THE

THEORY OF SOUND

BY

JOHN WILLIAM STRUTT, BARON RAYLEIGH, Sc.D., F.R.S.

HONORARY FELLOW OF TRINITY COLLEGE, CAMBRIDGE

WITH A HISTORICAL INTRODUCTION BY

ROBERT BRUCE LINDSAY

HAZARD PROFESSOR OF PHYSICS IN BROWN UNIVERSITY

IN TWO VOLUMES

VOLUME II

SECOND EDITION REVISED AND ENLARGED

..._W YORK

DOVER PUBLICATIONS

Published in Canada by General Publishing Company, Ltd., 30 Lesmill Road, Don Mills, Toronto, Ontario.
Published in the United Kingdom by Constable and Company, Ltd., 10 Orange Street, London WC 2.

This Dover edition, first published in 1945, is an unabridged republication of the second revised and enlarged edition of 1896. It is reprinted through special arrangement with The Macmillan Company, publishers of the original edition.

Standard Book Number: 486-60293-1
Library of Congress Catalog Card Number: 45-8804

Manufactured in the United States of America
Dover Publications, Inc.
180 Varick Street
New York, N. Y. 10014

PREFACE.

THE appearance of this second and concluding volume has been delayed by pressure of other work that could not well be postponed. As in Vol. I. the additions down to § 348 are indicated by square brackets, or by letters following the number of the section. From that point onwards the matter is new with the exception of § 381, which appeared in the first edition as § 348.

The additions to Chapter XIX. deal with aerial vibrations in narrow tubes where the influence of viscosity and heat conduction are important, and with certain phenomena of the second order dependent upon viscosity. Chapter XX. is devoted to capillary vibrations, and the explanation thereby of many beautiful observations due to Savart and other physicists. The sensitiveness of flames and smoke jets, a very interesting department of acoustics, is considered in Chapter XXI., and an attempt is made to lay the foundations of a theoretical treatment by the solution of problems respecting the stability, or otherwise, of stratified fluid motion. §§ 371, 372 deal with "bird-calls," investigated by Sondhauss, and with aeolian tones. In Chapter XXII. a slight sketch is given of the theory of the vibrations of elastic solids, especially as regards the propagation of plane waves, and the disturbance due to a harmonic force operative at one point of an infinite solid. The important problems of the vibrations of plates, cylinders and spheres, are perhaps best dealt with in works devoted specially to the theory of elasticity

The concluding chapter on the facts and theories of audition could not well have been omitted, but it has entailed labour out of proportion to the results. A large part of our knowledge upon this subject is due to Helmholtz, but most of the workers who have since published their researches entertain divergent views, in some cases, it would seem, without recognizing how fundamental their objections really are. And on several points the observations recorded by well qualified observers are so discrepant that no satisfactory conclusion can be drawn at the present time. The future may possibly shew that the differences are more nominal than real. In any case I would desire to impress upon the student of this part of our subject the importance of studying Helmholtz's views at first hand. In such a book as the present an imperfect outline of them is all that can be attempted. Only one thoroughly familiar with the *Tonempfindungen* is in a position to appreciate many of the observations and criticisms of subsequent writers.

TERLING PLACE, WITHAM.
February, 1896

EDITORIAL NOTE.

THE present re-issue has a few small corrections noted in the author's copy, and an addition [§ 335a] on the maximum disturbance that can be produced by an infinitesimal resonator exposed to plane waves. The author had written this out for inclusion.

May, 1926.

CONTENTS.

CHAPTER XI.

CHAPTER XII.

CHAPTER XIII.

CHAPTER XIV.

CHAPTER XV.

CHAPTER XVI.

CHAPTER XVII.

CHAPTER XVIII.

CHAPTER XIX.

CHAPTER XX.

CHAPTER XXI.

CHAPTER XXII.

CHAPTER XXIII.

[1] Appears now for the first time.
[2] Appeared in the First Edition.

ERRATUM.

Vol. I. p. 407, footnote. Add reference to Chree, *Camb. Phil. Trans.*, Vol. XIV. p. 250, 1887.

THE
THEORY OF SOUND

CHAPTER XI.

AERIAL VIBRATIONS.

236. SINCE the atmosphere is the almost universal vehicle of sound, the investigation of the vibrations of a gaseous medium has always been considered the peculiar problem of Physical Acoustics; but in all, except a few specially simple questions, chiefly relating to the propagation of sound in one dimension, the mathematical difficulties are such that progress has been very slow. Even when a theoretical result is obtained, it often happens that it cannot be submitted to the test of experiment, in default of accurate methods of measuring the intensity of vibrations. In some parts of the subject all that we can do is to solve those problems whose mathematical conditions are sufficiently simple to admit of solution, and to trust to them and to general principles not to leave us quite in the dark with respect to other questions in which we may be interested.

In the present chapter we shall regard fluids as perfect, that is to say, we shall assume that the mutual action between any two portions separated by an ideal surface is *normal to that surface*. Hereafter we shall say something about fluid friction; but, in general, acoustical phenomena are not materially disturbed by such deviation from perfect fluidity as exists in the case of air and other gases.

The equality of pressure in all directions about a given point is a necessary consequence of perfect fluidity, whether there be rest or motion, as is proved by considering the equilibrium of a small tetrahedron under the operation of the fluid pressures, the impressed forces, and the reactions against acceleration. In the limit, when the tetrahedron is taken indefinitely small, the fluid pressures on its sides become paramount, and equilibrium requires that their whole magnitudes be proportional to the areas of the faces over which they act. The pressure at the point x, y, z will be denoted by p.

237. If $\rho X dV$, $\rho Y dV$, $\rho Z dV$, denote the impressed forces acting on the element of mass ρdV, the equation of equilibrium is

$$dp = \rho \left(X dx + Y dy + Z dz \right),$$

where dp denotes the variation of pressure corresponding to changes dx, dy, dz in the co-ordinates of the point at which the pressure is estimated. This equation is readily established by considering the equilibrium of a small cylinder with flat ends, the projections of whose axis on those of co-ordinates are respectively dx, dy, dz. To obtain the equations of motion we have, in accordance with D'Alembert's Principle, merely to replace X, &c. by $X - Du/Dt$, &c., where Du/Dt, &c. denote the accelerations of the particle of fluid considered. Thus

$$\left. \begin{array}{l} \dfrac{dp}{dx} = \rho \left(X - \dfrac{Du}{Dt} \right) \\[2mm] \dfrac{dp}{dy} = \rho \left(Y - \dfrac{Dv}{Dt} \right) \\[2mm] \dfrac{dp}{dz} = \rho \left(Z - \dfrac{Dw}{Dt} \right) \end{array} \right\} \quad \dots\dots\dots\dots\dots (1).$$

In hydrodynamical investigations it is usual to express the velocities of the fluid u, v, w in terms of x, y, z and t. They then denote the velocities of the particle, whichever it may be, that at the time t is found at the point x, y, z. After a small interval of time dt, a new particle has reached x, y, z; $du/dt \cdot dt$ expresses the excess of its velocity over that of the first particle, while $Du/Dt \cdot dt$ on the other hand expresses the change in the velocity of the *original* particle in the same time, or the change of velocity at a point, which is not fixed in space, but moves with the fluid. To this notation we shall adhere. In the change contemplated in d/dt, the position in space (determined by the values of x, y, z) is retained invariable, while in D/Dt it is a certain particle of the fluid on which attention is fixed. The relation between the two kinds of differentiation with respect to time is expressed by

$$\frac{D}{Dt} = \frac{d}{dt} + u\frac{d}{dx} + v\frac{d}{dy} + w\frac{d}{dz} \ldots\ldots\ldots\ldots(2),$$

and must be clearly conceived, though in a large class of important problems with which we shall be occupied in the sequel, the distinction practically disappears. Whenever the motion is very small, the terms $u\,d/dx$, &c. diminish in relative importance, and ultimately $D/Dt = d/dt$.

238. We have further to express the condition that there is no creation or annihilation of matter in the interior of the fluid. If α, β, γ be the edges of a small rectangular parallelepiped parallel to the axes of co-ordinates, the quantity of matter which passes out of the included space in time dt in excess of that which enters is

$$\left\{\frac{d(\rho u)}{dx} + \frac{d(\rho v)}{dy} + \frac{d(\rho w)}{dz}\right\}\alpha\beta\gamma dt;$$

and this must be equal to the actual loss sustained, or

$$-\frac{d\rho}{dt}\alpha\beta\gamma dt.$$

Hence

$$\frac{d\rho}{dt} + \frac{d(\rho u)}{dx} + \frac{d(\rho v)}{dy} + \frac{d(\rho w)}{dz} = 0\ldots\ldots\ldots\ldots(1),$$

the so-called equation of continuity. When ρ is constant (with respect to both time and space), the equation assumes the simple form

$$\frac{du}{dx} + \frac{dv}{dy} + \frac{dw}{dz} = 0\ldots\ldots\ldots\ldots\ldots(2).$$

In problems connected with sound, the velocities and the variation of density are usually treated as small quantities. Putting $\rho = \rho_0(1+s)$, where s, called the *condensation*, is small, and neglecting the products $u\,ds/dx$, &c., we find

$$\frac{ds}{dt} + \frac{du}{dx} + \frac{dv}{dy} + \frac{dw}{dz} = 0\ldots\ldots\ldots\ldots(3).$$

In special cases these equations take even simpler forms. In the case of an incompressible fluid whose motion is entirely parallel to the plane of xy,

$$\frac{du}{dx} + \frac{dv}{dy} = 0 \ldots\ldots\ldots\ldots\ldots(4),$$

from which we infer that the expression $u\,dy - v\,dx$ is a perfect differential. Calling it $d\psi$, we have as the equivalent of (4)

$$u = \frac{d\psi}{dy}, \qquad v = -\frac{d\psi}{dx} \dots\dots\dots\dots\dots (5),$$

where ψ is a function of the co-ordinates which so far is perfectly arbitrary. The function ψ is called the *stream*-function, since the motion of the fluid is everywhere in the direction of the curves $\psi = $ constant. When the motion is steady, that is, always the same at the same point of space, the curves $\psi = $ constant mark out a system of pipes or channels in which the fluid may be supposed to flow. Analytically, the substitution of *one* function ψ for the *two* functions u and v is often a step of great consequence.

Another case of importance is when there is symmetry round an axis, for example, that of x. Everything is then expressible in terms of x and r, where $r = \sqrt{(y^2 + z^2)}$, and the motion takes place in planes passing through the axis of symmetry. If the velocities respectively parallel and perpendicular to the axis of symmetry be u and q, the equation of continuity is

$$\frac{d(ru)}{dx} + \frac{d(rq)}{dr} = 0 \dots\dots\dots\dots\dots (6),$$

which, as before, is equivalent to

$$ru = \frac{d\psi}{dr}, \qquad rq = -\frac{d\psi}{dx} \dots\dots\dots\dots(7),$$

ψ being the stream-function.

239. In almost all the cases with which we shall have to deal, the hydrodynamical equations undergo a remarkable simplification in virtue of a proposition first enunciated by Lagrange. If for any part of a fluid mass $u\,dx + v\,dy + w\,dz$ be at one moment a perfect differential $d\phi$, it will remain so for all subsequent time. In particular, if a fluid be originally at rest, and be then set in motion by conservative forces and pressures transmitted from the exterior, the quantities

$$\frac{dv}{dz} - \frac{dw}{dy}, \qquad \frac{dw}{dx} - \frac{du}{dz}, \qquad \frac{du}{dy} - \frac{dv}{dx},$$

(which we shall denote by ξ, η, ζ) can never depart from zero.

We assume that ρ is a function of p, and we shall write for brevity

$$\varpi = \int \frac{dp}{\rho} \dots\dots\dots\dots\dots\dots\dots(1).$$

The equations of motion obtained from (1), (2), § 237, are

$$\frac{d\varpi}{dx} = X - \frac{du}{dt} - u\frac{du}{dx} - v\frac{du}{dy} - w\frac{du}{dz} \dots\dots\dots(2),$$

with two others of the same form relating to y and z. By hypothesis,

$$\frac{dX}{dy} = \frac{dY}{dx};$$

so that by differentiating the first of the above equations with respect to y and the second with respect to x, and subtracting, we eliminate ϖ and the impressed forces, obtaining equations which may be put into the form

$$\frac{D\zeta}{Dt} = \frac{du}{dz}\xi + \frac{dv}{dz}\eta - \left(\frac{du}{dx} + \frac{dv}{dy}\right)\zeta \dots\dots\dots\dots(3),$$

with two others of the same form giving $D\xi/Dt$, $D\eta/Dt$.

In the case of an incompressible fluid, we may substitute for $du/dx + dv/dy$ its equivalent $-dw/dz$, and thus obtain

$$\frac{D\zeta}{Dt} = \frac{du}{dz}\xi + \frac{dv}{dz}\eta + \frac{dw}{dz}\zeta, \text{ \&c.} \dots\dots\dots\dots(4),$$

which are the equations used by Helmholtz as the foundation of his theorems respecting vortices.

If the motion be continuous, the coefficients of ξ, η, ζ in the above equations are all finite. Let L denote their greatest numerical value, and Ω the sum of the numerical values of ξ, η, ζ. By hypothesis, Ω is initially zero; the question is whether in the course of time it can become finite. The preceding equations shew that it cannot; for its rate of increase for a given particle is at any time less than $3L\Omega$, all the quantities concerned being positive. Now even if its rate of increase were as great as $3L\Omega$, Ω would never become finite, as appears from the solution of the equation

$$\frac{D\Omega}{Dt} = 3L\Omega \dots\dots\dots\dots\dots\dots (5).$$

A fortiori in the actual case, Ω cannot depart from zero, and the same must be true of ξ, η, ζ.

It is worth notice that this conclusion would not be disturbed by the presence of frictional forces acting on each particle proportional to its velocity, as may be seen by substituting $X - \kappa u$, $Y - \kappa v$, $Z - \kappa w$, for X, Y, Z in (2)[1]. But it is otherwise with the frictional forces which actually exist in fluids, and are dependent on the *relative* velocities of their parts.

The first satisfactory demonstration of the important proposition now under discussion was given by Cauchy; but that sketched above is due to Stokes[2]. It is not sufficient merely to shew that if, and whenever, ξ, η, ζ vanish, their differential coefficients $D\xi/Dt$, &c. vanish also, though this is a point that is often overlooked. When a body falls from rest under the action of gravity, $s \propto s^{\frac{1}{2}}$; but it does not follow that s never becomes finite. To justify that conclusion it would be necessary to prove that s vanishes in the limit, not merely to the first order, but to all orders of the small quantity t; which, of course, cannot be done in the case of a falling body. If, however, the equation had been $\dot{s} \propto s$, all the differential coefficients of s with respect to t would vanish with t, if s did so, and then it might be inferred legitimately that s could never vary from zero.

By a theorem due to Stokes, the moments of momentum about the axes of co-ordinates of any infinitesimal spherical portion of fluid are equal to ξ, η, ζ, multiplied by the moment of inertia of the mass; and thus these quantities may be regarded as the component rotatory velocities of the fluid at the point to which they refer.

If ξ, η, ζ vanish throughout a space occupied by moving fluid, any small spherical portion of the fluid if suddenly solidified would retain only a motion of translation. A proof of this proposition in a generalised form will be given a little later. Lagrange's theorem thus consists in the assertion that particles of fluid at any time destitute of rotation can never acquire it.

---

[1] By introducing such forces and neglecting the terms dependent on inertia, we should obtain equations applicable to the motion of electricity through uniform conductors.

[2] *Cambridge Trans.* Vol. VIII. p. 307, 1845. B. A. Report on Hydrodynamics, 1847.

240. A somewhat different mode of investigation has been adopted by Thomson, which affords a highly instructive view of the whole subject[1].

By the fundamental equations

$$d\varpi = X\,dx + Y\,dy + Z\,dz - \frac{Du}{Dt}\,dx - \frac{Dv}{Dt}\,dy - \frac{Dw}{Dt}\,dz.$$

Now $X\,dx + Y\,dy + Z\,dz = dR$, if the forces be conservative, and

$$\frac{Du}{Dt}\,dx + \frac{Dv}{Dt}\,dy + \frac{Dw}{Dt}\,dz$$

$$= \frac{D}{Dt}(u\,dx + v\,dy + w\,dz) - u\,\frac{D\,dx}{Dt} - v\,\frac{D\,dy}{Dt} - w\,\frac{D\,dz}{Dt},$$

in which

$$\frac{D\,dx}{Dt} = d\,\frac{Dx}{Dt} = du, \ \&c.$$

Thus, if $U^2 = u^2 + v^2 + w^2$, we have

$$d\varpi = dR - \frac{D}{Dt}(u\,dx + v\,dy + w\,dz) + \tfrac{1}{2}dU^2 \ldots\ldots\ldots(1),$$

or

$$\frac{D}{Dt}(u\,dx + v\,dy + w\,dz) = d\left(R + \tfrac{1}{2}U^2 - \varpi\right) \ \ldots\ldots(2).$$

Integrating this equation along any finite arc P_1P_2, moving with the fluid, we have

$$\frac{D}{Dt}\int(u\,dx + v\,dy + w\,dz) = \left(R + \tfrac{1}{2}U^2 - \varpi\right)_2 - \left(R + \tfrac{1}{2}U^2 - \varpi\right)_1 \ldots(3),$$

in which suffixes denote the values of the bracketed function at the points P_2 and P_1 respectively. If the arc be a complete circuit,

$$\frac{D}{Dt}\int(u\,dx + v\,dy + w\,dz) = 0 \ \ldots\ldots\ldots\ldots\ldots(4);$$

or, in words,

The line-integral of the tangential component velocity round any closed curve of a moving fluid remains constant throughout all time.

The line-integral in question is appropriately called the *circulation*, and the proposition may be stated:—

The circulation in any closed line moving with the fluid remains constant.

[1] *Vortex Motion.* *Edinburgh Transactions*, 1869.

In a state of rest the circulation is of course zero, so that, if a fluid be set in motion by pressures transmitted from the outside or by conservative forces, the circulation along any closed line must ever remain zero, which requires that $u\,dx + v\,dy + w\,dz$ be a complete differential.

But it does not follow conversely that in irrotational motion there can never be circulation, unless it be known that ϕ is single-valued; for otherwise $\int d\phi$ need not vanish round a closed circuit. In such a case all that can be said is that there is no circulation round any closed curve capable of being contracted to a point without passing out of space occupied by irrotationally moving fluid, or more generally, that the circulation is the same in all mutually reconcilable closed curves. Two curves are said to be reconcilable, when one can be obtained from the other by continuous deformation, without passing out of the irrotationally moving fluid.

Within an oval space, such as that included by an ellipsoid, all circuits are reconcilable, and therefore if a mass of fluid of that form move irrotationally, there can be no circulation along any closed curve drawn within it. Such spaces are called simply-connected. But in an annular space like that bounded by the surface of an anchor ring, a closed curve going round the ring is not continuously reducible to a point, and therefore there may be circulation along it, even although the motion be irrotational throughout the whole volume included. But the circulation is zero for every closed curve which does not pass round the ring, and has the same constant value for all those that do.

[In the above theorems "circulation" is defined without reference to mass. If the fluid be of uniform density, the *momentum* reckoned round a closed circuit is proportional to circulation, but in the case of a compressible fluid a distinction must be drawn. The existence of a velocity-potential does not then imply evanescence of the integral momentum reckoned round a closed circuit.]

241. When $u\,dx + v\,dy + w\,dz$ is an exact differential $d\phi$, the velocity in any direction is expressed by the corresponding rate of change of ϕ, which is called the velocity-potential, and

$$\frac{du}{dx} + \frac{dv}{dy} + \frac{dw}{dz}$$

may be replaced by

$$\frac{d^2\phi}{dx^2} + \frac{d^2\phi}{dy^2} + \frac{d^2\phi}{dz^2}.$$

If S denote any closed surface, the rate of flow outwards across the element dS is expressed by $dS \cdot d\phi/dn$, where $d\phi/dn$ is the rate of variation of ϕ in proceeding outwards along the normal. In the case of constant density, the total loss of fluid in time dt is thus

$$\iint \frac{d\phi}{dn}\, dS \cdot dt,$$

the integration ranging over the whole surface of S. If the space S be full both at the beginning and at the end of the time dt, the loss must vanish; and thus

$$\iint \frac{d\phi}{dn}\, dS = 0 \dots\dots\dots\dots\dots\dots(1).$$

The application of this equation to the element $dx\,dy\,dz$ gives for the equation of continuity of an incompressible fluid

$$\frac{d^2\phi}{dx^2} + \frac{d^2\phi}{dy^2} + \frac{d^2\phi}{dz^2} = 0 \dots\dots\dots\dots\dots\dots(2),$$

or, as it is generally written,

$$\nabla^2\phi = 0 \dots\dots\dots\dots\dots\dots (3);$$

when it is desired to work with polar co-ordinates, the transformed equation is more readily obtained directly by applying (1) to the corresponding element of volume, than by transforming (2) in accordance with the analytical rules for effecting changes in the independent variables.

Thus, if we take polar co-ordinates in the plane xy, so that

$$x = r\cos\theta, \quad y = r\sin\theta,$$

we find

$$\nabla^2\phi = \frac{d^2\phi}{dr^2} + \frac{1}{r}\frac{d\phi}{dr} + \frac{1}{r^2}\frac{d^2\phi}{d\theta^2} + \frac{d^2\phi}{dz^2} \dots\dots\dots\dots(4);$$

or, if we take polar co-ordinates in space,

$$x = r\sin\theta\cos\omega, \quad y = r\sin\theta\sin\omega, \quad z = r\cos\theta,$$

$$\nabla^2\phi = \frac{d^2\phi}{dr^2} + \frac{2}{r}\frac{d\phi}{dr} + \frac{1}{r^2\sin\theta}\frac{d}{d\theta}\left(\sin\theta\frac{d\phi}{d\theta}\right) + \frac{1}{r^2\sin^2\theta}\frac{d^2\phi}{d\omega^2} \dots(5).$$

Simpler forms are assumed in special cases, such, for example, as that of symmetry round z in (5).

When the fluid is compressible, and the motion such that the squares of small quantities may be neglected, the equation of continuity is by (3), § 238,

$$\frac{ds}{dt} + \nabla^2\phi = 0\ldots\ldots\ldots\ldots\ldots\ldots\ldots(6),$$

where any form of $\nabla^2\phi$ may be used that may be most convenient for the problem in hand.

242. The irrotational motion of incompressible fluid within any simply-connected closed space S is completely determined by the normal velocities over the surface of S. If S be a material envelope, it is evident that an arbitrary normal velocity may be impressed upon its surface, which normal velocity must be shared by the fluid immediately in contact, provided that the whole volume inclosed remain unaltered. If the fluid be previously at rest, it can acquire no molecular rotation under the operation of the fluid pressures, which shews that it must be possible to determine a function ϕ, such that $\nabla^2\phi = 0$ throughout the space inclosed by S, while over the surface $d\phi/dn$ has a prescribed value, limited only by the condition

$$\iint \frac{d\phi}{dn}\, dS = 0\ldots\ldots\ldots\ldots\ \ldots\ldots\ldots(1).$$

An analytical proof of this important proposition is indicated in Thomson and Tait's *Natural Philosophy*, § 317.

There is no difficulty in proving that but one solution of the problem is possible. By Green's theorem, if $\nabla^2\phi = 0$,

$$\iiint \left(\frac{d\phi^2}{dx^2} + \frac{d\phi^2}{dy^2} + \frac{d\phi^2}{dz^2}\right) dV = \iint \phi\, \frac{d\phi}{dn}\, dS\ldots\ldots\ldots(2),$$

the integration on the left-hand side ranging over the volume, and on the right over the surface of S. Now if ϕ and $\phi + \Delta\phi$ be two functions, satisfying Laplace's equation, and giving prescribed surface-values of $d\phi/dn$, their difference $\Delta\phi$ is a function also satisfying Laplace's equation, and making $d\Delta\phi/dn$ vanish over the surface of S. Under these circumstances the double integral in (2) vanishes, and we infer that at every point of S $d\Delta\phi/dx$, $d\Delta\phi/dy$, $d\Delta\phi/dz$ must be equal to zero. In other words $\Delta\phi$ must be constant, and the two motions identical. As a particular case, there can be no motion of the irrotational kind within the volume S, independently of a motion of the surface. The restriction to simply-connected spaces is rendered necessary by the failure of Green's theorem, which, as was first pointed out by Helmholtz, is otherwise possible.

When the space S is multiply-connected, the irrotational motion is still determinate, if besides the normal velocity at every point of S there be given the values of the constant circulations in all the possible irreconcilable circuits. For a complete discussion of this question we must refer to Thomson's original memoir, and content ourselves here with the case of a doubly-connected space, which will suffice for illustration.

Let $ABCD$ be an endless tube within which fluid moves irrotationally. For this motion there must exist a velocity-potential, whose differential coefficients, expressing, as they do, the component velocities, are necessarily single-valued, but which need not itself be single-valued. The simplest way of attacking the difficulty presented by the ambiguity of ϕ, is to conceive a barrier AB taken across the ring, so as to close the passage. The space $ABCDBAEF$ is then simply continuous, and Green's theorem applies to it without modification, if allowance be made for a possible finite difference in the value of ϕ on the two sides of the barrier. This difference, if it exist, is necessarily the same at all points of AB, and in the hydrodynamical application expresses the *circulation* round the ring.

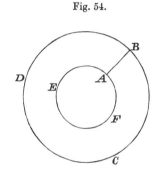

Fig. 54.

In applying the equation

$$\iiint\left(\frac{d\phi^2}{dx^2} + \frac{d\phi^2}{dy^2} + \frac{d\phi^2}{dz^2}\right) dV = \iint \phi \frac{d\phi}{dn} dS \ldots\ldots\ldots (2),$$

we have to calculate the double integral over the two faces of the barrier as well as over the original surface of the ring. Now since $\dfrac{d\phi}{dn}$ has the same value on the two sides,

$$\iint \phi \frac{d\phi}{dn} dS \text{ (over two faces of } AB) = \iint \frac{d\phi}{dn} \kappa \, dS = \kappa \iint \frac{d\phi}{dn} dS,$$

if κ denote the constant difference of ϕ. Thus, if κ vanish, or there be no circulation round the ring, we infer, just as for a simply-connected space, that ϕ is completely determined by the surface-values of $d\phi/dn$. If there be circulation, ϕ is still determined, if the amount of the circulation be given. For, if ϕ and $\phi + \Delta\phi$ be two functions satisfying Laplace's equation and giving the same amount of circulation and the same normal velocities at S, their difference $\Delta\phi$ also satisfies Laplace's equation and the condition that there shall be neither circulation nor normal velocities over S. But, as we have just seen, under these circumstances $\Delta\phi$ vanishes at every point.

Although in a doubly-connected space irrotational motion is possible independently of surface normal velocities, yet such a motion cannot be generated by conservative forces nor by motions imposed (at any previous time) on the bounding surface, for we have proved that if the fluid be originally at rest, there can never be circulation along any closed curve. Hence, for multiply-connected as well as simply-connected spaces, if a fluid be set in motion by arbitrary deformation of the boundary, the whole mass comes to rest so soon as the motion of the boundary ceases.

If in a fluid moving without circulation all the fluid outside a reentrant tube-like surface of uniform section become instantaneously solid, then also at the same moment all the fluid within the tube comes to rest. This mechanical interpretation, however unpractical, will help the student to understand more clearly what is meant by a fluid having no circulation, and it leads to an extension of Stokes' theorem with respect to molecular rotation. For, if all the fluid (moving subject to a velocity-potential) outside a spherical cavity of any radius become suddenly solid, the fluid inside the cavity can retain no motion. Or, as we may also state it, any spherical portion of an irrotationally moving [incompressible] fluid becoming suddenly solid would possess only a motion of translation, *without rotation*[1].

A similar proposition will apply to a cylinder disc, or cylinder with flat ends, in the case of fluid moving irrotationally in two dimensions only.

---

[1] Thomson on *Vortex Motion, loc. cit.*

The motion of an incompressible fluid which has been once at rest partakes of the remarkable property (§ 79) common to that of all systems which are set in motion with prescribed velocities, namely, that the energy is the least possible. If any other motion be proposed satisfying the equation of continuity and the boundary conditions, its energy is necessarily greater than that of the motion which would be generated from rest[1].

243. The fact that the irrotational motion of incompressible fluid depends upon a velocity-potential satisfying Laplace's equation, is the foundation of a far-reaching analogy between the motion of such a fluid, and that of electricity or heat in a uniform conductor, which it is often of great service to bear in mind. The same may be said of the connection between all the branches of Physics which depend mathematically on a potential, for it often happens that the analogous theorems are far from equally obvious. For example, the analytical theorem that, if $\nabla^2\phi = 0$,

$$\iint \frac{d\phi}{dn}\, dS = 0$$

over a closed surface, is most readily suggested by the fluid interpretation, but once obtained may be interpreted for electric or magnetic forces.

Again, in the theory of the conduction of heat or electricity, it is obvious that there can be no steady motion in the interior of S, without transmission across some part of the bounding surface, but this, when interpreted for incompressible fluids, gives an important and rather recondite law.

244. When a velocity-potential exists, the equation to determine the pressure may be put into a simpler form. We have from (1), § 240,

$$d\varpi = dR - \frac{D}{Dt}\, d\phi + \tfrac{1}{2} dU^2 \dots\dots\dots\dots\dots(1),$$

whence by integration

$$\varpi = \int \frac{dp}{\rho} = R - \frac{D\phi}{Dt} + \tfrac{1}{2} U^2.$$

[1] [The reader who wishes to pursue the study of general hydrodynamics is referred to the treatises of Lamb and Basset.]

Now
$$\frac{D\phi}{Dt} = \frac{d\phi}{dt} + u^2 + v^2 + w^2;$$
so that
$$\int \frac{dp}{\rho} = R - \frac{d\phi}{dt} - \tfrac{1}{2} U^2 \dots\dots\dots\dots\dots(2),$$

which is the form ordinarily given.

If ρ be constant, $\int \dfrac{dp}{\rho}$ is replaced, of course, by $\dfrac{p}{\rho}$.

The relation between p and ϕ in the case of impulsive motion from rest may be deduced from (2) by integration. We see that

$$\frac{1}{\rho} \int p\, dt = -\phi \quad \text{ultimately.}$$

The same conclusion may be arrived at by a direct application of mechanical principles to the circumstances of impulsive motion.

If $p = \kappa\rho$, equation (2) takes the form

$$\kappa \log \rho = R - \frac{d\phi}{dt} - \tfrac{1}{2} U^2 \dots\dots\dots\dots\dots(3).$$

If the motion be such that the component velocities are always the same at the same point of space, it is called *steady*, and ϕ becomes independent of the time. The equation of pressure is then

$$\int \frac{dp}{\rho} = R - \tfrac{1}{2} U^2 \dots\dots\dots\dots\dots(4),$$

or in the case when there are no impressed forces,

$$\int \frac{dp}{\rho} = C - \tfrac{1}{2} U^2 \dots\dots\dots\dots\dots(5).$$

In most acoustical applications of (2), the velocities and condensation are small, and then we may neglect the term $\tfrac{1}{2} U^2$, and substitute $\dfrac{\delta p}{\rho_0}$ for $\int \dfrac{dp}{\rho}$, if δp denote the small variable part of p; thus

$$\frac{\delta p}{\rho_0} = R - \frac{d\phi}{dt} \dots\dots\dots\dots\dots(6),$$
which with
$$\frac{ds}{dt} + \nabla^2\phi = 0 \dots\dots\dots\dots\dots(7)$$

are the equations by means of which the small vibrations of an elastic fluid are to be investigated.

If $a^2 = dp/d\rho$, so that $\delta p = a^2\rho_0 s$, (6) becomes

$$a^2 s = R - \frac{d\phi}{dt} \dots\dots\dots\dots\dots\dots(8),$$

and we get on elimination of s,

$$\frac{d^2\phi}{dt^2} = \frac{dR}{dt} + a^2\nabla^2\phi\dots\dots\dots\dots\dots\dots(9).$$

245. The simplest kind of wave-motion is that in which the excursions of every particle are parallel to a fixed line, and are the same in all planes perpendicular to that line. Let us therefore (assuming that $R = 0$) suppose that ϕ is a function of x (and t) only. Our equation (9) § 244 becomes

$$\frac{d^2\phi}{dt^2} = a^2\frac{d^2\phi}{dx^2} \dots\dots\dots\dots\dots\dots\dots(1),$$

the same as that already considered in the chapter on Strings. We there found that the general solution is

$$\phi = f(x - at) + F(x + at) \dots\dots\dots\dots\dots(2),$$

representing the propagation of independent waves in the positive and negative directions with the common velocity a.

Within such limits as allow the application of the approximate equation (1), the velocity of sound is entirely independent of the form of the wave, being, for example, the same for simple waves

$$\phi = A \cos \frac{2\pi}{\lambda} (x - at),$$

whatever the wave-length may be. The condition satisfied by the positive wave, and therefore by the initial disturbance if a positive wave alone be generated, is

$$a\frac{d\phi}{dx} + \frac{d\phi}{dt} = 0,$$

or by (8) § 244

$$u - as = 0 \dots\dots\dots\dots\dots\dots\dots(3).$$

Similarly, for a negative wave

$$u + as = 0 \dots\dots\dots\dots\dots\dots\dots(4).$$

Whatever the initial disturbance may be (and u and s are both arbitrary), it can always be divided into two parts, satisfying respectively (3) and (4), which are propagated undisturbed. In each component wave the direction of propagation is the same as that of the motion of the *condensed* parts of the fluid.

The rate at which energy is transmitted across unit of area of a plane parallel to the front of a progressive wave may be regarded as the mechanical measure of the intensity of the radiation. In the case of a simple wave, for which

$$\phi = A \cos \frac{2\pi}{\lambda} (x - at) \dots\dots\dots\dots(5),$$

the velocity $\dot{\xi}$ of the particle at x (equal to $d\phi/dx$) is given by

$$\dot{\xi} = -\frac{2\pi}{\lambda} A \sin \frac{2\pi}{\lambda} (x - at) \dots\dots\dots(6),$$

and the displacement ξ is given by

$$\xi = -\frac{A}{a} \cos \frac{2\pi}{\lambda} (x - at) \dots\dots\dots\dots(7).$$

The pressure $p = p_0 + \delta p$, where by (6) § **244**

$$\delta p = -\frac{2\pi}{\lambda} \rho_0 aA \sin \frac{2\pi}{\lambda} (x - at)\dots\dots\dots(8).$$

Hence, if W denote the work transmitted across unit area of the plane x in time t,

$$\frac{dW}{dt} = (p_0 + \delta p) \dot{\xi} = \tfrac{1}{2}\rho_0 a \left(\frac{2\pi}{\lambda}\right)^2 A^2 + \text{periodic terms}.$$

If the integration with respect to time extend over any number of complete periods, or practically whenever its range is sufficiently long, the periodic terms may be omitted, and we may take

$$W : t = \tfrac{1}{2}\rho_0 a \left(\frac{2\pi}{\lambda}\right)^2 A^2\dots\dots\dots\dots(9);$$

or by (3) and (6), if $\dot{\xi}$ now denote the maximum value of the velocity and s the maximum value of the condensation,

$$W = \tfrac{1}{2}\rho_0 \dot{\xi}^2 at = \tfrac{1}{2}\rho_0 a^3 s^2 t \dots\dots\dots\dots(10).$$

Thus the work consumed in generating waves of harmonic type is the same as would be required to give the maximum velocity $\dot{\xi}$ to the whole mass of air through which the waves extend[1].

---

[1] The earliest statement of the principle embodied in equation (10) that I have met with is in a paper by Sir W. Thomson, "On the possible density of the luminiferous medium, and on the mechanical value of a cubic mile of sun-light." *Phil. Mag.* IX. p. 36. 1855.

In terms of the maximum excursion ξ by (7) and (9)

$$W = 2\pi^2 \rho_0 \frac{a^3}{\lambda^2} \xi^2 t = 2\pi^2 \rho_0 a t \frac{\xi^2}{\tau^2} \dots\dots\dots\dots(11)^1,$$

where $\tau (= \lambda/a)$ is the periodic time. In a *given medium* the mechanical measure of the intensity is proportional to the square of the amplitude directly, and to the square of the periodic time inversely. The reader, however, must be on his guard against supposing that the mechanical measure of intensity of undulations of different wave lengths is a proper measure of the loudness of the corresponding sounds, as perceived by the ear.

In any plane progressive wave, whether the type be harmonic or not, the whole energy is equally divided between the potential and kinetic forms. Perhaps the simplest road to this result is to consider the formation of positive and negative waves from an initial disturbance, whose energy is wholly potential[2]. The total energies of the two derived progressive waves are evidently equal, and make up together the energy of the original disturbance. Moreover, in each progressive wave the condensation (or rare-faction) is one-half of that which existed at the corresponding point initially, so that the *potential* energy of each progressive wave is *one-quarter* of that of the original disturbance. Since, as we have just seen, the *whole* energy is *one-half* of the same quantity, it follows that in a progressive wave of any type one-half of the energy is potential and one-half is kinetic.

The same conclusion may also be drawn from the general expressions for the potential and kinetic energies and the relations between velocity and condensation expressed in (3) and (4). The potential energy of the element of volume dV is the work that would be gained during the expansion of the corresponding quantity of gas from its actual to its normal volume, the expansion being opposed throughout by the normal pressure p_0. At any stage of the expansion, when the condensation is s', the effective pressure δp is by § 244 $a^2 \rho_0 s'$, which pressure has to be multiplied by the corresponding increment of volume $dV . ds'$. The whole work gained during the expansion from dV to $dV(1 + s)$ is therefore $a^2 \rho_0 dV . \int_0^s s' ds'$ or $\frac{1}{2} a^2 \rho_0 dV . s^2$. The general expressions for the potential and kinetic energies are accordingly

[1] Bosanquet, *Phil. Mag.* XLV. p. 173. 1873.

[2] *Phil. Mag.* (5) I. p. 260. 1876.

$$\text{potential energy} = \tfrac{1}{2} a^2 \rho_0 \iiint s^2 \, dV \quad \ldots\ldots\ldots\ldots(12),$$

$$\text{kinetic energy} \quad = \quad \tfrac{1}{2} \rho_0 \iiint u^2 \, dV \quad \ldots\ldots\ldots\ldots(13),$$

and these are equal in the case of plane progressive waves for which

$$u = \pm \, as.$$

If the plane progressive waves be of harmonic type, u and s at any moment of time are circular functions of one of the space co-ordinates (x), and therefore the mean value of their squares is one-half of the maximum value. Hence the total energy of the waves is equal to the kinetic energy of the whole mass of air concerned, moving with the maximum velocity to be found in the waves, or to the potential energy of the same mass of air when condensed to the maximum density of the waves.

[It may be worthy of notice that when terms of the second order are retained, a purely periodic value of u does not correspond to a purely periodic motion. The quantity of fluid which passes unit of area at point x in time dt is $\rho u \, dt$, or $\rho_0 (1 + s) u \, dt$. If u be periodic, $\int u \, dt = 0$, but $\int s u \, dt$ may be finite. Thus in a positive progressive wave

$$\int s u \, dt = a \int s^2 dt,$$

and there is a transference of fluid in the direction of wave propagation.]

246. The first theoretical investigation of the velocity of sound was made by Newton, who assumed that the relation between pressure and density was that formulated in Boyle's law. If we assume $p = \kappa \rho$, we see that the velocity of sound is expressed by $\sqrt{\kappa}$, or $\sqrt{p} \div \sqrt{\rho}$, in which the dimensions of p $(= \text{force} \div \text{area})$ are $[M][L]^{-1}[T]^{-2}$, and those of ρ $(= \text{mass} \div \text{volume})$ are $[M][L]^{-3}$. Newton expressed the result in terms of the '*height of the homogeneous atmosphere*,' defined by the equation

$$g \rho h = p \quad \ldots\ldots\ldots\ldots\ldots\ldots\ldots\ldots\ldots(1),$$

where p and ρ refer to the pressure and the density at the earth's surface. The velocity of sound is thus $\sqrt{(gh)}$, or the velocity which would be acquired by a body falling freely under the action of gravity through half the height of the homogeneous atmosphere.

To obtain a numerical result we require to know a pair of simultaneous values of p and ρ.

[It is found by experiment[1] that at 0° Cent. under the pressure due (at Paris) to 760 mm. of mercury at 0° the density of dry air is ·0012933 gms. per cubic centimetre. If we assume as the density of mercury at 0° 13·5953[2], and $g = 980·939$, we have in C.G.S. measure

$$p = 760 \times 13·5953 \times 980·939, \quad \rho = ·0012933,$$

whence $\qquad a = \sqrt{(p/\rho)} = 27994·5 ;$

so that the velocity of sound at 0° would be 279·945 metres per second, falling short of the result of direct observation by about a sixth part.]

Newton's investigation established that the velocity of sound should be independent of the amplitude of the vibration, and also of the pitch, but the discrepancy between his calculated value (published in 1687) and the experimental value was not explained until Laplace pointed out that the use of Boyle's law involved the assumption that in the condensations and rarefactions accompanying sound the temperature remains constant, in contradiction to the known fact that, when air is suddenly compressed, its temperature rises. The laws of Boyle and Charles supply only one relation between the three quantities, pressure, volume, and temperature, of a gas, viz.

$$pv = R\theta \dots\dots\dots\dots\dots\dots\dots\dots(2),$$

where the temperature θ is measured from the zero of the gas thermometer; and therefore without some auxiliary assumption it is impossible to specify the connection between p and v (or ρ). Laplace considered that the condensations and rarefactions concerned in the propagation of sound take place with such rapidity that the heat and cold produced have not time to pass away, and that therefore the relation between volume and pressure is sensibly the same as if the air were confined in an absolutely non-conducting vessel. Under these circumstances the change of pressure corresponding to a given condensation or rarefaction is greater than on the hypothesis of constant temperature, and the velocity of sound is accordingly increased.

---

[1] On the Densities of the Principal Gases, *Proc. Roy. Soc.* vol. LIII. p. 147, 1893.

[2] Volkmann, *Wied. Ann.* vol. XIII. p. 221, 1881.

In equation (2) let v denote the volume and p the pressure of the unit of mass, and let θ be expressed in centigrade degrees reckoned from the absolute zero[1]. The condition of the gas (if uniform) is defined by any two of the three quantities p, v, θ, and the third may be expressed in terms of them. The relation between the simultaneous variations of the three quantities is

$$\frac{d\theta}{\theta} = \frac{dp}{p} + \frac{dv}{v} \quad \dots(3).$$

In order to effect the change specified by dp and dv, it is in general necessary to communicate heat to the gas. Calling the necessary quantity of heat dQ, we may write

$$dQ = \left(\frac{dQ}{dv}\right) dv + \left(\frac{dQ}{dp}\right) dp \quad \dotsi(4).$$

Suppose now (a) that $dp = 0$. Equations (3) and (4) give

$$\frac{dQ}{d\theta} \, (p \text{ const.}) = \left(\frac{dQ}{dv}\right) \frac{v}{\theta},$$

where $\dfrac{dQ}{d\theta}$ (p const.) expresses the specific heat of the gas under a constant pressure. This being denoted by κ_p, we have

$$\kappa_p = \left(\frac{dQ}{dv}\right) \frac{v}{\theta} \dotsi(5).$$

Again, suppose (b) that $dv = 0$. We find in a similar manner that, if κ_v denote the specific heat under a constant volume,

$$\kappa_v = \left(\frac{dQ}{dp}\right) \frac{p}{\theta} \dotsi(6).$$

In order to obtain the relation between dp and dv when there is no communication of heat, we have only to put $dQ = 0$. Thus

$$\left(\frac{dQ}{dv}\right) dv + \left(\frac{dQ}{dp}\right) dp = 0,$$

or, on substituting for the differential coefficients of Q their values in terms of κ_v, κ_p,

$$\kappa_p \frac{dv}{v} + \kappa_v \frac{dp}{p} = 0 \dotsi(7).$$

Since $v = 1/\rho$, $\qquad dv/v = -d\rho/\rho$;

so that $\qquad a^2 = \dfrac{dp}{d\rho} = \dfrac{p}{\rho} \dfrac{\kappa_p}{\kappa_v} = \dfrac{p}{\rho} \gamma \dotsi(8).$

[1] On the ordinary centigrade scale the absolute zero is about $-273°$.

if, as usual, the ratio of the specific heats be denoted by γ. Laplace's value of the velocity of sound is therefore greater than Newton's in the ratio of $\sqrt{\gamma} : 1$.

By integration of (8), we obtain for the relation between p and ρ, on the supposition of no communication of heat,

$$\frac{p}{p_0} = \left(\frac{\rho}{\rho_0}\right)^{\gamma} \dots\dots\dots\dots\dots\dots\dots(9)[1],$$

where p_0, ρ_0 are two simultaneous values. Under the same circumstances the relation between pressure and temperature is by (3)

$$\frac{p}{p_0} = \left(\frac{\theta}{\theta_0}\right)^{\frac{\gamma}{\gamma-1}} \dots\dots\dots\dots\dots\dots(10).$$

The magnitude of γ cannot be determined with accuracy by direct experiment, but an approximate value may be obtained by a method of which the following is the principle. Air is compressed into a reservoir capable of being put into communication with the external atmosphere by opening a wide valve. At first the temperature of the compressed air is raised, but after a time the superfluous heat passes away and the whole mass assumes the temperature of the atmosphere Θ. Let the pressure (measured by a manometer) be p. The valve is now opened for as short a time as is sufficient to permit the equilibrium of pressure to be completely established, that is, until the internal pressure has become equal to that of the atmosphere P. If the experiment be properly arranged, this operation is so quick that the air in the vessel has not sufficient time to receive heat from the sides, and therefore expands nearly according to the law expressed in (9). Its temperature θ at the moment the operation is complete is therefore determined by

$$\frac{p}{P} = \left(\frac{\Theta}{\theta}\right)^{\frac{\gamma}{\gamma-1}} \dots\dots\dots\dots\dots\dots(11).$$

The enclosed air is next allowed to absorb heat until it has regained the atmospheric temperature Θ, and its pressure (p') is then observed. During the last change the volume is constant, and therefore the relation between pressure and temperature gives

$$\frac{P}{p'} = \frac{\theta}{\Theta} \dots\dots\dots\dots\dots\dots\dots(12);$$

[1] It is here assumed that γ is constant. This equation appears to have been given first by Poisson.

so that by elimination of θ/Θ,

$$\frac{p}{P} = \left(\frac{p'}{P}\right)^{\frac{\gamma}{\gamma-1}},$$

whence

$$\gamma = \frac{\log p - \log P}{\log p - \log p'}, \quad \dots\dots\dots\dots\dots(13).$$

By experiments of this nature Clement and Desormes determined $\gamma = 1\cdot3492$; but the method is obviously not susceptible of any great accuracy. The value of γ required to reconcile the calculated and observed velocities of sound is $1\cdot408$, of the substantial correctness of which there can be little doubt.

We are not, however, dependent on the phenomena of sound for our knowledge of the magnitude of γ. The value of κ_p —the specific heat at constant pressure—has been determined experimentally by Regnault; and although on account of inherent difficulties the experimental method[1] may fail to yield a satisfactory result for κ_v, the information sought for may be obtained indirectly by means of a relation between the two specific heats, brought to light by the modern science of Thermodynamics.

If from the equations

$$\left.\begin{array}{l} \dfrac{dQ}{\theta} = \kappa_p \dfrac{dv}{v} + \kappa_v \dfrac{dp}{p} \\[2mm] \dfrac{d\theta}{\theta} = \dfrac{dv}{v} + \dfrac{dp}{p} \end{array}\right\} \quad \dots\dots\dots\dots\dots(14)$$

we eliminate dp, there results

$$dQ = (\kappa_p - \kappa_v)\frac{p\,dv}{R} + \kappa_v d\theta \quad \dots\dots\dots\dots(15).$$

Let us suppose that $dQ = 0$, or that there is no communication of heat. It is known that the heat developed during the compression of an approximately perfect gas, such as air, is almost exactly the thermal equivalent of the work done in compressing it. This important principle was assumed by Mayer in his celebrated memoir on the dynamical theory of heat, though on grounds which can hardly be considered adequate. However that may be, the principle itself is very nearly true, as has since been proved by the experiments of Joule and Thomson.

If we measure heat in dynamical units, Mayer's principle may be expressed $-\kappa_v d\theta = p\,dv$ on the understanding that there is

---

[1] [See, however, Joly, *Phil. Trans.* vol. CLXXXII. A, 1891.]

no communication of heat. Comparing this with (15), we see that

$$\kappa_p - \kappa_v = R \quad \dotfill (16),$$

and therefore

$$\gamma = \frac{\kappa_p}{\kappa_v} = \frac{\kappa_p}{\kappa_p - R} \quad \dotfill (17).$$

The value of pv in gravitation measure (gramme, centimetre) is $1033 \div \cdot001293$, at $0°$ Cent. so that

$$R = \frac{1033}{\cdot001293 \times 272\cdot85}.$$

By Regnault's experiments the specific heat of air is $\cdot2379$ of that of water; and in order to raise a gramme of water one degree Cent., 42350 gramme-centimetres of work must be done on it. Hence with the same units as for R,

$$\kappa_p = \cdot2379 \times 42350.$$

Calculating from these data, we find $\gamma = 1\cdot410$, agreeing almost exactly with the value deduced from the velocity of sound. This investigation is due to Rankine, who employed in it 1850 to calculate the specific heat of air, taking Joule's equivalent and the observed velocity of sound as data. In this way he anticipated the result of Regnault's experiments, which were not published until 1853.

247. Laplace's theory has often been the subject of mis-apprehension among students, and a stumblingblock to those remarkable persons, called by De Morgan 'paradoxers.' But there can be no reasonable doubt that, antecedently to all calculation, the hypothesis of no communication of heat is greatly to be preferred to the equally special hypothesis of constant temperature. There would be a real difficulty if the velocity of sound were not decidedly in excess of Newton's value, and the wonder is rather that the cause of the excess remained so long undiscovered.

The only question which can possibly be considered open, is whether a small part of the heat and cold developed may not escape by conduction or radiation before producing its full effect. Everything must depend on the rapidity of the alternations. Below a certain limit of slowness, the heat in excess, or defect, would have time to adjust itself, and the temperature would remain sensibly constant. In this case the relation between pressure and density would be that which leads to Newton's value of the velocity of sound. On the other hand, above a certain limit of quickness, the gas would behave as if confined in a non-conducting vessel, as supposed in Laplace's theory. Now although the circumstances of the actual problem are better represented by the latter than by the former supposition, there may still (it may be said) be a sensible deviation from the law of pressure and density involved in Laplace's theory, entailing a somewhat slower velocity of propagation of sound. This question has been carefully discussed by Stokes in a paper published in 1851[1], of which the following is an outline.

The mechanical equations for the *small* motion of air are

$$\frac{dp}{dx} = -\rho \frac{du}{dt} \text{ &c. } \dots\dots\dots\dots\dots\dots(1),$$

with the equation of continuity

$$\frac{ds}{dt} + \frac{du}{dx} + \frac{dv}{dy} + \frac{dw}{dz} = 0 \dots\dots\dots\dots\dots\dots(2).$$

The temperature is supposed to be uniform except in so far as it is disturbed by the vibrations themselves, so that if θ denote the *excess* of temperature,

$$p = \kappa\rho \left(1 + s + \alpha\theta\right) \dots\dots\dots\dots\dots\dots(3).$$

The effect of a small sudden condensation s is to produce an elevation of temperature, which may be denoted by βs. Let dQ be the quantity of heat entering the element of volume in time dt, measured by the rise of temperature that it would produce, if there were no condensation. Then (the distinction between D/Dt and d/dt being neglected)

$$\frac{d\theta}{dt} = \beta \frac{ds}{dt} + \frac{dQ}{dt} \dots\dots\dots\dots\dots\dots(4),$$

dQ/dt being a function of θ and its differential coefficients with respect to space, dependent on the special character of the dissipation. Two extreme cases may be mentioned; the first when the tendency to equalisation of temperature is due to conduction, the second when the operating cause is radiation, and the transparency of the medium such that radiant heat is

[1] *Phil. Mag.* (4) I. 305.

not sensibly absorbed within a distance of several wave-lengths. In the former case $dQ/dt \propto \nabla^2\theta$, and in the latter, which is that selected by Stokes for analytical investigation, $dQ/dt \propto (-\theta)$, Newton's law of radiation being assumed as a sufficient approximation to the truth. We have then

$$\frac{d\theta}{dt} = \beta \frac{ds}{dt} - q\theta \quad \dots\dots\dots\dots\dots\dots(5).$$

In the case of plane waves, to which we shall confine our attention, v and w vanish, while u, p, s, θ are functions of x (and t) only. Eliminating p and u between (1), (2) and (3), we find

$$\frac{d^2s}{dt^2} = \kappa \left(\frac{d^2s}{dx^2} + \alpha \frac{d^2\theta}{dx^2} \right),$$

from which and (5) we get

$$\left(\frac{d}{dt} + q \right) \frac{d^2s}{dt^2} = \kappa \left(\gamma \frac{d}{dt} + q \right) \frac{d^2s}{dx^2} \quad \dots\dots\dots\dots(6),$$

if γ be written (in the same sense as before) for $1 + \alpha\beta$.

If the vibrations be harmonic, we may suppose that s varies as e^{int}, and the equation becomes

$$\frac{d^2s}{dx^2} + \frac{n^2}{\kappa} \cdot \frac{q + in}{q + i\gamma n} \cdot s = 0 \quad \dots\dots\dots\dots\dots(7).$$

Let the coefficient of s in (7) be put into the form $\mu^2 e^{-2i\psi}$ where

$$\mu^4 = \frac{n^4}{\kappa^2} \cdot \frac{q^2 + n^2}{q^2 + \gamma^2 n^2} \quad \dots\dots\dots\dots\dots\dots(8),$$

and

$$2\psi = \tan^{-1}\frac{\gamma n}{q} - \tan^{-1}\frac{n}{q} = \tan^{-1}\frac{(\gamma - 1)\,nq}{\gamma n^2 + q^2} \quad \dots\dots(9).$$

Equation (7) is then satisfied by terms of the form

$$e^{\pm i\mu(\cos\psi - i\sin\psi)x},$$

but (μ being positive, and ψ less than $\frac{1}{2}\pi$) if we wish for the expression of the wave travelling in the positive direction, we must take the lower sign. Discarding the imaginary part, we find as the appropriate solution

$$s = A e^{-\mu\sin\psi x} \cos(nt - \mu\cos\psi\,x) \quad \dots\dots\dots\dots(10).$$

The first thing to be noticed is that the sound cannot be propagated to a distance unless $\sin \psi$ be insensible.

The velocity of propagation (V) is

$$V = n \mu^{-1} \sec \psi \ \dots\dots\dots\dots\dots(11),$$

which, when $\sin \psi$ is insensible, reduces to

$$V = n \mu^{-1} \ \dots\dots\dots\dots\dots(12).$$

Now from (9) we see that ψ cannot be insensible, unless q/n is either very great, or very small. On the first supposition from (11), or directly from (7), we have approximately, $V = \sqrt{\kappa}$ (Newton); and on the second, $V = \sqrt{(\kappa\gamma)}$, (Laplace), as ought evidently to be the case, when the meaning of q in (5) is considered. What we now learn is that, if q and n were comparable, the effect would be not merely a deviation of V from either of the limiting values, but a rapid stifling of the sound, which we know does not take place in nature.

Of this theoretical result we may convince ourselves, as Stokes explains, without the use of analysis. Imagine a mass of air to be confined within a closed cylinder, in which a piston is worked with a reciprocating motion. If the period of the motion be very long, the temperature of the air remains nearly constant, the heat developed by compression having time to escape by conduction or radiation. Under these circumstances the pressure is a function of volume, and whatever work has to be expended in producing a given compression is refunded when the piston passes through the same position in the reverse direction; no work is consumed in the long run. Next suppose that the motion is so rapid that there is no time for the heat and cold developed by the condensations and rarefactions to escape. The pressure is still a function of volume, and no work is dissipated. The only difference is that now the variations of pressure are more considerable than before in comparison with the variations of volume. We see how it is that both on Newton's and on Laplace's hypothesis the waves travel without dissipation, though with different velocities.

But in intermediate cases, when the motion of the piston is neither so slow that the temperature remains constant nor so quick that the heat has no time to adjust itself, the result is different. The work expended in producing a small condensation is no longer completely refunded during the corresponding rarefaction on account of the diminished temperature, part of the heat developed by the compression having in the meantime escaped. In fact the passage of heat by conduction or radiation from a warmer to a finitely colder body always involves dissipation, a principle which occupies a fundamental position in the science of Thermodynamics. In order therefore to maintain the motion of the piston, energy must be supplied from without, and if there be only a limited store to be drawn from, the motion must ultimately subside.

Another point to be noticed is that, if q and n were comparable, V would depend upon n, viz. on the pitch of the sound, a state of things which from experiment we have no reason to suspect. On the contrary the evidence of observation goes to prove that there is no such connection.

From (10) we see that the falling off in the intensity, estimated per wave-length, is a maximum with $\tan \psi$, or ψ; and by (9) ψ is a maximum when $q : n = \sqrt{\gamma}$. In this case

$$\mu = n\kappa^{-\frac{1}{2}}\gamma^{-\frac{1}{4}}, \qquad 2\psi = \tan^{-1}\gamma^{\frac{1}{2}} - \tan^{-1}\gamma^{-\frac{1}{2}}\ldots\ldots(13),$$

whence, if we take $\gamma = 1\cdot36$, $\quad 2\psi = 8°\,47'$.

Calculating from these data, we find that for each wave-length of advance, the amplitude of the vibration would be diminished in the ratio $\cdot6172$.

To take a numerical example, let

$\tau = \frac{1}{300}$ of a second, $\quad \lambda = $ wave-length $= 44$ inches [112 cm.].

In 20 yards [1828 cm.] the intensity would be diminished in the ratio of about 7 millions to one.

Corresponding to this,

$$q = 2198\ldots\ldots\ldots\ldots\ldots\ldots(14).$$

If the value of q were actually that just written, sounds of the pitch in question would be very rapidly stifled. We therefore infer that q is in fact either much greater or else much less. But even so large a value as 2000 is utterly inadmissible, as we may convince ourselves by considering the significance of equation (5).

Suppose that by a rigid envelope transparent to radiant heat, the volume of a small mass of gas were maintained constant, then the equation to determine its thermal condition at any time is

$$\frac{d\theta}{dt} + q\theta = 0,$$

whence
$$\theta = Ae^{-qt} \dots\dots\dots\dots\dots\dots(15),$$

where A denotes the initial excess of temperature, proving that after a time $1/q$ the excess of temperature would fall to less than half its original value. To suppose that this could happen in a two thousandth of a second of time would be in contradiction to the most superficial observation.

We are therefore justified in assuming that q is very small in comparison with n, and our equations then become approximately

$$\mu = \frac{n}{\kappa^{\frac{1}{2}}\gamma^{\frac{1}{2}}}, \qquad 2\psi = \frac{\gamma-1}{\gamma}\frac{q}{n}, \qquad V = n\mu^{-1} = \kappa^{\frac{1}{2}}\gamma^{\frac{1}{2}},$$

$$s = A\,e^{-(1-\gamma^{-1})\,qx/2V}\,\cos\frac{2\pi}{\lambda}\,(Vt-x)\,\dots\dots\dots(16).$$

The effects of a small radiation of heat are to be sought for rather in a damping of the vibration than in an altered velocity of propagation.

Stokes calculates that if $\gamma = 1\cdot414$, $V = 1100$, the ratio $(N : 1)$ in which the intensity is diminished in passing over a distance x, is given by $\log_{10} N = \cdot0001156\,qx$ in foot-second measure. Although we are not able to make precise measurements of the intensity of sound, yet the fact that audible vibrations can be propagated for many miles excludes any such value of q as could appreciably affect the velocity of transmission.

Neither is it possible to attribute to the air such a conducting power as could materially disturb the application of Laplace's theory. In order to trace the effects of conduction, we have only to replace q in (5) by $-q'd^2/dx^2$. Assuming as a particular solution

$$s = Ae^{i(nt+mx)},$$

we find
$$m^2 in\kappa\gamma = in^3 + q'n^2m^2 - \kappa q'm^4,$$

whence, if q' be relatively small,

$$m = \frac{-n}{\sqrt{(\kappa\gamma)}} \left(1 - \frac{\gamma-1}{\gamma} \frac{q'n}{2\kappa\gamma} i\right) \quad\dots\dots\dots\dots(17).$$

Thus the solution in real quantities is

$$s = A \cdot Exp\left(-\frac{\gamma-1}{\gamma} \frac{q'n^2x}{2\,(\kappa\gamma)^{\frac{3}{2}}}\right) \cdot \cos\left(nt - \frac{nx}{\sqrt{(\kappa\gamma)}}\right)\dots\dots(18),$$

leaving the velocity of propagation to this order of approximation still equal to $\sqrt{(\kappa\gamma)}$.

From (18) it appears that the first effect of conduction, as of radiation, is on the amplitude rather than on the velocity of propagation. In truth the conducting power of gases is so feeble, and in the case of audible sounds at any rate the time during which conduction can take place is so short, that disturbance from this cause is not to be looked for.

In the preceding discussions the waves are supposed to be propagated in an open space. When the air is confined within a tube, whose diameter is small in comparison with the wave-length, the conditions of the problem are altered, at least in the case of conduction. What we have to say on this head will, however, come more conveniently in another place.

248. From the expression $\sqrt{(p\gamma/\rho)}$, we see that in the same gas the velocity of sound is independent of the density, because if the temperature be constant, p varies as ρ $(p = R\rho\theta)$. On the other hand the velocity of sound is proportional to the square root of the absolute temperature, so that if a_0 be its value at 0° Cent.

$$a = a_0 \sqrt{1 + \frac{\theta'}{273}} \quad\dots\dots\dots\dots\dots\dots(1),$$

where the temperature is measured in the ordinary manner from the freezing point of water.

The most conspicuous effect of the dependence of the velocity of sound on temperature is the variability of the pitch of organ pipes. We shall see in the following chapters that the period of the note of a flue organ-pipe is the time occupied by a pulse in running over a distance which is a definite multiple of the length of the pipe, and therefore varies inversely as the velocity of propagation. The inconvenience arising from this alteration of pitch is aggravated by the fact that the reed pipes are not similarly affected; so that a change of temperature puts an organ out of tune with itself.

Prof. Mayer[1] has proposed to make the connection between temperature and wave-length the foundation of a pyrometric method, but I am not aware whether the experiment has ever been carried out.

The correctness of (1) as regards air at the temperatures of $0°$ and $100°$ has been verified experimentally by Kundt. See § 260.

In different gases at given temperature and pressure a is inversely proportional to the square roots of the densities, at least if γ be constant[2]. For the non-condensable gases γ does not sensibly vary from its value for air. [Thus in the case of hydrogen the velocity is greater than for air in the ratio

$$\sqrt{(1\cdot2933)} : \sqrt{(\cdot08993)},$$

or $$3\cdot792 : 1.]$$

The velocity of sound is not entirely independent of the degree of dryness of the air, since at a given pressure moist air is somewhat lighter than dry air. It is calculated that at $50°$ F. [$10°$ C.], air saturated with moisture would propagate sound between 2 and 3 feet per second faster than if it were perfectly dry. [1 foot $= 30\cdot5$ cm.]

The formula $a^2 = dp/d\rho$ may be applied to calculate the velocity of sound in liquids, or, if that be known, to infer conversely the coefficient of compressibility. In the case of water it is found by experiment that the compression per atmosphere is $\cdot0000457$. Thus, if $dp = 1033 \times 981$ in absolute C.G.S. units,

$$d\rho = \cdot0000457, \quad \text{since } \rho = 1.$$

Hence $$a = 1489 \text{ metres per second,}$$

which does not differ much from the observed value (1435).

249. In the preceding sections the theory of plane waves has been derived from the general equations of motion. We

---

[1] On an Acoustic Pyrometer. *Phil. Mag.* XLV. p. 18, 1873.

[2] According to the kinetic theory of gases, the velocity of sound is determined solely by, and is proportional to, the mean velocity of the molecules. Preston, *Phil. Mag.* (5) III. p. 441, 1877. [See also Waterston (1846), *Phil. Trans.* vol. CLXXXIII. A, p. 1, 1892.]

now proceed to an independent investigation in which the motion is expressed in terms of the actual position of the layers of air instead of by means of the velocity-potential, whose aid is no longer necessary inasmuch as in one dimension there can be no question of molecular rotation.

If y, $y + dy/dx \cdot dx$, define the actual positions at time t of neighbouring layers of air whose equilibrium positions are defined by x and $x + dx$, the density ρ of the included slice is given by

$$\rho \,:\, \rho_0 = 1 \,:\, \frac{dy}{dx} \quad \dots \dots \dots \dots \dots \dots \dots (1),$$

whence by (9) § 246,

$$p \,:\, p_0 = 1 \,:\, \left(\frac{dy}{dx}\right)^{\gamma} \quad \dots \dots \dots \dots \dots \dots (2),$$

the expansions and condensations being supposed to take place according to the adiabatic law. The mass of unit of area of the slice is $\rho_0 dx$, and the corresponding moving force is

$$- dp/dx \cdot dx,$$

giving for the equation of motion

$$\rho_0 \frac{d^2 y}{dt^2} + \frac{dp}{dx} = 0 \quad \dots \dots \dots \dots \dots \dots \dots (3).$$

Between (2) and (3) p is to be eliminated. Thus,

$$\left(\frac{dy}{dx}\right)^{\gamma+1} \frac{d^2 y}{dt^2} = \frac{p_0 \gamma}{\rho_0} \frac{d^2 y}{dx^2} \quad \dots \dots \dots \dots (4).$$

Equation (4) is an *exact* equation defining the actual abscissa y in terms of the equilibrium abscissa x and the time. If the motion be assumed to be small, we may replace $(dy/dx)^{\gamma+1}$, which occurs as the coefficient of the small quantity d^2y/dt^2, by its approximate value unity; and (4) then becomes

$$\frac{d^2 y}{dt^2} = \frac{p_0 \gamma}{\rho_0} \frac{d^2 y}{dx^2} \quad \dots \dots \dots \dots \dots \dots (5),$$

the ordinary approximate equation.

If the expansion be isothermal, as in Newton's theory, the equations corresponding to (4) and (5) are obtained by merely putting $\gamma = 1$.

Whatever may be the relation between p and ρ, depending on the constitution of the medium, the equation of motion is by (1) and (3)

$$\left(\frac{dy}{dx}\right)^2 \frac{d^2y}{dt^2} = \frac{dp}{d\rho} \frac{d^2y}{dx^2} \dots\dots\dots\dots\dots(6),$$

from which ρ, occurring in $dp/d\rho$, is to be eliminated by means of the relation between ρ and dy/dx expressed in (1).

250. In the preceding investigations of aerial waves we have supposed that the air is at rest except in so far as it is disturbed by the vibrations of sound, but we are of course at liberty to attribute to the whole mass of air concerned any common motion. If we suppose that the air is moving in the direction contrary to that of the waves and with the same actual velocity, the wave form, if permanent, is stationary in space, and the motion is *steady*. In the present section we will consider the problem under this aspect, as it is important to obtain all possible clearness in our views on the mechanics of wave propagation.

If u_0, p_0, ρ_0 denote respectively the velocity, pressure, and density of the fluid in its undisturbed state, and if u, p, ρ be the corresponding quantities at a point in the wave, we have for the equation of continuity

$$\rho u = \rho_0 u_0 \dots\dots\dots\dots\dots\dots(1),$$

and by (5) § 244 for the equation of energy

$$\int_{p_0}^{p} \frac{dp}{\rho} = \tfrac{1}{2} u_0^2 - \tfrac{1}{2} u^2 \dots\dots\dots\dots\dots(2).$$

Eliminating u, we get

$$\int_{p_0}^{p} \frac{dp}{\rho} = \tfrac{1}{2} u_0^2 \left(1 - \frac{\rho_0^2}{\rho^2}\right) \dots\dots\dots\dots(3),$$

determining the law of pressure under which alone it is possible for a stationary wave to maintain itself in fluid moving with velocity u_0. From (3)

$$\frac{dp}{d\rho} = u_0^2 \frac{\rho_0^2}{\rho^2} \dots\dots\dots\dots\dots(4),$$

or

$$p = \text{constant} - \frac{u_0^2 \rho_0^2}{\rho} \dots\dots\dots\dots(5).$$

Since the relation between the pressure and the density of actual gases is not that expressed in (5), we conclude that a self-maintaining stationary aerial wave is an impossibility, whatever may be the velocity u_0 of the general current, or in other words that a wave cannot be propagated relatively to the undisturbed parts of the gas without undergoing an alteration of type. Nevertheless, when the changes of density concerned are small, (5) may be satisfied approximately; and we see from (4) that the velocity of stream necessary to keep the wave stationary is given by

$$u_0 = \sqrt{\left(\frac{dp}{d\rho}\right)} \dots\dots\dots\dots\dots\dots(6),$$

which is the same as the velocity of the wave estimated relatively to the fluid.

This method of regarding the subject shews, perhaps more clearly than any other, the nature of the relation between velocity and condensation § 245 (3), (4). In a stationary wave-form a loss of velocity accompanies an augmented density according to the principle of energy, and therefore the fluid composing the condensed parts of a wave moves forward more slowly than the undisturbed portions. Relatively to the fluid therefore the motion of the condensed parts is in the same direction as that in which the waves are propagated.

When the relation between pressure and density is other than that expressed in (5), a stationary wave can be maintained only by the aid of an impressed force. By (1) and (2) § 237 we have, on the supposition that the motion is steady,

$$X = u \frac{du}{dx} + \frac{1}{\rho} \frac{dp}{dx} \dots\dots\dots\dots\dots\dots (7),$$

while the relation between u and ρ is given by (1). If we suppose that $p = a^2 \rho$, (7) becomes

$$X = (u^2 - a^2) \frac{d \log u}{dx} \dots\dots\dots\dots\dots(8),$$

shewing that an impressed force is necessary at every place where u is variable and unequal to a.

251. The reason of the change of type which ensues when a wave is left to itself is not difficult to understand. From the ordinary theory we know that an infinitely small disturbance is propagated with a certain velocity a, which velocity is relative to the parts of the medium undisturbed by the wave. Let us consider now the case of a wave so long that the variations of velocity and density are insensible for a considerable distance along it, and at a place where the velocity (u) is finite let us imagine a small secondary wave to be superposed. The velocity with which the secondary wave is propagated through the medium is a, but on account of the local motion of the medium itself the whole velocity of advance is $a + u$, and depends upon the part of the long wave at which the small wave is placed. What has been said of a secondary wave applies also to the parts of the long wave itself, and thus we see that after a time t the place, where a certain velocity u is to be found, is in advance of its original position by a distance equal, not to at, but to $(a + u) t$; or, as we may express it, u is propagated with a velocity $a + u$. In symbolical notation $u = f\{x - (a + u) t\}$, where f is an arbitrary function, an equation first obtained by Poisson[1].

From the argument just employed it might appear at first sight that alteration of type was a necessary incident in the progress of a wave, independently of any particular supposition as to the relation between pressure and density, and yet it was proved in § 250 that in the case of one particular law of pressure there would be no alteration of type. We have, however, tacitly assumed in the present section that a is constant, which is tantamount to a restriction to Boyle's law. Under any other law of pressure $\sqrt{(dp/d\rho)}$ is a function of ρ, and therefore, as we shall see presently, of u. In the case of the law expressed in (5) § 250, the relation between u and ρ for a progressive wave is such that $\sqrt{(dp/d\rho)} + u$ is constant, as much advance being lost by slower propagation due to augmented density as is gained by superposition of the velocity u.

So far as the constitution of the medium itself is concerned there is nothing to prevent our ascribing arbitrary values to both u and ρ, but in a progressive wave a relation between these two quantities must be satisfied. We know already (§ 245) that this is the case when the disturbance is small, and the following argument will not only shew that such a relation is to be expected in cases where the square of the motion must be retained, but will even define the form of the relation.

Whatever may be the law of pressure, the velocity of propagation of small disturbances is by § 245 equal to $\sqrt{(dp/d\rho)}$, and in

[1] Mémoire sur la Théorie du Son. *Journal de l'école polytechnique*, t. VII. p. 319. 1808.

a positive progressive wave the relation between velocity and condensation is

$$u : s = \sqrt{\left(\frac{dp}{d\rho}\right)} \quad \dots \dots \dots \dots \dots \dots (1).$$

If this relation be violated at any point, a wave will emerge, travelling in the negative direction. Let us now picture to ourselves the case of a positive progressive wave in which the changes of velocity and density are very gradual but become important by accumulation, and let us inquire what conditions must be satisfied in order to prevent the formation of a negative wave. It is clear that the answer to the question whether, or not, a negative wave will be generated at any point will depend upon the state of things in the immediate neighbourhood of the point, and not upon the state of things at a distance from it, and will therefore be determined by the criterion applicable to small disturbances. In applying this criterion we are to consider the velocities and condensations, not absolutely, but relatively to those prevailing in the neighbouring parts of the medium, so that the form of (1) proper for the present purpose is

$$du = \sqrt{\left(\frac{dp}{d\rho}\right)} \cdot \frac{d\rho}{\rho} \quad \dots \dots \dots \dots \dots \dots (2);$$

whence

$$u = \int \sqrt{\left(\frac{dp}{d\rho}\right)} \cdot \frac{d\rho}{\rho} \quad \dots \dots \dots \dots \dots \dots (3),$$

which is the relation between u and ρ necessary for a positive progressive wave. Equation (2) was obtained analytically by Earnshaw[1].

In the case of Boyle's law, $\sqrt{(dp/d\rho)}$ is constant, and the relation between velocity and density, given first, I believe, by Helmholtz[2], is

$$u = a \log \frac{\rho}{\rho_0} \quad \dots \dots \dots \dots \dots \dots (4),$$

if ρ_0 be the density corresponding to $u = 0$.

In this case Poisson's integral allows us to form a definite idea of the change of type accompanying the earlier stages of the progress of the wave, and it finally leads us to a difficulty which has not as yet been surmounted[3]. If we draw a curve to represent

[1] *Phil. Trans.* 1859, p. 146.

[2] *Fortschritte der Physik*, IV. p. 106. 1852.

[3] Stokes, "On a difficulty in the Theory of Sound." *Phil. Mag.* Nov. 1848.

the distribution of velocity, taking x for abscissa and u for ordinate, we may find the corresponding curve after the lapse of time t by the following construction. Through any point on the original curve draw a straight line in the positive direction parallel to x, and of length equal to $(a + u) t$, or, as we are concerned with the shape of the curve only, equal to $u t$. The locus of the ends of these lines is the velocity curve after a time t.

But this law of derivation cannot hold good indefinitely. The crests of the velocity curve gain continually on the troughs and must at last overtake them. After this the curve would indicate two values of u for one value of x, ceasing to represent anything that could actually take place. In fact we are not at liberty to push the application of the integral beyond the point at which the velocity becomes discontinuous, or the velocity curve has a vertical tangent. In order to find when this happens let us take two neighbouring points on any part of the curve which slopes downwards in the positive direction, and inquire after what time this part of the curve becomes vertical. If the difference of abscissæ be dx, the hinder point will overtake the forward point in the time $dx \div (- du)$. Thus the motion, as determined by Poisson's equation, becomes discontinuous after a time equal to the reciprocal, taken positively, of the greatest negative value of du/dx.

For example, let us suppose that

$$u = U \cos \frac{2\pi}{\lambda} \{x - (a + u) t\},$$

where U is the greatest initial velocity. When $t = 0$, the greatest negative value of du/dx is $- 2\pi U/\lambda$; so that discontinuity will commence at the time $t = \lambda/2\pi U$.

When discontinuity sets in, a state of things exists to which the usual differential equations are inapplicable; and the subsequent progress of the motion has not been determined. It is probable, as suggested by Stokes, that some sort of reflection would ensue. In regard to this matter we must be careful to keep purely mathematical questions distinct from physical ones. In practice we have to do with spherical waves, whose divergency may of itself be sufficient to hold in check the tendency to discontinuity. In actual gases too it is certain that before discontinuity could enter, the law of pressure would begin to change its form, and the influence of viscosity could no longer be neglected. But these considerations have nothing to do with the mathematical problem of determining what would happen to waves of finite amplitude in a medium, free from viscosity, whose pressure is under all circumstances exactly proportional to its density; and this problem has not been solved.

It is worthy of remark that, although we may of course conceive a wave of finite disturbance to exist at any moment, there is a limit to the duration of its previous independent existence. By drawing lines in the negative instead of in the positive direction we may trace the history of the velocity curve; and we see that as we push our inquiry further and further into past time the forward slopes become easier and the backward slopes steeper. At a time, equal to the greatest positive value of dx/du, antecedent to that at which the curve is first contemplated, the velocity would be discontinuous.

252. The complete integration of the exact equations (4) and (6) § 249 in the case of a progressive wave was first effected by Earnshaw[1]. Finding reason for thinking that in a sound wave the equation

$$\frac{dy}{dt} = F\left(\frac{dy}{dx}\right) \dots\dots\dots\dots\dots\dots(1)$$

must always be satisfied, he observed that the result of differentiating (1) with respect to t, viz.

$$\frac{d^2y}{dt^2} = \left\{F'\left(\frac{dy}{dx}\right)\right\}^2 \frac{d^2y}{dx^2} \dots\dots\dots\dots\dots(2),$$

can by means of the arbitrary function F be made to coincide with any dynamical equation in which the ratio of d^2y/dt^2 and d^2y/dx^2 is expressed in terms of dy/dx. The form of the function F being thus determined, the solution may be completed by the usual process applicable to such cases[2].

Writing for brevity α in place of dy/dx, we have

$$dy = \frac{dy}{dx}dx + \frac{dy}{dt}dt = \alpha\,dx + F(\alpha)\,dt,$$

and the integral is to be found by eliminating α between the equations

$$\left. \begin{array}{l} y = \alpha x + F(\alpha)\,t + \phi(\alpha) \\ 0 = \ x + F'(\alpha)\,t + \phi'(\alpha) \end{array} \right\} \dots\dots\dots\dots\dots(3),$$

α being equal to ρ_0/ρ, and ϕ being an arbitrary function.

[1] *Proceedings of the Royal Society*, Jan. 6, 1859. *Phil. Trans.* 1860, p. 133.

[2] Boole's *Differential Equations*, Ch. xiv.

If $p = a^2 \rho$, the exact equation (6 § 249) is

$$\left(\frac{dy}{dx}\right)^2 \frac{d^2y}{dt^2} = a^2 \frac{d^2y}{dx^2} \dots\dots\dots\dots\dots(4),$$

by comparison of which with (2) we see that

$$F'(\alpha) = \frac{\pm a}{\alpha} \dots\dots\dots\dots\dots(5),$$

or on integration

$$F(\alpha) = C \pm a \log \alpha \dots\dots\dots\dots\dots(6),$$

as might also have been inferred from (4) § 251. The constant C vanishes, if $F(\alpha)$, viz. u, vanish when $\alpha = 1$, or $\rho = \rho_0$; otherwise it represents a velocity of the medium as a whole, having nothing to do with the wave as such. For a *positive* progressive wave the lower signs in the ambiguities are to be used. Thus in place of (3), we have

$$\left.\begin{array}{l} y = \alpha x - a \log \alpha\, t + \phi(\alpha) \\ 0 = \alpha x - a t \quad + \alpha\, \phi'(\alpha) \end{array}\right\} \dots\dots\dots\dots(7),$$

and

$$u = - a \log \alpha = a \log \frac{\rho}{\rho_0} \dots\dots\dots\dots\dots(8).$$

If we subtract the second of equations (7) from the first, we get

$$y - at + at \log \alpha = \phi(\alpha) - \alpha\, \phi'(\alpha),$$

from which by (8) we see that $y - (a + u)\, t$ is an arbitrary function of α, or of u. Conversely therefore u is an arbitrary function of $y - (a + u)\, t$, and we may write

$$u = f\{y - (a + u)\, t\} \dots\dots\dots\dots\dots(9).$$

Equation (9) is Poisson's integral, considered in the preceding section, where the symbol x has the same meaning as here attaches to y.

253. The problem of plane waves of finite amplitude attracted also the attention of Riemann, whose memoir was communicated to the Royal Society of Göttingen on the 28th of November, 1859[1]. Riemann's investigation is founded on the general hydrodynamical equations investigated in §§ 237, 238, and is not restricted to any particular law of pressure. In order, however, not unduly to

---

[1] Ueber die Fortpflanzung ebener Luftwellen von endlicher Schwingungsweite. Göttingen, *Abhandlungen*, t. VIII. 1860. See also an excellent abstract in the *Fortschritte der Physik*, xv. p. 123. [Reference may be made also to a paper by C. V. Burton, *Phil. Mag.* xxxv. p. 317, 1893.]

extend the discussion of this part of our subject, already perhaps treated at greater length than its acoustical importance would warrant, we shall here confine ourselves to the case of Boyle's law of pressure.

Applying equations (1), (2) of § 237 and (1) of § 238 to the circumstances of the present problem, we get

$$\frac{du}{dt} + u\frac{du}{dx} = -a^2\frac{d\log\rho}{dx} \quad\dots\dots\dots\dots\dots(1),$$

$$\frac{d\log\rho}{dt} + u\frac{d\log\rho}{dx} = -\frac{du}{dx} \quad\dots\dots\dots\dots(2).$$

If we multiply (2) by $\pm a$, and afterwards add it to (1), we obtain

$$\frac{dP}{dt} = -(u+a)\frac{dP}{dx}, \qquad \frac{dQ}{dt} = -(u-a)\frac{dQ}{dx} \quad\dots\dots(3),$$

where $\qquad P = a\log\rho + u, \qquad Q = a\log\rho - u\dots\dots\dots\dots(4).$

Thus $\qquad dP = \frac{dP}{dx}\{dx - (u+a)\,dt\} \quad\dots\dots\dots\dots\dots(5),$

$$dQ = \frac{dQ}{dx}\{dx - (u-a)\,dt\} \quad\dots\dots\dots\dots(6).$$

These equations are more general than Poisson's and Earnshaw's in that they are not limited to the case of a single positive, or negative, progressive wave. From (5) we learn that whatever may be the value of P corresponding to the point x and the time t, the same value of P corresponds to the point $x + (u+a)\,dt$ at the time $t + dt$; and in the same way from (6) we see that Q remains unchanged when x and t acquire the increments $(u-a)\,dt$ and dt respectively. If P and Q be given at a certain instant of time as functions of x, and the representative curves be drawn, we may deduce the corresponding value of u by (4), and thus, as in § 251, construct the curves representing the values of P and Q after the small interval of time dt, from which the new values of u and ρ in their turn become known, and the process can be repeated.

The element of the fluid, to which the values of P and Q at any moment belong, is itself moving with the velocity u, so that the velocities of P and Q relatively to the element are numerically the same, and equal to a, that of P being in the positive direction and that of Q in the negative direction.

We are now in a position to trace the consequences of an initial disturbance which is confined to a finite portion of the medium, e.g. between $x = \alpha$ and $x = \beta$, outside which the medium is at rest and at its normal density, so that the values of P and Q are $a \log \rho_0$. Each value of P propagates itself in turn to the elements of fluid which lie in front of it, and each value of Q to those that lie behind it. The hinder limit of the region in which P is variable, viz. the place where P first attains the constant value $a \log \rho_0$, comes into contact first with the variable values of Q, and moves accordingly with a variable[1] velocity. At a definite time, requiring for its determination a solution of the differential equations, the hinder (left hand) limit of the region through which P varies, meets the hinder (right hand) limit of the region through which Q varies, after which the two regions separate themselves, and include between them a portion of fluid in its equilibrium condition, as appears from the fact that the values of P and Q are both $a \log \rho_0$. In the positive wave Q has the constant value $a \log \rho_0$, so that $u = a \log (\rho/\rho_0)$, as in (4) § 251; in the negative wave P has the same constant value, giving as the relation between u and ρ, $u = - a \log (\rho/\rho_0)$. Since in each progressive wave, when isolated, a law prevails connecting the quantities u and ρ, we see that in the positive wave du vanishes with dP, and in the negative wave du vanishes with dQ. Thus from (5) we learn that in a positive progressive wave du vanishes, if the increments of x and t be such as to satisfy the equation $dx - (u + a)\, dt = 0$, from which Poisson's integral immediately follows.

It would lead us too far to follow out the analytical development of Riemann's method, for which the reader must be referred to the original memoir; but it would be improper to pass over in silence an error on the subject of discontinuous motion into which Riemann and other writers have fallen. It has been held that a state of motion is possible in which the fluid is divided into two parts by a surface of discontinuity propagating itself with constant velocity, all the fluid on one side of the surface of discontinuity being in one uniform condition as to density and velocity, and on the other side in a second uniform condition in the same respects. Now, if this motion were possible, a motion of the same kind in which the surface of discontinuity is at rest would also be

---

[1] At this point an error seems to have crept into Riemann's work, which is corrected in the abstract of the *Fortschritte der Physik*.

possible, as we may see by supposing a velocity equal and opposite to that with which the surface of discontinuity at first moves, to be impressed upon the whole mass of fluid. In order to find the relations that must subsist between the velocity and density on the one side (u_1, ρ_1) and the velocity and density on the other side (u_2, ρ_2), we notice in the first place that by the principle of conservation of matter $\rho_2 u_2 = \rho_1 u_1$. Again, if we consider the momentum of a slice bounded by parallel planes and including the surface of discontinuity, we see that the momentum leaving the slice in the unit of time is for each unit of area $(\rho_2 u_2 = \rho_1 u_1) u_2$, while the momentum entering it is $\rho_1 u_1^2$. The difference of momentum must be balanced by the pressures acting at the boundaries of the slice, so that

$$\rho_1 u_1 (u_2 - u_1) = p_1 - p_2 = a^2 (\rho_1 - \rho_2),$$

whence

$$u_1 = a \sqrt{\left(\frac{\rho_2}{\rho_1}\right)}, \qquad u_2 = a \sqrt{\left(\frac{\rho_1}{\rho_2}\right)} \dots \dots \dots \dots (7).$$

The motion thus determined is, however, not possible; it satisfies indeed the conditions of mass and momentum, but it violates the condition of energy (§ 244) expressed by the equation

$$\tfrac{1}{2} u_2^2 - \tfrac{1}{2} u_1^2 = a^2 \log \rho_1 - a^2 \log \rho_2 \dots \dots \dots \dots (8).$$

This argument has been already given in another form in § 250, which would alone justify us in rejecting the assumed motion, since it appears that no steady motion is possible except under the law of density there determined. From equation (8) of that section we can find what impressed forces would be necessary to maintain the motion defined by (7). It appears that the force X, though confined to the place of discontinuity, is made up of two parts of opposite signs, since by (7) u passes *through* the value a. The whole moving force, viz. $\int X \rho \, dx$, vanishes, and this explains how it is that the condition relating to momentum is satisfied by (7), though the force X be ignored altogether.

253 a. Among the phenomena of the second order which admit of a ready explanation, a prominent place must be assigned to the repulsion of resonators discovered independently by Dvořák[1] and Mayer[2]. These observers found that an air resonator of any kind (Ch. XVI.) when exposed to a powerful source

---

[1] *Pogg. Ann.* CLVII. p. 42, 1876; *Wied. Ann.* III. p. 328, 1878.

[2] *Phil. Mag.* vol. VI. p. 225, 1878.

of sound experiences a force directed inwards from the mouth, somewhat after the manner of a rocket. A combination of four light resonators, mounted anemometer fashion upon a steel point, may be caused to revolve continuously.

If there be no impressed forces, equation (2) § 244 gives

$$\varpi = \int \frac{dp}{\rho} = -\frac{d\phi}{dt} - \tfrac{1}{2} U^2 \quad \dots\dots\dots\dots(1).$$

Distinguishing the values of the quantities at two points of space by suffixes, we may write

$$\varpi_1 - \varpi_0 = \frac{d}{dt}(\phi_0 - \phi_1) + \tfrac{1}{2}U_0^2 - \tfrac{1}{2}U_1^2\dots\dots\dots(2).$$

This equation holds good at every instant. Integrating it over a long range of time we obtain as applicable to every case of fluid motion in which the flow between the two points does not continually increase

$$\int\varpi_1 dt - \int\varpi_0 dt = \tfrac{1}{2}\int U_0^2 dt - \tfrac{1}{2}\int U_1^2 dt \dots\dots\dots(3).$$

The first point (with suffix 0) is now to be chosen at such a distance that the variation of pressure and the velocity are there insensible. Accordingly

$$\int\varpi_1 dt = -\tfrac{1}{2}\int U_1^2 dt\dots\dots\dots\dots(4).$$

This equation is true wherever the second point be taken. If it be in the interior of a resonator, or at a corner where three fixed walls meet, $U_1 = 0$, and therefore

$$\int(\varpi_1 - \varpi_0)\, dt = 0 \dots\dots\dots\dots(5),$$

or the mean value of ϖ in the interior is the same as at a distance outside.

By (9) § 246, if the expansions and contractions be adiabatic, $p \propto \rho^\gamma$; and $\varpi = p^{(\gamma-1)/\gamma}$. Thus

$$\sqrt{\left\{\left(\frac{p_1}{p_0}\right)^{(\gamma-1)/\gamma} - 1\right\}} dt = 0\dots\dots\dots\dots(6).$$

If in (6) we suppose that the difference between p_1 and p_0 is comparatively small, we may expand the function there contained by the binomial theorem. The approximate result may be expressed

$$\int\frac{p_1 - p_0}{p_0}\, dt = \frac{1}{2\gamma}\int\left(\frac{p_1 - p_0}{p_0}\right)^2 dt \quad \dots\dots\dots(7),$$

shewing that the mean value of $(p_1 - p_0)$ is positive, or in other words that the mean pressure in the resonator is *in excess* of the atmospheric pressure[1]. The resonator therefore tends to move as if impelled by a force acting normally over the area of its aperture and directed *inwards*.

The experiment may be made (after Dvořák) with a Helmholtz resonator by connecting the nipple with a horizontal and not too narrow glass tube in which moves a piston of ether. When a fork of suitable pitch, e.g. 256 or 512, is vigorously excited and presented to the mouth of the resonator, the movement of the ether shews an augmentation of pressure, while the similar presentation of the non-vibrating fork is without effect.

If to the first order of small quantities

$$(p - p_0)/p_0 = P \cos nt \dots\dots\dots\dots\dots(8),$$

its mean value correct to the second order is $P^2/4\gamma$, in which for air and the principal gases $\gamma = 1\cdot4$.

If the expansions and contractions be supposed to take place isothermally, the corresponding result is arrived at by putting $\gamma = 1$ in (7).

253 *b*. In § 253 *a* the effect to be explained is intimately connected with the compressibility of the fluid which occupies the interior of the resonator. In the class of phenomena now to be considered the compressibility of the fluid is of secondary importance, and the leading features of the explanation may be given upon the supposition that the fluid retains a constant density throughout.

If ρ be constant, (4) § 253 *a* may be written

$$\int (p_1 - p_0)\, dt = -\tfrac{1}{2}\rho \int U_1{}^2 dt \dots\dots\dots\dots(1),$$

shewing that the mean pressure at a place where there is motion is less than in the undisturbed parts of the fluid—a theorem due to Kelvin[2], and applied by him to the explanation of the attractions observed by Guthrie and other experimenters. Thus a vibrating tuning-fork, presented to a delicately suspended rectangle of paper, appears to exercise an attraction, the mean value of U^2 being greater on the face exposed to the fork than upon the back.

[1] *Phil. Mag.* vol. VI. p. 270, 1878.
[2] *Proc. Roy. Soc.* vol. XIX. p. 271, 1887.

In the above experiment the action depends upon the proximity of the source of disturbance. When the flow of fluid, whether steady or alternating, is uniform over a large region, the effect upon an obstacle introduced therein is a question of shape. In the case of a sphere there is manifestly no tendency to turn; and since the flow is symmetrical on the up-stream and down-stream sides, the mean pressures given by (1) balance one another. Accordingly a sphere experiences neither force nor couple. It is otherwise when the form of the body is elongated or flattened. That a flat obstacle tends to turn its flat side to the stream[1] may be inferred from the general character of the lines of flow round it. The pressures at the various points of the surface BC (Fig. 54 *a*) depend upon the velocities of the fluid there obtaining. The full pressure due to the complete stoppage of the stream is to be found at two points,

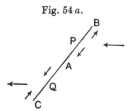

Fig. 54 *a*.

where the current divides. It is pretty evident that upon the up-stream side this lies (P) on AB, and upon the down-stream side upon AC at the corresponding point Q. The resultant of the pressures thus tends to turn AB so as to face the stream.

When the obstacle is in the form of an ellipsoid, the mathematical calculation of the forces can be effected; but it must suffice here to refer to the particular case of a thin circular disc, whose normal makes an angle θ with the direction of the undisturbed stream. It may be proved[2] that the moment M of the couple tending to diminish θ has the value given by

$$M = \tfrac{4}{3}\rho a^3 W^2 \sin 2\theta \quad\dots\dots\dots\dots\dots(2),$$

a being the radius of the disc and W the velocity of the stream. If the stream be alternating instead of steady, we have merely to employ the mean value of W^2, as appears from (1).

The observation that a delicately suspended disc sets itself across the direction of alternating currents of air originated in the attempt to explain certain anomalies in the behaviour of a magnetometer mirror[3]. In illustration, " a small disc of paper, about the size of a sixpence, was hung by a fine silk fibre across

[1] Thomson and Tait's *Natural Philosophy*, § 336, 1867.

[2] W. König, *Wied. Ann.* t. XLIII. p. 51, 1891.

[3] *Proc. Roy. Soc.* vol. XXXII. p. 110, 1881.

the mouth of a resonator of pitch 128. When a sound of this pitch is excited in the neighbourhood, there is a powerful rush of air into and out of the resonator, and the disc sets itself promptly across the passage. A fork of pitch 128 may be held near the resonator, but it is better to use a second resonator at a little distance in order to avoid any possible disturbance due to the neighbourhood of the vibrating prongs. The experiment, though rather less striking, was also successful with forks and resonators of pitch 256."

Upon this principle an instrument may be constructed for measuring the intensities of aerial vibrations of selected pitch[1]. A tube, measuring three quarters of a wave length, is open at one end and at the other is closed air-tight by a plate of glass. At one quarter of a wave length's distance from the closed end is hung by a silk fibre a light mirror with attached magnet, such as is used for reflecting galvanometers. In its undisturbed condition the plane of the mirror makes an angle of 45° with the axis of the tube. At the side is provided a glass window, through which light, entering along the axis and reflected by the mirror, is able to escape from the tube and to form a suitable image upon a divided scale. The tube as a whole acts as a resonator, and the alternating currents at the loop (§ 255) deflect the mirror through an angle which is read in the usual manner.

In an instrument constructed by Boys[2] the sensitiveness is exalted to an extraordinary degree. This is effected partly by the use of a very light mirror with suspension of quartz fibre, and partly by the adoption of double resonance. The large resonator is a heavy brass tube of about 10 cm. diameter, closed at one end, and of such length as to resound to *e'*. The mirror is hung in a short lateral tube forming a communication between the large resonator and a small glass bulb of suitable capacity. The external vibrations may be regarded as magnified first by the large resonator and then *again* by the small one, so that the mirror is affected by powerful alternating currents of air. The selection of pitch is so definite that there is hardly any response to sounds which are a semi-tone too high or too low.

Perhaps the most striking of all the effects of alternating aerial currents is the rib-like structure assumed by cork filings in

[1] *Phil. Mag.* vol. XIV. p. 186, 1882.
[2] *Nature*, vol. XLII. p. 604, 1890.

Kundt's experiment § 260. Close observation, while the vibrations are in progress, shews that the filings are disposed in thin laminæ transverse to the tube and extending upwards to a certain distance from the bottom. The effect is a maximum at the loops, and disappears in the neighbourhood of the nodes. When the vibrations stop, the laminæ necessarily fall, and in so doing lose much of their sharpness, but they remain visible as transverse streaks.

The explanation of this peculiar behaviour has been given by W. König[1]. We have seen that a single spherical obstacle experiences no force from an alternating current. But this condition of things is disturbed by the presence of a neighbour. Consider for simplicity the case of two spheres at a moderate distance apart, and so situated that the line of centres is either parallel to the stream, Fig. 54 *b*, or perpendicular to it, Fig. 54 *c*. It is easy to recognise that the velocity between the spheres will be less in the first case and greater in the second than on the averted hemispheres. Since the pressure increases as the velocity diminishes, it follows

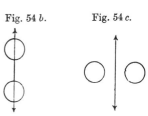

Fig. 54 *b*. Fig. 54 *c*.

that in the first position the spheres will repel one another, and that in the second position they will attract one another. The result of these forces between neighbours is plainly a tendency to aggregate in laminæ. The case may be contrasted with that of iron filings in a magnetic field, whose direction is parallel to that of the aerial current. There is then attraction in the first position and repulsion in the second, and the result is a tendency to aggregate in *filaments*.

On the foundation of the analysis of Kirchhoff, König has calculated the forces operative in the case of two spheres which are not too close together. If a_1, a_2 be the radii of the spheres, r their distance asunder, θ the angle between the line of centres and the direction of the current taken as axis of z (Fig. 54 *d*), W the velocity of the current, then the components of force upon the sphere B in the direction of z and of x

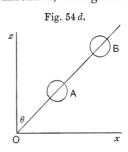

Fig. 54 *d*.

[1] *Wied. Ann.* t. XLII. pp. 353, 549, 1891.

drawn perpendicular to z in the plane containing z and the line of centres, are given by

$$Z = - \frac{3\pi\rho a_1^3 a_2^3 W^2}{r^4} \cos\theta \, (3 - 5\cos^2\theta)\ldots\ldots\ldots(3),$$

$$X = - \frac{3\pi\rho a_1^3 a_2^3 W^2}{r^4} \sin\theta \, (1 - 5\cos^2\theta)\ldots\ldots\ldots(4),$$

the third component Y vanishing by virtue of the symmetry. In the case of Fig. 54 b $\theta = 0$, and there is repulsion equal to $6\pi\rho a_1^3 a_2^3 W^2/r^4$; in the case of Fig. 54 c $\theta = \frac{1}{2}\pi$, and the force is an attraction $3\pi\rho a_1^3 a_2^3 W^2/r^4$. In oblique positions the direction of the force does not coincide with the line of centres.

If the spheres be rigidly connected, the forces upon the system reduce to a couple (tending to increase θ) of moment given by

$$-Z\sin\theta + X\cos\theta = \frac{3\pi\rho a_1^3 a_2^3 W^2}{r^3} \sin 2\theta \ldots\ldots\ldots(5).$$

When the current is alternating, we are to take the *mean* value of W^2 in (3), (4), (5).

254. The exact experimental determination of the velocity of sound is a matter of greater difficulty than might have been expected. Observations in the open air are liable to errors from the effects of wind, and from uncertainty with respect to the exact condition of the atmosphere as to temperature and dryness. On the other hand when sound is propagated through air contained in pipes, disturbance arises from friction and from transfer of heat; and, although no great errors from these sources are to be feared in the case of tubes of considerable diameter, such as some of those employed by Regnault, it is difficult to feel sure that the ideal plane waves of theory are nearly enough realized.

The following Table[1] contains a list of the principal experimental determinations which have been made hitherto.

Names of Observers.	Velocity of Sound at 0° Cent. in Metres.
Académie des Sciences (1738)	332
Benzenberg (1811)	333·7 / 332·3
Goldingham (1821)	331·1
Bureau des Longitudes (1822)	330·6
Moll and van Beek	332·2

[1] Bosanquet, *Phil. Mag.* April, 1877.

Names of Observers.	Velocity of Sound at 0° Cent. in Metres.
Stampfer and Myrback............................	332·4
Bravais and Martins (1844)	332·4
Wertheim	331·6
Stone (1871)	332·4
Le Roux...	330·7
Regnault	330·7

In Stone's experiments[1] the course over which the sound was timed commenced at a distance of 640 feet from the source, so that any errors arising from excessive disturbance were to a great extent avoided.

A method has been proposed by Bosscha[2] for determining the velocity of sound without the use of great distances. It depends upon the precision with which the ear is able to decide whether short ticks are simultaneous, or not. In König's[3] form of the experiment, two small electro-magnetic counters are controlled by a fork-interrupter (§ 64), whose period is one-tenth of a second, and give synchronous ticks of the same period. When the counters are close together the audible ticks coincide, but as one counter is gradually removed from the ear, the two series of ticks fall asunder. When the difference of distances is about 34 metres, coincidence again takes place, proving that 34 metres is about the distance traversed by sound in a tenth part of a second.

[On the basis of experiments made in pipes Violle and Vautier[4] give 331·10 as applicable in free air. The result includes a correction, amounting to 0·68, which is of a more or less theoretical character, representing the presumed influence of the pipe (0·7^m in diameter).]

[1] *Phil. Trans.* 1872, p. 1.   [2] *Pogg. Ann.* xcii. 486. 1854.
[3] *Pogg. Ann.* cxviii. 610. 1863.   [4] *Ann. de Chim.* t. xix.; 1890.

CHAPTER XII.

255. WE have already (§ 245) considered the solution of our fundamental equation, when the velocity-potential, in an unlimited fluid, is a function of one space co-ordinate only. In the absence of friction no change would be caused by the introduction of any number of fixed cylindrical surfaces, whose generating lines are parallel to the co-ordinate in question; for even when the surfaces are absent the fluid has no tendency to move across them. If one of the cylindrical surfaces be closed (in respect to its transverse section), we have the important problem of the axial motion of air within a cylindrical pipe, which, when once the mechanical conditions at the ends are given, is independent of anything that may happen outside the pipe.

Considering a simple harmonic vibration, we know (§ 245) that, if ϕ varies as e^{int},

$$\frac{d^2\phi}{dx^2} + k^2\phi = 0 \dots\dots\dots\dots\dots\dots(1),$$

where

$$k = \frac{2\pi}{\lambda} = \frac{n}{a}\dots\dots\dots\dots\dots\dots\dots(2).$$

The solution may be written in two forms—

$$\left.\begin{array}{l}\phi = (A \cos kx + B \sin kx)\,e^{int} \\ \phi = (A\,e^{ikx} + B\,e^{-ikx})\,e^{int}\end{array}\right\} \dots\dots\dots\dots\dots\dots(3),$$

of which finally only the real parts will be retained. The first form will be most convenient when the vibration is stationary, or nearly so, and the second when the motion reduces itself to a positive, or negative, progressive undulation. The constants A and B in the symbolical solution may be complex, and thus the final expression in terms of real quantities will involve *four* arbitrary constants. If we wish to use real quantities throughout, we must take

$$\phi = (A \cos kx + B \sin kx) \cos nt$$
$$+ (C \cos kx + D \sin kx) \sin nt \dots\dots\dots\dots (4),$$

but the analytical work would generally be longer. When no ambiguity can arise, we shall sometimes for the sake of brevity drop, or restore, the factor involving the time without express mention. Equations such as (1) are of course equally true whether the factor be understood or not.

Taking the first form in (3), we have

$$\left. \begin{aligned} \phi &= \quad\ A \cos kx + \quad B \sin kx \\ \frac{d\phi}{dx} &= -kA \sin kx + kB \cos kx \end{aligned} \right\} \dots\dots\dots\dots(5).$$

If there be any point at which either ϕ or $d\phi/dx$ is permanently zero, the ratio $A : B$ must be real, and then the vibration is *stationary*, that is, the same in phase at all points simultaneously.

Let us suppose that there is a node at the origin. Then when $x = 0$, $d\phi/dx$ vanishes, the condition of which is $B = 0$. Thus

$$\phi = A \cos kx\, e^{int}; \quad \frac{d\phi}{dx} = -kA \sin kx\, e^{int} \dots\dots\dots(6),$$

from which, if we substitute $Pe^{i\theta}$ for A, and throw away the imaginary part,

$$\left. \begin{aligned} \phi &= \quad\ P \cos kx \cos (nt + \theta) \\ \frac{d\phi}{dx} &= -kP \sin kx \cos (nt + \theta) \end{aligned} \right\} \dots\dots\dots\dots (7).$$

From these equations we learn that $d\phi/dx$ vanishes wherever $\sin kx = 0$; that is, that besides the origin there are nodes at the points $x = \frac{1}{2}m\lambda$, m being any positive or negative integer. At any of these places infinitely thin rigid plane barriers normal to x might be stretched across the tube without in any way altering the motion. Midway between each pair of consecutive nodes there is a *loop*, or place of no pressure variation, since $\delta p = -\rho\dot{\phi}$ (6) § 244. At any of these loops a communication with the external atmosphere might be opened, without causing any disturbance of the motion from air passing in or out. The loops are the places of maximum velocity, and the nodes those of maximum pressure variation. At intervals of λ everything is exactly repeated.

If there be a node at $x = l$, as well as at the origin, $\sin kl = 0$, or $\lambda = 2l/m$, where m is a positive integer. The gravest tone which can be sounded by air contained in a doubly closed pipe of length l is therefore that which has a wave-length equal to $2l$. This statement, it will be observed, holds good whatever be the gas with which the pipe is filled; but the frequency, or the place of the tone in the musical scale, depends also on the nature of the particular gas. The periodic time is given by

$$\tau = \frac{\lambda}{a} = \frac{2l}{a} \dots\dots\dots\dots\dots\dots (8).$$

The other tones possible for a doubly closed pipe have periods which are submultiples of that of the gravest tone, and the whole system forms a harmonic scale.

Let us now suppose, without stopping for the moment to inquire how such a condition of things can be secured, that there is a loop instead of a node at the point $x = l$. Equation (6) gives $\cos kl = 0$, whence $\lambda = 4l \div (2m + 1)$, where m is zero or a positive integer. In this case the gravest tone has a wave-length equal to four times the length of the pipe reckoned from the node to the loop, and the other tones form with it a harmonic scale, from which, however, all the members of even order are missing.

256. By means of a rigid barrier there is no difficulty in securing a node at any desired point of a tube, but the condition for a loop, i.e. that under no circumstances shall the pressure vary, can only be realized approximately. In most cases the variation of pressure at any point of a pipe may be made small by allowing a free communication with the external air. Thus Euler and Lagrange assumed constancy of pressure as the condition to be satisfied at the end of an open pipe. We shall afterwards return to the problem of the open pipe, and investigate by a rigorous process the conditions to be satisfied at the end. For our immediate purpose it will be sufficient to know, what is indeed tolerably obvious, that the open end of a pipe may be treated as a loop, if the diameter of the pipe be neglected in comparison with the wave-length, provided the external pressure in the neighbourhood of the open end be not itself variable from some cause independent of the motion within the pipe. When there is an independent source of sound, the pressure at the end of the pipe is the same as it would be in the same place, if the pipe were away. The impediment to securing the fulfilment of the condition for a loop at any desired point lies in the inertia of the machinery required to sustain the pressure. For theoretical purposes we may overlook this difficulty, and imagine a massless piston backed by a compressed spring also without mass. The assumption of a loop at an open end of a pipe is tantamount to neglecting the inertia of the outside air.

We have seen that, if a node exist at any point of a pipe, there must be a series, ranged at equal intervals $\frac{1}{2}\lambda$, that midway between each pair of consecutive nodes there must be a loop, and that the whole vibration must be stationary. The same conclusion follows if there be at any point a loop; but it may perfectly well happen that there are neither nodes nor loops, as for example in the case when the motion reduces to a positive or negative progressive wave. In stationary vibration there is no transference of energy along the tube in either direction, for energy cannot pass a node or a loop.

257. The relations between the lengths of an open or closed pipe and the wave-lengths of the included column of air may also be investigated by following the motion of a *pulse*, by which is understood a wave confined within narrow limits and composed of uniformly condensed or rarefied fluid. In looking at the matter from this point of view it is necessary to take into account carefully the circumstances under which the various reflections take place. Let us first suppose that a condensed pulse travels in the positive direction towards a barrier fixed across the tube. Since the energy contained in the wave cannot escape from the tube, there must be a reflected wave, and that this reflected wave is also a wave of condensation appears from the fact that there is no loss of fluid. The same conclusion may be arrived at in another way. The effect of the barrier may be imitated by the introduction of a similar and equidistant wave of condensation moving in the negative direction. Since the two waves are both condensed and are propagated in contrary directions, the velocities of the fluid composing them are equal and opposite, and therefore neutralise one another when the waves are superposed.

If the progress of the negative reflected wave be interrupted by a second barrier, a similar reflection takes place, and the wave, still remaining condensed, regains its positive character. When a distance has been travelled equal to twice the length of the pipe, the original state of things is completely restored, and the same cycle of events repeats itself indefinitely. We learn therefore that the period within a doubly closed pipe is the time occupied by a pulse in travelling twice the length of the pipe.

The case of an open end is somewhat different. The supplementary negative wave necessary to imitate the effect of the open end must evidently be a wave of rarefaction capable of neutralising the positive pressure of the condensed primary wave, and thus in the act of reflection a wave changes its character from condensed to rarefied, or from rarefied to condensed. Another way of considering the matter is to observe that in a positive condensed pulse the momentum of the motion is forwards, and in the absence of the necessary forces cannot be changed by the reflection. But forward motion in the reflected negative wave is indissolubly connected with the rarefied condition.

When both ends of a tube are open, a pulse travelling backwards and forwards within it is completely restored to its original state after traversing twice the length of the tube, suffering in the process two reflections, and thus the relation between length and period is the same as in the case of a tube, whose ends are both closed; but when one end of a tube is open and the other closed, a double passage is not sufficient to close the cycle of changes. The original condensed or rarefied character cannot be recovered until after two reflections from the open end, and accordingly in the case contemplated the period is the time required by the pulse to travel over *four* times the length of the pipe.

258. After the full discussion of the corresponding problems in the chapter on Strings, it will not be necessary to say much on the compound vibrations of columns of air. As a simple example we may take the case of a pipe open at one end and closed at the other, which is suddenly brought to rest at the time $t = 0$, after being for some time in motion with a uniform velocity parallel to its length. The initial state of the contained air is then one of uniform velocity u_0 parallel to x, and of freedom from compression and rarefaction. If we suppose that the origin is at the closed end, the general solution is by (7) § 255,

$$\phi = (A_1 \cos n_1 t + B_1 \sin n_1 t) \cos k_1 x$$
$$+ (A_2 \cos n_2 t + B_2 \sin n_2 t) \cos k_2 x$$
$$+ \dots \dots \dots \dots \dots \dots \dots \dots \dots \dots \dots \dots (1),$$

where $k_r = (r - \tfrac{1}{2}) \pi / l$, $n_r = a k_r$, and A_1, B_1, A_2, $B_2 \dots$ are arbitrary constants.

Since $\dot{\phi}$ is to be zero initially for all values of x, the coefficients B must vanish; the coefficients A are to be determined by the condition that for all values of x between 0 and l,

$$\Sigma \, k_r A_r \sin k_r x = - u_0 \dots \dots \dots \dots \dots (2),$$

where the summation extends to all integral values of r from 1 to ∞. The determination of the coefficients A from (2) is effected in the usual way. Multiplying by $\sin k_r x \, dx$, and integrating from 0 to l, we get

$$\tfrac{1}{2} l \, k_r A_r = - u_0 / k_r,$$

or

$$A_r = - \frac{2u_0}{k_r^2 l} \dots \dots \dots \dots \dots \dots (3).$$

The complete solution is therefore

$$\phi = - \frac{2u_0}{l} \sum_{r=1}^{r=\infty} \frac{\cos k_r x}{k_r^2} \cos n_r t \dots \dots \dots \dots (4).$$

259. In the case of a tube stopped at the origin and open at $x = l$, let $\phi = \cos nt$ be the value of the potential at the open end due to an external source of sound. Determining P and θ in equation (7) § 255, we find

$$\phi = \frac{\cos kx}{\cos kl} \cos nt \dots \dots \dots \dots \dots (1).$$

It appears that the vibration within the tube is a minimum, when $\cos kl = \pm 1$, that is when l is a multiple of $\tfrac{1}{2} \lambda$, in which case there is a node at $x = l$. When l is an odd multiple of $\tfrac{1}{4} \lambda$, $\cos kl$ vanishes, and then according to (1) the motion would become infinite. In this case the supposition that the pressure at the open end is independent of what happens within the tube breaks down; and we can only infer that the vibration is very large, in consequence of the isochronism. Since there is a node at $x = 0$, there must be a loop when x is an odd multiple of $\frac{1}{4}\lambda$, and we conclude that in the case of isochronism the variation of pressure at the open end of the tube due to the external cause is exactly neutralised by the variation of pressure due to the motion within the tube itself. If there were really at the open end a variation of pressure on the whole, the motion must increase without limit in the absence of dissipative forces.

If we suppose that the origin is a loop instead of a node, the solution is

$$\phi = \frac{\sin kx}{\sin kl} \cos nt \dots\dots\dots\dots\dots\dots(2),$$

where $\phi = \cos nt$ is the given value of ϕ at the open end $x = l$. In this case the expression becomes infinite, when $kl = m\pi$, or $l = \frac{1}{2}m\lambda$.

We will next consider the case of a tube, whose ends are both open and exposed to disturbances of the same period, making ϕ equal to He^{int}, Ke^{int} respectively. Unless the disturbances at the ends are in the same phase, one at least of the coefficients H, K must be complex.

Taking the first form in (3) § 255, we have as the general expression for ϕ

$$\phi = e^{int}(A \cos kx + B \sin kx).$$

If we take the origin in the middle of the tube, and assume that the values He^{int}, Ke^{int} correspond respectively to $x = l$, $x = -l$, we get to determine A and B,

$$H = A \cos kl + B \sin kl,$$
$$K = A \cos kl - B \sin kl,$$

whence

$$A = \frac{H + K}{2 \cos kl}, \qquad B = \frac{H - K}{2 \sin kl} \dots\dots\dots\dots (3),$$

giving

$$\phi = e^{int} \frac{H \sin k(l + x) + K \sin k(l - x)}{\sin 2kl} \dots\dots\dots(4).$$

This result might also be deduced from (2), if we consider that the required motion arises from the superposition of the motion, which is due to the disturbance He^{int} calculated on the hypothesis that the other end $x = -l$ is a loop, on the motion, which is due to Ke^{int} on the hypothesis that the end $x = l$ is a loop.

The vibration expressed by (4) cannot be *stationary*, unless the ratio $H : K$ be real, that is unless the disturbances at the ends be in similar, or in opposite, phases. Hence, except in the cases reserved, there is no loop anywhere, and therefore no place at which a branch tube can be connected along which sound will not be propagated[1].

At the middle of the tube, for which $x = 0$,

$$\phi = \frac{H + K}{2 \cos kl} e^{int} \dotfill (5),$$

shewing that the variation of pressure (proportional to $\dot{\phi}$) vanishes if $H + K = 0$, that is, if the disturbances at the ends be equal and in *opposite* phases. Unless this condition be satisfied, the expression becomes infinite when $2l = \frac{1}{2} (2m + 1) \lambda$.

At a point distant $\frac{1}{4}\lambda$ from the middle of the tube the expression for ϕ is

$$\phi = \frac{H - K}{2 \sin kl} e^{int} \dotfill (6),$$

vanishing when $H = K$, that is, when the disturbances at the ends are equal and in the *same* phase. In general ϕ becomes infinite, when $\sin kl = 0$, or $2l = m\lambda$.

If at one end of an unlimited tube there be a variation of pressure due to an external source, a train of progressive waves will be propagated inwards from that end. Thus, if the length along the tube measured from the open end be y, the velocity-potential is expressed by $\phi = \cos (nt - ny/a)$, corresponding to $\phi = \cos nt$ at $y = 0$; so that, if the cause of the disturbance within the tube be the passage of a train of progressive waves across the open end, the intensity within the tube will be the same as in the space outside. It must not be forgotten that the diameter of the tube is supposed to be infinitely small in comparison with the length of a wave.

---

[1] An arrangement of this kind has been proposed by Prof. Mayer (*Phil. Mag.* XLV. p. 90, 1873) for comparing the intensities of sources of sound of the same pitch. Each end of the tube is exposed to the action of one of the sources to be compared, and the distances are adjusted until the amplitudes of the vibrations denoted by H and K are equal. The branch tube is led to a manometric capsule (§ 262), and the method assumes that by varying the point of junction the disturbance of the flame can be stopped. From the discussion in the text it appears that this assumption is not theoretically correct.

Let us next suppose that the source of the motion is within the tube itself, due for example to the inexorable motion of a piston at the origin[1]. The constants in (5) § 255 are to be determined by the conditions that when $x = 0$, $d\phi/dx = \cos nt$ (say), and that, when $x = l$, $\phi = 0$. Thus $kA = -\tan kl$, $kB = 1$, and the expression for ϕ is

$$\phi = \frac{\sin k(x - l)}{k \cos kl} \dots\dots\dots\dots\dots\dots (7).$$

The motion is a minimum, when $\cos kl = \pm 1$, that is, when the length of the tube is a multiple of $\frac{1}{2}\lambda$.

When l is an odd multiple of $\frac{1}{4}\lambda$, the place occupied by the piston would be a node, if the open end were really a loop, but in this case the solution fails. The escape of energy from the tube prevents the energy from accumulating beyond a certain point; but no account can be taken of this so long as the open end is treated rigorously as a loop. We shall resume the question of resonance after we have considered in greater detail the theory of the open end, when we shall be able to deal with it more satisfactorily.

In like manner if the point $x = l$ be a node, instead of a loop, the expression for ϕ is

$$\phi = \frac{\cos k(l - x)}{k \sin kl} \dots\dots\dots\dots\dots\dots (8);$$

and thus the motion is a minimum when l is an odd multiple of $\frac{1}{4}\lambda$, in which case the origin is a loop. When l is an even multiple of $\frac{1}{4}\lambda$, the origin should be a node, which is forbidden by the conditions of the question. In this case according to (8) the motion becomes infinite, which means that in the absence of dissipative forces the vibration would increase without limit.

260. The experimental investigation of aerial waves within pipes has been effected with considerable success by Kundt[2]. To generate waves is easy enough; but it is not so easy to invent a method by which they can be effectually examined. Kundt discovered that the nodes of stationary waves can be made evident by dust. A little fine sand or lycopodium seed, shaken over the interior of a glass tube containing a vibrating column of air

---

[1] These problems are considered by Poisson, *Mém. de l'Institut*, t. II. p. 305, 1819.

[2] *Pogg. Ann.* t. CXXXV. p. 337, 1868.

disposes itself in recurring patterns, by means of which it is easy to determine the positions of the nodes and to measure the intervals between them. In Kundt's experiments the origin of the sound was in the longitudinal vibration of a glass tube called the sounding-tube, and the dust-figures were formed in a second and larger tube, called the wave-tube, the latter being provided with a moveable stopper for the purpose of adjusting its length. The other end of the wave-tube was fitted with a cork through which the sounding-tube passed half way. By suitable friction the sounding-tube was caused to vibrate in its gravest mode, so that the central point was nodal, and its interior extremity (closed with a cork) excited aerial vibrations in the wave-tube. By means of the stopper the length of the column of air could be adjusted so as to make the vibrations as vigorous as possible, which happens when the interval between the stopper and the end of the sounding-tube is a multiple of half the wave-length of the sound.

With this apparatus Kundt was able to compare the wave-lengths of the same sound in various gases, from which the relative velocities of propagation are at once deducible, but the results were not entirely satisfactory. It was found that the intervals of recurrence of the dust-patterns were not strictly equal, and, what was worse, that the pitch of the sound was not constant from one experiment to another. These defects were traced to a communication of motion to the wave-tube through the cork, by which the dust-figures were disturbed, and the pitch made irregular in consequence of unavoidable variations in the mounting of the apparatus. To obviate them, Kundt replaced the cork, which formed too stiff a connection between the tubes, by layers of sheet indiarubber tied round with silk, obtaining in this way a flexible and perfectly air-tight joint; and in order to avoid any risk of the comparison of wave-lengths being vitiated by an alteration of pitch, the apparatus was modified so as to make it possible to excite the two systems of dust-figures simultaneously and in response to the same sound. A collateral advantage of the new method consisted in the elimination of temperature-corrections.

In the improved "Double Apparatus" the sounding-tube was caused to vibrate *in its second mode* by friction applied near the middle; and thus the nodes were formed at the points distant from the ends by one-fourth of the length of the tube. At each of these points connection was made with an independent wave-tube, provided with an adjustable stopper, and with branch tubes and stop-cocks suitable for admitting the various gases to be experimented upon. It is evident that dust-figures formed in the two tubes correspond rigorously to the same pitch, and that therefore a comparison of the intervals of recurrence leads to a correct determination of the velocities of propagation, under the circumstances of the experiment, for the two gases with which the tubes are filled.

The results at which Kundt arrived were as follows:—

(a) The velocity of sound in a tube diminishes with the diameter. Above a certain diameter, however, the change is not perceptible.

(b) The diminution of velocity increases with the wave-length of the tone employed.

(c) Powder, scattered in a tube, diminishes the velocity of sound in narrow tubes, but in wide ones is without effect.

(d) In narrow tubes the effect of powder increases, when it is very finely divided, and is strongly agitated in consequence.

(e) Roughening the interior of a narrow tube, or increasing its surface, diminishes the velocity.

(f) In wide tubes these changes of velocity are of no importance, so that the method may be used in spite of them for exact determinations.

(g) The influence of the intensity of sound on the velocity cannot be proved.

(h) With the exception of the first, the wave-lengths of a tone as shewn by dust are not affected by the mode of excitation.

(i) In wide tubes the velocity is independent of pressure, but in small tubes the velocity increases with the pressure.

(j) All the observed changes in the velocity were due to friction and especially to exchange of heat between the air and the sides·of the tube.

(k) The velocity of sound at 100° agrees exactly with that given by theory[1].

---

[1] From some expressions in the memoir already cited, from which the notice in the text is principally derived, Kundt appears to have contemplated a continuation of his investigations; but I am unable to find any later publication on the subject.

We shall return to the question of the propagation of sound in narrow tubes as affected by the causes mentioned above (*j*), and shall then investigate the formulæ given by Helmholtz and Kirchhoff.

[The genesis of the peculiar transverse striation which forms a leading feature of the dust-figures has already been considered § 253 *b*. According to the observations of Dvořák[1] the powerful vibrations which occur in a Kundt's tube are accompanied by certain mean motions of the gas. Thus near the walls there is a flow from the loops to the nodes, and in the interior a return flow from the nodes to the loops. This is a consequence of viscosity acting with peculiar advantage upon the parts of the fluid contiguous to the walls[2]. We may perhaps return to this subject in a later chapter.]

261. In the experiments described in the preceding section the aerial vibrations are *forced*, the pitch being determined by the external source, and not (in any appreciable degree) by the length of the column of air. Indeed, strictly speaking, all sustained vibrations are forced, as it is not in the power of free vibrations to maintain themselves, except in the ideal case when there is absolutely no friction. Nevertheless there is an important practical distinction between the vibrations of a column of air as excited by a longitudinally vibrating rod or by a tuning-fork, and such vibrations as those of the organ-pipe or chemical harmonicon. In the latter cases the pitch of the sound depends principally on the length of the aerial column, the function of the wind or of the flame[3] being merely to restore the energy lost by friction and by communication to the external air. The air in an organ-pipe is to be considered as a column swinging almost freely, the lower end, across which the wind sweeps, being treated roughly as open, and the upper end as closed, or open, as the case may be. Thus the wave-length of the principal tone of a stopped pipe is four times the length of the pipe; and, except at the extremities, there is neither node nor loop. The overtones of the pipe are the *odd*

---

[1] *Pogg. Ann.* t. CLVII. p. 61, 1876.

[2] On the Circulation of Air observed in Kundt's Tubes, and on some allied Acoustical Problems, *Phil. Trans.* vol. CLXXV. p. 1, 1884.

[3] The subject of sensitive flames with and without pipes is treated in considerable detail by Prof. Tyndall in his work on Sound; but the mechanics of this class of phenomena is still very imperfectly understood. We shall return to it in a subsequent chapter.

harmonics, twelfth, higher third, &c., corresponding to the various subdivisions of the column of air. In the case of the twelfth, for example, there is a node at the point of trisection nearest to the open end, and a loop at the other point of trisection midway between the first and the stopped end of the pipe.

In the case of the open organ-pipe both ends are loops, and there must be at least one internal node. The wave-length of the principal tone is twice the length of the pipe, which is divided into two similar parts by a node in the middle. From this we see the foundation of the ordinary rule that the pitch of an open pipe is the same as that of a stopped pipe of half its length. For reasons to be more fully explained in a subsequent chapter, connected with our present imperfect treatment of the open end, the rule is only approximately correct. The open pipe, differing in this respect from the stopped pipe, is capable of sounding the whole series of tones forming the harmonic scale founded upon its principal tone. In the case of the octave there is a loop at the centre of the pipe and nodes at the points midway between the centre and the extremities.

Since the frequency of the vibration in a pipe is proportional to the velocity of propagation of sound in the gas with which the pipe is filled, the comparison of the pitches of the notes obtained from the same pipe in different gases is an obvious method of determining the velocity of propagation, in cases where the impossibility of obtaining a sufficiently long column of the gas precludes the use of the direct method. In this application Chladni with his usual sagacity led the way. The subject was resumed at a later date by Dulong[1] and by Wertheim[2], who obtained fairly satisfactory results.

262. The condition of the air in the interior of an organ-pipe was investigated experimentally by Savart[3], who lowered into the pipe a small stretched membrane on which a little sand was scattered. In the neighbourhood of a node the sand remained sensibly undisturbed, but, as a loop was approached, it danced with more and more vigour. But by far the most striking form of the

[1] Recherches sur les chaleurs spécifiques des fluides élastiques. *Ann. de Chim.*, t. XLI. p. 113, 1829.

[2] *Ann. de Chim.*, 3ième série, t. XXIII. p. 434, 1848.

[3] *Ann. de Chim.*, t. XXIV. p. 56, 1823.

experiment is that invented by König.   In this method the vibration is indicated by a small gas flame, fed through a tube which is in communication with a cavity called a manometric capsule. This cavity is bounded on one side by a membrane on which the vibrating air acts.   As the membrane vibrates, rendering the capacity of the capsule variable, the supply of gas becomes unsteady and the flame intermittent.   The period is of course too small for the intermittence to manifest itself as such when the flame is looked at steadily.   By shaking the head, or with the aid of a moveable mirror, the resolution into more or less detached images may be effected; but even without resolution the altered character of the flame is evident from its general appearance.   In the application to organ-pipes, one or more capsules are mounted on a pipe in such a manner that the membranes are in contact with the vibrating column of air; and the difference in the flame is very marked, according as the associated capsule is situated at a node or at a loop.

263.   Hitherto we have supposed the pipe to be straight, but it will readily be anticipated that, when the cross section is small and does not vary in area, straightness is not a matter of importance.   Conceive a curved axis of x running along the middle of the pipe, and let the constant section perpendicular to this axis be S. When the greatest diameter of S is very small in comparison with the wave-length of the sound, the velocity-potential ϕ becomes nearly invariable over the section; applying Green's theorem to the space bounded by the interior of the pipe and by two cross sections, we get

$$\iiint \nabla^2 \phi \, dV = S \cdot \Delta \left(\frac{d\phi}{dx} \right).$$

Now by the general equation of motion

$$\iiint \nabla^2 \phi \, dV = \frac{1}{a^2} \iiint \ddot{\phi} \, dV = \frac{1}{a^2} \frac{d^2}{dt^2} \iiint \phi \, dV = \frac{S}{a^2} \frac{d^2}{dt^2} \int \phi \, dx,$$

and in the limit, when the distance between the sections is made to vanish,

$$\int \phi \, dx = \phi \, dx, \qquad \Delta \left(\frac{d\phi}{dx} \right) = \frac{d^2\phi}{dx^2} \, dx;$$

so that

$$\frac{d^2\phi}{dt^2} = a^2 \frac{d^2\phi}{dx^2} \dots\dots\dots\dots\dots\dots\dots\dots(1),$$

shewing that ϕ depends upon x in the same way as if the pipe were straight. By means of equation (1) the vibrations of air in curved pipes of uniform section may be easily investigated, and the results are the rigorous consequences of our fundamental equations (which take no account of friction), when the section is supposed to be infinitely small. In the case of thin tubes such as would be used in experiment, they suffice at any rate to give a very good representation of what actually happens.

264. We now pass on to the consideration of certain cases of connected tubes. In the accompanying figure AD represents a thin pipe, which divides at D into two branches DB, DC. At E the branches reunite and form a single tube EF. The sections of the single tubes and of the branches are assumed to be uniform as well as very small.

Fig. 55.

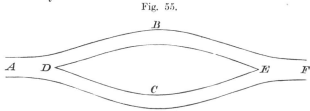

In the first instance let us suppose that a positive wave of arbitrary type is advancing in A. On its arrival at the fork D, it will give rise to positive waves in B and C, and, unless a certain condition be satisfied, to a negative reflected wave in A. Let the potential of the positive waves be denoted by f_A, f_B, f_C, f being in each case a function of $x - at$: and let the reflected wave be $F(x + at)$. Then the conditions to be satisfied at D are first that the pressures shall be the same for the three pipes, and secondly that the whole velocity of the fluid in A shall be equal to the sum of the whole velocities of the fluid in B and C. Thus, using A, B, C to denote the areas of the sections, we have, § 244,

$$\left.\begin{array}{l} f_A' - F' = f_B' = f_C' \\ A\,(f_A' + F') = B f_B' + C f_C' \end{array}\right\} \quad \dots\dots\dots\dots\dots\dots\dots(1);$$

whence
$$F' = \frac{B + C - A}{B + C + A}\, f_A' \quad \dots\dots\dots\dots\dots\dots(2),$$

$$f_B' = f_C' = \frac{2A}{B + C + A}\, f_A' \quad \dots\dots\dots\dots\dots(3)[1].$$

[1] These formulæ, as applied to determine the reflected and refracted waves at the junction of two tubes of sections $B + C$, and A respectively, are given by

It appears that f_B and f_C are always the same. There is no reflection, if

$$B + C = A \dots\dots\dots(4),$$

that is, if the combined sections of the branches be equal to the section of the trunk; and, when this condition is satisfied,

$$f_B = f_C = f_A \dots\dots\dots(5).$$

The wave then advances in B and C exactly as it would have done in A, had there been no break. If the lengths of the branches between D and E be equal, and the section of F be equal to that of A, the waves on arrival at E combine into a wave propagated along F, and again there is no reflection. The division of the tube has thus been absolutely without effect; and since the same would be true for a negative wave passing from F to A, we may conclude generally that a tube may be divided into two, or more, branches, all of the same length, without in any way influencing the law of aërial vibration, provided that the whole section remain constant. If the lengths of the branches from D to E be unequal, the result is different. Besides the positive wave in F, there will be in general negative reflected waves in B and C. The most interesting case is when the wave is of harmonic type and one of the branches is longer than the other by a multiple of $\frac{1}{2}\lambda$. If the difference be an *even* multiple of $\frac{1}{2}\lambda$, the result will be the same as if the branches were of equal length, and no reflection will ensue. But suppose that, while B and C are equal in section, one of them is longer than the other by an *odd* multiple of $\frac{1}{2}\lambda$. Since the waves arrive at E in opposite phases, it follows from symmetry that the positive wave in F must vanish, and that the pressure at E, which is necessarily the same for all the tubes, must be constant. The waves in B and C are thus reflected as from an open end. That the conditions of the question are thus satisfied may also be seen by supposing a barrier taken across the tube F in the neighbourhood of E in such a way that the tubes B and C communicate without a change of section. The wave in each tube will then pass on into the other without interruption, and the pressure-variation at E, being the resultant of equal and opposite components, will vanish. This being so, the barrier may be removed without altering the conditions, and no wave will be propagated along F, whatever its section may be. The arrange-

Poisson, *Mém. de l'Institut*, t. II. p. 305, 1819. The reader will not forget that both diameters must be small in comparison with the wave-length.

ment now under consideration was invented by Herschel, and has been employed by Quincke and others for experimental purposes,—an application that we shall afterwards have occasion to describe. The phenomenon itself is often referred to as an example of interference, to which there can be no objection, but the same cannot be said when the reader is led to suppose that the positive waves neutralise each other in F, and that there the matter ends. It must never be forgotten that there is no loss of energy in interference, but only a different distribution; when energy is diverted from one place, it reappears in another. In the present case the positive wave in A conveys energy with it. If there is no wave along F, there are two possible alternatives. Either energy accumulates in the branches, or else it passes back along A in the form of a negative wave. In order to see what really happens, let us trace the progress of the waves reflected back at E.

These waves are equal in magnitude and start from E in opposite phases; in the passage from E to D one has to travel a greater distance than the other by an odd multiple of $\frac{1}{2}\lambda$; and therefore on arrival at D they will be in complete accordance. Under these circumstances they combine into a single wave, which travels negatively along A, and there is no reflection. When the negative wave reaches the end of the tube A, or is otherwise disturbed in its course, the whole or a part may be reflected, and then the process is repeated. But however often this may happen there will be no wave along F, unless by accumulation, in consequence of a coincidence of periods, the vibration in the branches becomes so great that a small fraction of it can no longer be neglected.

Or we may reason thus. Suppose the tube F cut off by a barrier as before. The motion in the ring being due to forces acting at D is necessarily symmetrical with respect to D, and D'—the point which divides $DBCD$ into equal parts. Hence D' is a node, and the vibration is stationary. This being the case, at a point E distant $\frac{1}{4}\lambda$ from D' on either side, there must be a loop; and if the barrier be removed there will still be no tendency to produce vibration in F. If the perimeter of the ring be a multiple of λ, there may be

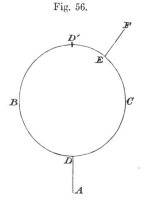

Fig. 56.

vibration within it of the period in question, independently of any lateral openings.

Any combination of connected tubes may be treated in a similar manner. The general

Fig. 57.

principle is that at any junction a space can be taken large enough to include all the region through which the want of uniformity affects the law of the waves, and yet so small that its longest dimension may be neglected in comparison with λ. Under these circumstances the fluid within the space in question may be treated as if the wave-length were infinite, or the fluid itself incompressible, in which case its velocity-potential would satisfy $\nabla^2\phi = 0$, following the same laws as electricity.

265. When the section of a pipe is variable, the problem of the vibrations of air within it cannot generally be solved. The case of conical pipes will be treated on a future page. At present we will investigate an approximate expression for the pitch of a nearly cylindrical pipe, taking first the case where both ends are closed. The method that will be employed is similar to that used for a string whose density is not quite constant, §§ 91, 140, depending on the principle that the period of a free vibration fulfils the stationary condition, and may therefore be calculated from the potential and kinetic energies of any hypothetical motion not departing far from the actual type. In accordance with this plan we shall assume that the velocity normal to any section S is constant over the section, as must be very nearly the case when the variation of S is slow. Let X represent the total transfer of fluid at time t across the section at x, reckoned from the equilibrium condition; then $\dot{X}$ represents the total velocity of the current, and $\dot{X} \div S$ represents the actual velocity of the particles of fluid, so that the kinetic energy of the motion within the tube is expressed by

$$T = \tfrac{1}{2}\rho \int \frac{\dot{X}^2}{S}\,dx \quad \dots\dots\dots\dots\dots\dots(1).$$

The potential energy § 245 (12) is expressed in general by

$$V = \tfrac{1}{2} a^2 \rho \iiint s^2\, dV,$$

or, since $dV = Sdx$, by

$$V = \tfrac{1}{2} a^2 \rho \int S s^2 \, dx \dots\dots\dots\dots\dots(2).$$

Again, by the condition of continuity,

$$-s = \frac{1}{S} \frac{dX}{dx} \dots\dots\dots\dots\dots\dots(3),$$

and thus

$$V = \tfrac{1}{2} a^2 \rho \int \frac{1}{S} \left(\frac{dX}{dx}\right)^2 dx \dots\dots\dots\dots\dots(4).$$

If we now assume for X an expression of the same form as would obtain if S were constant, viz.

$$X = \sin \frac{\pi x}{l} \cos nt \dots\dots\dots\dots\dots(5),$$

we obtain from the values of T and V in (1) and (4),

$$n^2 = \frac{a^2 \pi^2}{l^2} \int_0^l \cos^2 \frac{\pi x}{l} \frac{dx}{S} \div \int_0^l \sin^2 \frac{\pi x}{l} \frac{dx}{S} \dots\dots\dots(6),$$

or, if we write $S = S_0 + \Delta S$ and neglect the square of ΔS,

$$n^2 = \frac{a^2 \pi^2}{l^2} \left\{ 1 - 2 \int_0^l \cos \frac{2\pi x}{l} \frac{\Delta S}{S_0} \frac{dx}{l} \right\} \dots\dots\dots(7).$$

The result may be expressed conveniently in terms of Δl, the correction that must be made to l in order that the pitch may be calculated from the ordinary formula, as if S were constant. For the value of Δl we have

$$\Delta l = \int_0^l \cos \frac{2\pi x}{l} \frac{\Delta S}{S_0} \, dx \dots\dots\dots\dots\dots(8).$$

The effect of a variation of section is greatest near a node or near a loop. An enlargement of section in the first case lowers the pitch, and in the second case raises it. At the points midway between the nodes and loops a slight variation of section is without effect. The pitch is thus decidedly altered by an enlargement or contraction near the middle of the tube, but the influence of a slight conicality would be much less.

The expression for Δl given by (8) is applicable as it stands to the gravest tone only; but we may apply it to the mth tone of the harmonic scale, if we modify it by the substitution of $\cos(2m\pi x/l)$ for $\cos(2\pi x/l)$.

In the case of a tube *open* at both ends (5) is replaced by

$$X = \cos \frac{\pi x}{l} \cos nt \dots\dots\dots\dots\dots\dots(9),$$

which leads to

$$\Delta l = -\int_0^l \cos \frac{2\pi x}{l} \frac{\Delta S}{S_0} dx \dots\dots\dots\dots(10),$$

instead of (8). The pitch of the sound is now raised by an enlargement at the ends, or by a contraction at the middle, of the tube; and, as before, it is unaffected by a slight general conicality (§ 281).

266. The case of progressive waves moving in a tube of variable section is also interesting. In its general form the problem would be one of great difficulty; but where the change of section is very gradual, so that no considerable alteration occurs within a distance of a great many wave-lengths, the principle of energy will guide us to an approximate solution. It is not difficult to see that in the case supposed there will be no sensible reflection of the wave at any part of its course, and that therefore the energy of the motion must remain unchanged[1]. Now we know, § 245, that for a given area of wave-front, the energy of a train of simple waves is as the square of the amplitude, from which it follows that as the waves advance the amplitude of vibration varies inversely as the square root of the section of the tube. In all other respects the type of vibration remains absolutely unchanged. From these results we may get a general idea of the action of an ear-trumpet. It appears that according to the ordinary approximate equations, there is no limit to the concentration of sound producible in a tube of gradually diminishing section.

The same method is applicable, when the density of the medium varies slowly from point to point. For example, the amplitude of a sound-wave moving upwards in the atmosphere may be determined by the condition that the energy remains unchanged. From § 245 it appears that the amplitude is inversely as the square root of the density[2].

---

[1] *Phil. Mag.* (5) i. p. 261, 1876.

[2] A delicate question arises as to the ultimate fate of sonorous waves propagated upwards. It should be remarked that in rare air the deadening influence of viscosity is much increased.

CHAPTER XIII.

SPECIAL PROBLEMS. REFLECTION AND REFRACTION OF PLANE WAVES.

267. BEFORE undertaking the discussion of the general equations for aërial vibrations we may conveniently turn our attention to a few special problems, relating principally to motion in two dimensions, which are susceptible of rigorous and yet comparatively simple solution. In this way the reader, to whom the subject is new, will acquire some familiarity with the ideas and methods employed before attacking more formidable difficulties.

In the previous chapter (§ 255) we investigated the vibrations in one dimension, which may take place parallel to the axis of a tube, of which both ends are closed. We will now inquire what vibrations are possible within a closed rectangular box, dispensing with the restriction that the motion is to be in one dimension only. For each simple vibration of which the system is capable, ϕ varies as a circular function of the time, say $\cos kat$, where k is some constant; hence $\ddot{\phi} = -k^2 a^2 \phi$, and therefore by the general differential equation (9) § 244

$$\nabla^2 \phi + k^2 \phi = 0 \quad \ldots \ldots \ldots \ldots \ldots \ldots \ldots \ldots (1).$$

Equation (1) must be satisfied throughout the whole of the included volume. The surface condition to be satisfied over the six sides of the box is simply

$$\frac{d\phi}{dn} = 0 \quad \ldots \ldots \ldots \ldots \ldots \ldots \ldots \ldots (2),$$

where dn represents an element of the normal to the surface. It is only for special values of k that it is possible to satisfy (1) and (2) simultaneously.

Taking three edges which meet as axes of rectangular co-ordinates, and supposing that the lengths of the edges are respectively α, β, γ, we know (§ 255) that

$$\phi = \cos\left(p\,\frac{\pi x}{\alpha}\right), \quad \phi = \cos\left(q\,\frac{\pi y}{\beta}\right), \quad \phi = \cos\left(r\,\frac{\pi z}{\gamma}\right),$$

where p, q, r are integers, are particular solutions of the problem. By any of these forms equation (2) is satisfied, and provided that k be equal to $p\pi/\alpha$, $q\pi/\beta$, or $r\pi/\gamma$, as the case may be, (1) is also satisfied. It is equally evident that the boundary equation (2) is satisfied over all the surface by the form

$$\phi = \cos\left(p\,\frac{\pi x}{\alpha}\right)\cos\left(q\,\frac{\pi y}{\beta}\right)\cos\left(r\,\frac{\pi z}{\gamma}\right) \ \ldots\ldots\ldots\ (3),$$

a form which also satisfies (1), if k be taken such that

$$k^2 = \pi^2\left(\frac{p^2}{\alpha^2} + \frac{q^2}{\beta^2} + \frac{r^2}{\gamma^2}\right) \ \ldots\ldots\ldots\ldots\ldots\ldots\ (4),$$

where as before p, q, r are integers[1].

The general solution, obtained by compounding all particular solutions included under (3), is

$$\phi = \Sigma\,\Sigma\,\Sigma\,(A\cos kat + B\sin kat)$$
$$\times \cos\left(p\,\frac{\pi x}{\alpha}\right)\cos\left(q\,\frac{\pi y}{\beta}\right)\cos\left(r\,\frac{\pi z}{\gamma}\right) \ \ldots\ldots\ldots\ldots(5),$$

in which A and B are arbitrary constants, and the summation is extended to all integral values of p, q, r.

This solution is sufficiently general to cover the case of any initial state of things within the box, not involving molecular rotation. The initial distribution of velocities depends upon the initial value of ϕ, or $\int(u_0 dx + v_0 dy + w_0 dz)$, and by Fourier's theorem can be represented by (5), suitable values being ascribed to the coefficients A. In like manner an arbitrary initial distribution of condensation (or rarefaction), depending on the initial value of $\dot{\phi}$, can be represented by ascribing suitable values to the coefficients B.

The investigation might be presented somewhat differently by commencing with assuming in accordance with Fourier's

[1] Duhamel, *Liouville Journ. Math.*, vol. XIV. p. 84, 1849.

theorem that the general value of ϕ at time t can be expressed in the form

$$\phi = \Sigma \Sigma \Sigma \, C \cos\left(p \, \frac{\pi x}{\alpha}\right) \cos\left(q \, \frac{\pi y}{\beta}\right) \cos\left(r \, \frac{\pi z}{\gamma}\right),$$

in which the coefficients C may depend upon t, but not upon x, y, z. The expressions for T and V would then be formed, and shewn to involve only the squares of the coefficients C, and from these expressions would follow the normal equations of motion connecting each normal co-ordinate C with the time.

The gravest mode of vibration is that in which the entire motion is parallel to the longest dimension of the box, and there is no internal node. Thus, if α be the greatest of the three sides α, β, γ, we are to take $p = 1$, $q = 0$, $r = 0$.

In the case of a cubical box, $\alpha = \beta = \gamma$, and then instead of (4) we have

$$k^2 = \frac{\pi^2}{\alpha^2} \left(p^2 + q^2 + r^2\right) \dots\dots\dots\dots\dots\dots (6),$$

or, if λ be the wave-length of plane waves of the same period,

$$\lambda = 2\alpha \div \sqrt{(p^2 + q^2 + r^2)}\dots\dots\dots\dots\dots(7).$$

For the gravest mode $p = 1$, $q = 0$, $r = 0$, or $p = 0$, $q = 1$, $r = 0$, &c., and $\lambda = 2\alpha$. The next gravest is when $p = 1$, $q = 1$, $r = 0$, &c., and then $\lambda = \sqrt{2}\,\alpha$. When $p = 1$, $q = 1$, $r = 1$, $\lambda = 2\alpha/\sqrt{3}$. For the fourth gravest mode $p = 2$, $q = 0$, $r = 0$, &c., and then $\lambda = 4\alpha$.

As in the case of the membrane (§ 197), when two or more primitive modes have the same period of vibration, other modes of like period may be derived by composition.

The trebly infinite series of possible simple component vibrations is not necessarily completely represented in particular cases of compound vibrations. If, for example, we suppose the contents of the box in its initial condition to be neither condensed nor rarefied in any part, and to have a uniform velocity, whose components parallel to the axes of co-ordinates are respectively u_0, v_0, w_0, no simple vibrations are generated for which more than one of the three numbers p, q, r is finite. In fact each component initial velocity may be considered separately, and the problem is similar to that solved in § 258.

In future chapters we shall meet with other examples of the vibrations of air within completely closed vessels.

Some of the natural notes of the air contained within a room may generally be detected on singing the scale. Probably it is somewhat in this way that blind people are able to estimate the size of rooms[1].

In long and narrow passages the vibrations parallel to the length are too slow to affect the ear, but notes due to transverse vibrations may often be heard. The relative proportions of the various overtones depend upon the place at which the disturbance is created[2].

In some cases of this kind the pitch of the vibrations, whose direction is principally transverse, is influenced by the occurrence of longitudinal motion. Suppose, for example, in (3) and (4), that $q = 1$, $r = 0$, and that α is much greater than β. For the principal transverse vibration $p = 0$, and $k = \pi/\beta$. But besides this there are other modes of vibration in which the motion is principally transverse, obtained by ascribing to p small integral values. Thus, when $p = 1$,

$$k^2 = \pi^2 \left(\frac{1}{\alpha^2} + \frac{1}{\beta^2} \right),$$

shewing that the pitch is nearly the same as before[3].

268. If we suppose γ to become infinitely great, the box of the preceding section is transformed into an infinite rectangular tube, whose sides are α and β. Whatever may be the motion of the air within this tube, its velocity-potential may be expressed by Fourier's theorem in the series

$$\phi = \Sigma \, \Sigma \, A_{pq} \cos \frac{p\pi x}{\alpha} \cos \frac{q\pi y}{\beta} \quad \ldots\ldots\ldots\ldots\ldots (1),$$

where the coefficients A are independent of x and y. By the use of this form we secure the fulfilment of the boundary condition

---

[1] A remarkable instance is quoted in Young's *Natural Philosophy*, II. p. 272, from Darwin's *Zoonomia*, II. 487. "The late blind Justice Fielding walked for the first time into my room, when he once visited me, and after speaking a few words said, 'This room is about 22 feet long, 18 wide, and 12 high'; all which he guessed by the ear with great accuracy."

[2] Oppel, *Die harmonischen Obertöne des durch parallele Wände erregten Reflexionstones. Fortschritte der Physik*, xx. p. 130.

[3] There is an underground passage in my house in which it is possible, by singing the right note, to excite free vibrations of many seconds' duration, and it often happens that the resonant note is affected with distinct beats. The breadth of the passage is about 4 feet, and the height about $6\frac{1}{2}$ feet.

that there is to be no velocity across the sides of the tube; the nature of A as a function of z and t depends upon the other conditions of the problem.

Let us consider the case in which the motion at every point is harmonic, and due to a normal motion imposed upon a barrier stretching across the tube at $z = 0$. Assuming ϕ to be proportional to e^{ikat} at all points, we have the usual differential equation

$$\frac{d^2\phi}{dx^2} + \frac{d^2\phi}{dy^2} + \frac{d^2\phi}{dz^2} + k^2\phi = 0 \quad \dots\dots\dots\dots\dots(2),$$

which by the conjugate property of the functions must be satisfied separately by each term of (1). Thus to determine A_{pq} as a function of z, we get

$$\frac{d^2A_{pq}}{dz^2} + \left[k^2 - \pi^2 \left(\frac{p^2}{\alpha^2} + \frac{q^2}{\beta^2} \right) \right] A_{pq} = 0 \quad \dots\dots\dots\dots(3).$$

The solution of this equation differs in form according to the sign of the coefficient of A_{pq}. When p and q are both zero, the coefficient is necessarily positive, but as p and q increase the coefficient changes sign. If the coefficient be positive and be called μ^2, the general value of A_{pq} may be written

$$A_{pq} = B_{pq}\, e^{i(kat+\mu z)} + C_{pq}\, e^{i(kat-\mu z)} \quad \dots\dots\dots\dots\dots(4),$$

where, as the factor e^{ikat} is expressed, B_{pq}, C_{pq} are absolute constants. However, the first term in (4) expresses a motion propagated in the negative direction, which is excluded by the conditions of the problem, and thus we are to take simply as the term corresponding to p, q,

$$\phi = C_{pq} \cos\frac{p\pi x}{\alpha} \cos\frac{q\pi y}{\beta}\, e^{i(kat-\mu z)}.$$

In this expression C_{pq} may be complex; passing to real quantities and taking two new real arbitrary constants, we obtain

$$\phi = [D_{pq} \cos(kat - \mu z) + E_{pq} \sin(kat - \mu z)] \cos\frac{p\pi x}{\alpha} \cos\frac{q\pi y}{\beta} \dots(5).$$

We have now to consider the form of the solution in cases where the coefficient of A_{pq} in (3) is negative. If we call it $-\nu^2$, the solution corresponding to (4) is

$$A_{pq} = e^{ikat}(B_{pq}\, e^{\nu z} + C_{pq}\, e^{-\nu z}) \dots\dots\dots\dots\dots(6),$$

of which the first term is to be rejected as becoming infinite with z. We thus obtain corresponding to (5)

$$\phi = e^{-\nu z} \left[D_{pq} \cos kat + E_{pq} \sin kat \right] \cos \frac{p\pi x}{\alpha} \cos \frac{q\pi y}{\beta} \quad \ldots\ldots(7).$$

The solution obtained by combining all the particular solutions given by (5) and (7) is the general solution of the problem, and allows of a value of $d\phi/dz$ over the section $z = 0$, arbitrary at every point in both amplitude and phase.

At a great distance from the source the terms given in (7) become insensible, and the motion is represented by the terms of (5) alone. The effect of the terms involving high values of p and q is thus confined to the neighbourhood of the source, and at moderate distances any sudden variations or discontinuities in the motion at $z = 0$ are gradually eased off and obliterated.

If we fix our attention on any particular simple mode of vibration (for which p and q do not both vanish), and conceive the frequency of vibration to increase from zero upwards, we see that the effect, at first confined to the neighbourhood of the source, gradually extends further and further and, after a certain value is passed, propagates itself to an infinite distance, the critical frequency being that of the two dimensional free vibrations of the corresponding mode. Below the critical point no work is required to *maintain* the motion; above it as much work must be done at $z = 0$ as is carried off to infinity in the same time.

268 a. If in the general formulæ of § 267 we suppose that $r = 0$, we fall back upon the case of a motion purely two-dimensional. The third dimension (γ) of the chamber is then a matter of indifference; and the problem may be supposed to be that of the vibrations of a rectangular plate of air bounded, for example, by two parallel plates of glass, and confined at the rectangular boundary. In this form it has been treated both theoretically and experimentally by Kundt[1]. The velocity-potential is simply

$$\phi = \cos \left(p \frac{\pi x}{\alpha} \right) \cos \left(q \frac{\pi y}{\beta} \right) \ldots\ldots\ldots\ldots\ldots(1),$$

where p and q are integers; and the frequency is determined by

$$k^2 = \pi^2 (p^2/\alpha^2 + q^2/\beta^2) \ldots\ldots\ldots\ldots\ldots(2).$$

[1] *Pogg. Ann.* vol. XL. pp. 177, 337, 1873.

If the plate be *open* at the boundary, an approximate solution may be obtained by supposing that ϕ is there evanescent. In this case the expression for ϕ is derived from (1) by writing sines instead of cosines, while the frequency equation retains the same form (2). This has already been discussed under the head of membranes in § 197. If $\alpha = \beta$, so that the rectangle becomes a square, the various normal modes of the same pitch may be combined, as explained in § 197.

In Kundt's experiments the vibrations were excited through a perforation in one of the glass plates, to which was applied the extremity of a suitably tuned rod vibrating longitudinally, and the division into segments was indicated by the behaviour of cork filings. As regards pitch there was a good agreement with calculation in the case of plates *closed* at the boundary. When the rectangular boundary was *open*, the observed frequencies were too small, a discrepancy to be attributed to the merely approximate character of the assumption that the pressure is there invariable (see § 307).

The theory of the circular plate of air depends upon Bessel's functions, and is considered in § 339.

269. We will now examine the result of the composition of two trains of plane waves of harmonic type, whose amplitudes and wave-lengths are equal, but whose directions of propagation are inclined to one another at an angle 2α. The problem is one of two dimensions only, inasmuch as everything is the same in planes perpendicular to the lines of intersection of the two sets of wave-fronts.

At any moment of time the positions of the planes of maximum condensation for each train of waves may be represented by parallel lines drawn at equal intervals λ on the plane of the paper, and these lines must be supposed to move with a velocity a in a direction perpendicular to their length. If both sets of lines be drawn, the paper will be divided into a system of equal parallelograms, which advance in the direction of one set of diagonals. At each corner of a parallelogram the condensation is doubled by the superposition of the two trains of waves, and in the centre of each parallelogram the rarefaction is a maximum for the same reason. On each diagonal there is therefore a series of maxima and minima condensations, advancing without change of relative position and with velocity $a/\cos\alpha$. Between each adjacent pair of lines of maxima and minima there is a parallel line of zero condensation, on which the two trains of waves neutralize one another. It is especially remarkable that, if the wave-pattern were visible (like the corresponding water wave-pattern to which the whole of the preceding argument is applicable), it would appear to move forwards without change of type in a direction different from that of either component train, and with a velocity different from that with which both component trains move.

In order to express the result analytically, let us suppose that the two directions of propagation are equally inclined at an angle α to the axis of x. The condensations themselves may be denoted by

$$\cos\frac{2\pi}{\lambda}(a\,t - x\cos\alpha - y\sin\alpha)$$

and

$$\cos\frac{2\pi}{\lambda}(a\,t - x\cos\alpha + y\sin\alpha)$$

respectively, and thus the expression for the resultant is

$$s = \cos\frac{2\pi}{\lambda}(a\,t - x\cos\alpha - y\sin\alpha) + \cos\frac{2\pi}{\lambda}(a\,t - x\cos\alpha + y\sin\alpha)$$

$$= 2\cos\frac{2\pi}{\lambda}(a\,t - x\cos\alpha)\,\cos\frac{2\pi}{\lambda}(y\sin\alpha)\ldots\ldots\ldots(1).$$

It appears from (1) that the distribution of s on the plane xy advances parallel to the axis of x, unchanged in type, and with a uniform velocity $a/\cos\alpha$. Considered as depending on y, s is a maximum, when $y\sin\alpha$ is equal to 0, λ, 2λ, 3λ, &c., while for the intermediate values, viz. $\frac{1}{2}\lambda$, $\frac{3}{2}\lambda$, &c., s vanishes.

If $\alpha = \frac{1}{2}\pi$, so that the two trains of waves meet one another directly, the velocity of propagation parallel to x becomes infinite, and (1) assumes the form

$$s = 2\cos\left(\frac{2\pi}{\lambda}at\right)\cos\left(\frac{2\pi}{\lambda}y\right)\ldots\ldots\ldots\ldots(2);$$

which represents *stationary* waves.

The problem that we have just been considering is in reality the same as that of the reflection of a train of plane waves by an infinite plane wall. Since the expression on the right-hand side of equation (1) is an even function of y, s is symmetrical with respect to the axis of x, and consequently there is no motion across that axis. Under these circumstances it is evident that the motion could in no way be altered by the introduction along the axis of x of an absolutely immovable wall. If α be the angle between the surface and the direction of propagation of the incident waves, the velocity with which the places of maximum condensation (corresponding to the greatest elevation of water-waves) move along the wall is $a/\cos\alpha$. It may be noticed that the aërial pressures have no tendency to move the wall as a whole, except in the case of absolutely perpendicular incidence, since they are at any moment as much negative as positive.

269 a. When sound waves proceeding from a distant source are reflected perpendicularly by a solid wall, the superposition of the direct and reflected waves gives rise to a system of nodes and loops, exactly as in the case of a tube considered in § 255. The nodal planes, viz. the surfaces of evanescent motion, occur at distances from the wall which are even multiples of the quarter wave length, and the loops bisect the intervals between the nodes. In exploring experimentally it is usually best to seek the places of minimum effect, but whether these will be nodes or loops depends upon the apparatus employed, a consideration of which the neglect has led to some confusion[1]. Thus a resonator will cease to respond when its mouth coincides with a loop, so that this method of experimenting gives the *loops* whether the resonator be in connection with the ear or with a "manometric capsule" (§ 282). The same conclusion applies also to the use of the unaided ear, except that in this case the head is an obstacle large enough to disturb sensibly the original distribution of the loop and nodes[2]. If on the other hand the indicating apparatus be a small stretched membrane exposed upon both sides, or a sensitive smoke jet or flame, the places of vanishing disturbance are the *nodes*[3].

The complete establishment of stationary vibrations with nodes and loops occupies a certain time during which the sound is to be maintained. When a harmonium reed is sounding steadily in a room free from carpets and curtains, it is easy, listening with a resonator, to find places where the principal tone is almost entirely subdued. But at the first moment of putting down the

---

[1] N. Savart, *Ann. d. Chim.* LXXI. p. 20, 1839 ; XL. p. 385, 1845.

[2] *Phil. Mag.* VII. p. 150, 1879.

[3] *Phil. Mag. loc. cit.* p. 153.

key, or immediately after letting it go, the tone in question asserts itself, often with surprising vigour.

The formation of stationary nodes and loops in front of a reflecting wall may be turned to good account when it is desired to determine the wave-lengths of aërial vibrations. The method is especially valuable in the case of very acute sounds and of vibrations of frequency so high as to be inaudible. With the aid of a high pressure sensitive flame vibrations produced by small "bird-calls" may be traced down to a complete wave-length of 6 mm., corresponding to a frequency of about 55,000 per second.

270. So long as the medium which is the vehicle of sound continues of unbroken uniformity, plane waves may be propagated in any direction with constant velocity and with type unchanged; but a disturbance ensues when the waves reach any part where the mechanical properties of the medium undergo a change. The general problem of the vibrations of a variable medium is probably quite beyond the grasp of our present mathematics, but many of the points of physical interest are raised in the case of plane waves. Let us suppose that the medium is uniform above and below a certain infinite plane ($x = 0$), but that in crossing that plane there is an abrupt variation in the mechanical properties on which the propagation of sound depends—namely the *compressibility* and the *density*. On the upper side of the plane (which for distinctness of conception we may suppose horizontal) a train of plane waves advances so as to meet it more or less obliquely; the problem is to determine the (refracted) wave which is propagated onwards within the second medium, and also that thrown back into the first medium, or reflected. We have in the first place to form the equations of motion and to express the boundary conditions.

In the upper medium, if ρ be the natural density and s the condensation,
$$\text{density} = \rho\,(1 + s),$$
and
$$\text{pressure} = P\,(1 + As),$$
where A is a coefficient depending on the compressibility, and P is the undisturbed pressure. In like manner in the lower medium
$$\text{density} = \rho_1\,(1 + s_1),$$
$$\text{pressure} = P\,(1 + A_1 s_1),$$

the undisturbed pressure being the same on both sides of $x = 0$. Taking the axis of z parallel to the line of intersection of the plane of the waves with the surface of separation $x = 0$, we have for the upper medium (§ 244),

$$\frac{d^2\phi}{dt^2} = V^2 \left(\frac{d^2\phi}{dx^2} + \frac{d^2\phi}{dy^2} \right) \dots\dots\dots\dots\dots(1),$$

and

$$\frac{d\phi}{dt} + V^2 s = 0 \dots\dots\dots\dots\dots\dots(2),$$

where

$$V^2 = PA \div \rho \dots\dots\dots\dots\dots\dots(3).$$

Similarly, in the lower medium,

$$\frac{d^2\phi_1}{dt^2} = V_1^2 \left(\frac{d^2\phi_1}{dx^2} + \frac{d^2\phi_1}{dy^2} \right) \dots\dots\dots\dots\dots(4),$$

and

$$\frac{d\phi_1}{dt} + V_1^2 s_1 = 0 \dots\dots\dots\dots\dots\dots(5),$$

where

$$V_1^2 = PA_1 \div \rho_1 \dots\dots\dots\dots\dots\dots(6).$$

These equations must be satisfied at all points of the fluid. Further the boundary conditions require (i) that at all points of the surface of separation the velocities perpendicular to the surface shall be the same for the two fluids, or

$$d\phi/dx = d\phi_1/dx, \quad \text{when } x = 0 \dots\dots\dots\dots(7);$$

(ii) that the pressures shall be the same, whence $A_1 s_1 = A s$, or by (2), (3), (5) and (6),

$$\rho \, d\phi/dt = \rho_1 \, d\phi_1/dt, \quad \text{when } x = 0 \dots\dots\dots\dots(8).$$

In order to represent a train of waves of harmonic type, we may assume ϕ and ϕ_1 to be proportional to $e^{i(ax+by+ct)}$, where $ax + by = \text{const.}$ gives the direction of the plane of the waves. If we assume for the incident wave,

$$\phi = \phi' e^{i(ax+by+ct)} \dots\dots\dots\dots\dots(9),$$

the reflected and refracted waves may be represented respectively by

$$\phi = \phi'' e^{i(-ax+by+ct)} \dots\dots\dots\dots\dots(10),$$

$$\phi_1 = \phi_1 e^{i(a_1 x+by+ct)} \dots\dots\dots\dots\dots(11).$$

The coefficient of t is necessarily the same in all three waves on account of the periodicity, and the coefficient of y must be the same, since the traces of all the waves on the plane of separation must move together. With regard to the coefficient of x, it appears by substitution in the differential equations that its sign is changed in passing from the incident to the reflected wave ; in fact

$$c^2 = V^2 [(\pm a)^2 + b^2] = V_1^2 [a_1^2 + b^2] \ldots \ldots \ldots (12).$$

Now $b \div \sqrt{(a^2 + b^2)}$ is the sine of the angle included between the axis of x and the normal to the plane of the waves—in optical language, the sine of the angle of incidence, and $b \div \sqrt{(a_1^2 + b^2)}$ is in like manner the sine of the angle of refraction. If these angles be called θ, θ_1, (12) asserts that $\sin \theta : \sin \theta_1$ is equal to the constant ratio $V : V_1$,—the well-known law of sines. The laws of refraction and reflection follow simply from the fact that the velocity of propagation normal to the wave-fronts is constant in each medium, that is to say, independent of the *direction* of the wave-front, taken in connection with the equal velocities of the traces of all the waves on the plane of separation ($V \div \sin \theta = V_1 \div \sin \theta_1$). It remains to satisfy the boundary conditions (7) and (8).

These give

$$\left. \begin{array}{l} a (\phi' - \phi'') = a_1 \phi_1 \\ \rho (\phi' + \phi'') = \rho_1 \phi_1 \end{array} \right\} \ldots \ldots \ldots \ldots \ldots (13),$$

whence

$$2\phi' = \left(\frac{\rho_1}{\rho} + \frac{a_1}{a} \right) \phi_1 ; \qquad 2\phi'' = \left(\frac{\rho_1}{\rho} - \frac{a_1}{a} \right) \phi_1 \ldots \ldots \ldots (14).$$

This completes the symbolical solution. If a_1 (and θ_1) be real, we see that if the incident wave be

$$\phi = \cos (ax + by + ct),$$

or in terms of V, λ, and θ,

$$\phi = \cos \frac{2\pi}{\lambda} (x \cos \theta + y \sin \theta + Vt) \ldots \ldots \ldots (15),$$

the reflected wave is

$$\phi = \frac{\dfrac{\rho_1}{\rho} - \dfrac{\cot \theta_1}{\cot \theta}}{\dfrac{\rho_1}{\rho} + \dfrac{\cot \theta_1}{\cot \theta}} \cos \frac{2\pi}{\lambda} (- x \cos \theta + y \sin \theta + Vt) \ldots (16),$$

and the refracted wave is

$$\phi_1 = \frac{2}{\dfrac{\rho_1}{\rho} + \dfrac{\cot \theta_1}{\cot \theta}} \cos \frac{2\pi}{\lambda_1} (x \cos \theta_1 + y \sin \theta_1 + V_1 t) \ldots (17).$$

The formula for the amplitude of the reflected wave, viz.

$$\frac{\phi''}{\phi'} = \frac{\dfrac{\rho_1}{\rho} - \dfrac{\cot\theta_1}{\cot\theta}}{\dfrac{\rho_1}{\rho} + \dfrac{\cot\theta_1}{\cot\theta}} \dots\dots\dots\dots\dots(18),$$

is here obtained on the supposition that the waves are of harmonic type; but since it does not involve λ, and there is no change of phase, it may be extended by Fourier's theorem to waves of any type whatever.

If there be no reflected wave, $\cot\theta_1 : \cot\theta = \rho_1 : \rho$, from which and $(1 + \cot^2\theta_1) : (1 + \cot^2\theta) = V^2 : V_1^2$, we deduce

$$\left(\frac{\rho_1^2}{\rho^2} - \frac{V^2}{V_1^2}\right)\cot^2\theta = \frac{V^2}{V_1^2} - 1 \dots\dots\dots\dots(19),$$

which shews that, provided the refractive index $V_1 : V$ be intermediate in value between unity and $\rho : \rho_1$, there is always an angle of incidence at which the wave is completely intromitted; but otherwise there is no such angle.

Since (18) is not altered (except as to sign) by an interchange of θ, θ_1; ρ, ρ_1; &c., we infer that a wave incident in the second medium at an angle θ_1 is reflected in the same proportion as a wave incident in the first medium at an angle θ.

As a numerical example let us suppose that the upper medium is air at atmospheric pressure, and the lower medium water. Substituting for $\cot\theta_1$ its value in terms of θ and the refractive index, we get

$$\frac{\cot\theta_1}{\cot\theta} = \frac{V}{V_1}\sqrt{1 - \left(\frac{V_1^2}{V^2} - 1\right)\tan^2\theta}\dots\dots\dots(20),$$

or, since $V_1 : V = 4\cdot3$ approximately,

$$\cot\theta_1/\cot\theta = \cdot23\sqrt{(1 - 17\cdot5\tan^2\theta)},$$

which shews that the ratio of cotangents diminishes to zero, as θ increases from zero to about $13°$, after which it becomes imaginary, indicating total reflection, as we shall see presently. It must be remembered that in applying optical terms to acoustics, it is the *water* that must be conceived to be the 'rare' medium. The ratio of densities is about $770 : 1$; so that

$$\frac{\phi''}{\phi'} = \frac{1 - \cdot0003\sqrt{(1 - 17\cdot5\tan^2\theta)}}{1 + \cdot0003\sqrt{(1 - 17\cdot5\tan^2\theta)}}$$

$$= 1 - \cdot0006\sqrt{(1 - 17\cdot5\tan^2\theta)} \text{ very nearly.}$$

Even at perpendicular incidence the reflection is sensibly perfect.

If both media be gaseous, $A_1 = A$, if the temperature be constant; and even if the development of heat by compression be taken into account, there will be no sensible difference between A and A_1 in the case of the simple gases. Now, if $A_1 = A$, $\rho_1 : \rho = \sin^2 \theta : \sin^2 \theta_1$, and the formula for the intensity of the reflected wave becomes

$$\frac{\phi''}{\phi'} = \frac{\sin 2\theta - \sin 2\theta_1}{\sin 2\theta + \sin 2\theta_1} = \frac{\tan (\theta - \theta_1)}{\tan (\theta + \theta_1)} \dots\dots\dots (21),$$

coinciding with that given by Fresnel for light polarized perpendicularly to the plane of incidence. In accordance with Brewster's law the reflection vanishes at the angle of incidence, whose tangent is V/V_1.

But, if on the other hand $\rho_1 = \rho$, the cause of disturbance being the change of compressibility, we have

$$\frac{\phi''}{\phi'} = \frac{\tan \theta_1 - \tan \theta}{\tan \theta_1 + \tan \theta} = \frac{\sin (\theta_1 - \theta)}{\sin (\theta_1 + \theta)} \dots\dots\dots\dots(22),$$

agreeing with Fresnel's formula for light polarized in the plane of incidence. In this case the reflected wave does not vanish at any angle of incidence.

In general, when $\theta = 0$,

$$\phi'' : \phi' = \frac{\rho_1}{\rho} - \frac{V}{V_1} : \frac{\rho_1}{\rho} + \frac{V}{V_1} \dots\dots\dots\dots(23);$$

so that there is no reflection, if $\rho_1 : \rho = V : V_1$. In the case of gases $V^2 : V_1^2 = \rho_1 : \rho$, and then

$$\frac{\phi''}{\phi'} = \frac{\sqrt{\rho_1} - \sqrt{\rho}}{\sqrt{\rho_1} + \sqrt{\rho}} = \frac{V - V_1}{V + V_1} \dots\dots\dots\dots(24).$$

Suppose, for example, that after perpendicular incidence reflection takes place at a surface separating air and hydrogen. We have

$$\rho = \cdot001276, \quad \rho_1 = \cdot00008837 \cdot$$

whence $\sqrt{\rho} : \sqrt{\rho_1} = 3\cdot800$, giving

$$\phi'' = - \cdot5833 \, \phi'.$$

The ratio of intensities, which is as the square of the amplitudes, is $\cdot3402 : 1$, so that about one-third part is reflected.

If the difference between the two media be very small, and we write $V_1 = V + \delta V$, (24) becomes

$$\frac{\phi''}{\phi'} = - \tfrac{1}{2} \frac{\delta V}{V} \dots\dots\dots\dots\dots(25).$$

If the first medium be air at 0° Cent., and the second medium be air at $t°$ Cent., $V + \delta V = V\sqrt{(1 + \cdot00366\,t)}$; so that

$$\phi''/\phi' = - \cdot00091\,t.$$

The ratio of the intensities of the reflected and incident sounds is therefore $\cdot83 \times 10^{-6} \times t^2 : 1$.

As another example of the same kind we may take the case in which the first medium is dry air and the second is air of the same temperature saturated with moisture. At 10° Cent. air saturated with moisture is lighter than dry air by about one part in 220, so that $\delta V = \frac{1}{440} V$ nearly. Hence we conclude from (25) that the reflected sound is only about one 774,000th part of the incident sound.

From these calculations we see that reflections from warm or moist air must generally be very small, though of course the effect may accumulate by repetition. It must also be remembered that in practice the transition from one state of things to the other would be gradual, and not abrupt, as the present theory supposes. If the space occupied by the transition amount to a considerable fraction of the wave-length, the reflection would be materially lessened. On this account we might expect grave sounds to travel through a heterogeneous medium less freely than acute sounds.

The reflection of sound from surfaces separating portions of gas of different densities has engaged the attention of Tyndall, who has devised several striking experiments in illustration of the subject[1]. For example, sound from a high-pitched reed was conducted through a tin tube towards a sensitive flame, which served as an indicator. By the interposition of a coal-gas flame issuing from an ordinary bat's-wing burner between the tube and the sensitive flame, the greater part of the effect could be cut off. Not only so, but by holding the flame at a suitable angle, the sound could be reflected through another tube in sufficient quantity to excite a second sensitive flame, which but for the interposition of the reflecting flame would have remained undisturbed.

[The refraction of Sound has been demonstrated experimentally by Sondhauss[2] with the aid of a collodion balloon charged with carbonic acid.]

---

[1] *Sound*, 3rd edition, p. 282, 1875.

[2] *Pogg. Ann.* t. 85, p. 378, 1852. *Phil. Mag.* vol. v. p. 73, 1853.

The preceding expressions (16), (17), (18) hold good in every case of reflection from a 'denser' medium; but if the velocity of sound be greater in the lower medium, and the angle of incidence exceed the critical angle, a_1 becomes imaginary, and the formulæ require modification. In the latter case it is impossible that a refracted wave should exist, since, even if the angle of refraction were 90°, its trace on the plane of separation must necessarily outrun the trace of the incident wave.

If $-i a_1'$ be written in place of a_1, the symbolical equations are

Incident wave
$$\phi = e^{i(ax+by+ct)},$$

Reflected wave
$$\phi = \frac{\dfrac{\rho_1}{\rho} + i\,\dfrac{a_1'}{a}}{\dfrac{\rho_1}{\rho} - i\,\dfrac{a_1'}{a}}\, e^{i(-ax+by+ct)},$$

Refracted wave
$$\phi_1 = \frac{2}{\dfrac{\rho_1}{\rho} - i\,\dfrac{a_1'}{a}}\, e^{i(-ia_1'x+by+ct)};$$

from which by discarding the imaginary parts, we obtain

Incident wave
$$\phi = \cos(ax + by + ct) \quad\text{.......................}(26),$$

Reflected wave
$$\phi = \cos(-ax + by + ct + 2\epsilon) \quad\text{................}(27),$$

Refracted wave
$$\phi = \frac{2}{\left(\dfrac{\rho_1^2}{\rho^2} + \dfrac{a_1'^2}{a^2}\right)^{\frac{1}{2}}}\, e^{a_1'x} \cos(by + ct + \epsilon) \quad\text{.......}(28),$$

where
$$\tan \epsilon = \frac{a_1'\rho}{a\,\rho_1} \quad\text{.......................}(29).$$

These formulæ indicate total reflection. The disturbance in the second medium is not a wave at all in the ordinary sense, and at a short distance from the surface of separation (x negative) becomes insensible. Calculating a_1' from (12) and expressing it in terms of θ and λ, we find

$$a_1' = \frac{2\pi}{\lambda}\sqrt{\sin^2\theta - \frac{V^2}{V_1^2}} \quad\text{................}(30),$$

shewing that the disturbance does not penetrate into the second medium more than a few wave-lengths.

The difference of phase between the reflected and the incident waves is 2ϵ, where

$$\tan \epsilon = \frac{\rho}{\rho_1} \sqrt{\tan^2 \theta - \frac{V^2}{V_1^2} \sec^2 \theta} \dots\dots\dots\dots(31).$$

If the media have the same compressibilities, $\rho : \rho_1 = V_1^2 : V^2$, and

$$\tan \epsilon = \frac{V_1}{V} \sqrt{\frac{V_1^2}{V^2} \tan^2 \theta - \sec^2 \theta} \dots\dots\dots\dots(32).$$

Since there is no loss of energy in reflection and refraction, the work transmitted in any time across any area of the front of the incident wave must be equal to the work transmitted in the same time across corresponding areas of the reflected and refracted waves. These corresponding areas are plainly in the ratio

$$\cos \theta : \cos \theta : \cos \theta_1 ;$$

and thus by § 245 (τ being the same for all the waves),

$$\cos \theta \frac{\rho}{V} (\phi'^2 - \phi''^2) = \cos \theta_1 \frac{\rho_1}{V_1} \phi_1^2,$$

or since $V : V_1 = \sin \theta : \sin \theta_1,$

$$\rho \cot \theta (\phi'^2 - \phi''^2) = \rho_1 \cot \theta_1 \phi_1^2 \dots\dots\dots\dots(33),$$

which is the energy condition, and agrees with the result of multiplying together the two boundary equations (13).

When the velocity of propagation is greater in the lower than in the upper medium, and the angle of incidence exceeds the critical angle, no energy is transmitted into the second medium; in other words the reflection is total.

The method of the present investigation is substantially the same as that employed by Green in a paper on the Reflection and Refraction of Sound[1]. The case of perpendicular incidence was first investigated by Poisson[2], who obtained formulæ corresponding to (23) and (24), which had however been already given by Young for the reflection of Light. In a subsequent memoir[3] Poisson considered the general case of oblique incidence, limiting himself, however, to gaseous media for which Boyle's law holds good, and by a very complicated analysis arrived at a result equivalent to

---

[1] *Cambridge Transactions*, vol. VI. p. 403, 1838.

[2] *Mém. de l'Institut*, t. II. p. 305. 1819.

[3] "Mémoire sur le mouvement de deux fluides élastiques superposés." *Mém. de l'Institut*, t. X. p. 317. 1831.

(21). He also verified that the energies of the reflected and refracted waves make up that of the incident wave[1].

271. If the second medium be indefinitely extended downwards with complete uniformity in its mechanical properties, the transmitted wave is propagated onwards continually. But if at $x = -l$ there be a further change in the compressibility, or density, or both, part of the wave will be thrown back, and on arrival at the first surface ($x = 0$) will be divided into two parts, one transmitted into the first medium, and one reflected back, to be again divided at $x = -l$, and so on. By following the progress of these waves the solution of the problem may be obtained, the resultant reflected and transmitted waves being compounded of an infinite convergent series of components, all parallel and harmonic. This is the method usually adopted in Optics for the corresponding problem, and is quite rigorous, though perhaps not always sufficiently explained; but it does not appear to have any advantage over a more straightforward analysis. In the following investigation we shall confine ourselves to the case where the third medium is similar in its properties to the first medium.

In the first medium
$$\phi = \phi' e^{i(ax+by+ct)} + \phi'' e^{i(-ax+by+ct)}.$$
In the second medium
$$\psi = \psi' e^{i(a_1 x+by+ct)} + \psi'' e^{i(-a_1 x+by+ct)}.$$
In the third medium
$$\phi = \phi_1 e^{i(ax+by+ct)},$$
with the conditions
$$c^2 = V^2(a^2 + b^2) = V_1^2(a_1^2 + b^2) \dots\dots\dots\dots(1).$$

At the two surfaces of separation we have to secure the equality of normal motions and pressures; for $x = 0$,
$$\left. \begin{array}{l} a(\phi' - \phi'') = a_1(\psi' - \psi'') \\ \rho(\phi' + \phi'') = \rho_1(\psi' + \psi'') \end{array} \right\} \dots\dots\dots\dots(2);$$
for $x = -l$,
$$\left. \begin{array}{l} a_1(\psi' e^{-ia_1 l} - \psi'' e^{ia_1 l}) = a\phi_1 e^{-ial} \\ \rho_1(\psi' e^{-ia_1 l} + \psi'' e^{ia_1 l}) = \rho\phi_1 e^{-ial} \end{array} \right\} \dots\dots\dots\dots(3),$$

---

[1] [It is interesting and encouraging to note Laplace's remark in a correspondence with T. Young. The great analyst writes (1817) "Je persiste à croire que le problème de la propagation des ondes, lorsqu'elles traversent différens milieux, n'a jamais été résolu, et qu'il surpasse peut-être les forces actuelles de l'analyse" (Young's *Works*, vol. I. p. 374).]

from which ψ' and ψ'' are to be eliminated.   We get

$$\left.\begin{aligned}(\phi' - \phi'') \cos a_1 l - i\frac{a_1 \rho}{a \rho_1} (\phi' + \phi'') \sin a_1 l = \phi_1 e^{-ial}\\[2mm](\phi' + \phi'') \cos a_1 l - i\frac{a \rho_1}{a_1 \rho} (\phi' - \phi'') \sin a_1 l = \phi_1 e^{-ial}\end{aligned}\right\}\dots\dots(4);$$

and from these, if for brevity $a\rho_1/a_1\rho = \alpha$,

$$\frac{\phi''}{\phi'} = \frac{\alpha - \alpha^{-1}}{\alpha + \alpha^{-1} - 2i \cot a_1 l}\dots\dots\dots\dots(5),$$

$$\frac{\phi_1}{\phi'} = \frac{2\,e^{ial}}{2 \cos a_1 l + i \sin a_1 l\,(\alpha + \alpha^{-1})}\dots\dots\dots\dots(6).$$

In order to pass to real quantities, these expressions must be put into the form $Re^{i\theta}$.   If a_1 be real, we find corresponding to the incident wave

$$\phi = \cos (ax + by + ct),$$

the reflected wave

$$\phi = \frac{(\alpha^{-1} - \alpha) \sin (-ax + by + ct - \epsilon)}{\sqrt{\{4 \cot^2 a_1 l + (\alpha + \alpha^{-1})^2\}}}\dots\dots\dots(7),$$

and the transmitted wave

$$\phi = \frac{2 \cos (ax + by + ct + al - \epsilon)}{\sqrt{\{4 \cos^2 a_1 l + \sin^2 a_1 l\,(\alpha + \alpha^{-1})^2\}}}\dots\dots\dots(8),$$

where

$$\tan \epsilon = \tfrac{1}{2}(\alpha + \alpha^{-1}) \tan a_1 l\dots\dots\dots\dots(9).$$

If $\alpha = \rho_1 \cot \theta / \rho \cot \theta_1 = 1$, there is no reflected wave, and the transmitted wave is represented by

$$\phi = \cos (ax + by + ct + al - a_1 l),$$

shewing that, except for the alteration of phase, the whole of the medium might as well have been uniform.

If l be small, we have approximately for the reflected wave

$$\phi = \tfrac{1}{2}a_1 l\,(\alpha^{-1} - \alpha)\,\sin (-ax + by + ct),$$

a formula applying when the plate is thin in comparison with the wave-length.   Since $a_1 = (2\pi/\lambda_1) \cos \theta_1$, it appears that for a given angle of incidence the amplitude varies inversely as λ_1, or as λ.

In any case the reflection vanishes, if $\cot^2 a_1 l = \infty$, that is, if

$$2l \cos \theta_1 = m\lambda_1,$$

m being an integer.   The wave is then wholly transmitted.

At perpendicular incidence, the intensity of the reflection is expressed by

$$\left(\frac{V\rho}{V_1\rho_1} - \frac{V_1\rho_1}{V\rho} \right) \div \sqrt{4 \cot^2 \frac{2\pi l}{V_1\tau} + \left(\frac{V\rho}{V_1\rho_1} + \frac{V_1\rho_1}{V\rho} \right)^2} \ldots\ldots\ldots(10).$$

Let us now suppose that the second medium is incompressible, so that $V_1 = \infty$; our expression becomes

$$-\frac{\pi\rho_1 l/\rho\lambda}{\sqrt{\{1 + \pi^2 (\rho_1 l/\rho\lambda)^2\}}} \ldots\ldots\ldots\ldots\ldots\ldots(11),$$

shewing how the amount of reflection depends upon the relative masses of such quantities of the media as have volumes in the ratio of $l : \lambda$. It is obvious that the second medium behaves like a rigid body and acts only in virtue of its inertia. If this be sufficient, the reflection may become sensibly total.

We have now to consider the case in which a_1 is imaginary. In the symbolical expressions (5) and (6) $\cos a_1 l$ and $i \sin a_1 l$ are real, while α, $\alpha + \alpha^{-1}$, $\alpha - \alpha^{-1}$ are pure imaginaries. Thus, if we suppose that $a_1 = i a_1'$, $\alpha = i \alpha'$, and introduce the notation of the hyperbolic sine and cosine (§ 170), we get

$$\frac{\phi''}{\phi'} = \frac{-i (\alpha' + \alpha'^{-1}) \sinh a_1' l}{2 \cosh a_1' l - i (\alpha' - \alpha'^{-1}) \sinh a_1' l},$$

$$\frac{\phi_1}{\phi'} = \frac{2 e^{ial}}{2 \cosh a_1' l - i (\alpha' - \alpha'^{-1}) \sinh a_1' l}.$$

Hence, if the incident wave be

$$\phi = \cos (ax + by + ct),$$

the reflected wave is expressed by

$$\phi = \frac{(\alpha' + \alpha'^{-1}) \sinh a_1' l \cos (- ax + by + ct + \epsilon)}{\sqrt{\{4 \cosh^2 a_1' l + (\alpha' - \alpha'^{-1})^2 \sinh^2 a_1' l\}}} \ldots\ldots\ldots(12),$$

where
$$\cot \epsilon = \tfrac{1}{2} (\alpha'^{-1} - \alpha') \tanh a_1' l \ldots\ldots\ldots\ldots(13),$$

and the transmitted wave is expressed by

$$\phi = \frac{2 \sin (ax + by + ct + al + \epsilon)}{\sqrt{\{4 \cosh^2 a_1' l + (\alpha' - \alpha'^{-1})^2 \sinh^2 a_1' l\}}} \ldots\ldots\ldots(14).$$

It is easy to verify that the energies of the reflected and transmitted waves account for the whole energy of the incident wave. Since in the present case the corresponding areas of wave-front are equal for all three waves, it is only necessary to add the squares of the amplitudes given in equations (7), (8), or in equations (12), (14).

272. These calculations of reflection and refraction under various circumstances might be carried further, but their interest would be rather optical than acoustical. It is important to bear in mind that no energy is destroyed by any number of reflections and refractions, whether partial or total, what is lost in one direction always reappearing in another.

On account of the great difference of densities reflection is usually nearly total at the boundary between air and any solid or liquid matter. Sounds produced in air are not easily communicated to water, and *vice versâ* sounds, whose origin is under water, are heard with difficulty in air. A beam of wood, or a metallic wire, acts like a speaking tube, conveying sounds to considerable distances with very little loss.

272 a. In preceding sections the surface of separation, at which reflection takes place, is supposed to be absolutely plane. It is of interest, both from an acoustical and from an optical point of view, to inquire what effect would be produced by roughnesses, or corrugations, in the reflecting surface ; and the problem thus presented may be solved without difficulty to a certain extent by the method of § 268, especially if we limit ourselves to the case of perpendicular incidence. The equation of the reflecting surface will be supposed to be $z = \zeta$, where ζ is a periodic function of x whose mean value is zero. As a particular case we may take

$$\zeta = c \cos px \quad\dots\dots\dots\dots\dots\dots\dots(1);$$

but in general we should have to supplement the first term of the series expressed in (1) by cosines and sines of the multiples of px. The velocity-potential of the incident wave (of amplitude unity) may be written

$$\phi = e^{ik(at+z)} \quad\dots\dots\dots\dots\dots\dots\dots(2).$$

For the regularly reflected wave we have $\phi = A_0 e^{-ikz}$, the time factor being dropped for the sake of brevity ; but to this must be added terms in $\cos px$, $\cos 2px$, &c. Thus, as the complete value of ϕ in the upper medium,

$$\phi = e^{ikz} + A_0 e^{-ikz} + A_1 e^{-i\mu_1 z} \cos px + A_2 e^{-i\mu_2 z} \cos 2px + \dots \ \dots(3),$$

in which

$$\mu_1^2 = k^2 - p^2, \quad \mu_2^2 = k^2 - 4p^2, \quad\dots\dots\dots\dots\dots(4).$$

The expression (3), in which for simplicity sines of multiples of px have been omitted from the first, would be sufficiently general even though cosines of multiples of px accompanied $c \cos px$ in (1).

As explained in § 268, much turns upon whether the quantities $\mu_1, \mu_2, \ldots$ are real or imaginary. In the latter case the corresponding terms are sensible only in the neighbourhood of $z = 0$. If all the values of μ be imaginary, as happens when $p > k$, the reflected wave soon reduces itself to its first term.

For any real value of μ, say μ_r, the corresponding part of the velocity-potential is

$$\phi = \tfrac{1}{2} A_r \left\{ e^{-i(\mu_r z - rpx)} + e^{-i(\mu_r z + rpx)} \right\},$$

representing plane waves inclined to z at angles whose sines are $\pm rp/k$. These are known in Optics as the spectra of the rth order. When the wave-length of the corrugation is less than that of the vibration, there are no lateral spectra.

In the lower medium we have

$$\phi_1 = B_0 e^{ik_1 z} + B_1 e^{i\mu_1' z} \cos px + B_2 e^{i\mu_2' z} \cos 2px + \ldots \ldots (5),$$

where $$\mu_1'^2 = k_1^2 - p^2, \qquad \mu_2'^2 = k_1^2 - 4p^2, \ldots \ldots \ldots \ldots (6).$$

In each exponential the coefficient of z is to be taken positive; if it be imaginary, because the wave is propagated in the negative direction; if it be real, because the disturbance must decrease, and not increase, in penetrating the second medium.

The conditions to be satisfied at the boundary are (§ 270) that

$$\rho \phi = \rho_1 \phi_1 \ldots \ldots \ldots \ldots \ldots \ldots \ldots (7),$$

and that $d\phi/dn = d\phi_1/dn$, where dn is perpendicular to the surface $z = \zeta$. Hence

$$\frac{d(\phi - \phi_1)}{dz} - \frac{d(\phi - \phi_1)}{dx} \frac{d\zeta}{dx} = 0 \ldots \ldots \ldots \ldots (8).$$

Thus far there is no limitation upon either the amplitude (c) or the wave-length ($2\pi/p$) of the corrugation. We will now suppose that the wave-length is very large, so that p^2 may be neglected throughout. Under these conditions, (8) reduces to

$$d(\phi - \phi_1)/dz = 0 \ldots \ldots \ldots \ldots \ldots (9).$$

In the differentiation of (3) and (5) with respect to z, the various terms are multiplied by the coefficients $\mu_1, \mu_2, \ldots \mu_1', \mu_2', \ldots$;

but when p^2 is neglected these quantities may be identified with k, k_1 respectively. Thus at the boundary

$$\frac{d\phi}{dz} = ik\{e^{ik\zeta} - A_0 e^{-ik\zeta} - A_1 e^{-ik\zeta}\cos px - \ldots\ldots\ldots\ldots\ldots\};$$

and

$$\frac{d\phi_1}{dz} = ik_1\phi_1 = \frac{ik_1\rho\phi}{\rho_1},$$

by (7). Accordingly,

$$k_1\rho\{e^{ik\zeta} + A_0 e^{-ik\zeta} + A_1 e^{-ik\zeta}\cos px + \ldots\ldots\ldots\ldots\}$$

$$= k\rho_1\{e^{ik\zeta} - A_0 e^{-ik\zeta} - A_1 e^{-ik\zeta}\cos px - \ldots\ldots\ldots\ldots\},$$

or $$\frac{k_1\rho - k\rho_1}{k_1\rho + k\rho_1}e^{2ik\zeta} + A_0 + A_1\cos px + A_2\cos 2px + \ldots = 0\ldots\ldots(10).$$

By this equation A_0, A_1, &c. are determined when ζ is known.

If we put $\zeta = 0$, we fall back on previous results (23) § 270 for a truly plane surface. Thus A_1, $A_2,\ldots$ vanish, while

$$A_0 = \frac{k\rho_1 - k_1\rho}{k\rho_1 + k_1\rho}\ldots\ldots\ldots\ldots\ldots\ldots(11),$$

expressing the amplitude of the wave regularly reflected.

We will now apply (10) to the case of a simple corrugation, as expressed in (1), and for brevity we will denote the right hand member of (11) by R. The determination of A_0, $A_1,\ldots$ requires the expression of $e^{2ik\zeta}$ in Fourier's series. We have (compare § 343)

$$e^{2ikc\cos px} = J_0(2kc) - 2J_2(2kc)\cos 2px + 2J_4(2kc)\cos 4px + \ldots$$

$$+ i\{2J_1(2kc)\cos px - 2J_3(2kc)\cos 3px + 2J_5(2kc)\cos 5px - \ldots\}$$

$$\ldots\ldots\ldots(12),$$

where J_0, $J_1,\ldots$ are the Bessel's functions of the various orders. Thus

$$\left.\begin{array}{ll}
A_0/R = J_0(2kc), & A_1/R = 2iJ_1(2kc), \\
A_2/R = -2J_2(2kc), & A_3/R = -2iJ_3(2kc), \\
A_4/R = 2J_4(2kc), & A_5/R = 2iJ_5(2kc),
\end{array}\right\}\ldots(13),$$

the coefficients of even order being real, and those of odd order pure imaginaries. The complete solution of the problem of reflection, under the restriction that p is small, is then obtained by substitution in (3); and it may be remarked that it is the same as would be furnished by the usual optical methods, which take account only of phase retardations. Thus, as regards the wave reflected parallel to z, the retardation at any point of the surface due to the corrugation is 2ζ, or $2c \cos px$. The influence of the corrugations is therefore to change the amplitude of the reflected vibration in the ratio

$$\int \cos (2kc \cos px)\, dx : \int dx, \quad \text{or} \quad J_0 (2kc).$$

In like manner the amplitude of each of the lateral spectra of the first order is $J_1 (2kc)$, and so on. The sum of the intensities of all the reflected waves is

$$R^2 \{J_0^2 + 2J_1^2 + 2J_2^2 + \ldots\} = R^2 \ldots\ldots\ldots\ldots(14)$$

by a known theorem; so that, in the case supposed (of p infinitely small), the fraction of the whole energy thrown back is the same as if the surface were smooth.

It should be remarked that in this theory there is no limitation upon the value of $2kc$. If $2kc$ be small, only the earlier terms of the series are sensible, the Bessel's function $J_n (2kc)$ being of order $(2kc)^n$. When on the other hand $2kc$ is large, the early terms are small, while the series is less convergent. The values of J_0 and J_1 are tabulated in § 200. For certain values of $2kc$ individual reflected waves vanish. In the case of the regularly reflected wave, or spectrum of zero order, this first occurs when $2kc = 2\cdot404$, § 206, or $c = \cdot2\lambda$.

The full solution of the problem of the present section would require the determination of the reflection when k is given for all values of c and for all values of p. We have considered the case of p infinitely small, and we shall presently deal with the case where $p > k$. For intermediate values of p the problem is more difficult, and in considering them we shall limit ourselves to the simpler boundary conditions which obtain when no energy penetrates the second medium. The simplest case of all arises when $\rho_1 = 0$, so that the boundary equation (7) reduces to

$$\phi = 0 \ldots\ldots\ldots\ldots\ldots\ldots(15),$$

the condition for an "open end," § 256. We may also refer to the case of a rigid wall, or "closed" end, where the surface condition is

$$d\phi/dn = 0 \ldots\ldots\ldots\ldots\ldots\ldots(16).$$

By (3) and (15) the condition to be satisfied at the surface is

$$e^{2ikz} + A_0 + A_1 e^{i(k - \mu_1)z} \cos px + A_2 e^{i(k - \mu_2)z} \cos 2px + \ldots = 0 \ldots(16).$$

In our problem z is given by (1) as a function of x; and the equations of condition are to be found by equating to zero the coefficients of the various terms involving $\cos px$, $\cos 2px$, &c., when the left hand member of (16) is expanded in Fourier's series. The development of the various exponentials is effected as in (12); and the resulting equations are

$$J_0(2k) + A_0 + iA_1 J_1(k - \mu_1) - A_2 J_2(k - \mu_2) - \ldots = 0 \ldots (17),$$

$$2iJ_1(2k) + A_1\{J_0(k - \mu_1) - J_2(k - \mu_1)\}$$
$$+ A_2\{iJ_1(k - \mu_2) - iJ_3(k - \mu_2)\} + \ldots = 0 \ldots\ldots (18),$$

$$-2J_2(2k) + A_1\{iJ_1(k - \mu_1) - iJ_3(k - \mu_1)\}$$
$$+ A_2\{J_0(k - \mu_2) + J_4(k - \mu_2)\} + \ldots = 0 \ldots\ldots(19),$$

and so on, where for the sake of brevity c has been made equal to unity. So far as $(k - \mu)$ may be treated as real, as happens for a large number of terms when p is small relatively to k, the various Bessel's functions are all real, and thus the A's of even order are real and the A's of odd order are pure imaginaries. Accordingly the phase of the perpendicularly reflected wave is the same as if $c = 0$; but it must be remembered that this conclusion is in reality only approximate, because, however small p may be, the μ's end by becoming imaginary.

From the above equations it is easy to obtain the value of A_0 as far as the term in p^4. From (19)

$$A_2 = 2J_2(2k);$$

from (18)

$$iA_1 = 2J_1(2k) + (k - \mu_2)J_2(2k);$$

and finally from (17)

$$-A_0 = J_0(2k) + (k - \mu_1)J_1(2k)$$
$$+ \{\tfrac{1}{2}(k - \mu_1)(k - \mu_2) - \tfrac{1}{4}(k - \mu_2)^2\}J_2(2k) + \ldots \ldots (20).$$

From (4)

$$k - \mu_1 = \frac{p^2}{2k} + \frac{p^4}{8k^3} + \ldots;$$

so that, as expanded in powers of p with reintroduction of c,

$$-A_0 = J_0(2kc) + \frac{p^2}{k^2}.\tfrac{1}{2}kc.J_1(2kc)$$
$$+ \frac{p^4}{k^4}\{\tfrac{1}{8}kc.J_1(2kc) - \tfrac{1}{2}k^2c^2.J_2(2kc)\}\ldots\ldots\ldots(21)[1].$$

[1] Brit. Ass. Rep. 1893, p. 691.

This gives the amplitude of the perpendicularly reflected wave, with omission of p^6 and higher powers of p.

The case of reflection from a fixed wall is a little more complicated. By (8) the boundary condition is

$$d\phi/dz + pc \sin px \,.\, d\phi/dx = 0,$$

which gives

$$e^{2ikz} - A_0 - \frac{\mu_1}{k} A_1 e^{i(k-\mu_1)z} \cos px - \frac{\mu_2}{k} A_2 e^{i(k-\mu_2)z} \cos 2px - \ldots$$

$$- \frac{p^2 c \sin px}{ik} \{A_1 e^{i(k-\mu_1)z} \sin px + 4 A_2 e^{i(k-\mu_2)z} \sin 2px + \ldots\} = 0$$

$$\ldots\ldots\ldots(22)$$

as the equation to be satisfied when $z = c \cos px$. The first approximation to A_1 gives

$$A_1 = 2i J_1(2kc) \ldots\ldots\ldots\ldots\ldots\ldots(23);$$

whence to a second approximation

$$A_0 = J_0(2kc) + \left\{-\tfrac{1}{2}(k - \mu_1) + \frac{p^2 c}{2k}\right\} i A_1$$

$$= J_0(2kc) - \frac{p^2}{2k^2} \,.\, kc \,.\, J_1(2kc) \ldots\ldots\ldots\ldots(24).$$

The first approximation to the various coefficients may be found by putting $R = +1$ in (13).

When $p > k$, there are no diffracted spectra, and the whole energy of the wave incident upon an impenetrable medium must be represented in the wave directly reflected. The modulus of A_0 is therefore unity. When $p < k$, the energy is divided between the various spectra, including that of zero order. There is thus a relation between the squares of the moduli of $A_0, A_1, A_2, \ldots$, the series being continued as long as μ is real.

A more analytical investigation may be based upon v. Helmholtz's theorem (§ 293), according to which

$$\int \left\{\psi \frac{d\chi}{dn} - \chi \frac{d\psi}{dn}\right\} dS = 0,$$

where S is any closed surface, and ψ and χ satisfy the equation

$$\nabla^2 + k^2 = 0.$$

In order to apply this we take for ψ and χ the real and imaginary parts respectively of ϕ as given by (3). Thus representing each complex coefficient A_n in the form $C_n + iD_n$, we get

$$\psi = \cos kz + C_0\cos kz + D_0 \sin kz$$
$$+ (C_1 \cos \mu_1 z + D_1\sin \mu_1 z)\cos px +\ldots\ldots\ldots\ldots\ldots(25),$$

$$\chi = \sin kz - C_0 \sin kz + D_0\cos kz$$
$$+(- C_1\sin \mu_1 z + D_1\cos \mu_1 z)\cos px +\ldots\ldots\ldots\ldots\ldots(26).$$

In (25), (26), when the series are carried sufficiently far, the terms change their form on account of μ becoming imaginary; but for the present purpose these terms will not be required, as they disappear when z is very great. The surface of integration S is made up of the reflecting surface and of a plane parallel to it at a great distance. Although this surface is not strictly closed, it may be treated as such, since the part still remaining open laterally at infinity does not contribute sensibly to the result. Now the part of the integral corresponding to the reflecting surface vanishes, either because

$$\psi = \chi = 0,$$

or else because $d\psi/dn = d\chi/dn = 0 ;$

and we conclude that when z is great

$$\int \left\{ \psi \frac{d\chi}{dz} - \chi \frac{d\psi}{dz} \right\} dx = 0 \ \ldots\ldots\ldots\ldots\ldots(27).$$

The application of (27) to the values of ψ and χ in (25), (26) gives

$$C_0{}^2 + D_0{}^2 + \frac{\mu_1}{2k}(C_1{}^2 + D_1{}^2) + \frac{\mu_2}{2k}(C_2{}^2 + D_2{}^2) + \ldots = 1 \ \ldots\ldots(28),$$

the series in (28) being continued so far as to include every *real* value of μ.

In (28) $\frac{1}{4}(C_n{}^2 + D_n{}^2)$ represents the intensity of each spectrum of the nth order.

The coefficient μ_n/k is equal to $\cos \theta_n$, where θ_n is the obliquity of the diffracted rays. The meaning of this factor will be evident when it is remarked that to each unit of area of the waves incident and directly reflected, there corresponds an area $\cos \theta_n$ of the waves which constitute the spectrum of the nth order.

If all the values of μ are imaginary, as happens when $p > k$, (28) reduces to

$$C_0{}^2 + D_0{}^2 = 1 \ \ldots\ldots\ldots\ldots\ldots\ldots\ldots(29),$$

or the intensity of the wave directly reflected is unity. It is of importance to notice the full significance of this result. However deep the corrugations may be, if only they are periodic in a period less than the wave-length of the vibration, the regular reflection is total. An extremely rough wall will thus reflect sound waves of moderate pitch as well as if it were theoretically smooth.

The above investigation is limited to the case where the second medium is impenetrable, so that the whole energy of the incident wave is thrown back in the regularly reflected wave and in the diffracted spectra. It is an interesting question whether the conclusion that corrugations of period less than λ have no effect can be extended so as to apply when there is a wave regularly transmitted. It is evident that the principle of energy does not suffice to decide the question, but it is probable that the answer should be in the negative. If we suppose the corrugations of given period to become very deep and involved, it would seem that the condition of things would at last approach that of a very gradual transition between the media, in which case (§ 148 b) the reflection tends to vanish.

Our limits will not allow us to treat at length the problem of oblique incidence upon a corrugated surface; but one or two remarks may be made.

If p^2 may be neglected, the solution corresponding to (13) is

$$A_0 = R J_0 (2kc \cos \theta) \dots\dots\dots\dots\dots(30),$$

θ being the angle of incidence and reflection, and R the value of A_0, § 270, corresponding to $c = 0$. The factor expressing the effect of the corrugations is thus a function of $c \cos \theta$; so that a deep corrugation when θ is large may have the same effect as a shallow one when θ is small.

Whatever be the angle of incidence, there are no reflected spectra (except of zero order) when the wave-length of the corrugation is less than the *half* of that of the vibrations. Hence, if the second medium be impenetrable, the regular reflection under the above condition is total.

The reader who wishes to pursue the study of the theory of gratings is referred to treatises on optics, and to papers by the Author[1], and by Prof. Rowland[2].

---

[1] The Manufacture and Theory of Diffraction Gratings, *Phil. Mag.* vol. XLVII. pp. 81, 193, 1874; On Copying Diffraction Gratings, and on some Phenomena connected therewith, *Phil. Mag.* vol. XI. p. 196, 1881; *Enc. Brit.* Wave Theory of Light.

[2] Gratings in Theory and Practice, *Phil. Mag.* vol. XXXV. p. 397, 1893.

CHAPTER XIV.

273. In connection with the general problem of aërial vibrations in three dimensions one of the first questions, which naturally offers itself, is the determination of the motion in an unlimited atmosphere consequent upon arbitrary initial disturbances. It will be assumed that the disturbance is *small*, so that the ordinary approximate equations are applicable, and further that the initial velocities are such as can be derived from a velocity-potential, or (§ 240) that there is no *circulation*. If the latter condition be violated, the problem is one of vortex motion, on which we do not enter. We shall also suppose in the first place that no external forces act upon the fluid, so that the motion to be investigated is due solely to a disturbance actually existing at a time ($t = 0$), previous to which we do not push our inquiries. The method that we shall employ is not very different from that of Poisson[1], by whom the problem was first successfully attacked.

If u_0, v_0, w_0 be the initial velocities at the point x, y, z, and s_0 the initial condensation, we have (§ 244),

$$\phi_0 = \int (u_0 dx + v_0 dy + w_0 dz) \dots\dots\dots\dots(1),$$

$$\dot{\phi}_0 = -a^2 s_0 \dots\dots\dots\dots\dots\dots\dots\dots (2),$$

by which the initial values of the velocity-potential ϕ and of its differential coefficient with respect to time $\dot{\phi}$ are determined. The problem before us is to determine ϕ at time t from the above

---

[1] Sur l'intégration de quelques équations linéaires aux différences partielles, et particulièrement de l'équation générale du mouvement des fluides élastiques. *Mém. de l'Institut*, t. III. p. 121. 1820.

initial values, and the general equation applicable at all times and places,

$$\left(\frac{d^2}{dt^2} - a^2 \nabla^2\right) \phi = 0 \quad \dots\dots\dots\dots\dots\dots \quad (3).$$

When ϕ is known, its derivatives give the component velocities at any point.

The symbolical solution of (3) may be written

$$\phi = \sin (ia\nabla t).\,\theta + \cos (ia\nabla t).\,\chi \dots\dots\dots\dots\dots(4),$$

where θ and χ are two arbitrary functions of x, y, z and $i = \sqrt{(-1)}$. To connect θ and χ with the initial values of ϕ and $\dot{\phi}$, which we shall denote by f and F respectively, it is only necessary to observe that when $t = 0$, (4) gives

$$\phi_0 = \chi, \qquad \dot{\phi}_0 = ia\nabla.\,\theta \,;$$

so that our result may be expressed

$$\phi = \cos (ia\nabla t).\,f + \frac{\sin (ia\nabla t)}{ia\nabla}.\,F \,\dots\dots\dots\dots\dots(5),$$

in which equation the question of the interpretation of odd powers of ∇ need not be considered, as both the symbolic functions are wholly even.

In the case where ϕ was a function of x only, we saw (§ 245) that its value for any point x at time t depended on the initial values of ϕ and $\dot{\phi}$ at the points whose co-ordinates were $x - at$ and $x + at$, and was wholly independent of the initial circumstances at all other points. In the present case the simplest supposition open to us is that the value of ϕ at a point O depends on the initial values of ϕ and $\dot{\phi}$ at points situated on the surface of the sphere, whose centre is O and radius at; and, as there can be no reason for giving one direction a preference over another, we are thus led to investigate the expression for the mean value of a function over a spherical surface in terms of the successive differential coefficients of the function at the centre.

By the symbolical form of Maclaurin's theorem the value of $F(x, y, z)$ at any point P on the surface of the sphere of radius r may be written

$$F(x, y, z) = e^{x\frac{d}{dx_0} + y\frac{d}{dy_0} + z\frac{d}{dz_0}}.\,F(x_0, y_0, z_0),$$

the centre of the sphere O being the origin of co-ordinates. In the integration over the surface of the sphere d/dx_0, d/dy_0, d/dz_0 behave as constants; we may denote them temporarily by l, m, n, so that $\nabla^2 = l^2 + m^2 + n^2$.

Thus, r being the radius of the sphere, and dS an element of its surface, since, by the symmetry of the sphere, we may replace any function of $\dfrac{lx + my + nz}{\sqrt{(l^2 + m^2 + n^2)}}$ by the same function of z without altering the result of the integration,

$$\iint e^{lx+my+nz}\,dS = \iint (e^{\nabla})^{\frac{lx+my+nz}{\sqrt{(l^2+m^2+n^2)}}}\,dS$$

$$= \iint e^{\nabla z}\,dS = 2\pi r \int_{-r}^{+r} e^{\nabla z}\,dz = \frac{2\pi r}{\nabla}\left(e^{\nabla r} - e^{-\nabla r}\right) = 4\pi r^2 \frac{\sin(i\nabla r)}{i\nabla r}.$$

The mean value of F over the surface of the sphere of radius r is thus expressed by the result of the operation on F of the symbol $\sin(i\nabla r)/i\nabla r$, or, if $\iint d\sigma$ denote integration with respect to angular space,

$$\frac{1}{4\pi}\iint F(r)\,d\sigma = \frac{\sin(i\nabla r)}{i\nabla r}\,.\,F \quad\ldots\ldots\ldots\ldots\ldots(6).$$

By comparison with (5) we now see that so far as ϕ depends on the initial values of $\dot\phi$, it is expressed by

$$\phi = \frac{t}{4\pi}\iint F(at)\,d\sigma \quad\ldots\ldots\ldots\ldots\ldots\ldots(7),$$

or in words, ϕ at any point at time t is the mean of the initial values of $\dot\phi$ over the surface of the sphere described round the point in question with radius at, the whole multiplied by t.

By Stokes' rule (§ 95), or by simple inspection of (5), we see that the part of ϕ depending on the initial values of ϕ may be derived from that just written by differentiating with respect to t and changing the arbitrary function. The complete value of ϕ at time t is therefore

$$\phi = \frac{t}{4\pi}\iint F(at)\,d\sigma + \frac{1}{4\pi}\frac{d}{dt}\,t\iint f(at)\,d\sigma \ldots\ldots\ldots(8),$$

which is Poisson's result [1].

On account of the importance of the present problem, it may

---

[1] Another investigation will be found in Kirchhoff's *Vorlesungen über Mathematische Physik*, p. 317. 1876. [See also Note to § 273 at the end of this volume.]

be well to verify the solution *a posteriori*. We have first to prove that it satisfies the general differential equation (3). Taking for the present the first term only, and bearing in mind the general symbolic equation

$$\frac{d^2}{dt^2} t = \frac{1}{t} \frac{d}{dt} t^2 \frac{d}{dt} \quad \dots\dots\dots\dots\dots\dots (9),$$

we find from (8)

$$\frac{d^2\phi}{dt^2} = \frac{1}{4\pi t} \frac{d}{dt} t^2 \iint \frac{d}{dt} F(at)\, d\sigma = \frac{1}{4\pi at} \frac{d}{dt} \iint \frac{d\,F(at)}{d\,(at)}\, dS,$$

dS being the surface element of the sphere $r = at$.

But by Green's theorem

$$\iint \frac{d\,F(r)}{dr}\, dS = \iiint \nabla^2 F\, dV \quad (r < at);$$

and thus

$$\frac{d^2\phi}{dt^2} = \frac{1}{4\pi at} \frac{d}{dt} \cdot \iiint \nabla^2 F\, dV \quad (r < at)$$

$$= \frac{1}{4\pi t} \iint \nabla^2 F\, dS \quad (r = at) = \frac{a^2 t}{4\pi} \iint \nabla^2 F\, d\sigma.$$

Now $\iint \nabla^2 F\, d\sigma$ is the same as $\nabla^2 \iint F\, d\sigma$, and thus (3) is in fact satisfied.

Since the second part of ϕ is obtained from the first by differentiation, it also must satisfy the fundamental equation.

With respect to the initial conditions we see that when t is made equal to zero in (8),

$$\phi = \frac{1}{4\pi} \iint f(at)\, d\sigma \quad (t = 0) = f(0) ;$$

$$\dot{\phi} = \frac{1}{4\pi} \iint F(at)\, d\sigma \quad (t = 0) + \frac{1}{4\pi} \frac{d^2}{dt^2} t \iint f(at)\, d\sigma \quad (t = 0),$$

of which the first term becomes in the limit $F(0)$. When $t = 0$,

$$\frac{d^2}{dt^2} t \iint f(at)\, d\sigma = 2\frac{d}{dt} \iint f(at)\, d\sigma \quad (t = 0)$$

$$= 2a \iint f'(at)\, d\sigma \quad (t = 0) = 0,$$

since the oppositely situated elements cancel in the limit, when the radius of the spherical surface is indefinitely diminished. The expression in (8) therefore satisfies the prescribed initial conditions as well as the general differential equation.

274. If the initial disturbance be confined to a space T, the integrals in (8) § 273 are zero, unless some part of the surface of the sphere $r = at$ be included within T. Let O be a point external to T, r_1 and r_2 the radii of the least and greatest spheres described about O which cut it. Then so long as $at < r_1$, ϕ remains equal to zero. When at lies between r_1 and r_2, ϕ may be finite, but for values greater than r_2 ϕ is again zero. The disturbance is thus at any moment confined to those parts of space for which at is intermediate between r_1 and r_2. The limit of the wave is the envelope of spheres with radius at, whose centres are situated on the surface of T. "When t is small, this system of spheres will have an exterior envelope of two sheets, the outer of these sheets being exterior, and the inner interior to the shell formed by the assemblage of the spheres The outer sheet forms the outer limit to the portion of the medium in which the dilatation is different from zero. As t increases, the inner sheet contracts, and at last its opposite sides cross, and it changes its character from being exterior, with reference to the spheres, to interior. It then expands, and forms the inner boundary of the shell in which the wave of condensation is comprised [1]." The successive positions of the boundaries of the wave are thus a series of parallel surfaces, and each boundary is propagated normally with a velocity equal to a.

If at the time $t = 0$ there be no motion, so that the initial disturbance consists merely in a variation of density, the subsequent condition of things is expressed by the first term of (8) § 273. Let us suppose that the original disturbance, still limited to a finite region T, consists of condensation only, without rarefaction. It might be thought that the same peculiarity would attach to the resulting wave throughout the whole of its subsequent course; but, as Prof. Stokes has remarked, such a conclusion would be erroneous. For values of the time less than r_1/a the potential at O is zero; it then becomes negative (s_0 being positive), and continues negative until it vanishes again when $t = r_2/a$, after which it always remains equal to zero. While ϕ is diminishing, the medium at O is in a state of condensation, but as ϕ increases again to zero, the state of the medium at O is one of rarefaction. The wave propagated outwards consists therefore of two parts at least, of which the first is condensed and the last rarefied. Whatever may be the character of the original disturbance within T, the final value of ϕ

[1] Stokes, "Dynamical Theory of Diffraction." *Camb. Trans.* IX. p. 15, 1849.

at any external point O is the same as the initial value, and there-fore, since $a^2s = -\dot{\phi}$, the mean condensation during the passage of the wave, depending on the integral $\int s\, dt$, is zero. Under the head of spherical waves we shall have occasion to return to this subject (§ 279).

The general solution embodied in (8) § 273 must of course embrace the particular case of plane waves, but a few words on this application may not be superfluous, for it might appear at first sight that the effect at a given point of a disturbance initially confined to a slice of the medium enclosed between two parallel planes would not pass off in any finite time, as we know it ought to do. Let us suppose for simplicity that ϕ_0 is zero throughout, and that within the slice in question the initial value $\dot{\phi}_0$ is constant. From the theory of plane waves we know that at any arbitrary point the disturbance will finally cease after the lapse of a time t, such that at is equal to the distance (d) of the point under consideration from the further boundary of the initially disturbed region; while on the other hand, since the sphere of radius at continues to cut the region, it would appear from the general formula that the disturbance continues. It is true indeed that ϕ remains finite, but this is not inconsistent with rest. It will in fact appear on examination that the mean value of $\dot{\phi}_0$ multiplied by the radius of the sphere is the same whatever may be the position and size of the sphere, provided only that it cut completely through the region of original disturbance. If $at > d$, ϕ is thus constant with respect both to space and time, and accordingly the medium is at rest.

[The same principles may find an application to the phenomena of *thunder*. Along the path of the lightning we may perhaps suppose that the generation of heat is uniform, equivalent to a uniform initial distribution of condensation. It appears that the value of ϕ at O the point of observation can change rapidly only when the sphere $r = at$ meets the path of the discharge at its extremities or very obliquely.]

275. In two dimensions, when ϕ is independent of z, it might be supposed that the corresponding formula would be obtained by simply substituting for the sphere of radius at the circle of equal radius. This, however, is not the case. It may be proved that the mean value of a function $F(x, y)$ over the circumference of a circle of radius r is $J_0(ir\nabla)F_0$, where $i = \sqrt{(-1)}$,

$$\nabla^2 = d^2/dx_0^2 + d^2/dy_0^2,$$

and J_0 is *Bessel's* function of zero order; so that

$$\frac{1}{2\pi r}\int F(x, y)\, ds = \left(1 + \frac{r^2\nabla^2}{2^2} + \frac{r^4\nabla^4}{2^2 \cdot 4^2} + \ldots\right)F,$$

differing from what is required to satisfy the fundamental equation.

The correct result applicable to two dimensions may be obtained from the general formula. The element of spherical surface dS may be replaced by $r\, dr\, d\theta/\cos\psi$, where r, θ are plane polar co-ordinates, and ψ is the angle between the tangent plane and that in which the motion takes place. Thus

$$\cos\psi = \frac{\sqrt{(a^2t^2 - r^2)}}{at},$$

$F(at)$ is replaced by $F(r, \theta)$, and so

$$\phi = \iint \frac{F(r, \theta)\, r\, dr\, d\theta}{4\pi a \sqrt{(a^2t^2 - r^2)}} \ldots\ldots\ldots\ldots\ldots\ldots(1),$$

where the integration extends over the area of the circle $r = at$. The other term might be obtained by Stokes' rule.

This solution is applicable to the motion of a layer of gas between two parallel planes, or to that of an unlimited stretched membrane, which depends upon the same fundamental equation.

276. From the solution in terms of initial conditions we may, as usual (§ 66), deduce the effect of a continually renewed disturbance. Let us suppose that throughout the space T (which will ultimately be made to vanish), a uniform disturbance $\dot{\phi}$, equal to $\Phi(t')\, dt'$, is communicated at time t'. The resulting value of ϕ at time t is

$$\frac{S}{4\pi a^2(t - t')}\Phi(t')\, dt',$$

where S denotes the part of the surface of the sphere $r = a(t - t')$ intercepted within T, a quantity which vanishes, unless $a(t - t')$ be comprised between the narrow limits r_1 and r_2. Ultimately $t - t'$ may be replaced by r/a, and $\Phi(t')$ by $\Phi(t - r/a)$; and the result of the integration with respect to dt' is found by writing T (the volume) for $\int a S\, dt'$. Hence

$$\phi = \frac{T}{4\pi a^2 r}\Phi\left(t - \frac{r}{a}\right)\ldots\ldots\ldots\ldots\ldots\ldots\ldots(1),$$

shewing that the disturbance originating at any point spreads itself symmetrically in all directions with velocity a, and with amplitude varying inversely as the distance. Since any number of particular solutions may be superposed, the general solution of the equation

$$\ddot{\phi} = a^2 \nabla^2 \phi + \Phi \dots\dots\dots\dots\dots\dots(2)$$

may be written

$$\phi = \frac{1}{4\pi a^2} \iiint \Phi \left(t - \frac{r}{a} \right) \frac{dV}{r} \dots\dots\dots\dots\dots(3),$$

r denoting the distance of the element dV situated at x, y, z from O (at which ϕ is estimated), and $\Phi(t - r/a)$ the value of Φ for the point x, y, z at the time $t - r/a$. Complementary terms, satisfying through all space the equation $\ddot{\phi} = a^2 \nabla^2 \phi$, may of course occur independently.

In our previous notation (§ 244)

$$\Phi = \frac{d}{dt} \int (X\,dx + Y\,dy + Z\,dz) ;$$

and it is assumed that $X\,dx + Y\,dy + Z\,dz$ is a complete differential. Forces, under whose action the medium could not adjust itself to equilibrium, are excluded ; as for instance, a force uniform in magnitude and direction within a space T, and vanishing outside that space. The nature of the disturbance denoted by Φ is perhaps best seen by considering the extreme case when Φ vanishes except through a small volume, which is supposed to diminish without limit, while the magnitude of Φ increases in such a manner that the whole effect remains finite. If then we integrate equation (2) through a small space including the point at which Φ is ultimately concentrated, we find in the limit

$$0 = a^2 \iint \frac{d\phi}{dn}\,dS + \iiint \Phi\,dV \dots\dots\dots\dots(4),$$

shewing that the effect of Φ may be represented by a proportional introduction or abstraction of fluid at the place in question. The simplest source of sound is thus analogous to a focus in the theory of conduction of heat, or to an electrode in the theory of electricity.

277. The preceding expressions are general in respect of the relation to time of the functions concerned ; but in almost all the applications that we shall have to make, it will be convenient to analyse the motion by Fourier's theorem and treat separately the simple harmonic motions of various periods, afterwards, if necessary, compounding the results. The values of ϕ and Φ, if simple harmonic at every point of space, may be expressed in the form $R \cos(nt + \epsilon)$, R and ϵ being independent of time, but variable from point to point. But as in such cases it often conduces to simplicity to add the term $iR \sin(nt + \epsilon)$, making altogether $Re^{i(nt+\epsilon)}$, or $Re^{i\epsilon} \cdot e^{int}$, we will assume simply that all the functions which enter into a problem are proportional to e^{int}, the coefficients being in general complex. After our operations are completed, the real and imaginary parts of the expressions can be separated, either of them by itself constituting a solution of the question.

Since ϕ is proportional to e^{int}, $\ddot{\phi} = -n^2 \phi$; and the differential equation becomes

$$\nabla^2\phi + k^2\phi + a^{-2}\Phi = 0 \ldots\ldots\ldots\ldots\ldots\ldots\ldots(1),$$

where, for the sake of brevity, k is written in place of n/a. If λ denote the *wave-length* of the vibration of the period in question,

$$k = n/a = 2\pi/\lambda \ldots\ldots\ldots\ldots\ldots\ldots\ldots (2).$$

To adapt (3) of the preceding section to the present case, it is only necessary to remark that the substitution of $t - r/a$ for t is effected by introducing the factor $e^{-inr/a}$, or e^{-ikr}: thus

$$\Phi(t - r/a) = e^{-ikr} \Phi(t),$$

and the solution of (1) is

$$\phi = \frac{1}{4\pi a^2} \iiint \frac{e^{-ikr}}{r} \Phi \, dV \ldots\ldots\ldots\ldots\ldots(3),$$

to which may be added any solution of $\nabla^2\phi + k^2\phi = 0$.

If the disturbing forces be all in the same phase, and the region through which they act be very small in comparison with the wave-length, e^{-ikr} may be removed from under the integral sign, and at a sufficient distance we may take

$$\phi = \frac{e^{-ikr}}{4\pi a^2 r} \iiint \Phi \, dV,$$

or in real quantities, on restoring the time factor and replacing $\iiint \Phi \, dV$ by Φ_1,

$$\phi = \Phi_1 \frac{\cos(nt - kr + \epsilon)}{4\pi a^2 r} \ldots\ldots\ldots\ldots\ldots (4).$$

In order to verify that (3) satisfies the differential equation (1), we may proceed as in the theory of the common potential. Considering one element of the integral at a time, we have first to shew that

$$\phi = \frac{e^{-ikr}}{r} \dots\dots\dots\dots\dots\dots\dots(5)$$

satisfies $\nabla^2\phi + k^2\phi = 0$, at points for which r is finite. The simplest course is to express ∇^2 in polar co-ordinates referred to the element itself as pole, when it appears that

$$\nabla^2 \frac{e^{-ikr}}{r} = \left(\frac{d^2}{dr^2} + \frac{2}{r}\frac{d}{dr}\right)\frac{e^{-ikr}}{r} = \frac{1}{r}\frac{d^2}{dr^2} r \cdot \frac{e^{-ikr}}{r} = -k^2 \frac{e^{-ikr}}{r}.$$

We infer that (3) satisfies $\nabla^2\phi + k^2\phi = 0$, at all points for which Φ vanishes. In the case of a point at which Φ does not vanish, we may put out of account all the elements situated at a finite distance (as contributing only terms satisfying $\nabla^2\phi + k^2\phi = 0$), and for the element at an infinitesimal distance replace e^{-ikr} by unity. Thus on the whole

$$(\nabla^2 + k^2)\,\phi = \frac{1}{4\pi a^2} \nabla^2 \iiint \Phi \frac{dV}{r} = \frac{1}{a^2}\,\Phi,$$

exactly as in Poisson's theorem for the common potential[1].

278. The effect of a force Φ_1 distributed over a surface S may be obtained as a limiting case from (3) § 277. $\Phi\,dV$ is replaced by $\Phi\,b\,dS$, b denoting the thickness of the layer; and in the limit we may write $\Phi\,b = \Phi_1$. Thus

$$\phi = \frac{1}{4\pi a^2} \iint \Phi_1 \frac{e^{-ikr}}{r}\,dS\dots\dots\dots\dots\dots(1).$$

The value of ϕ is the same on the two sides of S, but there is discontinuity in its derivatives. If dn be drawn outwards from S normally, (4) § 276 gives

$$\left(\frac{d\phi}{dn}\right)_1 + \left(\frac{d\phi}{dn}\right)_2 = -\frac{1}{a^2}\,\Phi_1 \dots\dots\dots\dots(2)[2].$$

If the surface S be plane, the integral in (1) is evidently symmetrical with respect to it, and therefore

$$(d\phi/dn)_1 = (d\phi/dn)_2.$$

[1] See Thomson and Tait's *Natural Philosophy*, § 491.
[2] Helmholtz. *Crelle*, t. 57, p. 21, 1860.

Hence, if $d\phi/dn$ be the given normal velocity of the fluid in contact with the plane, the value of ϕ is determined by

$$\phi = -\frac{1}{2\pi} \iint \frac{d\phi}{dn} \frac{e^{-ikr}}{r} dS \dots \dots (3),$$

which is a result of considerable importance. To exhibit it in terms of real quantities, we may take

$$d\phi/dn = Pe^{i(nt+\epsilon)} \dots \dots (4),$$

P and ϵ being real functions of the position of dS. The symbolical solution then becomes

$$\phi = -\frac{1}{2\pi} \iint P\, e^{i(nt-kr+\epsilon)} \frac{dS}{r} \dots \dots (5),$$

from which, if the imaginary part be rejected, we obtain

$$\phi = -\frac{1}{2\pi} \iint P \frac{\cos(nt-kr+\epsilon)}{r} dS \dots \dots (6),$$

corresponding to

$$d\phi/dn = P\cos(nt+\epsilon) \dots \dots (7).$$

The same method is applicable to the general case when the motion is not restricted to be simple harmonic. We have

$$\phi = -\frac{1}{2\pi} \iint V\left(t - \frac{r}{a}\right) \cdot \frac{dS}{r} \dots \dots (8),$$

where by $V(t-r/a)$ is denoted the normal velocity at the plane for the element dS at the time $t-r/a$, that is to say, at a time r/a antecedent to that at which ϕ is estimated.

In order to complete the solution of the problem for the unlimited mass of fluid lying on one side of an infinite plane, we have to add the most general value of ϕ, consistent with $V=0$. This part of the question is identical with the general problem of reflection from an infinite rigid plane[1].

It is evident that the effect of the constraint will be represented by the introduction on the other side of the plane of fictitious initial displacements and forces, forming in conjunction with those actually existing on the first side a system perfectly symmetrical with respect to the plane. Whatever the initial values of ϕ and $\dot{\phi}$ may be belonging to any point on the first side, the same must be ascribed to its *image*, and in like manner whatever function of

[1] Poisson, *Journal de l'école polytechnique*, t. VII. 1808.

the time Φ may be at the first point, it must be conceived to be the same function of the time at the other. Under these circumstances it is clear that for all future time ϕ will be symmetrical with respect to the plane, and therefore the normal velocity zero. So far then as the motion on the first side is concerned, there will be no change if the plane be removed, and the fluid continued indefinitely in all directions, provided the circumstances on the second side are the exact reflection of those on the first. This being understood, the general solution of the problem for a fluid bounded by an infinite plane is contained in the formulæ (8) § 273, (3) § 277, and (8) of the present section. They give the result of arbitrary initial conditions (ϕ_0 and $\dot{\phi}_0$), arbitrary applied forces (Φ), and arbitrary motion of the plane (V).

Measured by the resulting potential, a source of given magnitude, i.e. a source at which a given introduction and withdrawal of fluid takes place, is thus twice as effective when close to a rigid plane, as if it were situated in the open; and the result is ultimately the same, whether the source be concentrated in a point close to the plane, or be due to a corresponding normal motion of the surface of the plane itself.

The operation of the plane is to double the effective pressures which oppose the expansion and contraction at the source, and therefore to double the total energy emitted; and since this energy is diffused through only the half of angular space, the intensity of the sound is quadrupled, which corresponds to a doubled amplitude, or potential (§ 245).

We will now suppose that instead of $d\phi/dn = 0$, the prescribed condition at the infinite plane is that $\phi = 0$. In this case the fictitious distribution of ϕ_0, $\dot{\phi}_0$, Φ, on the second side of the plane must be the *opposite* of that on the first side, so that the sum of the values at two corresponding points is always zero. This secures that on the plane of symmetry itself ϕ shall vanish throughout.

Let us next suppose that there are two parallel surfaces S_1 S_2, separated by the infinitely small interval dn, and that the value of Φ_1 on the second surface is equal and opposite to the value of Φ_1 on the first. In crossing S_1, there is by (2) a finite change in the value of $d\phi/dn$ to the amount of Φ_1/a^2, but in crossing S_2 the same finite change occurs in the reverse direction. When dn is reduced without limit, and $\Phi_1 dn$ replaced by Φ_{11}, $d\phi/dn$ will be the same on the two sides of the double sheet, but there will be discontinuity in the value of ϕ to the amount of Φ_{11}/a^2. At the same time (1) becomes

$$\phi = \frac{1}{4\pi a^2} \iint \frac{d}{dn}\left(\frac{e^{-ikr}}{r}\right) \Phi_{11} \, dS \ldots\ldots\ldots\ldots\ldots(9).$$

If the surface S be plane, the values of ϕ on the two sides of it are numerically equal, and therefore close to the surface itself

$$\phi = \pm \tfrac{1}{2} a^{-2} \Phi_{11}.$$

Hence (9) may be written

$$\phi = -\frac{1}{2\pi} \iint \frac{d}{dn}\left(\frac{e^{-ikr}}{r}\right) \phi \, dS \ldots\ldots\ldots\ldots\ldots(10),$$

where ϕ under the integral sign represents the surface-potential, positive on the one side and negative on the other, due to the action of the forces at S. The direction of dn must be understood to be *towards* the side at which ϕ is to be estimated.

279. The problem of spherical waves diverging from a point has already been forced upon us and in some degree considered, but on account of its importance it demands a more detailed treatment. If the centre of symmetry be taken as pole the velocity-potential is a function of r only, and (§ 241) ∇^2 reduces to $\dfrac{d^2}{dr^2} + \dfrac{2}{r}\dfrac{d}{dr}$, or to $\dfrac{1}{r}\dfrac{d^2}{dr^2}r$. The equation of free motion (3) § 273 thus becomes

$$\frac{d^2(r\phi)}{dt^2} = a^2 \frac{d^2(r\phi)}{dr^2} \ldots\ldots\ldots\ldots\ldots\ldots(1),$$

whence, as in § 245,

$$r\phi = f(at - r) + F(at + r)\ldots\ldots\ldots\ldots(2).$$

The values of the velocity and condensation are to be found by differentiation in accordance with the formulæ

$$u = \frac{d\phi}{dr}, \qquad s = -\frac{1}{a^2}\frac{d\phi}{dt}\ldots\ldots\ldots\ldots\ldots(3).$$

As in the case of one dimension, the first term represents a wave advancing in the direction of r increasing, that is to say, a divergent wave, and the second term represents a wave converging upon the pole. The latter does not in itself possess much interest. If we confine our attention to the divergent wave, we have

$$u = -\frac{f(at - r)}{r^2} - \frac{f'(at - r)}{r}; \qquad as = -\frac{f'(at - r)}{r}\ldots\ldots(4).$$

When r is very great the term divided by r^2 may be neglected, and then approximately

$$u = as \dots\dots\dots\dots\dots\dots\dots\dots\dots(5),$$

the same relation as obtains in the case of a plane wave, as might have been expected.

If the type be harmonic,

$$r\phi = A\, e^{ik(at-r+\theta)} \dots\dots\dots\dots\dots\dots\dots(6),$$

or, if only the real part be retained,

$$r\phi = A \cos \frac{2\pi}{\lambda} (at + \theta - r) \dots\dots\dots\dots\dots(7).$$

If a divergent disturbance be confined to a spherical shell, within and without which there is neither condensation nor velocity, the character of the wave is limited by a remarkable relation, first pointed out by Stokes[1]. From equations (4) we have

$$(as - u)\, r^2 = f(at - r),$$

shewing that the value of $f(at - r)$ is the same, viz. zero, both inside and outside the shell to which the wave is limited. Hence by (4), if α and β be radii less and greater than the extreme radii of the shell,

$$\int_{\alpha}^{\beta} s\, r\, dr = 0 \dots\dots\dots\dots\dots\dots\dots(8),$$

which is the expression of the relation referred to. As in § 274, we see that a condensed or a rarefied wave cannot exist alone. When the radius becomes great in comparison with the thickness, the variation of r in the integral may be neglected, and (8) then expresses that the *mean* condensation is zero.

[Availing himself of Foucault's method for rendering visible minute optical differences, Töpler[2] succeeded in observing spherical sonorous waves originating in small electric sparks, and their reflection from a plane wall. Subsequently photographic records of similar phenomena have been obtained by Mach[3].]

In applying the general solution (2) to deduce the motion resulting from arbitrary initial circumstances, we must remember that in its present form it is too general for the purpose, since it covers the case in which the pole is itself a source, or place where

---

[1] *Phil. Mag.* xxxiv. p. 52. 1849.

[2] *Pogg. Ann.* vol. cxxxi. pp. 33, 180. 1867.

[3] *Sitzber. der Wiener Akad.*, 1889.

fluid is introduced or withdrawn in violation of the equation of continuity. The total current across the surface of a sphere of radius r is $4\pi r^2 u$, or by (2) and (3)

$$-4\pi \{f(at-r) + F(at+r)\} + 4\pi r \{F'(at+r) - f'(at-r)\},$$

so that, if the pole be not a source, $f(at-r) + F(at+r)$, or $r\phi$, must vanish with r. Thus

$$f(at) + F(at) = 0 \dots\dots\dots\dots\dots\dots (9),$$

an equation which must hold good for all positive values of the argument[1].

By the known initial circumstances the values of u and s are determined for the time $t = 0$, and for all (positive) values of r. If these initial values be represented by u_0 and s_0, we obtain from (2) and (3)

$$\left. \begin{array}{l} f(-r) + F(r) = r \int u_0 \, dr \\ f(-r) - F(r) = a \int s_0 \, r \, dr \end{array} \right\} \dots\dots\dots\dots (10),$$

by which the function f is determined for all negative arguments, and the function F for all positive arguments. The form of f for positive arguments follows by means of (9), and then the whole subsequent motion is determined by (2). The form of F for negative arguments is not required.

The initial disturbance divides itself into two parts, travelling in opposite directions, in each of which $r\phi$ is propagated with constant velocity a, and the inwards travelling wave is continually reflected at the pole. Since the condition to be there satisfied is $r\phi = 0$, the case is somewhat similar to that of a parallel tube terminated by an *open* end, and we may thus perhaps better understand why the condensed wave, arising from the liberation of a mass of condensed air round the pole, is followed immediately by a wave of rarefaction.

[The composite character of the wave resulting from an initial condensation may be invoked to explain a phenomenon which has often occasioned surprise. When windows are broken by a violent explosion in their neighbourhood, they are frequently observed to

---

[1] The solution for spherical vibrations may be obtained without the use of (1) by superposition of trains of plane waves, related similarly to the pole, and travelling outwards in all directions symmetrically.

have fallen *outwards* as if from exposure to a wave of rarefaction. This effect may be attributed to the second part of the compound wave; but it may be asked why should the second part preponderate over the first? If the window were freely suspended, the momentum acquired from the waves of condensation and rarefaction would be equal. But under the actual conditions it may well happen that the force of the condensed wave is spent in overcoming the resistance of the supports, and then the rarefied wave is left free to produce its full effect.]

280. Returning now to the case of a train of harmonic waves travelling outwards continually from the pole as source, let us investigate the connection between the velocity-potential and the quantity of fluid which must be supposed to be introduced and withdrawn alternately. If the velocity-potential be

$$\phi = - \frac{A}{4\pi r} \cos k \, (at - r) \dots\dots\dots\dots\dots(1),$$

we have, as in the preceding section, for the total current crossing a sphere of radius r,

$$4\pi r^2 \frac{d\phi}{dr} = A \left\{ \cos k \, (at - r) - kr \sin k \, (at - r) \right\} = A \cos kat,$$

where r is small enough. If the maximum rate of introduction of fluid be denoted by A, the corresponding potential is given by (1).

It will be observed that when the source, as measured by A, is finite, the potential and the pressure-variation (proportional to $\dot{\phi}$) are infinite at the pole. But this does not, as might for a moment be supposed, imply an infinite emission of energy. If the pressure be divided into two parts, one of which has the same phase as the velocity, and the other the same phase as the acceleration, it will be found that the former part, on which the work depends, is finite. The infinite part of the. pressure does no work on the whole, but merely keeps up the vibration of the air immediately round the source, whose effective inertia is indefinitely great.

We will now investigate the energy emitted from a simple source of given magnitude, supposing for the sake of greater generality that the source is situated at the vertex of a rigid cone of solid angle ω. If the rate of introduction of fluid at the source be $A \cos kat$, we have

$$\omega r^2 \, d\phi/dr = A \cos kat$$

ultimately, corresponding to

$$\phi = -\frac{A}{\omega r} \cos k\,(at-r)\dots\dots\dots\dots(2);$$

whence

$$\dot{\phi} = \frac{kaA}{\omega r} \sin k\,(at-r)\dots\dots\dots\dots(3),$$

and

$$\omega r^2 \frac{d\phi}{dr} = A\{\cos k\,(at-r) - kr \sin k\,(at-r)\}\dots\dots(4).$$

Thus, as in § 245, if dW be the work transmitted in time dt, we get, since $\delta p = -\rho\,\dot{\phi}$,

$$\frac{dW}{dt} = -\frac{\rho kaA^2}{\omega r} \sin k\,(at-r) \cos k\,(at-r)$$
$$+ \rho\,\frac{k^2 aA^2}{\omega} \sin^2 k\,(at-r).$$

Of the right-hand member the first term is entirely periodic, and in the second the mean value of $\sin^2 k\,(at-r)$ is $\frac{1}{2}$. Thus in the long run

$$W = \frac{\rho k^2 aA^2}{2\omega}\,t\dots\dots\dots\dots\dots(5)[1].$$

It will be remarked that when the source is given, the amplitude varies inversely as ω, and therefore the intensity inversely as ω^2. For an acute cone the intensity is greater, not only on account of the diminution in the solid angle through which the sound is distributed, but also because the total energy emitted from the source is itself increased.

When the source is in the open, we have only to put $\omega = 4\pi$, and when it is close to a rigid plane, $\omega = 2\pi$.

The results of this article find an interesting application in the theory of the speaking trumpet, or (by the law of reciprocity §§ 109, 294) hearing trumpet. If the diameter of the large open end be small in comparison with the wave-length, the waves on arrival suffer copious reflection, and the ultimate result, which must depend largely on the precise relative lengths of the tube and of the wave, requires to be determined by a different process. But by sufficiently prolonging the cone, this reflection may be diminished, and it will tend to cease when the diameter of the open end includes a large number of wave-lengths. Apart from friction it would therefore be possible by diminishing ω to obtain from a given source any desired amount of energy, and at the

<hr>

[1] Cambridge Mathematical Tripos Examination, 1876.

same time by lengthening the cone to secure the unimpeded transference of this energy from the tube to the surrounding air.

From the theory of diffraction it appears that the sound will not fall off to any great extent in a lateral direction, unless the diameter at the large end exceed half a wave-length. The ordinary explanation of the effect of a common trumpet, depending on a supposed concentration of *rays* in the axial direction, is thus untenable.

281. By means of Euler's equation,

$$\frac{d^2(r\phi)}{dt^2} = a^2 \frac{d^2(r\phi)}{dr^2} \dots \dots \dots (1),$$

we may easily establish a theory for conical pipes with open ends, analogous to that of Bernoulli for parallel tubes, subject to the same limitation as to the smallness of the diameter of the tubes in comparison with the wave-length of the sound[1]. Assuming that the vibration is stationary, so that $r\phi$ is everywhere proportional to $\cos kat$, we get from (1)

$$\frac{d^2(r\phi)}{dr^2} + k^2 \cdot r\phi = 0 \dots \dots \dots (2),$$

of which the general solution is

$$r\phi = A \cos kr + B \sin kr \dots \dots \dots (3).$$

The condition to be satisfied at an open end, viz., that there is to be no condensation or rarefaction, gives $r\phi = 0$, so that, if the extreme radii of the tube be r_1 and r_2, we have

$$A \cos kr_1 + B \sin kr_1 = 0, \quad A \cos kr_2 + B \sin kr_2 = 0,$$

whence by elimination of $A : B$, $\sin k(r_2 - r_1) = 0$, or $r_2 - r_1 = \frac{1}{2} m\lambda$, where m is an integer. In fact since the form of the general solution (3) and the condition for an open end are the same as for a parallel tube, the result that the length of the tube is a multiple of the half wave-length is necessarily also the same.

A cone, which is complete as far as the vertex, may be treated as if the vertex were an open end, since, as we saw in § 279, the condition $r\phi = 0$ is there satisfied.

The resemblance to the case of parallel tubes does not extend to the position of the nodes. In the case of the gravest vibration

---

[1] D. Bernoulli, *Mém. d. l'Acad. d. Sci.* 1762; Duhamel, Liouville *Journ. Math.* vol. XIV. p. 98, 1849.

of a parallel tube open at both ends, the node occupies a central position, and the two halves vibrate synchronously as tubes open at one end and stopped at the other. But if a conical tube were divided by a partition at its centre, the two parts would have different periods, as is evident, because the one part differs from a parallel tube by being contracted at its open end where the effect of a contraction is to depress the pitch, while the other part is contracted at its stopped end, where the effect is to raise the pitch. In order that the two periods may be the same, the partition must approach nearer to the narrower end of the tube. Its actual position may be determined analytically from (3) by equating to zero the value of $d\phi/dr$.

When both ends of a conical pipe are closed, the corresponding notes are determined by eliminating $A : B$ between the equations,

$$A (\cos kr_1 + kr_1 \sin kr_1) + B (\sin kr_1 - kr_1 \cos kr_1) = 0,$$

$$A (\cos kr_2 + kr_2 \sin kr_2) + B (\sin kr_2 - kr_2 \cos kr_2) = 0,$$

of which the result may be put into the form

$$kr_2 - \tan^{-1} kr_2 = kr_1 - \tan^{-1} kr_1 \ldots\ldots\ldots\ldots (4).$$

If $r_1 = 0$, we have simply

$$\tan kr_2 = kr_2 \ldots\ldots\ldots\ldots\ldots\ldots (5)[1];$$

if r_1 and r_2 be very great, $\tan^{-1} kr_1$ and $\tan^{-1} kr_2$ are both odd multiples of $\frac{1}{2}\pi$, so that $r_2 - r_1$ is a multiple of $\frac{1}{2}\lambda$, as the theory of parallel tubes requires.

[If $r_2 - r_1 = l$, $r_2 + r_1 = r$, (4) may be written

$$\tan kl = \frac{kl}{1 + \frac{1}{4}k^2(r^2 - l^2)} \ldots\ldots\ldots\ldots (6).$$

When r is great in comparison with l, the approximate solution of (6) gives

$$\lambda = \frac{2l}{m}\left(1 - \frac{4l^2}{m^2\pi^2 r^2}\right) \ldots\ldots\ldots\ldots (7),$$

m being an integer. The influence of conicality upon the pitch is thus of the second order.

Experiments upon conical pipes have been made by Boutet[2] and by Blaikley[3].]

[1] For the roots of this equation see § 207.

[2] *Ann. d. Chim.* vol. xxi. p. 150, 1870.

[3] *Phil. Mag.* vi. p. 119, 1878.

282. If there be two distinct sources of sound of the same pitch, situated at O_1 and O_2, the velocity-potential ϕ at a point P whose distances from O_1, O_2 are r_1 and r_2, may be expressed

$$\phi = A\,\frac{\cos k\,(at - r_1)}{r_1} + B\,\frac{\cos k\,(at - r_2 - \alpha)}{r_2}\ldots\ldots(1),$$

where A and B are coefficients representing the magnitudes of the sources (which without loss of generality may be supposed to have the same sign), and α represents the retardation (considered as a distance) of the second source relatively to the first. The two trains of spherical waves are in agreement at any point P, if $r_2 + \alpha - r_1 = \pm\, m\lambda$, where m is an integer, that is, if P lie on any one of a system of hyperboloids of revolution having foci at O_1 and O_2. At points lying on the intermediate hyperboloids, represented by $r_2 + \alpha - r_1 = \pm\, \frac{1}{2}(2m+1)\lambda$, the two sets of waves are opposed in phase, and neutralize one another as far as their actual magnitudes permit. The neutralization is complete, if $r_1 : r_2 = A : B$, and then the density at P continues permanently unchanged. The intersections of this sphere with the system of hyperboloids will thus mark out in most cases several circles of absolute silence. If the distance $O_1 O_2$ between the sources be great in comparison with the length of a wave, and the sources themselves be not very unequal in power, it will be possible to depart from the sphere $r_1 : r_2 = A : B$ for a distance of several wave-lengths, without appreciably disturbing the equality of intensities, and thus to obtain over finite surfaces several alternations of sound and of almost complete silence.

There is some difficulty in actually realising a satisfactory interference of two independent sounds. Unless the unison be extraordinarily perfect, the silences are only momentary and are consequently difficult to appreciate. It is therefore best to employ sources which are mechanically connected in such a way that the relative phases of the sounds issuing from them cannot vary. The simplest plan is to repeat the first sound by reflection from a flat wall (§§ 269, 278), but the experiment then loses something in directness owing to the fictitious character of the second source. Perhaps the most satisfactory form of the experiment is that described in the Philosophical Magazine for June 1877 by myself. "An intermittent electric current, obtained from a fork interrupter making 128 vibrations per second, excited by means of electromagnets two other forks, whose frequency was 256, (§§ 63, 64).

These latter forks were placed at a distance of about ten yards apart, and were provided with suitably tuned resonators, by which their sounds were reinforced. The pitch of the forks was necessarily identical, since the vibrations were forced by electromagnetic forces of absolutely the same period. With one ear closed it was found possible to define the places of silence with considerable accuracy, a motion of about an inch being sufficient to produce a marked revival of sound. At a point of silence, from which the line joining the forks subtended an angle of about 60°, the apparent striking up of one fork, when the other was stopped, had a very peculiar effect."

Another method is to duplicate a sound coming along a tube by means of branch tubes, whose open ends act as sources. But the experiment in this form is not a very easy one.

It often happens that considerations of symmetry are sufficient to indicate the existence of places of silence. For example, it is evident that there can be no variation of density in the continuation of the plane of a vibrating plate, nor in the equatorial plane of a symmetrical solid of revolution vibrating in the direction of its axis. More generally, any plane is a plane of silence, with respect to which the sources are symmetrical in such a manner that at any point and at its image in the plane there are sources of equal intensities and of opposite phases, or, as it is often more conveniently expressed, of the same phase and of opposite amplitudes.

If any number of sources in the same phase, whose amplitudes are on the whole as much negative as positive, be placed on the circumference of a circle, they will give rise to no disturbance of pressure at points on the straight line which passes through the centre of the circle and is directed at right angles to its plane. This is the case of the symmetrical bell (§ 232), which emits no sound in the direction of its axis[1].

The accurate experimental investigation of aërial vibrations is beset with considerable difficulties, which have been only partially surmounted hitherto. In order to avoid unwished for reflections it is generally necessary to work in the open air, where delicate apparatus, such as a sensitive flame, is difficult of management. Another impediment arises from the presence of the experimenter himself, whose person is large enough to disturb materially the

---

[1] *Phil. Mag.* (5), III. p. 460. 1877.

state of things which he wishes to examine. Among indicators of sound may be mentioned membranes stretched over cups, the agitation being made apparent by sand, or by small pendulums resting lightly against them. If a membrane be simply stretched across a hoop, both its faces are acted upon by nearly the same forces, and consequently the motion is much diminished, unless the membrane be large enough to cast a sensible shadow, in which its hinder face may be protected. Probably the best method of examining the intensity of sound at any point in the air is to divert a portion of it by means of a tube ending in a small cone or resonator, the sound so diverted being led to the ear, or to a manometric capsule. In this way it is not difficult to determine places of silence with considerable precision.

By means of the same kind of apparatus it is possible to examine even the *phase* of the vibration at any point in air, and to trace out the surfaces on which the phase does not vary[1]. If the interior of a resonator be connected by flexible tubing with a manometric capsule, which influences a small gas flame, the motion of the flame is related in an invariable manner (depending on the apparatus itself) to the variation of pressure at the mouth of the resonator; and in particular the interval between the lowest drop of the flame and the lowest pressure at the resonator is independent of the absolute time at which these effects occur. In Mayer's experiment two flames were employed, placed close together in one vertical line, and were examined with a revolving mirror. So long as the associated resonators were undisturbed, the serrations of the two flames occupied a fixed relative position, and this relative position was also maintained when one resonator was moved about so as to trace out a surface of invariable phase. For further details the reader must be referred to the original paper.

283. When waves of sound impinge upon an obstacle, a portion of the motion is thrown back as an echo, and under cover of the obstacle there is formed a sort of sound shadow. In order, however, to produce shadows in anything like optical perfection, the dimensions of the intervening body must be considerable. The standard of comparison proper to the subject is the wavelength of the vibration; it requires almost as extreme conditions to produce *rays* in the case of sound, as it requires in optics to avoid producing them. Still, sound shadows thrown by hills, or

[1] Mayer, *Phil. Mag.* (4), XLIV. p. 321. 1872.

buildings, are often tolerably complete, and must be within the experience of all.

For closer examination let us take first the case of plane waves of harmonic type impinging upon an immovable plane screen, of infinitesimal thickness, in which there is an aperture of any form, the plane of the screen ($x = 0$) being parallel to the fronts of the waves. The velocity-potential of the undisturbed train of waves may be taken,

$$\phi = \cos (nt - kx)\ldots\ldots\ldots\ldots\ldots\ldots\ldots(1).$$

If the value of $d\phi/dx$ over the apertui be known, formulæ (6) and (7) § 278 allow us to calculate the value of ϕ at any point on the further side. In the ordinary theory of diffraction, as given in works on optics, it is assumed that the disturbance in the plane of the aperture is the same as if the screen were away. This hypothesis, though it can never be rigorously exact, will suffice when the aperture is very large in comparison with the wavelength, as is usually the case in optics.

For the undisturbed wave we have

$$\frac{d\phi}{dx}(x = 0) = k \sin nt \ldots\ldots\ldots\ldots\ldots\ldots(2),$$

and therefore on the further side, we get

$$\phi = -\frac{k}{2\pi} \iint \frac{\sin (nt - kr)}{r}\, dS \ldots\ldots\ldots\ldots\ldots(3),$$

the integration extending over the area of the aperture. Since $k = 2\pi/\lambda$, we see by comparison with (1) that in supposing a primary wave broken up, with the view of applying Huygens' principle, dS must be divided by λr, and the phase must be accelerated by a quarter of a period.

When r is large in comparison with the dimensions of the aperture, the composition of the integral is best studied by the aid of Fresnel's[1] zones. With the point O, for which ϕ is to be estimated, as centre describe a series of spheres of radii increasing by the constant difference $\frac{1}{2}\lambda$, the first sphere of the series being of such radius (c) as to touch the plane of the screen. On this plane are thus marked out a series of circles, whose radii ρ are

---

[1] [These zones are usually spoken of as Huygens' zones by optical writers (e.g. Billet, *Traité d'Optique physique*, vol. I. p. 102, Paris, 1858); but, as has been pointed out by Schuster (*Phil. Mag.* vol. XXXI. p. 85, 1891), it is more correct to name them after Fresnel.]

given by $\rho^2 + c^2 = (c + \frac{1}{2}n\lambda)^2$, or $\rho^2 = nc\lambda$, very nearly; so that the rings into which the plane is divided, being of approximately equal area, make contributions to ϕ which are approximately equal in numerical magnitude and alternately opposite in sign. If O lie decidedly within the projection of the area, the first term of the series representing the integral is finite, and the terms which follow are alternately opposite in sign and of numerical magnitude at first nearly constant, but afterwards diminishing gradually to zero, as the parts of the rings intercepted within the aperture become less and less. The case of an aperture, whose boundary is equidistant from O, is excepted.

In a series of this description any term after the first is neutralized almost exactly [that is, so far as first differences are concerned] by half the sum of those which immediately precede and follow it, so that the sum of the whole series is represented approximately by half the first term, which stands over uncompensated. We see that, provided a sufficient number of zones be included within the aperture, the value of ϕ at the point O is independent of the nature of the aperture, and is therefore the same as if there had been no screen at all. Or we may calculate directly the effect of the circle with which the system of zones begins; a course which will have the advantage of bringing out more clearly the significance of the change of phase which we found it necessary to introduce when the primary wave was broken up. Thus, let us conceive the circle in question divided into infinitesimal rings of equal area. The parts of ϕ due to each of these rings are equal in amplitude and of phase ranging uniformly over half a complete period. The phase of the resultant is therefore midway between those of the extreme elements, that is to say, a quarter of a period behind that due to the element at the centre of the circle. The amplitude of the resultant will be less than if all its components had been in the same phase, in the ratio $\int_0^\pi \sin x \, dx : \pi$, or $2 : \pi$; and therefore since the area of the circle is $\pi\lambda c$, half the effect of the first zone is

$$\phi = -\frac{1}{2} \cdot \frac{2}{\pi} \cdot \frac{\sin(nt - kc - \frac{1}{2}\pi)}{\lambda c} \cdot \pi\lambda c = \cos(nt - kc),$$

the same as if the primary wave were to pass on undisturbed.

When the point O is well away from the projection of the aperture, the result is quite different. The series representing the integral then converges at both ends, and by the same reasoning as before its sum is seen to be approximately zero. We conclude that if the projection of O on the plane $x = 0$ fall within the aperture, and be nearer to O by a great many wave-lengths than the nearest point of the boundary of the aperture, then the disturbance at O is nearly the same as if there were no obstacle at all; but, if the projection of O fall outside the aperture and be nearer to O by a great many wave-lengths than the nearest point of the boundary, then the disturbance at O practically vanishes. This is the theory of sound rays in its simplest form.

The argument is not very different if the screen be oblique to the plane of the waves. As before, the motion on the further side of the screen may be regarded as due to the normal motion of the particles in the plane of the aperture, but this normal motion now varies in phase from point to point. If the primary waves proceed from a source at Q, Fresnel's zones for a point P are the series of ellipses represented by $r_1 + r_2 = PQ + \frac{1}{2} n \lambda$, where r_1 and r_2 are the distances of any point on the screen from Q and P respectively, and n is an integer. On account of the assumed smallness of λ in comparison with r_1 and r_2, the zones are at first of equal area and make equal and opposite contributions to the value of ϕ; and thus by the same reasoning as before we may conclude that at any point decidedly outside the geometrical projection of the aperture the disturbance vanishes, while at any point decidedly within the geometrical projection the disturbance is the same as if the primary wave had passed the screen unimpeded. It may be remarked that the increase of area of the Fresnel's zones due to obliquity is compensated in the calculation of the integral by the correspondingly diminished value of the normal velocity of the fluid. The enfeeblement of the primary wave between the screen and the point P due to divergency is represented by a diminution in the area of the Fresnel's zones below that corresponding to plane incident waves in the ratio $r_1 + r_2 : r_1$.

There is a simple relation between the transmission of sound through an aperture in a screen and its reflection from a plane reflector of the same form as the aperture, of which advantage may sometimes be taken in experiment. Let us imagine a source similar to Q and in the same phase to be placed at Q', the *image* of Q in the plane of the screen, and let us suppose that the screen is removed and replaced by a plate whose form and position is exactly that of the aperture; then we know that the effect at P of the two sources is uninfluenced by the presence of the plate, so that the vibration from Q' reflected from the plate and the vibration from Q transmitted round the plate together make up the same vibration as would be received from Q if there were no obstacle at all. Now according to the assumption which we made at the beginning of this section, the unimpeded vibration from Q may be regarded as composed of the vibration that finds its way round the plate and of that which would pass an aperture of the same form in an infinite screen, and thus the vibration from Q as transmitted through the aperture is equal to the vibration from Q' as reflected from the plate.

In order to obtain a nearly complete reflection it is not necessary that the reflecting plate include more than a small number of Fresnel's zones. In the case of direct reflection the radius ρ of the first zone is determined by the equation

$$\rho^2 (1/c_1 + 1/c_2) = \lambda \quad \ldots\ldots\ldots\ldots\ldots\ldots (4),$$

where c_1 and c_2 are the distances from the reflector of the source and of the point of observation. When the distances concerned are great, the zones become so large that ordinary walls are insufficient to give a complete reflection, but at more moderate distances echos are often nearly perfect. The area necessary for complete reflection depends also upon the wave-length; and thus it happens that a board or plate, which would be quite inadequate to reflect a grave musical note, may reflect very fairly a hiss or the sound of a high whistle. In experiments on reflection by screens of moderate size, the principal difficulty is to get rid sufficiently of the direct sound. The simplest plan is to reflect the sound from an electric bell, or other fairly steady source, round the corner of a large building[1].

284. In the preceding section we have applied Huygens' principle to the case where the primary wave is supposed to be broken up at the surface of an imaginary plane. If we really know what the normal motion at the plane is, we can calculate the disturbance at any point on the further side by a rigorous process. For surfaces other than the plane the problem has not been solved generally; nevertheless, it is not difficult to see that when the radii of curvature of the surface are very great in comparison with the wave-length, the effect of a normal motion of an

---

[1] *Phil. Mag.* (5), III. p. 458. 1877.

element of the surface must be very nearly the same as if the surface were plane. On this understanding we may employ the same integral as before to calculate the aggregate result. As a matter of convenience it is usually best to suppose the wave to be broken up at what is called in optics a *wave-surface*, that is, a surface at every point of which the *phase* of the disturbance is the same.

Let us consider the application of Huygens' principle to calculate the progress of a given divergent wave. With any point P, at which the disturbance is required, as centre, describe a series of spheres of radii continually increasing by the constant difference $\frac{1}{2}\lambda$, the first of the series being of such radius (c) as to touch the given wave-surface at C. If R be the radius of curvature of the surface in any plane through P and C, the corresponding radius ρ of the outer boundary of the nth zone is given by the equation

$$R + c = \sqrt{\{R^2 - \rho^2\}} + \sqrt{\{(c + \tfrac{1}{2} n\lambda)^2 - \rho^2\}},$$

from which we get approximately

$$\rho^2 = n\lambda \div \left(\frac{1}{R} + \frac{1}{c}\right)\dots\dots\dots\dots\dots\dots(1).$$

If the surface be one of revolution round PC, the area of the first n zones is $\pi\rho^2$, and since ρ^2 is proportional to n, it follows that the zones are of equal area. If the surface be not of revolution, the area of the first n zones is represented $\frac{1}{2}\int\rho^2 d\theta$, where θ is the azimuth of the plane in which ρ is measured, but it still remains true that the zones are of equal area. Since by hypothesis the normal motion does not vary rapidly over the wave-surface, the disturbances at P due to the various zones are nearly equal in magnitude and alternately opposite in sign, and we conclude that, as in the case of plane waves, the aggregate effect is the half of that due to the first zone. The phase at P is accordingly retarded behind that prevailing over the given wave-surface by an amount corresponding to the distance c.

The intensity of the disturbance at P depends upon the area of the first Fresnel's zone, and upon the distance c. In the case of symmetry, we have

$$\frac{\pi\rho^2}{c} = \frac{\pi\lambda R}{R + c},$$

which shews that the disturbance is less than if R were infinite in the ratio $R + c : R$. This diminution is the effect of divergency, and is the same as would be obtained on the supposition that the motion is limited by a conical tube whose vertex is at the centre of curvature (§ 266). When the surface is not of revolution, the value of $\frac{1}{2}\int_0^{2\pi}\rho^2 d\theta \div c$ may be expressed in terms of the principal radii of curvature R_1 and R_2, with which R is connected by the relation

$$1/R = \cos^2\theta/R_1 + \sin^2\theta/R_2.$$

We obtain on effecting the integration

$$\frac{1}{2c}\int_0^{2\pi}\rho^2 d\theta = \frac{\pi\lambda\sqrt{R_1 R_2}}{\sqrt{(R_1 + c)(R_2 + c)}} \dots\dots\dots\dots(2),$$

so that the amplitude is diminished by divergency in the ratio $\sqrt{(R_1 + c)(R_2 + c)} : \sqrt{R_1 R_2}$, a result which might be anticipated by supposing the motion limited to a tube formed by normals drawn through a small contour traced on the wave-surface.

Although we have spoken hitherto of diverging waves only, the preceding expressions may also be applied to waves converging in one or in both of the principal planes, if we attach suitable signs to R_1 and R_2. In such a case the area of the first Fresnel's zone is greater than if the wave were plane, and the intensity of the vibration is correspondingly increased. If the point P coincide with one of the principal centres of curvature, the expression (2) becomes infinite. The investigation, on which (2) was founded, is then insufficient; all that we are entitled to affirm is that the disturbance is much greater at P than at other points on the same normal, that the disproportion increases with the frequency, and that it would become infinite for notes of infinitely high pitch, whose wave-length would be negligible in comparison with the distances concerned.

285. Huygens' principle may also be applied to investigate the reflection of sound from curved surfaces. If the material surface of the reflector yielded so completely to the aërial pressures that the normal motion at every point were the same as it would have been in the absence of the reflector, then the sound waves would pass on undisturbed. The reflection which actually ensues when the surface is unyielding may therefore be regarded as due to a normal motion of each element of the reflector, equal and opposite to that of the primary waves at the same point, and may be investigated by the formula proper to plane surfaces in the manner of the preceding section, and subject to a similar limitation as to the relative magnitudes of the wave-length and of the other distances concerned.

The most interesting case of reflection occurs when the surface is so shaped as to cause a concentration of rays upon a particular point (P). If the sound issue originally from a simple source at Q, and the surface be an ellipsoid of revolution having its foci at P and Q, the concentration is complete, the vibration reflected from every element of the surface being in the same phase on arrival at Q. If Q be infinitely distant, so that the incident waves are plane, the surface becomes a paraboloid having its focus at P, and its axis parallel to the incident rays. We must not suppose, however, that a symmetrical wave diverging from Q is converted by reflection at the ellipsoidal surface into a spherical wave converging symmetrically upon P; in fact, it is easy to see that the intensity of the convergent wave must be different in different directions. Nevertheless, when the wave-length is very small in comparison with the radius, the different parts of the convergent wave become approximately independent of one another, and their progress is not materially affected by the failure of perfect symmetry.

The increase of loudness due to curvature depends upon the area of reflecting surface, from which disturbances of uniform phase arrive, as compared with the area of the first Fresnel's zone of a plane reflector in the same position. If the distances of the reflector from the source and from the point of observation be considerable, and the wave-length be not very small, the first Fresnel's zone is already rather large, and therefore in the case of a reflector of moderate dimensions but little is · gained by making it concave. On the other hand, in laboratory experiments, when the distances are moderate and the sounds employed are of high pitch, $e.g.$ the ticking of a watch or the cracking of electric sparks, concave reflectors are very efficient and give a distinct concentration of sound on particular spots.

286. We have seen that if a ray proceeding from Q passes after reflection at a plane or curved surface through P, the point R at which it meets the surface is determined by the condition that $QR + RP$ is a minimum (or in some cases a maximum). The point R is then the centre of the system of Fresnel's zones; the amplitude of the vibration at P depends upon the area of the first zone, and its phase depends upon the distance $QR + RP$. If there be no point on the surface of the reflector, for which $QR + RP$ is a maximum or a minimum, the system of Fresnel's zones has no centre, and there is no ray proceeding from Q which arrives at P after reflection from the surface. In like manner if sound be reflected more than once, the course of a ray is determined by the condition that its whole length between any two points is a maximum or a minimum.

The same principle may be applied to investigate the *refraction* of sound in a medium, whose mechanical properties vary gradually from point to point. The variation is supposed to be so slow that no sensible reflection occurs, and this is not inconsistent with decided refraction of the rays in travelling distances which include a very great number of wave-lengths. It is evident that what we are now concerned with is not merely the length of the ray, but also the velocity with which the wave travels along it, inasmuch as this velocity is no longer constant. The condition to be satisfied is that the *time* occupied by a wave in travelling along a ray between any two points shall be a maximum or a minimum; so that, if V be the velocity of propagation at any point, and ds an element of the length of the ray, the condition may be expressed, $\delta \int V^{-1} ds = 0$. This is Fermat's principle of least time.

The further developement of this part of the subject would lead us too far into the domain of geometrical optics. The fundamental assumption of the smallness of the wave-length, on which the doctrine of rays is built, having a far wider application to the phenomena of light than to those of sound, the task of developing its consequences may properly be left to the cultivators of the sister science. In the following sections the methods of optics are applied to one or two isolated questions, whose acoustical interest is sufficient to demand their consideration in the present work.

287.   One of the most striking of the phenomena connected with the propagation of sound within closed buildings is that presented by "whispering galleries," of which a good and easily accessible example is to be found in the circular gallery at the base of the dome of St Paul's cathedral. As to the precise mode of action acoustical authorities are not entirely agreed. In the opinion of the Astronomer Royal[1] the effect is to be ascribed to reflection from the surface of the dome overhead, and is to be observed at the point of the gallery diametrically opposite to the source of sound. Every ray proceeding from a radiant point and reflected from the surface of a spherical reflector, will after reflection intersect that diameter of the sphere which contains the radiant point. This diameter is in fact a degraded form of one of the two caustic surfaces touched by systems of rays in general, being the loci of the centres of principal curvature of the surface to which the rays are normal. The concentration of rays on one diameter thus effected, does not require the proximity of the radiant point to the reflecting surface.

Judging from some observations that I have made in St Paul's whispering gallery, I am disposed to think that the principal phenomenon is to be explained somewhat differently. The abnormal loudness with which a whisper is heard is not confined to the position diametrically opposite to that occupied by the whisperer, and therefore, it would appear, does not depend materially upon the symmetry of the dome. The whisper seems to creep round the gallery horizontally, not necessarily along the shorter arc, but rather along that arc towards which the whisperer faces. This is a consequence of the very unequal audibility of a whisper in front of and behind the speaker, a phenomenon which may easily be observed in the open air[2].

Let us consider the course of the rays diverging from a radiant point P, situated near the surface of a reflecting sphere, and let us denote the centre of the sphere by O, and the diameter passing through P by AA', so that A is the point on the surface nearest to P. If we fix our attention on a ray which issues from P at an angle $\pm \theta$ with the tangent plane at A, we see that after any number of reflections it continues to touch a concentric sphere of radius $OP \cos \theta$, so that the whole conical pencil of rays which originally make angles with the tangent plane at A numerically less than θ, is ever afterwards included between the reflecting surface and that of the concentric sphere of radius $OP \cos \theta$. The usual divergence in three dimensions entailing a diminishing intensity varying as r^{-2} is replaced by a divergence in two dimensions, like that of waves issuing from a source situated between

---

[1] Airy *On Sound*, 2nd edition, 1871, p. 145.

[2] *Phil. Mag.* (5), III. p. 458, 1877.

two parallel reflecting planes, with an intensity varying as r^{-1}. The less rapid enfeeblement of sound by distance than that usually experienced is the leading feature in the phenomena of whispering galleries.

The thickness of the sheet included between the two spheres becomes less and less as A approaches P, and in the limiting case of a radiant point situated on the surface of the reflector is expressed by $OA (1 - \cos \theta)$, or, if θ be small, $\frac{1}{2} OA . \theta^2$ approximately. The solid angle of the pencil, which determines the whole amount of radiation in the sheet, is $4\pi\theta$; so that as θ is diminished without limit the intensity becomes infinite, as compared with the intensity at a finite distance from a similar source in the open.

It is evident that this clinging, so to speak, of sound to the surface of a concave wall does not depend upon the exactness of the spherical form. But in the case of a true sphere, or rather of any surface symmetrical with respect to AA', there is in addition the other kind of concentration spoken of at the commencement of the present section which is peculiar to the point A' diametrically opposite to the source. It is probable that in the case of a nearly spherical dome like that of St Paul's a part of the observed effect depends upon the symmetry, though perhaps the greater part is referable simply to the general concavity of the walls.

The propagation of earthquake disturbances is probably affected by the curvature of the surface of the globe acting like a whispering gallery, and perhaps even sonorous vibrations generated at the surface of the land or water do not entirely escape the same kind of influence.

In connection with the acoustics of public buildings there are many points which still remain obscure. It is important to bear in mind that the loss of sound in a single reflection at a smooth wall is very small, whether the wall be plane or curved. In order to prevent reverberation it may often be necessary to introduce carpets or hangings to absorb the sound. In some cases the presence of an audience is found sufficient to produce the desired effect. In the absence of all deadening material the prolongation of sound may be very considerable, of which perhaps the most striking example is that afforded by the Baptistery at Pisa, where the notes of the common chord sung consecutively may be heard ringing on together for many seconds[1]. According to Henry[2] it is important to prevent the repeated reflection of sound backwards and forwards along the *length* of a hall intended for public speaking, which may be accomplished by suitably placed oblique surfaces. In this way the number of reflections in a given time is increased, and the undue prolongation of sound is checked.

288. Almost the only instance of acoustical refraction, which has a practical interest, is the deviation of sonorous rays from a rectilinear course due to heterogeneity of the atmosphere. The variation of pressure at different levels does not of itself give rise to refraction, since the velocity of sound is independent of density; but, as was first pointed out by Prof. Osborne Reynolds[3], the case is different with the variations of temperature which are usually to be met with. The temperature of the atmosphere is determined principally by the condensation or rarefaction, which any portion of air must undergo in its passage from one level to another, and its normal state is one of " convective equilibrium[4]," rather than of uniformity. According to this view the relation between pressure and density is that expressed in (9) § 246, and the velocity of sound is given by

$$V^2 = \frac{dp}{d\rho} = \gamma \frac{p_0}{\rho_0} \left(\frac{\rho}{\rho_0}\right)^{\gamma-1} \quad \dots\dots\dots\dots\dots (1).$$

To connect the pressure and density with the elevation (z), we have the hydrostatical equation

$$dp = -g\rho \, dz \dots\dots\dots\dots\dots\dots (2),$$

from which and (1) we find

$$V^2 = V_0^2 - (\gamma - 1) gz \dots\dots\dots\dots\dots (3),$$

if V_0 be the velocity at the surface. The corresponding relation between temperature and elevation obtained by means of equation (10) § 246 is

$$\frac{\theta}{\theta_0} = 1 - \frac{\gamma - 1}{V_0^2} gz \dots\dots\dots\dots\dots (4),$$

where θ_0 is the temperature at the surface.

---

[1] [Some observations of my own, made in 1883, gave the duration as 12 seconds. If a note changes pitch, both sounds are heard together and may give rise to a combination-tone, § 68. See Haberditzl, Ueber die von Dvořák beobachteten Variationston. Wien, *Akad. Sitzber.*, 77, p. 204, 1878.]

[2] *Amer. Assoc. Proc.* 1856, p. 119.

[3] *Proceedings of the Royal Society*, Vol. xxii. p. 531. 1874.

[4] Thomson, *On the convective equilibrium of temperature in the atmosphere.* *Manchester Memoirs*, 1861—62.

According to (4) the fall of temperature would be about 1° Cent. in 330 feet [100 m.], which does not differ much from the results of Glaisher's balloon observations. When the sky is clear, the fall of temperature during the day is more rapid than when the sky is cloudy, but towards sunset the temperature becomes approximately constant[1]. Probably on clear nights it is often warmer above than below.

The explanation of acoustical refraction as dependent upon a variation of temperature with height is almost exactly the same as that of the optical phenomenon of mirage. The curvature (ρ^{-1}) of a ray, whose course is approximately horizontal, is easily estimated by the method given by Prof. James Thomson[2]. Normal planes drawn at two consecutive points along the ray meet at the centre of curvature and are tangential to the wave-surface in its two consecutive positions. The portions of rays at elevations z and $z + \delta z$ respectively intercepted between the normal planes are to one another in the ratio $\rho : \rho - \delta z$, and also, since they are described in the same time, in the ratio $V : V + \delta V$. Hence in the limit

$$\frac{1}{\rho} = -\frac{d \log V}{dz} \quad\ldots\ldots\ldots\ldots\ldots\ldots\ldots (5).$$

In the normal state of the atmosphere a ray, which starts horizontally, turns gradually upwards, and at a sufficient distance passes over the head of an observer whose station is at the same level as the source. If the source be elevated, the sound is heard at the surface of the earth by means of a ray which starts with a downward inclination; but, if both the observer and the source be on the surface, there is no direct ray, and the sound is heard, if at all, by means of diffraction. The observer may then be said to be situated in a sound shadow, although there may be no obstacle in the direct line between himself and the source. According to (3)

$$2V\,dV/dz = -(\gamma - 1)g,$$

so that

$$\rho = \frac{2V^2}{(\gamma - 1)g} = \frac{4}{\gamma - 1} \cdot \frac{V^2}{2g} \quad\ldots\ldots\ldots\ldots (6);$$

or the radius of curvature of a horizontal ray is about ten times the height through which a body must fall under the action of

---

[1] *Nature*, Sept. 20, 1877.

[2] See Everett, *On the Optics of Mirage.* *Phil. Mag.* (4) XLV. pp. 161, 248.

gravity in order to acquire a velocity equal to the velocity of sound. If the elevations of the observer and of the source be z_1 and z_2, the greatest distance at which the sound can be heard otherwise than by diffraction is

$$\sqrt{(2z_1\rho)} + \sqrt{(2z_2\rho)} \dots\dots\dots\dots\dots\dots (7).$$

It is not to be supposed that the condition of the atmosphere is always such that the relation between velocity and elevation is that expressed in (3). When the sun is shining, the variation of temperature upwards is more rapid: on the other hand, as Prof. Reynolds has remarked, when rain is falling, a much slower variation is to be expected. In the arctic regions, where the nights are long and still, radiation may have more influence than convection in determining the equilibrium of temperature, and if so the propagation of sound in a horizontal direction would be favoured by the approximately isothermal condition of the atmosphere.

The general differential equation for the path of a ray, when the surfaces of equal velocity are parallel planes, is readily obtained from the law of sines. If θ be the angle of incidence, $V/\sin\theta$ is not altered by a refracting surface, and therefore in the case supposed remains constant along the whole course of a ray. If x be the horizontal co-ordinate, and the constant value of $V/\sin\theta$ be called c, we get $dx/dz = V/\sqrt{(c^2 - V^2)}$,

or
$$x = \int \frac{V dx}{\sqrt{c^2 - V^2}} \dots\dots\dots\dots\dots\dots (8).$$

If the law of velocity be that expressed in (3),

$$dz = -\frac{2V dV}{(\gamma - 1)g},$$

and thus
$$x = -\frac{2}{(\gamma - 1)g} \int \frac{V^2 dV}{\sqrt{c^2 - V^2}},$$

or, on effecting the integration,

$$(\gamma - 1)g\, x = \text{constant} + V\sqrt{(c^2 - V^2)} - c^2 \sin^{-1}(V/c) \dots\dots (9),$$

in which V may be expressed in terms of z by (3).

A simpler result will be obtained by taking an approximate form of (3), which will be accurate enough to represent the cases of practical interest. Neglecting the square and higher powers of z, we may take

$$V^{-1} = V_0^{-1} + \frac{g(\gamma - 1)z}{2V_0^3} \dots\dots\dots\dots (10).$$

Writing for brevity β in place of $\frac{1}{2}g(\gamma-1)/V_0^3$, we have

$$\beta \, dz = dV^{-1}.$$

By substitution in (8)

$$c\beta x = \int \frac{d(c/V)}{\sqrt{(c^2/V^2-1)}} = \log\left[\frac{c}{V} + \sqrt{(c^2/V^2-1)}\right]\ldots..(11),$$

the origin of x being taken so as to correspond with $V = c$, that is at the place where the ray is horizontal. Expressing V in terms of x, we find

$$2c/V = e^{c\beta x} + e^{-c\beta x},$$

whence $$\beta z = -V_0^{-1} + \frac{1}{2c}(e^{c\beta x} + e^{-c\beta x})\ldots\ldots\ldots\ldots(12).$$

The path of each ray is therefore a catenary whose vertex is downwards; the linear parameter is $\dfrac{2V_0^3}{g(\gamma-1)c}$, and varies from ray to ray.

289. Another cause of atmospheric refraction is to be found in the action of wind. It has long been known that sounds are generally better heard to leeward than to windward of the source; but the fact remained unexplained until Stokes[1] pointed out that the increasing velocity of the wind overhead must interfere with the rectilinear propagation of sound rays. From Fermat's law of least time it follows that the course of a ray in a moving, but otherwise homogeneous, medium, is the same as it would be in a medium, of which all the parts are at rest, if the velocity of propagation be increased at every point by the component of the wind-velocity in the direction of the ray. If the wind be horizontal, and do not vary in the same horizontal plane, the course of a ray, whose direction is everywhere but slightly inclined to that of the wind, may be calculated on the same principles as were applied in the preceding section to the case of a variable temperature, the normal velocity of propagation at any point being increased, or diminished, by the local wind-velocity, according as the motion of the sound is to leeward or to windward. Thus, when the wind increases overhead, which may be looked upon as the normal state of things, a horizontal ray travelling to windward is gradually bent upwards, and at a moderate distance passes over the head of an observer; rays travelling with the wind, on the

[1] *Brit. Assoc. Rep.* 1857, p. 22.

other hand, are bent downwards, so that an observer to leeward of the source hears by a direct ray which starts with a slight upward inclination, and has the advantage of being out of the way of obstructions for the greater part of its course.

The law of refraction at a horizontal surface, in crossing which the velocity of the wind changes discontinuously, is easily investigated. It will be sufficient to consider the case in which the direction of the wind and the ray are in the same vertical plane. If θ be the angle of incidence, which is also the angle between the plane of the wave and the surface of separation, U be the velocity of the air in that direction which makes the smaller angle with the ray, and V be the common velocity of propagation, the velocity of the trace of the plane of the wave on the surface of separation is

$$\frac{V}{\sin \theta} + U \dots\dots\dots\dots\dots\dots (1),$$

which quantity is unchanged by the refraction. If therefore U' be the velocity of the wind on the second side, and θ' be the angle of refraction,

$$\frac{V}{\sin \theta} + U = \frac{V}{\sin \theta'} + U' \dots\dots\dots\dots\dots(2),$$

which differs from the ordinary optical law. If the wind-velocity vary continuously, the course of a ray may be calculated from the condition that the expression (1) remains constant.

If we suppose that $U = 0$, the greatest admissible value of U' is

$$U' = V \{\operatorname{cosec} \theta - 1\} \dots\dots\dots\dots (3).$$

At a stratum where U' has this value, the direction of the ray which started at an angle θ has become parallel to the refracting surfaces, and a stratum where U' has a greater value cannot be penetrated at all. Thus a ray travelling upwards in still air at an inclination $(\frac{1}{2}\pi - \theta)$ to the horizon is reflected by a wind overhead of velocity exceeding that given in (3), and this independently of the velocities of intermediate strata. To take a numerical example, all rays whose upward inclination is less than 11°, are totally reflected by a wind of the same azimuth moving at the moderate speed of 15 miles per hour. The effects of such a wind on the propagation of sound cannot fail to be very important. Over the surface of still water sound moving to leeward, being confined between parallel reflecting planes, diverges in two dimensions only, and may therefore be heard at distances far greater than would otherwise be possible. Another possible effect of the reflector overhead is to render sounds audible which in still air would be intercepted by hills or other obstacles intervening. For the production of these phenomena it is not necessary that there be absence of wind at the source of sound, but, as appears at once from the form of (2), merely that the *difference* of velocities $U' - U$ attain a sufficient value.

The differential equation to the path of a ray, when the wind-velocity U is continuously variable, is

$$V \sqrt{1 + \left(\frac{dz}{dx}\right)^2} = c \pm U \ \dots\dots\dots\dots\dots (4),$$

whence

$$x = \int \frac{V \, dz}{\sqrt{\{(c \pm U)^2 - V^2\}}} \ \dots\dots\dots\dots\dots (5).$$

In comparing (5) with (8) of the preceding section, which is the corresponding equation for ordinary refraction, we must remember that V is now constant. If, for the sake of obtaining a definite result, we suppose that the law of variation of wind at different levels is that expressed by

$$U = \alpha + \beta z \ \dots\dots\dots\dots\dots (6),$$

we have

$$\beta x = V \int \frac{dU}{\sqrt{\{(c \pm U)^2 - V^2\}}} \ \dots\dots\dots\dots (7),$$

which is of the same form as (11) of the preceding section. The course of a ray is accordingly a catenary in the present case also, but there is a most important distinction between the two problems. When the refraction is of the ordinary kind, depending upon a variable velocity of propagation, the direction of a ray may be reversed. In the case of atmospheric refraction, due to a diminution of temperature upwards, the course of a ray is a catenary, whose vertex is downwards, in whichever direction the ray may be propagated. When the refraction is due to wind, whose velocity increases upwards, according to the law expressed in (6) with β positive, the path of a ray, whose direction is upwind, is also along a catenary with vertex downwards, but a ray whose direction is downwind cannot travel along this path. In the latter case the vertex of the catenary along which the ray travels is directed upwards.

290. In the paper by Reynolds already referred to, an account is given of some interesting experiments especially directed to test the theory of refraction by wind. It was found that " In the direction of the wind, when it was strong, the sound (of an electric bell) could be heard as well with the head on the ground as when raised, even when in a hollow with the bell hidden from view by the slope of the ground ; and no advantage whatever was gained either by ascending to an elevation or raising the bell. Thus, with the wind over the grass the sound could be heard 140 yards, and over snow 360 yards, either with the head lifted or on the ground ; whereas at right angles to the wind on all occasions the range was extended by raising either the observer or the bell."

" Elevation was found to affect the range of sound against the wind in a much more marked manner than at right angles."

" Over the grass no sound could be heard with the head on the ground at 20 yards from the bell, and at 30 yards it was lost with the head 3 feet from the ground, and its full intensity was lost when standing erect at 30 yards. At 70 yards, when standing erect, the sound was lost at long intervals, and was only faintly heard even then ; but it became continuous again when the ear was raised 9 feet from the ground, and it reached its full intensity at an elevation of 12 feet."

Prof. Reynolds thus sums up the results of his experiments :—

1. " When there is no wind, sound proceeding over a rough surface is more intense above than below."

2. " As long as the velocity of the wind is greater above than below, sound is lifted up to windward and is not destroyed."

3. " Under the same circumstances it is brought down to leeward, and hence its range extended at the surface of the ground."

Atmospheric refraction has an important bearing on the audibility of fog-signals, a subject which within the last few years has occupied the attention of two eminent physicists, Prof. Henry in America and Prof. Tyndall in this country. Henry[1] attributes almost all the vagaries of distant sounds to refraction, and has shewn how it is possible by various suppositions as to the motion of the air overhead to explain certain abnormal phenomena which have come under the notice of himself and other observers, while

[1] Report of the Lighthouse Board of the United States for the year 1874.

Tyndall[1], whose investigations have been equally extensive, considers the very limited distances to which sounds are sometimes audible to be due to an actual stopping of the sound by a flocculent condition of the atmosphere arising from unequal heating or moisture. That the latter cause is capable of operating in this direction to a certain extent cannot be doubted. Tyndall has proved by laboratory experiments that the sound of an electric bell may be sensibly intercepted by alternate layers of gases of different densities; and, although it must be admitted that the alternations of density were both more considerable and more abrupt than can well be supposed to occur in the open air, except perhaps in the immediate neighbourhood of the solid ground, some of the observations on fog-signals themselves seem to point directly to the explanation in question.

Thus it was found that the blast of a siren placed on the summit of a cliff overlooking the sea was followed by an echo of gradually diminishing intensity, whose duration sometimes amounted to as much as 15 seconds. This phenomenon was observed "when the sea was of glassy smoothness," and cannot apparently be attributed to any other cause than that assigned to it by Tyndall. It is therefore probable that refraction and acoustical opacity are both concerned in the capricious behaviour of fog-signals. *A priori* we should certainly be disposed to attach the greater importance to refraction, and Reynolds has shewn that some of Tyndall's own observations admit of explanation upon this principle. A failure in *reciprocity* can only be explained in accordance with theory by the action of wind (§ 111).

According to the hypothesis of acoustic clouds, a difference might be expected in the behaviour of sounds of long and of short duration, which it may be worth while to point out here, as it does not appear to have been noticed by any previous writer. Since energy is not lost in reflection and refraction, the intensity of radiation at a given distance from a continuous source of sound (or light) is not altered by an enveloping cloud of spherical form and of uniform density, the loss due to the intervening parts of the cloud being compensated by reflection from those which lie beyond the source. When, however, the sound is of short duration, the intensity at a distance may be very much diminished by the cloud on account of the different distances of its reflecting parts and the

[1] *Phil. Trans.* 1874. *Sound*, 3rd edition, Ch. VII.

consequent drawing out of the sound, although the whole intensity, as measured by the time-integral, may be the same as if there had been no cloud at all. This is perhaps the explanation of Tyndall's observation, that different kinds of signals do not always preserve the same order of effectiveness. In some states of the weather a " howitzer firing a 3-lb. charge commanded a larger range than the whistles, trumpets, or syren," while on other days " the inferiority of the gun to the syren was demonstrated in the clearest manner." It should be noticed, however, that in the same series of experiments it was found that the liability of the sound of a gun " to be quenched or deflected by an opposing wind, so as to be practically useless at a very short distance to windward, is very remarkable." The refraction proper must be the same for all kinds of sounds, but for the reason explained above, the diffraction round the edge of an obstacle may be less effective for the report of a gun than for the sustained note of a siren.

Another point examined by Tyndall was the influence of fog on the propagation of sound. In spite of isolated assertions to the contrary[1], it was generally believed on the authority of Derham that the influence of fog was prejudicial. Tyndall's observations prove satisfactorily that this opinion is erroneous, and that the passage of sound is favoured by the homogeneous condition of the atmosphere which is the usual concomitant of foggy weather. When the air is saturated with moisture, the fall of temperature with elevation according to the law of convective equilibrium is much less rapid than in the case of dry air, on account of the condensation of vapour which then accompanies expansion. From a calculation by Thomson[2] it appears that in warm fog the effect of evaporation and condensation would be to diminish the fall of temperature by one-half. The acoustical refraction due to temperature would thus be lessened, and in other respects no doubt the condition of the air would be favourable to the propagation of sound, provided no obstruction were offered by the suspended particles themselves. In a future chapter we shall investigate the disturbance of plane sonorous waves by a small obstacle, and we shall find that the effect depends upon the ratio of the diameter of the obstacle to the wave-length of the sound.

The reader who is desirous of pursuing this subject may

---

[1] See for example Desor, *Fortschritte der Physik*, XI. p. 217. 1855.

[2] *Manchester Memoirs*, 1861—62.

consult a paper by Reynolds " On the Refraction of Sound by the Atmosphere[1]," as well as the authorities already referred to. It may be mentioned that Reynolds agrees with Henry in considering refraction to be the really important cause of disturbance, but further observations are much needed. See also § 294.

291. On the assumption that the disturbance at an aperture in a screen is the same as it would have been at the same place in the absence of the screen, we may solve various problems respecting the diffraction of sound by the same methods as are employed for the corresponding problems in physical optics. For example, the disturbance at a distance on the further side of an infinite plane wall, pierced with a circular aperture on which plane waves of sound impinge directly, may be calculated as in the analogous problem of the diffraction pattern formed at the focus of a circular object-glass. Thus in the case of a symmetrical speaking trumpet the sound is a maximum along the axis of the instrument, where all the elementary disturbances issuing from the various points of the plane of the mouth are in one phase. In oblique directions the intensity is less; but it does not fall materially short of the maximum value until the obliquity is such that the difference of distances of the nearest and furthest points of the mouth amounts to about half a wave-length. At a somewhat greater obliquity the mouth may be divided into two parts, of which the nearer gives an aggregate effect equal in magnitude, but opposite in phase, to that of the further; so that the intensity in this direction vanishes. In directions still more oblique the sound revives, increases to an intensity equal to ·017 of that along the axis[2], again diminishes to zero, and so on, the alternations corresponding to the bright and dark rings which surround the central patch of light in the image of a star. If R denote the radius of the mouth, the angle, at which the first silence occurs, is $\sin^{-1}(\cdot 610\,\lambda/R)$. When the diameter of the mouth does not exceed $\frac{1}{2}\lambda$, the elementary disturbances combine without any considerable antagonism of phase, and the intensity is nearly uniform in all directions. It appears that concentration of sound along the axis requires that the ratio $R : \lambda$ should be large, a condition not usually satisfied in the ordinary use of speaking trumpets, whose efficiency depends rather upon an increase in the original volume

---

[1] *Phil. Trans.* Vol. 166, p. 315.   1876.
[2] Verdet, *Leçons d'optique physique*, t. I. p. 306.

of sound (§ 280).   When, however, the vibrations are of very short
wave-length, a trumpet of moderate size is capable of effecting a
considerable concentration along the axis, as I have myself verified
in the case of a hiss.

292.   Although such calculations as those referred to in the
preceding section are useful as giving us a general idea of the
phenomena of diffraction, it must not be forgotten that the
auxiliary assumption on which they are founded is by no means
strictly and generally true.   Thus in the case of a wave directly
incident upon a screen the normal velocity in the plane of the
aperture is not constant, as has been supposed, but increases from
the centre towards the edge, becoming infinite at the edge itself.
In order to investigate the conditions by which the actual velocity
is determined, let us for the moment suppose that the aperture is
filled up.   The incident wave $\phi = \cos(nt - kx)$ is then perfectly
reflected, and the velocity-potential on the negative side of the
screen $(x = 0)$ is

$$\phi = \cos(nt - kx) + \cos(nt + kx) \dots\dots\dots\dots(1),$$

giving, when $x = 0$, $\phi = 2 \cos nt$.   This corresponds to the vanishing of the normal velocity over the area of the aperture; the
completion of the problem requires us to determine a variable
normal velocity over the aperture such that the potential due to it
(§ 278) shall increase by the constant quantity $2 \cos nt$ in crossing
from the negative to the positive side; or, since the crossing
involves simply a change of sign, to determine a value of the
normal velocity over the area of the aperture which shall give on
the positive side $\phi = \cos nt$ over the same area.   The result of
superposing the two motions thus defined satisfies all the conditions of the problem, giving the same velocity and pressure on the
two sides of the aperture, and a vanishing normal velocity over the
remainder of the screen.

If $P \cos(nt + \epsilon)$ denote the value of $d\phi/dx$ at the various points
of the area (S) of the aperture, the condition for determining
P and ϵ is by (6) § 278,

$$-\frac{1}{2\pi} \iint P \frac{\cos(nt - kr + \epsilon)}{r} dS = \cos nt \dots\dots\dots(2),$$

where r denotes the distance between the element dS and any
fixed point in the aperture.   When P and ϵ are known, the

complete value of ϕ for any point on the positive side of the screen is given by

$$\phi = -\frac{1}{2\pi} \iint P \frac{\cos(nt - kr + \epsilon)}{r} \, dS \dots \dots \dots (3),$$

and for any point on the negative side by

$$\phi = +\frac{1}{2\pi} \iint P \frac{\cos(nt - kr + \epsilon)}{r} \, dS + 2 \cos nt \cos kx \dots (4).$$

The expression of P and ϵ for a finite aperture, even if of circular form, is probably beyond the power of known methods; but in the case where the dimensions are very small in comparison with the wave-length the solution of the problem may be effected for the circle and the ellipse. If r be the distance between two points, both of which are situated in the aperture, kr may be neglected, and we then obtain from (2)

$$\epsilon = 0, \quad 1 = -\frac{1}{2\pi} \iint P \frac{dS}{r} \dots \dots \dots (5),$$

shewing that $-P/2\pi$ is the density of the matter which must be distributed over S in order to produce there the constant potential unity. At a distance from the opening on the positive side we may consider r as constant, and take

$$\phi = M \frac{\cos(nt - kr)}{r} \dots \dots \dots (6),$$

where $M = -\frac{1}{2\pi} \iint P \, dS$, denoting the total quantity of matter which must be supposed to be distributed. It will be shewn on a future page (§ 306) that for an ellipse of semimajor axis a, and eccentricity e,

$$M = a \div F(e) \dots \dots \dots (7),$$

where F is the symbol of the complete elliptic function of the first kind. In the case of a circle, $F(e) = \frac{1}{2}\pi$, and

$$M = \frac{2a}{\pi} \dots \dots \dots (8).$$

This result is quite different from that which we should obtain on the hypothesis that the normal velocity in the aperture has the value proper to the primary wave. In that case by (3) § 283

$$\phi = -\frac{\pi a^2}{\lambda} \frac{\sin(nt - kr)}{r} \dots \dots \dots (9).$$

If there be several small apertures, whose distances apart are much greater than their dimensions, the same method gives

$$\phi = M_1 \frac{\cos(nt - kr_1)}{r_1} + M_2 \frac{\cos(nt - kr_2)}{r_2} + \ldots\ldots (10).$$

The diffraction of sound is a subject which has attracted but little attention either from mathematicians or experimentalists. Although the general character of the phenomena is well understood, and therefore no very startling discoveries are to be expected, the exact theoretical solution of a few of the simpler problems, which the subject presents, would be interesting; and, even with the present imperfect methods, something probably might be done in the way of experimental examination.

292 a. By means of a bird-call giving waves of about 1 cm. wave-length and a high pressure sensitive flame it is possible to imitate many interesting optical experiments. With this apparatus the shadow of an obstacle so small as the hand may be made apparent at a distance of several feet.

An experiment shewing the antagonism between the parts of a wave corresponding to the first and second Fresnel's zones (§ 283)

Fig. 57 a.

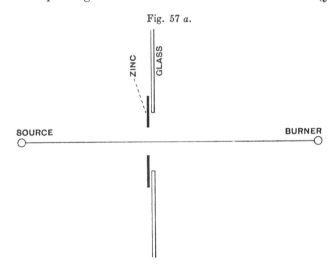

is very effective. A large glass screen (Fig. 57 a) is perforated with a circular hole 20 cm. in diameter, and is so situated between the source of sound and the burner that the aperture corresponds to the first two zones. By means of a zinc plate, held close to the glass, the aperture may be reduced to 14 cm., and then admits only the first zone. If the adjustments are well made, the flame, unaffected by the waves which penetrate the larger aperture, flares violently when the aperture is further restricted by the zinc plate. Or, as an alternative, the perforated plate may be replaced by a disc of 14 cm. diameter, which allows the second zone to be operative while the first is blocked off.

If a, b denote the distances of the screen from the source and from the point of observation, the external radius ρ of the nth zone is given by

$$\sqrt{(a^2 + \rho^2)} + \sqrt{(b^2 + \rho^2)} - a - b = \tfrac{1}{2}n\lambda,$$

or approximately

$$\rho^2 = n\lambda \frac{ab}{a + b} \dots\dots\dots\dots\dots\dots\dots(1).$$

When $a = b$,

$$\rho^2 = \tfrac{1}{2}n\lambda a \dots\dots\dots\dots\dots\dots\dots (2).$$

With the apertures specified above, $\rho^2 = 49$ for $n = 1$; $\rho^2 = 100$ for $n = 2$; so that

$$\lambda a = 100,$$

the measurements being in centimetres. This gives the suitable distances when λ is known. In an actual experiment $\lambda = 1\cdot2$, $a = 83$.

The process of augmenting the total effect by blocking out the alternate zones may be carried much further. Thus when a suitable circular grating, cut out of a sheet of zinc, is interposed between the source of sound and the flame, the effect is many times greater than when the screen is removed altogether[1]. As in Soret's corresponding optical experiment, the grating plays the part of a condensing lens.

The focal length of the lens is determined by (1), which may be written in the form

$$\frac{1}{f} = \frac{1}{a} + \frac{1}{b} = \frac{n\lambda}{\rho^2} \dots\dots\dots\dots\dots\dots (3);$$

so that

$$f = \rho^2/n\lambda \dots\dots \dots\dots\dots\dots (4).$$

In an actual grating constructed upon this plan eight zones—the first, third, fifth &c.—are occupied by metal. The radius of the first zone, or central circle, is $7\cdot6$ cm., so that $\rho^2/n = 58$. Thus, if $\lambda = 1\cdot2$ cm., $f = 48$ cm. If a and b are equal, each must be 96 cm

---

[1] "Diffraction of Sound," *Proc. Roy. Inst.* Jan. 20, 1888.

The condition of things at the centre of the shadow of a circular disc is still more easily investigated. If we construct in imagination a system of zones beginning with the circular edge of the disc, we see, as in § 283, that the total effect at a point upon the axis, being represented by the half of that of the first zone, is the same as if no obstacle at all were interposed. This analogue of a famous optical phenomenon is readily exhibited[1]. In one experiment a glass disc 38 cm. in diameter was employed, and its distances from the source and from the flame were respectively 70 cm. and 25 cm. A bird-call giving a pure tone ($\lambda = 1\cdot5$ cm.) is suitable, but may be replaced by a toy reed or other source giving short, though not necessarily simple, waves. In private work the ear furnished with a rubber tube may be used instead of a sensitive flame.

The region of no sensible shadow, though not confined to a mathematical point upon the axis, is of small dimensions, and a very moderate movement of the disc in its own plane suffices to reduce the flame to quiet. Immediately surrounding the central spot there is a ring of almost complete silence, and beyond that again a moderate revival of effect. The calculation of the intensity of sound at points off the axis of symmetry is too complicated to be entered upon here. The results obtained by Lommel[2] may be readily adapted to the acoustical problem. With the data specified above the diameter of the silent ring immediately surrounding the central region of activity is about 1·7 cm.

293. The value of a function ϕ which satisfies $\nabla^2\phi = 0$ throughout the interior of a simply-connected closed space S can be expressed as the potential of matter distributed over the surface of S. In a certain sense this is also true of the class of functions with which we are now occupied, which satisfy $\nabla^2\phi + k^2\phi = 0$. The following is Helmholtz's proof[3]. By Green's theorem, if ϕ and ψ denote any two functions of x, y, z,

$$\iint \phi \frac{d\psi}{dn} dS - \iiint \phi \nabla^2\psi \, dV = \iint \psi \frac{d\phi}{dn} dS - \iiint \psi \nabla^2\phi \, dV.$$

[1] "Acoustical Observations," *Phil. Mag.* Vol. IX. p. 281, 1880; *Proc. Roy. Inst.* loc. cit.

[2] *Abh. der bayer. Akad. der Wiss.* ii. Cl., xv. Bd., ii. Abth. See also *Encyclopædia Britannica*, Article "Wave Theory."

[3] *Theorie der Luftschwingungen in Röhren mit offenen Enden.* Crelle, Bd. LVII. p. 1. 1860.

To each side add $-\iiint k^2 \phi \psi \, dV$; then if

$$a^2 (\nabla^2 \phi + k^2 \phi) + \Phi = 0, \qquad a^2 (\nabla^2 \psi + k^2 \psi) + \Psi = 0,$$

$$a^2 \iint \phi \frac{d\psi}{dn} \, dS + \iiint \phi \, \Psi \, dV = a^2 \iint \psi \frac{d\phi}{dn} \, dS + \iiint \psi \, \Phi \, dV \ldots (1).$$

If Φ and Ψ vanish within S, we have simply

$$\iint \phi \frac{d\psi}{dn} \, dS = \iint \psi \frac{d\phi}{dn} \, dS \ldots \ldots \ldots \ldots \ldots (2).$$

Suppose, however, that

$$\phi = \frac{e^{-ikr}}{r} \ldots \ldots \ldots \ldots \ldots \ldots (3),$$

where r represents the distance of any point from a fixed origin O within S. At all points, except O, Φ vanishes; and the last term in (1) becomes

$$\iiint \psi \, \Phi \, dV = -a^2 \iiint \psi \, \nabla^2 \left(\frac{1}{r} \right) dV = 4\pi a^2 \psi,$$

ψ referring to the point O. Thus

$$4\pi \psi = \iint \frac{d\psi}{dn} \frac{e^{-ikr}}{r} \, dS - \iint \psi \frac{d}{dn} \left(\frac{e^{-ikr}}{r} \right) dS$$
$$+ \frac{1}{a^2} \iiint \psi \frac{e^{-ikr}}{r} \, dV \ldots \ldots \ldots (4),$$

in which, if Ψ vanish, we have an expression for the value of Ψ at any interior point O in terms of the surface values of ψ and of $d\psi/dn$. In the case of the common potential, on which we fall back by putting $k = 0$, ψ would be determined by the surface values of $d\psi/dn$ only. But with k finite, this law ceases to be universally true. For a given space S there is, as in the case investigated in § 267, a series of determinate values of k, corresponding to the periods of the possible modes of simple harmonic vibration which may take place within a closed rigid envelope having the form of S. With any of these values of k, it is obvious that ψ cannot be determined by its normal variation over S, and the fact that it satisfies throughout S the equation $\nabla^2 \psi + k^2 \psi = 0$. But if the supposed value of k do not coincide with one of the series, then the problem is determinate; for the difference of any two possible solutions, if finite, would satisfy the condition of giving no normal velocity over S, a condition which by hypothesis cannot be satisfied with the assumed value of k.

If the dimensions of the space S be very small in comparison with $\lambda (= 2\pi/k)$, e^{-ikr} may be replaced by unity; and we learn that ψ differs but little from a function which satisfies throughout S the equation $\nabla^2\phi = 0$.

294. On his extension of Green's theorem (1) Helmholtz founds his proof of the important theorem contained in the following statement: *If in a space filled with air which is partly bounded by finitely extended fixed bodies and is partly unbounded, sound waves be excited at any point* A, *the resulting velocity-potential at a second point* B *is the same both in magnitude and phase, as it would have been at* A, *had* B *been the source of the sound.*

If the equation

$$a^2 \iint \left(\phi \frac{d\psi}{dn} - \psi \frac{d\phi}{dn} \right) dS = \iiint (\psi\Phi - \phi\Psi) \, dV \ldots\ldots(1),$$

in which ϕ and ψ are arbitrary functions, and

$$\Phi = -a^2 (\nabla^2\phi + k^2\phi), \qquad \Psi = -a^2 (\nabla^2\psi + k^2\psi),$$

be applied to a space completely enclosed by a rigid boundary and containing any number of detached rigid fixed bodies, and if ϕ, ψ be velocity-potentials due to sources within S, we get

$$\iiint (\psi\Phi - \phi\Psi) \, dV = 0 \ldots\ldots\ldots\ldots\ldots(2).$$

Thus, if ϕ be due to a source concentrated in one point A, $\Phi = 0$ except at that point, and

$$\iiint \psi\Phi \, dV = \psi_A \iiint \Phi \, dV,$$

where $\iiint \Phi \, dV$ represents the intensity of the source. Similarly, if ψ be due to a source situated at B,

$$\iiint \phi\Psi \, dV = \phi_B \iiint \Psi \, dV.$$

Accordingly, if the sources be finite and equal, so that

$$\iiint \Phi \, dV = \iiint \Psi \, dV \ldots\ldots\ldots\ldots\ldots(3),$$

it follows that

$$\psi_A = \phi_B \ldots\ldots\ldots\ldots\ldots\ldots (4),$$

which is the symbolical statement of Helmholtz's theorem.

If the space S extend to infinity, the surface integral still vanishes, and the result is the same; but it is not necessary to go into detail here, as this theorem is included in the vastly more general principle of reciprocity established in Chapter V. The investigation there given shews that the principle remains true in the presence of dissipative forces, provided that these arise from resistances varying as the first power of the velocity, that the fluid need not be homogeneous, nor the neighbouring bodies rigid or fixed. In the application to infinite space, all obscurity is avoided by supposing the vibrations to be slowly dissipated after having escaped to a distance from A and B, the sources under contemplation.

The reader must carefully remember that in this theorem equal sources of sound are those produced by the periodic introduction and abstraction of equal quantities of fluid, or something whose effect is the same, and that equal sources do not necessarily evolve equal amounts of energy in equal times. For instance, a source close to the surface of a large obstacle emits twice as much energy as an equal source situated in the open.

As an example of the use of this theorem we may take the case of a hearing, or speaking, trumpet consisting of a conical tube, whose efficiency is thus seen to be the same, whether a sound produced at a point outside is observed at the vertex of the cone, or a source of equal strength situated at the vertex is observed at the external point.

It is important also to bear in mind that Helmholtz's form of the reciprocity theorem is applicable only to *simple* sources of sound, which in the absence of obstacles would generate symmetrical waves. As we shall see more clearly in a subsequent chapter, it is possible to have sources of sound, which, though concentrated in an infinitely small region, do not satisfy this condition. It will be sufficient here to consider the case of *double* sources, for which the modified reciprocal theorem has an interest of its own.

Let us suppose that A is a simple source, giving at a point B the potential $-\psi$, and that A' is an equal and opposite source situated at a neighbouring point, whose potential at B is $\psi + \Delta\psi$. If both sources be in operation simultaneously, the potential at B is $\Delta\psi$. Now let us suppose that there is a simple source at B, whose intensity and phase are the same as those of the sources at A and A'; the resulting potential at A is ψ, and at A' $\psi + \Delta\psi$. If the distance AA' be denoted by h, and be supposed to diminish without limit, the velocity of the fluid at A in the direction AA' is the limit of $\Delta\psi/h$. Hence, if we define a unit double source as the limit of two equal and opposite simple sources whose distance is diminished, and whose intensity is increased without limit in such a manner that the product of the intensity and the distance is the same as for two unit simple sources placed at the unit distance apart, we may say that the velocity of the fluid at A in direction AA' due to a unit simple source at B is numerically equal to the potential at B due to a unit *double* source at A, whose axis is in the direction AA'. This theorem, be it observed, is true in spite of any obstacles or reflectors that may exist in the neighbourhood of the sources.

Again, if AA' and BB' represent two unit double sources of the same phase, the velocity at B in direction BB' due to the source AA' is the same as the velocity at A in direction AA' due to the source BB'. These and other results of a like character may also be obtained on an immediate application of the general principle of § 108. These examples will be sufficient to shew that in applying the principle of reciprocity it is necessary to attend to the character of the sources. A double source, situated in an open space, is inaudible from any point in its equatorial plane, but it does not follow that a simple source in the equatorial plane is inaudible from the position of the double source. On this principle, I believe, may be explained a curious experiment by Tyndall[1], in which there was an apparent failure of reciprocity[2]. The source of sound employed was a reed of very high pitch, mounted in a tube, along whose axis the intensity was considerably greater than in oblique directions.

295. The kinetic energy T of the motion within a closed surface S is expressed by

$$T = \tfrac{1}{2}\rho_0 \iiint \Sigma \left(\frac{d\phi}{dx}\right)^2 dV \dots\dots\dots\dots(1);$$

[1] *Proceedings of the Royal Institution*, Jan. 1875. Also Tyndall, *On Sound*, 3rd edition, p. 405.

[2] See a note "On the Application of the Principle of Reciprocity to Acoustics." *Royal Society Proceedings*, Vol. xxv. p. 118, 1876, or *Phil. Mag.* (5), iii. p. 300.

so that $\quad \dfrac{dT}{dt} = \rho_0 \iiint \Sigma \dfrac{d\phi}{dx} \dfrac{d\dot{\phi}}{dx} \, dV$

$$= \rho_0 \iint \dot{\phi} \dfrac{d\phi}{dn} \, dS - \rho_0 \iiint \dot{\phi} \, \nabla^2 \phi \, dV \dots\dots\dots\dots\dots (2),$$

by Green's theorem. For the potential energy V_1 we have by (12) § 245

$$V_1 = \dfrac{\rho_0}{2a^2} \iiint \phi^2 \, dV \dots\dots\dots\dots\dots\dots(3),$$

whence $\quad \dfrac{dV_1}{dt} = \dfrac{\rho_0}{a^2} \iiint \phi \dot{\phi} \, dV = \dfrac{\rho_0}{a^2} \iiint \left\{ \dfrac{dR}{dt} + a^2 \nabla^2 \phi \right\} \dot{\phi} \, dV \dots(4),$

by the general equation of motion (9) § 244. Thus, if E denote the whole energy within the space S,

$$\dfrac{dE}{dt} = \rho_0 \iint \dot{\phi} \dfrac{d\phi}{dn} \, dS + \dfrac{\rho_0}{a^2} \iiint \dfrac{dR}{dt} \, \dot{\phi} \, dV \dots\dots\dots (5),$$

of which the first term represents the work transmitted across the boundary S, and the second represents the work done by internal sources of sound.

If the boundary S be a fixed rigid envelope, and there be no internal sources, E retains its initial value throughout the motion. This principle has been applied by Kirchhoff[1] to prove the determinateness of the motion resulting from given arbitrary initial conditions. Since every element of E is positive, there can be no motion within S, if E be zero. Now, if there were two motions possible corresponding to the same initial conditions, their difference would be a motion for which the initial value of E was zero; but by what has just been said such a motion cannot exist.

[1] *Vorlesungen über Math. Physik*, p. 311.

CHAPTER XV.

296. WHEN a train of plane waves, otherwise unimpeded, impinges upon a space occupied by matter, whose mechanical properties differ from those of the surrounding medium, secondary waves are thrown off, which may be regarded as a disturbance due to the change in the nature of the medium—a point of view more especially appropriate, when the *region of disturbance*, as well as the alteration of mechanical properties, is small. If the medium and the obstacle be fluid, the mechanical properties spoken of are two—the *compressibility* and the *density*: no account is here taken of friction or viscosity. In the chapter on spherical harmonic analysis we shall consider the problem here proposed on the supposition that the obstacle is spherical, without any restriction as to the smallness of the change of mechanical properties; in the present investigation the form of the obstacle is arbitrary, but we assume that the squares and higher powers of the changes of mechanical properties may be omitted.

If ξ, η, ζ denote the displacements parallel to the axes of co-ordinates of the particle, whose equilibrium position is defined by x, y, z, and if σ be the normal density, and m the constant of compressibility so that $\delta p = ms$, the equations of motion are

$$\sigma \frac{d^2\xi}{dt^2} + \frac{d(ms)}{dx} = 0 \quad \ldots\ldots\ldots\ldots\ldots(1),$$

and two similar equations in η and ζ. On the assumption that the whole motion is proportional to e^{ikat}, where as usual $k = 2\pi/\lambda$, and (§ 244) $a^2 = m/\sigma$, (1) may be written

$$\frac{d(ms)}{dx} - \sigma k^2 a^2 \xi = 0 \quad \ldots\ldots\ldots\ldots\ldots(2).$$

The relation between the condensation s, and the displacements ξ, η, ζ, obtained by integrating (3) § 238 with respect to the time, is

$$- s = \frac{d\xi}{dx} + \frac{d\eta}{dy} + \frac{d\zeta}{dz} \dotfill (3).$$

For the system of primary waves advancing in the direction of $- x$, η and ζ vanish; if ξ_0, s_0 be the values of ξ and s, and m_0, σ_0 be the mechanical constants for the undisturbed medium, we have as in (2)

$$\frac{d\,(m_0 s_0)}{dx} - \sigma_0 k^2 a^2 \xi_0 = 0 \dotfill (4);$$

but ξ_0, s_0 do not satisfy (2) at the region of disturbance on account of the variation in m and σ, which occurs there. Let us assume that the complete values are $\xi_0 + \xi$, η, ζ, $s_0 + s^1$, and substitute in (2). Then taking account of (4), we get

$$\frac{d\,(ms)}{dx} - \sigma k^2 a^2 \xi + (m - m_0) \frac{ds_0}{dx} + s_0 \frac{dm}{dx} - (\sigma - \sigma_0) k^2 a^2 \xi_0 = 0,$$

or, as it may also be written,

$$\frac{d}{dx} (ms) - \sigma k^2 a^2 \xi + \frac{d}{dx} (\Delta m . s_0) - \Delta \sigma . k^2 a^2 \xi_0 = 0 \dotfill (5),$$

if Δm, $\Delta \sigma$ stand respectively for $m - m_0$, $\sigma - \sigma_0$. The equations in η and ζ are in like manner

$$\left. \begin{array}{l} \dfrac{d}{dy} (ms) - \sigma k^2 a^2 \eta + \dfrac{d}{dy} (\Delta m . s_0) = 0 \\[2mm] \dfrac{d}{dz} (ms) - \sigma k^2 a^2 \zeta + \dfrac{d}{dz} (\Delta m . s_0) = 0 \end{array} \right\} \dotfill (6).$$

It is to be observed that Δm, $\Delta \sigma$ vanish, except through a small space, which is regarded as the region of disturbance; ξ, η, ζ, s, being the result of the disturbance are to be treated as small quantities of the order Δm, $\Delta \sigma$; so that in our approximate analysis the variations of m and σ in the first two terms of (5) and (6) are to be neglected, being there multiplied by small quantities. We thus obtain from (5) and (6) by differentiation and addition, with use of (3), as the differential equation in s,

$$\nabla^2 (ms) + k^2 ms = k^2 a^2 \frac{d}{dx} (\Delta \sigma . \xi_0) - \nabla^2 (\Delta m . s_0) \dotfill (7).$$

---

[1] [This notation was adopted for brevity. It might be clearer to take $\xi = \xi_0 + \Delta \xi$, $s = s_0 + \Delta s$, &c.; so that ξ, s, &c. should retain their former meanings.]

As in § 277, the solution of (7) is

$$4\pi m\, s = \iiint \frac{e^{-ikr}}{r} \left\{ \nabla^2 (\Delta m.s_0) - k^2 a^2 \frac{d}{dx} (\Delta\sigma.\xi_0) \right\} dV \ldots\ldots(8),$$

in which the integration extends over a volume completely including the region of disturbance. The integrals in (8) may be transformed with the aid of Green's theorem. Calling the two parts respectively P and Q, we have

$$P = \iiint \frac{e^{-ikr}}{r} \nabla^2 (\Delta m.s_0)\, dV = \iiint \Delta m.s_0 \nabla^2 \left(\frac{e^{-ikr}}{r}\right) dV$$

$$+ \iint \left\{ \frac{e^{-ikr}}{r} \frac{d}{dn} (\Delta m.s_0) - \Delta m.s_0 \frac{d}{dn} \left(\frac{e^{-ikr}}{r}\right) \right\} dS,$$

where S denotes the surface of the space through which the triple integration extends. Now on S, Δm and $\dfrac{d}{dn}(\Delta m.s_0)$ vanish, so that both the surface integrals disappear. Moreover

$$\nabla^2 \left(\frac{e^{-ikr}}{r}\right) = \frac{1}{r} \frac{d^2}{dr^2} e^{-ikr} = -k^2 \frac{e^{-ikr}}{r};$$

and thus

$$P = -k^2 \iiint \frac{e^{-ikr}}{r} \Delta m.s_0\, dV \ldots\ldots\ldots\ldots\ldots(9).$$

If the region of disturbance be small in comparison with λ, we may write

$$P = -k^2 s_0 \frac{e^{-ikr}}{r} \iiint \Delta m\, dV \ldots\ldots\ldots\ldots(10).$$

In like manner for the second integral in (8), we find

$$Q = -k^2 a^2 \iiint \frac{e^{-ikr}}{r} \frac{d}{dx} (\Delta\sigma.\xi_0)\, dV$$

$$= k^2 a^2 \iiint \Delta\sigma.\xi_0 \frac{d}{dx} \left(\frac{e^{-ikr}}{r}\right) dV = ik^3 a^2 \xi_0 \mu \frac{e^{-ikr}}{r} \iiint \Delta\sigma\, dV \ldots(11),$$

where μ denotes the cosine of the angle between x and r. The linear dimension of the region of disturbance is neglected in comparison with λ, and λ is neglected in comparison with r.

If T be the volume of the space through which Δm, $\Delta\sigma$ are sensible, we may write

$$\iiint \Delta m\, dV = T.\Delta m, \qquad \iiint \Delta\sigma\, dV = T.\Delta\sigma,$$

if on the right-hand sides Δm, $\Delta \sigma$ refer to the *mean values* of the variations in question. Thus from (8)

$$s = - \frac{k^2 T e^{-ikr}}{4\pi m \, r} \left\{ \Delta m . s_0 - ika^2 \Delta \sigma . \xi_0 \, \mu \right\} \dots \dots (12).$$

To express ξ_0 in terms of s_0, we have from (3), $\xi_0 = - \int s_0 \, dx$; and thus, if the condensation for the primary waves be $s_0 = e^{ik\,(at+x)}$, $ik\xi_0 = - s_0$, and (12) may be put into the form

$$s \, : \, s_0 = - \frac{\pi T e^{-ikr}}{\lambda^2 r} \left\{ \frac{\Delta m}{m} + \frac{\Delta \sigma}{\sigma} \, \mu \right\} \dots \dots \dots (13),$$

in which s_0 denotes the condensation of the primary waves at the place of disturbance at time t, and s denotes the condensation of the secondary waves at the same time at a distance r from the disturbance. Since the difference of phase represented by the factor e^{-ikr} corresponds simply to the distance r, we may consider that a simple reversal of phase occurs at the place of disturbance. The amplitude of the secondary waves is inversely proportional to the distance r, and to the *square* of the wave-length λ. Of the two terms expressed in (13) the first is symmetrical in all directions round the place of disturbance, while the second varies as the cosine of the angle between the primary and the secondary rays. Thus a place at which m varies behaves as a *simple* source, and a place at which σ varies behaves as a *double* source (§ 294).

That the secondary disturbance must vary as λ^{-2} may be proved immediately by the method of dimensions. Δm and $\Delta \sigma$ being given, the amplitude is necessarily proportional to T, and in accordance with the principle of energy must also vary inversely as r. Now the only quantities (dependent upon space, time, and mass) of which the ratio of amplitudes can be a function, are T, r, λ, a (the velocity of sound), and σ, of which the last cannot occur in the expression of a simple ratio, as it is the only one of the five which involves a reference to mass. Of the remaining four quantities T, r, λ, and a, the last is the only one which involves a reference to time, and is therefore excluded. We are left with T, r, and λ, of which the only combination varying as Tr^{-1}, and independent of the unit of length, is $Tr^{-1} \lambda^{-2}$.[1]

An interesting application of the results of this section may be made to explain what have been called *harmonic echoes*[2].

[1] " On the Light from the Sky," *Phil. Mag.* Feb. 1871, and " On the scattering of Light by small Particles," *Phil. Mag.* June, 1871.

[2] *Nature*, 1873, VIII. 319.

If the primary sound be a compound musical note, the various component tones are scattered in unlike proportions. The octave, for example, is sixteen times stronger relatively to the fundamental tone in the secondary than it was in the primary sound. There is thus no difficulty in understanding how it may happen that echoes returned from such reflecting bodies as groups of trees may be raised an octave. The phenomenon has also a complementary side. If a number of small bodies lie in the path of waves of sound, the vibrations which issue from them in all directions are at the expense of the energy of the main stream, and where the sound is compound, the exaltation of the higher harmonics in the scattered waves involves a proportional deficiency of them in the direct wave after passing the obstacles. This is perhaps the explanation of certain echoes which are said to return a sound graver than the original; for it is known that the pitch of a pure tone is apt to be estimated too low. But the evidence is conflicting, and the whole subject requires further careful experimental investigation; it may be commended to the attention of those who may have the necessary opportunities. While an alteration in the *character* of a sound is easily intelligible, and must indeed generally happen to a limited extent, a change in the pitch of a simple tone would be a violation of the law of forced vibrations, and hardly to be reconciled with theoretical ideas.

In obtaining (13) we have neglected the effect of the variable nature of the medium *on the disturbance.* When the disturbance on this supposition is thoroughly known, we might approximate again in the same manner. The additional terms so obtained would be necessarily of the second order in Δm, $\Delta \sigma$, so that our expressions are in all cases correct as far as the first powers of those quantities.

Even when the region of disturbance is not small in comparison with λ, the same method is applicable, provided the squares of Δm, $\Delta \sigma$ be really negligible. The total effect of any obstacle may then be calculated by integration from those of its parts. In this way we may trace the transition from a small region of disturbance whose *surface* does not come into consideration, to a thin plate of a few or of a great many square wavelengths in area, which will ultimately reflect according to the regular optical law. But if the obstacle be at all elongated in the direction of the primary rays, this method of calculation soon

ceases to be practically available, because, even although the
change of mechanical properties be very small, the interaction
of the various parts of the obstacle cannot be left out of account.
This caution is more especially needed in dealing with the case of
light, where the wave-length is so exceedingly small in comparison
with the dimensions of ordinary obstacles.

297.   In some degree similar to the effect produced by a
change in the mechanical properties of a small region of the fluid,
is that which ensues when the square of the motion rises anywhere to such importance that it can be no longer neglected.
$\nabla^2\phi + k^2\phi$ then acquires a finite value dependent upon the square
of the motion.   Such places therefore act like sources of sound;
the periods of the sources including the submultiples of the original period.   Thus any part of space, at which the intensity
accumulates to a sufficient extent, becomes itself a secondary
source, emitting the harmonic tones of the primary sound.   If
there be two primary sounds of sufficient intensity, the secondary
vibrations have frequencies which are the sums and differences of
the frequencies of the primaries (§ 68)[1].

298.   The pitch of a sound is liable to modification when the
source and the recipient are in relative motion.   It is clear, for
instance, that an observer approaching a fixed source will meet
the waves with a frequency exceeding that proper to the sound, by
the number of wave-lengths passed over in a second of time.   Thus
if v be the velocity of the observer and a that of sound, the
frequency is altered in the ratio $a \pm v : a$, according as the motion
is towards or from the source.   Since the alteration of pitch is
constant, a musical performance would still be heard in tune,
although in the second case, when a and v are nearly equal, the
fall in pitch would be so great as to destroy all musical character.
If we could suppose v to be greater than a, a sound produced after
the motion had begun would never reach the observer, but sounds
previously excited would be gradually overtaken and heard in the
reverse of the natural order.   If $v = 2a$, the observer would hear
a musical piece in correct time and tune, but *backwards*.

Corresponding results ensue when the source is in motion and
the observer at rest; the alteration depending only on the relative
motion in the line of hearing.   If the source and the observer move
with the same velocity there is no alteration of frequency, whether

[1] Helmholtz über Combinationstöne.   Pogg. *Ann.* Bd. xcix. s. 497. 1856.

the medium be in motion, or not. With a relative motion of 40 miles [64 kilometres] per hour the alteration of pitch is very conspicuous, amounting to about a semitone. The whistle of a locomotive is heard too high as it approaches, and too low as it recedes from an observer at a station, changing rather suddenly at the moment of passage.

The principle of the alteration of pitch by relative motion was first enunciated by Doppler[1], and is often called Doppler's principle. Strangely enough its legitimacy was disputed by Petzval[2], whose objection was the result of a confusion between two perfectly distinct cases, that in which there is a relative motion of the source and recipient, and that in which the medium is in motion while the source and the recipient are at rest. In the latter case the circumstances are mechanically the same as if the medium were at rest and the source and the recipient had a common motion, and therefore by Doppler's principle no change of pitch is to be expected.

Doppler's principle has been experimentally verified by Buijs Ballot[3] and Scott Russell, who examined the alterations of pitch of musical instruments carried on locomotives. A laboratory instrument for proving the change of pitch due to motion has been invented by Mach[4]. It consists of a tube six feet [183 cm.] in length, capable of turning about an axis at its centre. At one end is placed a small whistle or reed, which is blown by wind forced along the axis of the tube. An observer situated in the plane of rotation hears a note of fluctuating pitch, but if he places himself in the prolongation of the axis of rotation, the sound becomes steady. Perhaps the simplest experiment is that described by König[5]. Two c'' tuning-forks mounted on resonance cases are prepared to give with each other four beats per second. If the graver of the forks be made to approach the ear while the other remains at rest, one beat is *lost* for each two feet [61 cm.] of approach; if, however, it be the more acute of the two forks which approaches the ear, one beat is *gained* in the same distance.

[1] Theorie des farbigen Lichtes der Doppelsterne. Prag, 1842. See Pisko, *Die neueren Apparate der Akustik.* Wien, 1865.

[2] *Wien. Ber.* viii. 134. 1852. *Fortschritte der Physik*, viii. 167.

[3] Pogg. *Ann.* lxvi. p. 321.

[4] Pogg. *Ann.* cxii. p. 66, 1861, and cxvi. p. 333, 1862.

[5] König's *Catalogue des Appareils d'Acoustique.* Paris, 1865.

A modification of this experiment due to Mayer[1] may also be noticed. In this case one fork excites the vibrations of a second in unison with itself, the excitation being made apparent by a small pendulum, whose bob rests against the extremity of one of the prongs. If the exciting fork be at rest, the effect is apparent up to a distance of 60 feet [1830 cm.], but it ceases when the exciting fork is moved rapidly to or fro in the direction of the line joining the two forks.

There is some difficulty in treating mathematically the problem of a moving source, arising from the fact that any practical source acts also as an obstacle. Thus in the case of a bell carried through the air, we should require to solve a problem difficult enough without including the vibrations at all. But the solution of such a problem, even if it could be obtained, would throw no particular light on Doppler's law, and we may therefore advantageously simplify the question by idealizing the bell into a simple source of sound.

In § 147 we considered the problem of a moving source of disturbance in the case of a stretched string. The theory for aerial waves in one dimension is precisely similar, but for the general case of three dimensions some extension is necessary, in order to take account of the possibility of a motion across the direction of the sound rays. From §§ 273, 276 it appears that the effect at any point O of a source of sound is the same, whether the source be at rest, or whether it move in any manner on the surface of a sphere described about O as centre. If the source move in such a manner as to change its distance (r) from O, its effect is altered in two ways. Not only is the *phase* of the disturbance on arrival at O affected by the variation of distance, but the *amplitude* also undergoes a change. The latter complication however may be put out of account, if we limit ourselves to the case in which the source is sufficiently distant. On this understanding we may assert that the effect at O of a disturbance generated at time t and at distance r is the same as that of a similar disturbance generated at the time $t + \delta t$ and at the distance $r - a \delta t$. In the case of a periodic disturbance a velocity of approach (v) is equivalent to an increase of frequency in the ratio $a : a + v$.

299. We will now investigate the forced vibrations of the air contained within a rectangular chamber, due to internal sources

[1] *Phil. Mag.* (4), XLIII. p. 278, 1872.

of sound. By § 267 it appears that the result at time t of an initial condensation confined to the neighbourhood of the point ξ, η, ζ is

$$\dot{\phi} = \Sigma\Sigma\Sigma \, ka \, B_{pqr} \cos kat \cos\left(p\,\frac{\pi x}{\alpha}\right) \cos\left(q\,\frac{\pi y}{\beta}\right) \cos\left(r\,\frac{\pi z}{\gamma}\right) \quad \ldots(1),$$

where

$$ka\,B_{pqr} = \frac{8}{\alpha\beta\gamma} \cos\left(p\,\frac{\pi\xi}{\alpha}\right) \cos\left(q\,\frac{\pi\eta}{\beta}\right) \cos\left(r\,\frac{\pi\zeta}{\gamma}\right)\iiint\dot{\phi}_0\,dx\,dy\,dz\ldots(2),$$

from which the effect of an impressed force may be deduced, as in § 276. The disturbance $\iiint \dot{\phi}_{t'}\,dx\,dy\,dz$ communicated at time t' being denoted by $\iiint \Phi(t')\,dt'\,dx\,dy\,dz$, or $\Phi_1(t')\,dt'$, the resultant disturbance at time t is

$$\dot{\phi} = \frac{8}{\alpha\beta\gamma}\,\Sigma\Sigma\Sigma \cos\left(p\,\frac{\pi x}{\alpha}\right)\cos\left(q\,\frac{\pi y}{\beta}\right)\cos\left(r\,\frac{\pi z}{\gamma}\right) \times$$

$$\cos\left(p\,\frac{\pi\xi}{\alpha}\right)\cos\left(q\,\frac{\pi\eta}{\beta}\right)\cos\left(r\,\frac{\pi\zeta}{\gamma}\right)\int_{-\infty}^{t}\Phi_1(t')\cos ka(t-t')\,dt' \quad \ldots(3).$$

The symmetry of this expression with respect to x, y, z and ξ, η, ζ is an example of the principle of reciprocity (§ 107).

In the case of a harmonic force, for which $\Phi_1(t') = A \cos mat'$ we have to consider the value of

$$\int_{-\infty}^{t} \cos mat' \cos ka\,(t-t')\,dt' \quad \ldots\ldots\ldots\ldots\ldots(4).$$

Strictly speaking, this integral has no definite value; but, if we wish for the expression of the forced vibrations only, we must omit the integrated function at the lower limit, as may be seen by supposing the introduction of very small dissipative forces. We thus obtain

$$\int_{-\infty}^{t} \Phi_1(t')\cos ka\,(t-t')\,dt' = A\,\frac{ma\sin mat}{(m^2-k^2)\,a^2} \quad \ldots\ldots(5).$$

As might have been predicted, the expressions become infinite in case of a coincidence between the period of the source and one of the natural periods of the chamber. Any particular normal vibration will not be excited, if the source be situated on one of its loops.

The effect of a multiplicity of sources may readily be inferred by summation or integration.

300. When sound is excited within a cylindrical pipe, the simplest kind of excitation that we can suppose is by the forced vibration of a piston. In this case the waves are plane from the beginning. But it is important also to inquire what happens when the source, instead of being uniformly diffused over the section, is concentrated in one point of it. If we assume (what, however, is not unreservedly true) that at a sufficient distance from the source the waves become plane, the law of reciprocity is sufficient to guide us to the desired information.

Let A be a simple source in an unlimited tube, B, B' two points of the same normal section in the region of plane waves. *Ex hypothesi*, the potentials at B and B' due to the source A are the same, and accordingly by the law of reciprocity equal sources at B and B' would give the same potential at A. From this it follows that the effect of any source is the same at a distance, as if the source were uniformly diffused over the section which passes through it. For example, if B and B' were equal sources in opposite phases, the disturbance at A would be nil.

The energy emitted by a simple source situated within a tube may now be calculated. If the section of the tube be σ, and the source such that in the open the potential due to it would be

$$\phi = -\frac{A}{4\pi} \cdot \frac{\cos k\,(at - r)}{r} \quad \ldots\ldots\ldots\ldots(1),$$

the velocity-potential at a distance within the tube will be the same as if the cause of the disturbance were the motion of a piston at the origin, giving the same total displacement, and the energy emitted will also be the same. Now from (1)

$$2\pi r^2 \frac{d\phi}{dr} = \tfrac{1}{2} A \cos kat \quad \text{ultimately,}$$

and therefore if ψ be the velocity-potential of the plane waves in the tube (supposed parallel to z), we may take

$$\sigma \frac{d\psi}{dz} = \tfrac{1}{2} A \cos k\,(at - z) \ldots\ldots\ldots\ldots(2),$$

corresponding to which

$$\dot{\psi} = -\frac{aA}{2\sigma} \cos k\,(at - z) \ldots\ldots\ldots\ldots(3).$$

Hence, as in § 245, the energy (W) emitted *on each side* of the source is given by

$$\frac{dW}{dt} = \sigma\left(-\rho\,\dot\psi\,\frac{d\psi}{dz}\right)_{z=0} = \frac{\rho a A^2}{4\sigma}\cos^2 kat\,;$$

so that in the long run

$$W = \frac{\rho a A^2}{8\sigma}\,t \dots\dots\dots\dots\dots(4).$$

If the tube be stopped by an immovable piston placed close to the source, the whole energy is emitted in one direction; but this is not all. In consequence of the doubled pressure, twice as much energy as before is developed, and thus in this case

$$W = \frac{\rho a A^2}{2\sigma}\,t \dots\dots\dots\dots\dots(5).$$

The narrower the tube, the greater is the energy issuing from a given source. It is interesting to compare the efficiency of a source at the stopped end of a cylindrical tube with that of an equal source situated at the vertex of a cone. From § 280 we have in the latter case,

$$W' = \rho\,\frac{k^2 a A^2}{2\omega}\,t \dots\dots\dots\dots\dots(6),$$

so that

$$W : W' = \omega : k^2\sigma \dots\dots\dots\dots\dots(7).$$

The energies emitted in the two cases are the same when $\omega = k^2\sigma$, that is, when the section of the cylinder is equal to the area cut off by the cone from a sphere of radius k^{-1}.

301. We have now to examine how far it is true that vibrations within a cylindrical tube become approximately plane at a sufficient distance from their source. Taking the axis of z parallel to the generating lines of the cylinder, let us investigate the motion, whose potential varies as e^{ikat}, on the positive side of a source, situated at $z = 0$. If ϕ be the potential and ∇^2 stand for $d^2/dx^2 + d^2/dy^2$ the equation of the motion is

$$\left(\frac{d^2}{dz^2} + \nabla^2 + k^2\right)\phi = 0 \dots\dots\dots\dots\dots(1).$$

If ϕ be independent of z, it represents vibrations wholly transverse to the axis of the cylinder. If the potential be then proportional to e^{ipat}, it must satisfy

$$(\nabla^2 + p^2)\phi = 0 \dots\dots\dots\dots\dots(2),$$

as well as the condition that over the boundary of the section

$$\frac{d\phi}{dn} = 0 \dots\dots\dots\dots\dots\dots\dots(3).$$

In order that these equations may be compatible, p is restricted to certain definite values corresponding to the periods of the natural vibrations. A zero value of p gives $\phi =$ constant, which solution, though it is of no significance in the two dimension problem, we shall presently have to consider. For each admissible value of p, there is a definite normal function u of x and y (§ 92), such that a solution is

$$\phi = A u e^{ipat} \dots\dots\dots\dots\dots\dots(4).$$

Two functions u, u', corresponding to different values of p, are conjugate, viz. make

$$\iint u u' dx\, dy = 0 \dots\dots\dots\dots\dots\dots(5),$$

and any function of x and y may be expanded within the contour in the series

$$\phi = A_0 u_0 + A_1 u_1 + A_2 u_2 + \dots\dots\dots\dots\dots\dots(6),$$

in which u_0, corresponding to $p = 0$, is constant.

In the actual problem ϕ may still be expanded in the same series, provided that A_0, A_1, &c. be regarded as functions of z. By substitution in (1) we get, having regard to (2),

$$u_0 \left\{ \frac{d^2 A_0}{dz^2} + k^2 A_0 \right\} + u_1 \left\{ \frac{d^2 A_1}{dz^2} + (k^2 - p_1^2) A_1 \right\}$$

$$+ u_2 \left\{ \frac{d^2 A_2}{dz^2} + (k^2 - p_2^2) A_2 \right\} + \dots = 0 \dots\dots\dots\dots(7),$$

in which, by virtue of the conjugate property of the normal functions, each coefficient of u must vanish separately. Thus

$$\frac{d^2 A_0}{dz^2} + k^2 A_0 = 0, \qquad \frac{d^2 A}{dz^2} + (k^2 - p^2) A = 0 \dots\dots\dots(8).$$

The solution of the first of these equations is

$$A_0 = \alpha_0 e^{ikz} + \beta_0 e^{-ikz},$$

giving

$$\phi_0 = \alpha_0 u_0 e^{ik(at+z)} + \beta_0 u_0 e^{ik(at-z)} \dots\dots\dots\dots(9).$$

The solution of the general equation in A assumes a different form, according as $k^2 - p^2$ is positive or negative. If the forced vibration be graver in pitch than the gravest of the purely transverse natural vibrations, every finite value of p^2 is greater than k^2, or $k^2 - p^2$ is always negative.   Putting

$$k^2 - p^2 = - \mu^2 \quad\dots\dots\dots\dots\dots\dots\dots(10),$$

we have $A = \alpha e^{\mu z} + \beta e^{-\mu z},$

whence $\phi = (\alpha e^{\mu z} + \beta e^{-\mu z})\, u e^{ikat} \dots\dots\dots\dots\dots(11).$

Now under the circumstances supposed, it is evident that the motion does not become infinite with z, so that all the coefficients α vanish.   For a somewhat different reason the same is true of α_0, as there can be no wave in the negative direction.   We may therefore take

$$\phi = \beta_0 e^{ik(at-z)} + \beta_1 u_1 e^{-\mu_1 z}\, e^{ikat} + \beta_2 u_2 e^{-\mu_2 z}\, e^{ikat} + \dots\dots\dots(12),$$

an expression which reduces to its first term when z is sufficiently great.   We conclude that in all cases the waves ultimately become plane, *if the forced vibration be graver than the gravest of the natural transverse vibrations.*

In the case of a circular cylinder, of radius r, the gravest transverse vibration has a wave-length equal to $2\pi r \div 1.841 = 3.413\, r$ (§ 339).   If then the wave-length of the forced vibration exceed $3.413\, r$, the waves ultimately become plane.   It may happen however that the waves ultimately become plane, although the wave-length fall short of the above limit.   For example, if the source of vibration be symmetrical with respect to the axis of the tube, *e.g.* a simple source situated on the axis itself, the gravest transverse vibration with which we should have to deal would be more than an octave higher than in the general case, and the wave-length of the forced vibration might have less than half the above value.

From (12), when $z = 0$,

$$\frac{d\phi}{dz} = - ik\beta_0 e^{ikat} - \mu_1 \beta_1 u_1 e^{ikat} - \dots$$

whence $\iint \dfrac{d\phi}{dz}\, d\sigma = - ik\beta_0\, \sigma\, e^{ikat} \quad\dots\dots\dots\dots\dots(13),$

inasmuch as $\iint u_1\, d\sigma$, $\iint u_2\, d\sigma$, &c., all vanish.

It appears accordingly that the plane waves at a distance are the same as would be produced by a rigid piston at the origin, giving the same mean normal velocity as actually exists. Any normal motion of which the negative and positive parts are equal, produces ultimately no effect.

When there is no restriction on the character of the source, and when some of the transverse natural vibrations are graver than the actual one, some of the values of $k^2 - p^2$ are positive, and then terms enter of the form

$$\phi = \beta u\, e^{ikat}\, e^{-i\sqrt{(k^2-p^2)}z},$$

or in real quantities

$$\phi = \beta u \cos \{k\,at - \sqrt{(k^2 - p^2)}\,z\}.................(14),$$

indicating that the peculiarities of the source are propagated to an infinite distance.

The problem here considered may be regarded as a generalization of that of § 268. For the case of a circular cylinder it may be worked out completely with the aid of Bessel's functions, but this must be left to the reader.

302. In § 278 we have fully determined the motion of the air due to the normal periodic motion of a bounding plane plate of infinite extent. If $d\phi/dn$ be the given normal velocity at the element dS,

$$\phi = -\frac{1}{2\pi}\iint \frac{d\phi}{dn}\frac{e^{-ikr}}{r}\,dS ....(1)$$

gives the velocity-potential at any point P distant r from dS. The remainder of this chapter is devoted to the examination of the particular case of this problem which arises when the normal velocity has a given constant value over a circular area of radius R, while over the remainder of the plane it is zero. In particular we shall investigate what forces due to the reaction of the air will act on a rigid circular plate, vibrating with a simple harmonic motion in an equal circular aperture cut out of a rigid plane plate extending to infinity.

For the whole variation of pressure acting on the plate we have (§ 244)

$$\iint \delta p\,dS = -\sigma \iint \dot{\phi}\,dS = -ika\sigma \iint \phi\,dS,$$

where σ is the natural density, and ϕ varies as e^{ikat}. Thus by (1)

$$\iint \delta p \, dS = \frac{ika\sigma}{\pi} \frac{d\phi}{dn} \Sigma\Sigma \frac{e^{-ikr}}{r} \, dS dS' \dots\dots\dots\dots(2).$$

In the double sum

$$\Sigma\Sigma \frac{e^{-ikr}}{r} \, dS dS' \quad \dots\dots\dots\dots\dots(3),$$

which we have now to evaluate, each pair of elements is to be taken *once* only, and the product is to be summed after multiplication by the factor $r^{-1} e^{-ikr}$, depending on their mutual distance. The best method is that suggested by Prof. Maxwell for the common potential[1]. The quantity (3) is regarded as the work that would be consumed in the complete dissociation of the matter composing the disc, that is to say, in the removal of every element from the influence of every other, on the supposition that the potential of two elements is proportional to $r^{-1} e^{-ikr}$. The amount of work required, which depends only on the initial and final states, may be calculated by supposing the operation performed in any way that may be most convenient. For this purpose we suppose that the disc is divided into elementary rings, and that each ring is carried away to infinity before any of the interior rings are disturbed.

The first step is the calculation of the potential (V) at the edge of a disc of radius c. Taking polar co-ordinates (ρ, θ) with any point of the circumference for pole, we have

$$V = \iint \frac{e^{-ik\rho}}{\rho} \rho \, d\rho \, d\theta = \int_{-\frac{1}{2}\pi}^{+\frac{1}{2}\pi} \int_0^{2c \cos \theta} e^{-ik\rho} \, d\rho \, d\theta = \frac{2}{ik} \int_0^{\frac{1}{2}\pi} \{1 - e^{-2ikc \cos \theta}\} \, d\theta.$$

This quantity must be multiplied by $2\pi c \, dc$, and afterwards integrated with respect to c between the limits 0 and R. But it will be convenient first to effect a transformation. We have

$$\frac{2}{\pi} \int_0^{\frac{1}{2}\pi} e^{-2ikc \cos \theta} \, d\theta = \frac{2}{\pi} \int_0^{\frac{1}{2}\pi} e^{-2ikc \sin \theta} \, d\theta$$

$$= \frac{2}{\pi} \int_0^{\frac{1}{2}\pi} \cos (2kc \sin \theta) \, d\theta - \frac{2i}{\pi} \int_0^{\frac{1}{2}\pi} \sin (2kc \sin \theta) \, d\theta$$

$$= J_0(z) - i K(z) \dots\dots\dots\dots\dots\dots\dots\dots\dots(4),$$

where z is written for $2kc$. $J_0(z)$ is the Bessel's function of zero

---

[1] Theory of Resonance. *Phil. Trans.* 1870.

order (§ 200), and $K(z)$ is a function defined by the equation

$$K(z) = \frac{2}{\pi} \int_0^{\frac{1}{2}\pi} \sin(z \sin \theta) \, d\theta$$

$$= \frac{2}{\pi} \left\{ z - \frac{z^3}{1^2 \cdot 3^2} + \frac{z^5}{1^2 \cdot 3^2 \cdot 5^2} - \frac{z^7}{1^2 \cdot 3^2 \cdot 5^2 \cdot 7^2} + \ldots \ldots \right\} \ldots \ldots (5).$$

Deferring for the moment the further consideration of the function K, we have

$$V = \frac{\pi}{k} [K(z) - i \{1 - J_0(z)\}] \ldots \ldots \ldots (6),$$

and thus

$$\Sigma \Sigma \frac{e^{-ikr}}{r} \, dS dS' = \frac{\pi^2}{2k^3} \int_0^{2kR} z \, dz \, [K(z) - i \{1 - J_0(z)\}].$$

Now by (6) § 200 and (8) § 204

$$\int_0^z z \, dz \, J_0(z) = z \, J_1(z) \ldots \ldots \ldots \ldots (7);$$

and thus, if K_1 be defined by

$$K_1(z) = \int_0^z z \, dz \, K(z) \ldots \ldots \ldots (8),$$

we may write

$$\Sigma \Sigma \frac{e^{-ikr}}{r} \, dS dS' = \frac{\pi^2}{2k^3} K_1(2kR) - i \frac{\pi^2 R^2}{k} \left(1 - \frac{J_1(2kR)}{kR} \right) \ldots (9).$$

From this the total pressure is derived by introduction of the factor $\dfrac{ika\sigma}{\pi} \dfrac{d\phi}{dn}$, so that

$$\iint \delta p \, dS = a\sigma \cdot \pi R^2 \cdot \frac{d\phi}{dn} \left(1 - \frac{J_1(2kR)}{kR} \right) + i \frac{a\sigma\pi}{2k^2} \frac{d\phi}{dn} K_1(2kR) .. (10).$$

The reaction of the air on the disc may thus be divided into two parts, of which the first is proportional to the velocity of the disc, and the second to the acceleration. If ξ denote the displacement of the disc, so that $\dot{\xi} = \dfrac{d\phi}{dn}$, we have $\ddot{\xi} = ika \, \dot{\xi} = ika \dfrac{d\phi}{dn}$; and therefore in the equation of motion of the disc, the reaction of the air is represented by a frictional force $a\sigma \cdot \pi R^2 \cdot \dot{\xi} \left(1 - \dfrac{J_1(2kR)}{kR} \right)$ retarding the motion, and by an accession to the inertia equal to $\dfrac{\pi\sigma}{2k^3} K_1(2kR)$.

When kR is small, we have from the ascending series for J_1 (5) § 200,

$$1 - \frac{J_1(2kR)}{kR} = \frac{k^2 R^2}{1.2} - \frac{k^4 R^4}{1.2^2.3} + \frac{k^6 R^6}{1.2^2.3^2.4} - \frac{k^8 R^8}{1.2^2.3^2.4^2.5} + \ldots \quad (11),$$

so that the frictional term is approximately

$$\tfrac{1}{2} a\sigma . \pi R^2 . k^2 R^2 . \dot{\xi} \ldots\ldots\ldots\ldots\ldots\ldots\ldots (12).$$

From the nature of the case the coefficient of $\dot{\xi}$ must be positive, otherwise the reaction of the air would tend to augment, instead of to diminish, the motion. That $J_1(z)$ is in fact always less than $\tfrac{1}{2} z$ may be verified as follows. If θ lie between 0 and π, and z be positive, $\sin(z\sin\theta) - z\sin\theta$ is negative, and therefore also

$$\frac{1}{\pi}\int_0^\pi \{\sin(z\sin\theta) - z\sin\theta\}\sin\theta\, d\theta$$

is negative. But this integral is $J_1(z) - \tfrac{1}{2} z$, which is accordingly negative for all positive values of z.

When kR is great, $J_1(2kR)$ tends to vanish, and then the frictional term becomes simply $a\sigma . \pi R^2 . \dot{\xi}$. This result might have been expected; for when kR is very large, the wave motion in the neighbourhood of the disc becomes approximately plane. We have then by (6) and (8) § 245, $dp = a\rho_0\dot{\xi}$, in which ρ_0 is the density (σ); so that the retarding force is $\pi R^2 \delta p = a\sigma . \pi R^2 . \dot{\xi}$.

We have now to consider the term representing an alteration of inertia, and among other things to prove that this alteration is an increase, or that $K_1(z)$ is positive. By direct integration of the ascending series (5) for K (which is always convergent),

$$K_1(z) = \frac{2}{\pi}\left\{\frac{z^3}{1^2.3} - \frac{z^5}{1^2.3^2.5} + \frac{z^7}{1^2.3^2.5^2.7} \quad \ldots\ldots\right\} \ldots\ldots\ldots (13).$$

When therefore kR is small, we may take as the expression for the increase of inertia

$$\frac{8\sigma R^3}{3} = \sigma . \pi R^2 . \frac{8R}{3\pi} \ldots\ldots\ldots\ldots\ldots\ldots\ldots (14).$$

This part of the reaction of the air is therefore represented by supposing the vibrating plate to carry with it a mass of air equal to that contained in a cylinder whose base is the plate, and whose height is equal to $8R/3\pi$; so that, when the plate is sufficiently small, the mass to be added is independent of the period of vibration.

From the series (5) for $K(z)$, it may be proved immediately that

$$\frac{1}{z}\frac{d}{dz}\left(z\frac{d}{dz}\right)K(z) = \frac{2}{\pi z} - K(z) \quad \ldots\ldots\ldots\ldots(15),$$

or

$$\left(\frac{d^2}{dz^2} + \frac{1}{z}\frac{d}{dz} + 1\right)K(z) = \frac{2}{\pi z} \quad \ldots\ldots\ldots\ldots(16).$$

From the first form (15) it follows that

$$K_1(z) = \int_0^z K(z)\,z\,dz = \frac{2}{\pi}z - z\frac{dK(z)}{dz} \quad \ldots\ldots\ldots(17).$$

By means of this expression for $K_1(z)$ we may readily prove that the function is always positive. For

$$\frac{dK(z)}{dz} = \frac{d}{dz}\cdot\frac{2}{\pi}\int_0^{\frac{1}{2}\pi}\sin(z\sin\theta)\,d\theta = \frac{2}{\pi}\int_0^{\frac{1}{2}\pi}\cos(z\sin\theta)\sin\theta\,d\theta\ldots(18);$$

so that

$$K_1(z) = \frac{2z}{\pi}\left\{1 - \int_0^{\frac{1}{2}\pi}\cos(z\sin\theta)\sin\theta\,d\theta\right\}$$

$$= \frac{4z}{\pi}\int_0^{\frac{1}{2}\pi}\sin^2\left(\tfrac{1}{2}z\sin\theta\right)\sin\theta\,d\theta\ldots\ldots\ldots\ldots(19),$$

an integral of which every element is positive. When z is very large, $\cos(z\sin\theta)$ fluctuates with great rapidity, and thus $K_1(z)$ tends to the form

$$K_1(z) = \frac{2}{\pi}\cdot z \quad \ldots\ldots\ldots\ldots\ldots\ldots\ldots(20).$$

When z is great, the ascending series for K and K_1, though always ultimately convergent, become useless for practical calculation, and it is necessary to resort to other processes. It will be observed that the differential equation (16) satisfied by K is the same as that belonging to the Bessel's function J_0, with the exception of the term on the right-hand side, viz. $2/\pi z$. The function K is therefore included in the form obtained by adding to the general solution of Bessel's equation containing two arbitrary constants any particular solution of (16). Such a particular solution is

$$\tfrac{1}{2}\pi\cdot K(z) = z^{-1} - z^{-3} + 1^2\cdot3^2\cdot z^{-5} - 1^2\cdot3^2\cdot5^2\cdot z^{-7} + 1^2\cdot3^2\cdot5^2\cdot7^2\cdot z^{-9} - \ldots(21),$$

as may be readily verified on substitution. The series on the right of (21), notwithstanding its ultimate divergency, may be used successfully for computation when z is great. It is in fact the analytical equivalent of $\int_0^\infty e^{-\beta} (z^2 + \beta^2)^{-\frac{1}{2}} d\beta$, and we might take

$$K(z) = \frac{2}{\pi} \int_0^\infty \frac{e^{-\beta} d\beta}{\sqrt{(z^2 + \beta^2)}} + \text{Complementary Function},$$

determining the two arbitrary constants by an examination of the forms assumed when z is very great. But it is perhaps simpler to follow the method used by Lipschitz [1] for Bessel's functions.

By (4) we have

$$J_0(z) - i K(z) = \frac{2}{\pi} \int_0^{\frac{1}{2}\pi} e^{-iz \cos \theta} d\theta = \frac{2}{\pi} \int_0^1 \frac{e^{-izv} dv}{\sqrt{(1 - v^2)}} \quad \ldots\ldots(22).$$

Consider the integral $\int \dfrac{e^{-zw} dw}{\sqrt{(1 + w^2)}}$, where w is a complex variable of the form $u + iv$. Representing, as usual, simultaneous pairs of values of u and v by the co-ordinates of a point, we see that the value of the integral will be zero, if the integration with respect to w range round the rectangle, whose angular points are respectively 0, h, $h + i$, i, where h is any real positive quantity. Thus

$$\int_0^h \frac{e^{-zu} du}{\sqrt{1 + u^2}} + \int_0^i \frac{e^{-z(h+iv)} d(iv)}{\sqrt{1 + (h + iv)^2}} + \int_h^0 \frac{e^{-z(u+i)} du}{\sqrt{1 + (u + i)^2}} + \int_i^0 \frac{e^{-izv} d(iv)}{\sqrt{1 - v^2}} = 0,$$

from which, if we suppose that $h = \infty$,

$$\int_0^1 \frac{e^{-izv} dv}{\sqrt{1 - v^2}} = -i \int_0^\infty \frac{e^{-zu} du}{\sqrt{1 + u^2}} + i \int_0^\infty \frac{e^{-z(u+i)} du}{\sqrt{1 + (u + i)^2}} \ldots\ldots\ldots(23).$$

Replacing uz by β, we may write (23) in the form

$$\int_0^1 \frac{e^{-izv} dv}{\sqrt{(1 - v^2)}} = -i \int_0^\infty \frac{e^{-\beta} d\beta}{z\sqrt{(1 + \beta^2/z^2)}} + \frac{e^{-i(z - \frac{1}{4}\pi)}}{\sqrt{(2z)}} \int_0^\infty \frac{e^{-\beta} \beta^{-\frac{1}{2}} d\beta}{\sqrt{(1 - i\beta/2z)}} \ldots(24).$$

The first term on the right in (24) is entirely imaginary; it therefore follows by (22) that $\frac{1}{2}\pi J_0(z)$ is the real part of the second term. By expanding the binomial under the integral sign, and afterwards integrating by the formula

$$\int_0^\infty e^{-\beta} \beta^{q-\frac{1}{2}} d\beta = \Gamma(q + \tfrac{1}{2}),$$

we obtain as the expansion for $J_0(z)$ in negative powers of z,

$$J_0(z) = \sqrt{\left(\frac{2}{\pi z}\right)} \left\{ 1 - \frac{1^2 \cdot 3^2}{1 \cdot 2 \cdot (8z)^2} + \cdots \right\} \cos(z - \tfrac{1}{4}\pi)$$

$$+ \sqrt{\left(\frac{2}{\pi z}\right)} \left\{ \frac{1^2}{1 \cdot 8z} - \frac{1^2 \cdot 3^2 \cdot 5^2}{1 \cdot 2 \cdot 3 \cdot (8z)^3} + \cdots \right\} \sin(z - \tfrac{1}{4}\pi) \ldots\ldots(25).$$

---

[1] Crelle, Bd. LVI. 1859. Lömmel, *Studien über die Bessel'schen Functionen*, p. 59.

By stopping the expansion after any desired number of terms, and forming the expression for the remainder, it may be proved that the error committed by neglecting the remainder cannot exceed the last term retained (§ 200).

In like manner the imaginary part of the right-hand member of (24) is the equivalent of $-\frac{1}{2}i\pi K(z)$, so that

$$K(z) = \frac{2}{\pi}\left\{z^{-1} - z^{-3} + 1^2.3^2.z^{-5} - 1^2.3^2.5^2.z^{-7} + \ldots\ldots\right\}$$
$$+ \sqrt{\left(\frac{2}{\pi z}\right)}\left\{1 - \frac{1^2.3^2}{1.2.(8z)^2} + \frac{1^2.3^2.5^2.7^2}{1.2.3.4.(8z)^4} - \ldots\ldots\right\}\sin\left(z - \tfrac{1}{4}\pi\right)$$
$$- \sqrt{\left(\frac{2}{\pi z}\right)}\left\{\frac{1^2}{1.8z} - \frac{1^2.3^2.5^2}{1.2.3.(8z)^3} + \ldots\ldots\right\}\cos\left(z - \tfrac{1}{4}\pi\right)\ldots\ldots(26).$$

The value of $K_1(z)$ may now be determined by means of (17). We find

$$\frac{dK}{dz} = -\frac{2}{\pi}\left\{z^{-2} - 3.z^{-4} + 1^2.3^2.5.z^{-6} - 1^2.3^2.5^2.7.z^{-8} + \ldots\ldots\right\}$$
$$+ \sqrt{\left(\frac{2}{\pi z}\right)}\cos\left(z - \tfrac{1}{4}\pi\right)\left\{1 + \frac{3.5.1}{1.2.(8z)^2} - \frac{3.5.7.9.1.3.5}{1.2.3.4.(8z)^4} + \ldots\ldots\right\}$$
$$- \sqrt{\left(\frac{2}{\pi z}\right)}\sin\left(z - \tfrac{1}{4}\pi\right)\left\{\frac{3}{1.(8z)} - \frac{3.5.7.1.3}{1.2.3.(8z)^3}\right.$$
$$\left. + \frac{3.5.7.9.11.1.3.5.7}{1.2.3.4.5.(8z)^5} - \ldots\ldots\right\}\ldots\ldots(27).$$

The final expression for $K_1(z)$ may be put into the form

$$K_1(z) = \frac{2}{\pi}\left\{z + z^{-1} - 3.z^{-3} + 1^2.3^2.5.z^{-5} - 1^2.3^2.5^2.7.z^{-7} + \ldots\ldots\right\}$$
$$- \sqrt{\frac{2z}{\pi}}.\cos\left(z - \tfrac{1}{4}\pi\right)\left\{1 - \frac{(1^2-4)(3^2-4)}{1.2.(8z)^2}\right.$$
$$\left. + \frac{(1^2-4)(3^2-4)(5^2-4)(7^2-4)}{1.2.3.4.(8z)^4} - \ldots\ldots\right\}$$
$$- \sqrt{\frac{2z}{\pi}}.\sin\left(z - \tfrac{1}{4}\pi\right)\left\{\frac{1^2-4}{1.8z} - \frac{(1^2-4)(3^2-4)(5^2-4)}{1.2.3.(8z)^3} + \ldots\ldots\right\}^{1}\ldots(28).$$

It appears then that K_1 does not vanish when z is great, but approximates to $2z/\pi$. But although the accession to the inertia,

----

[1] As was to be expected, the series within brackets are the same as those that occur in the expression of the function $J_1(z)$.

which is proportional to K_1, becomes infinite with R, it vanishes ultimately when compared with the area of the disc, and with the other term which represents the dissipation. And this agrees with what we should anticipate from the theory of plane waves.

If, independently of the reaction of the air, the mass of the plate be M, and the force of restitution be $\mu\xi$, the equation of motion of the plate when acted on by an impressed force F, proportional to e^{ikat}, will be

$$\left\{M + \frac{\pi\sigma}{2k^3} K_1(2kR)\right\}\ddot{\xi} + a\sigma\pi R^2\left(1 - \frac{J_1(2kR)}{kR}\right)\dot{\xi} + \mu\xi = F \dots(29);$$

or by (13), if, as will be usual in practical applications, kR be small,

$$\left(M + \frac{8\sigma R^3}{3}\right)\ddot{\xi} + \frac{a\sigma\pi k^2 R^4}{2}\dot{\xi} + \mu\xi = F \dots\dots\dots(30).$$

Two particular cases of this problem deserve notice. First let M and μ vanish, so that the plate, itself devoid of mass, is subject to no other forces than F and those arising from aerial pressures. Since $\ddot{\xi} = ika\dot{\xi}$, the frictional term is relatively negligible, and we get when kR is very small,

$$a\sigma\pi R^2 . \frac{8kR}{3\pi}\dot{\xi} = -iF \dots\dots\dots\dots\dots(31).$$

Next let M and μ be such that the natural period of the plate, when subject to the reaction of the air, is the same as that imposed upon it. Under these circumstances

$$\left(M + \frac{8\sigma R^3}{3}\right)\ddot{\xi} + \mu\xi = 0,$$

and therefore

$$a\sigma\pi R^2 . \frac{k^2 R^2}{2} . \dot{\xi} = F \dots\dots\dots\dots(32).$$

Comparing with (31), we see that the amplitude of vibration is greater in the case when the inertia of the air is balanced, in the ratio of $16 : 3\pi kR$, shewing a large increase when kR is small. In the first case the phase of the motion is such that comparatively very little work is done by the force F; while in the second, the inertia of the air is compensated by the spring, and then F, being of the same phase as the velocity, does the maximum amount of work.

CHAPTER XVI.

303. In the pipe closed at one end and open at the other we had an example of a mass of air endowed with the property of vibrating in certain definite periods peculiar to itself in more or less complete independence of the external atmosphere. If the air beyond the open end were entirely without mass, the motion within the pipe would have no tendency to escape, and the contained column of air would behave like any other complex system not subject to dissipation. In actual experiment the inertia of the external air cannot, of course, be got rid of, but when the diameter of the pipe is small, the effect produced in the course of a few periods may be insignificant, and then vibrations once excited in the pipe have a certain degree of persistence. The narrower the channel of communication between the interior of a vessel and the external medium, the greater does the independence become. Such cavities constitute resonators; in the presence of an external source of sound, the contained air vibrates in unison, and with an amplitude dependent upon the relative magnitudes of the natural and forced periods, rising to great intensity in the case of approximate isochronism. When the original cause of sound ceases, the resonator yields back the vibrations stored up as it were within it, thus becoming itself for a short time a secondary source. The theory of resonators constitutes an important branch of our subject.

As an introduction to it we may take the simple case of a stopped cylinder, in which a piston moves without friction. On the further side of the piston the air is supposed to be devoid of inertia, so that the pressure is absolutely constant. If now the piston be set into vibration of very long period, it is clear that the contained air will be at any time very nearly in the equilibrium condition (of uniform density) corresponding to the momentary position of the piston. If the mass of the piston be very considerable in comparison with that of the included air, the natural vibrations resulting from a displacement will occur nearly as if the air had no inertia; and in deriving the period from the kinetic and potential energies, the former may be calculated without allowance for the inertia of the air, and the latter as if the rarefaction and condensation were uniform. Under the circumstances contemplated the air acts merely as a spring in virtue of its resistance to compression or dilatation; the form of the containing vessel is therefore immaterial, and the period of vibration remains the same, provided the capacity be not varied.

When a gas is compressed or rarefied, the mechanical value of the resulting displacement is found by multiplying each infinitesimal increment of volume by the corresponding pressure and integrating over the range required. In the present case it is of course only the difference of pressure on the two sides of the piston which is really operative, and this for a small change is proportional to the alteration of volume. The whole mechanical value of the small change is the same as if the expansion were opposed throughout by the *mean*, that is half the final, pressure; thus corresponding to a change of volume from S to $S + \delta S$, since $p = a^2 \rho$,

$$V = p \cdot \frac{\delta S}{2S} \cdot \delta S = \tfrac{1}{2} \rho a^2 \frac{(\delta S)^2}{S} \quad \ldots\ldots\ldots\ldots\ldots(1)[1].$$

If A denote the area of the piston, M its mass, and x its linear displacement, $\delta S = Ax$, and the equation of motion is

$$M\ddot{x} + \frac{\rho a^2 A^2}{S} x = 0 \ldots\ldots\ldots\ldots\ldots\ldots(2),$$

indicating vibrations, whose periodic time is

$$\tau = 2\pi \div aA \sqrt{\frac{\rho}{MS}} \quad \ldots\ldots\ldots\ldots\ldots(3).$$

Let us now imagine a vessel containing air, whose interior communicates with the external atmosphere by a narrow aperture or neck. It is not difficult to see that this system is capable of vibrations similar to those just considered, the air in the neighbourhood of the aperture supplying the place of the piston. By sufficiently increasing S, the period of the vibration may be made as long as we please, and we obtain finally a state of things in

---

[1] Compare (12) § 245.

which the kinetic energy of the motion may be neglected except in the neighbourhood of the aperture, and the potential energy may be calculated as if the density in the interior of the vessel were uniform. In flowing through the aperture under the operation of a difference of pressure on the two sides, or in virtue of its own inertia after such pressure has ceased, the air moves approximately as an incompressible fluid would do under like circumstances, provided that the space through which the kinetic energy is sensible be very small in comparison with the length of the wave. The suppositions on which we are about to proceed are not of course strictly correct as applied to actual resonators such as are used in experiment, but they are near enough to the mark to afford an instructive view of the subject and in many cases a foundation for a sufficiently accurate calculation of the pitch. They become rigorous only in the limit when the wave-length is indefinitely great in comparison with the dimensions of the vessel.

[On the above principles we may at once calculate the pitch of a resonator of volume S, whose cavity communicates with the external air by a long cylindrical neck of length L and area A. The mass of the aerial piston is $\rho A L$; so that (3) gives as the period of vibration

$$\tau = \frac{2\pi}{a}\sqrt{\left(\frac{LS}{A}\right)} \quad \dots\dots\dots\dots\dots\dots(4);$$

or, if λ be the length of plane waves of the same pitch,

$$\lambda = a\tau = 2\pi\sqrt{\left(\frac{LS}{A}\right)} \quad \dots\dots\dots\dots\dots(5).$$

If the cross-section of the neck be a circle of radius R, $A = \pi R^2$, and we obtain the formula (8) of § 307.]

304. The kinetic energy of the motion of an incompressible fluid through a given channel may be expressed in terms of the density ρ, and the rate of transfer, or current, $\dot{X}$, for under the circumstances contemplated the character of the motion is always the same. Since T necessarily varies as ρ and as $\dot{X}^2$, we may put

$$T = \tfrac{1}{2}\rho\,\frac{\dot{X}^2}{c} \quad \dots\dots\dots\dots\dots\dots\dots (1),$$

where the constant c, which depends only on the nature of the channel, is a linear quantity, as may be inferred from the fact that the dimensions of $\dot{X}$ are 3 in space and -1 in time. In fact, if ϕ be the velocity-potential,

$$T = \tfrac{1}{2}\rho \iiint \left[\frac{d\phi^2}{dx^2} + \frac{d\phi^2}{dy^2} + \frac{d\phi^2}{dz^2} \right] dx\, dy\, dz = \tfrac{1}{2}\rho \iint \phi\, \frac{d\phi}{dn}\, dS,$$

by Green's theorem, where the integration is to be extended over a surface including the whole region through which the motion is sensible. At a sufficient distance on either side of the aperture, ϕ becomes constant, and if the constant values be denoted by ϕ_1 and ϕ_2, and the integration be now limited to that half of S towards which the fluid flows, we have

$$T = \tfrac{1}{2}\rho\, (\phi_1 - \phi_2) \iint \frac{d\phi}{dn}\, dS = \tfrac{1}{2}\rho\, (\phi_1 - \phi_2)\, \dot{X}.$$

Now, since within S ϕ is determined linearly by its surface values, $\iint \dfrac{d\phi}{dn}\, dS$, or X, is proportional to $(\phi_1 - \phi_2)$. If we put $\dot{X} = c\, (\phi_1 - \phi_2)$, we get as before $T = \tfrac{1}{2}\rho \dot{X}^2/c$.

The nature of the constant c will be better understood by considering the electrical problem, whose conditions are mathematically identical with those of that under discussion. Let us suppose that the fluid is replaced by uniformly conducting material, and that the boundary of the channel or aperture is replaced by insulators. We know that if by battery power or otherwise, a difference of electric potential be maintained on the two sides, a steady current through the aperture of proportional magnitude will be generated. The ratio of the

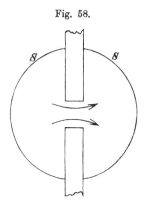

Fig. 58.

total current to the electromotive force is called the *conductivity* of the channel, and thus we see that our constant c represents simply this conductivity, on the supposition that the specific conducting power of the hypothetical substance is unity. The same thing may be otherwise expressed by saying that c is the side of the cube, whose resistance between opposite faces is the same as that of the channel. In the sequel we shall often avail ourselves of the electrical analogy.

When c is known, the proper tone of the resonator can be easily deduced. Since

$$V = \tfrac{1}{2}\rho a^2 \frac{X^2}{S}, \qquad T = \tfrac{1}{2}\rho \frac{\dot{X}^2}{c} \quad \dots\dots\dots\dots\dots \quad (2),$$

the equation of motion is

$$\ddot{X} + \frac{a^2 c}{S} X = 0 \dots\dots\dots\dots\dots\dots\dots(3),$$

indicating simple oscillations performed in a time

$$\tau = 2\pi \div \sqrt{\frac{a^2 c}{S}} \quad \dots\dots\dots\dots\dots\dots\dots(4).$$

If N be the frequency, or number of complete vibrations executed in the unit time,

$$N = \frac{a}{2\pi} \sqrt{\frac{c}{S}} \dots\dots\dots\dots\dots\dots\dots(5).$$

The wave-length λ, which is the quantity most closely connected with the dimensions of the cavity, is given by

$$\lambda = \frac{a}{N} = 2\pi \sqrt{\frac{S}{c}} \quad \dots\dots\dots\dots\dots\dots \quad (6),$$

and varies directly as the linear dimension. The wave-length, it will be observed, is a function of the size and shape of the resonator only, while the frequency depends also upon the nature of the gas; and it is important to remark that it is on the nature of the gas in and near the channel that the pitch depends and not on that occupying the interior of the vessel, for the inertia of the air in the latter situation does not come into play, while the compressibility of all gases is very approximately the same. Thus in the case of a pipe, the substitution of hydrogen for air in the neighbourhood of a node would make but little difference, but its effect in the neighbourhood of a loop would be considerable.

Hitherto we have spoken of the channel of communication as single, but if there be more than one channel, the problem is not essentially altered. The same formula for the frequency is still applicable, if as before we understand by c the whole conductivity between the interior and exterior of the vessel. When the channels are situated sufficiently far apart to act independently one of another, the resultant conductivity is the simple sum of those belonging to the separate channels; otherwise the resultant is less than that calculated by mere addition.

If there be two precisely similar channels, which do not interfere, and whose conductivity taken separately is c, we have

$$N = \sqrt{2} \times \frac{a}{2\pi} \sqrt{\frac{c}{S}} \quad\ldots\ldots\ldots\ldots\ldots\ldots (7),$$

shewing that the note is higher than if there were only one channel in the ratio $\sqrt{2} : 1$, or by rather less than a fifth—a law observed by Sondhauss and proved theoretically by Helmholtz in the case, where the channels of communication consist of simple holes in the infinitely thin sides of the reservoir.

305. The investigation of the conductivity for various kinds of channels is an important part of the theory of resonators; but in all except a very few cases the accurate solution of the problem is beyond the power of existing mathematics. Some general principles throwing light on the question may however be laid down, and in many cases of interest an approximate solution, sufficient for practical purposes, may be obtained.

We know (§§ 79, 242) that the energy of a fluid flowing through a channel cannot be greater than that of any fictitious motion giving the same total current. Hence, if the channel be narrowed in any way, or any obstruction be introduced, the conductivity is thereby diminished, because the alteration is of the nature of an additional constraint. Before the change the fluid was free to adopt the distribution of flow finally assumed. In cases where a rigorous solution cannot be obtained we may use the minimum property to estimate an inferior limit to the conductivity; the energy calculated from a hypothetical law of flow can never be less than the truth, and must exceed it unless the hypothetical and the actual motion coincide.

Another general principle, which is of frequent use, may be more conveniently stated in electrical language. The quantity with which we are concerned is the conductivity of a certain conductor composed of matter of unit specific conductivity. The principle is that if the conductivity of any part of the conductor be increased that of the whole is increased, and if the conductivity of any part be diminished that of the whole is diminished, exception being made of certain very particular cases, where no alteration ensues. In its passage through a conductor electricity distributes itself, so that the energy dissipated is for a given total current the least possible (§ 82). If now the specific resistance of any part be diminished, the total dissipation would be less than before, even if the distribution of currents remained unchanged. *A fortiori* will this be the case, when the currents redistribute themselves so as to make the dissipation a minimum. If an infinitely thin lamina of matter stretching across the channel be made perfectly conducting, the resistance of the whole will be diminished, unless the lamina coincide with one of the undisturbed equipotential surfaces. In the excepted case no effect will be produced.

306. Among different kinds of channels an important place must be assigned to those consisting of simple apertures in unlimited plane walls of infinitesimal thickness. In practical applications it is sufficient that a wall be very thin in proportion to the dimensions of the aperture, and approximately plane within a distance from the aperture large in proportion to the same quantity.

On account of the symmetry on the two sides of the wall, the motion of the fluid in the plane of the aperture must be normal, and therefore the velocity-potential must be constant; over the remainder of the plane the motion must be exclusively tangential, so that to determine ϕ on one side of the plane we have the conditions (i) $\phi = $ constant over the aperture, (ii) $d\phi/dn = 0$ over the rest of the plane of the wall, (iii) $\phi = $ constant at infinity.

Since we are concerned only with the differences of ϕ we may suppose that at infinity ϕ vanishes. It will be seen that conditions (ii) and (iii) are satisfied by supposing ϕ to be the potential of attracting matter distributed over the aperture; the remainder of the problem consists in determining the distribution of matter so that its potential may be constant over the same area. The problem is mathematically the same as that of determining the distribution of electricity on a charged conducting plate situated in an open space, whose form is that of the aperture under consideration, and the conductivity of the aperture may be expressed in terms of the *capacity* of the plate of the statical problem. If ϕ denote the constant potential in the aperture, the electrical resistance (for one side only) will be

$$\phi_1 \div \iint \frac{d\phi}{dn} d\sigma,$$

the integration extending over the area of the opening.

Now $\iint \dfrac{d\phi}{dn}\, d\sigma = 2\pi \times$ (whole quantity of matter distributed),

and thus, if M be the capacity, or charge corresponding to unit-potential, the total resistance is $(\pi M)^{-1}$. Accordingly for the conductivity, which is the reciprocal of the resistance,

$$c = \pi M \dots\dots\dots\dots\dots\dots\dots(1).$$

So far as I am aware, the ellipse is the only form of aperture for which c or M can be determined theoretically [1], in which case the result is included in the known solution of the problem of determining the distribution of charge on an ellipsoidal conductor. From the fact that a shell bounded by two concentric, similar and similarly situated ellipsoids exerts no force on an internal particle, it is easy to see that the superficial density at any point of an ellipsoid necessary to give a constant potential is proportional to the perpendicular (p) let fall from the centre upon the tangent plane at the point in question. Thus if ρ be the density, $\rho = \kappa p$; the whole quantity of matter Q is given by

$$Q = \iint \rho\, dS = \kappa \iint p\, dS = 4\pi\kappa abc \dots\dots\dots\dots(2),[2]$$

so that
$$\rho = \frac{Q p}{4\pi\, abc} \dots\dots\dots\dots\dots\dots\dots(3).$$

In the usual notation

$$p = 1 \div \sqrt{\frac{x^2}{a^4} + \frac{y^2}{b^4} + \frac{z^2}{c^4}},$$

or, since $\qquad z^2/c^2 = 1 - x^2/a^2 - y^2/b^2,$

$$p = c \div \sqrt{1 - \frac{x^2}{a^2} - \frac{y^2}{b^2} + \frac{c^2 x^2}{a^4} + \frac{c^2 y^2}{b^4}}.$$

If we now suppose that c is infinitely small, we obtain the particular case of an elliptic plate, and if we no longer distinguish between the two surfaces, we get

$$\rho = \frac{Q}{2\pi ab} \div \sqrt{1 - \frac{x^2}{a^2} - \frac{y^2}{b^2}}\dots\dots\dots\dots\dots(4).$$

---

[1] The case of a resonator with an elliptic aperture was considered by Helmholtz (Crelle, Bd. 57, 1860), whose result is equivalent to (8).

[2] $2c$ being for the moment the third principal axis of the ellipsoid.

We have next to find the value of the constant potential (P). By considering the value of P at the centre of the plate, we see that

$$P = \iint \frac{\rho \, dS}{r} = \iint \rho \, dr \, d\theta.$$

Integrating first with respect to r, we have

$$\int_0^r \rho \, dr = Q \div 4a \sqrt{(1 - e^2 \cos^2 \theta)},$$

e being the eccentricity; and thus

$$P = \frac{Q}{a} \int_0^{\frac{1}{2}\pi} \frac{d\theta}{\sqrt{(1 - e^2 \cos^2 \theta)}} = \frac{Q}{a} F(e),$$

where F is the symbol of the complete elliptic function of the first order. Putting $P = 1$, we find

$$\frac{c}{\pi} = M = \frac{a}{F(e)} \quad \dots\dots\dots\dots\dots\dots(5),$$

as the final expression for the capacity of an ellipse, whose semi-major axis is a and eccentricity is e. In the particular case of the circle, $e = 0$, $F(e) = \frac{1}{2}\pi$, and thus for a circle of radius R,

$$c = 2R \dots\dots\dots\dots\dots\dots\dots\dots(6).$$

If the capacity of the resonator be S, we find from (6) § 304

$$\lambda = \pi \sqrt{\left(\frac{2S}{R}\right)} \quad \dots\dots\dots\dots\dots\dots\dots (7).$$

The area of the ellipse (σ) is given by

$$\sigma = \pi a^2 \sqrt{(1 - e^2)},$$

and hence in terms of σ

$$\frac{1}{c} = \frac{1}{2} \sqrt{\left(\frac{\pi}{\sigma}\right)} \cdot \frac{2F(e) (1 - e^2)^{\frac{1}{4}}}{\pi} \quad \dots\dots\dots\dots(8).$$

When e is small, we obtain by expanding in powers of e previous to integration,

$$F(e) = \frac{1}{2}\pi \left\{ 1 + \frac{1^2}{2^2} e^2 + \frac{1^2 \cdot 3^2}{2^2 \cdot 4^2} e^4 + \frac{1^2 \cdot 3^2 \cdot 5^2}{2^2 \cdot 4^2 \cdot 6^2} e^6 + \dots \right\} \dots(9),$$

whence

$$\frac{2F(e) (1 - e^2)^{\frac{1}{4}}}{\pi} = 1 - \frac{e^4}{64} - \frac{e^6}{64} + \dots$$

Neglecting e^8 and higher powers, we have therefore

$$c = 2\sqrt{\left(\frac{\sigma}{\pi}\right)} \cdot \left(1 + \frac{e^4}{64} + \frac{e^6}{64} + \ldots \right) \ldots \ldots \ldots (10).$$

From this result we see that, if its eccentricity be small, the conductivity of an elliptic aperture is very nearly the same as that of a circular aperture *of equal area*. Among various forms of aperture of given area there must be one which has a minimum conductivity, and, though a formal proof might be difficult, it is easy to recognise that this can be no other than the circle. An inferior limit to the value of c is thus always afforded by the conductivity of the circle of equal area, that is $2\sqrt{(\sigma/\pi)}$, and when the true form is nearly circular, this limit may be taken as a close approximation to the real value.

The value of λ is then given by

$$\lambda = 2^{\frac{1}{2}} \pi^{\frac{5}{4}} \sigma^{-\frac{1}{4}} S^{\frac{1}{2}} \ldots \ldots \ldots \ldots \ldots (11).$$

In order to shew how slightly a moderate eccentricity affects the value of c, I have calculated the following short table with the aid of Legendre's values of $F(e)$. Putting $e = \sin\psi$, we have $\cos\psi$ as the ratio of axes, and for the conductivity

$$c = 2\sqrt{\left(\frac{\sigma}{\pi}\right)} \cdot \frac{\pi}{2\sqrt{(\cos\psi)} \cdot F(\sin\psi)}.$$

ψ.	$e = \sin\psi$.	$b : a = \cos\psi$.	$\pi \div 2F(e)(1-e^2)^{\frac{1}{4}}$.
0°	·00000	1·00000	1·0000
20°	·34204	·93969	1·0002
30°	·50000	·86603	1·0013
40°	·64279	·76604	1·0044
50°	·76604	·64279	1·0122
60°	·86603	·50000	1·0301
70°	·93969	·34202	1·0724
80°	·98481	·17365	1·1954
90°	1·00000	·00000	∞

The value of the last factor given in the fourth column is the ratio of the conductivity of the ellipse to *that of a circle of equal area*. It appears that even when the ellipse is so eccentric that the ratio of the axes is $2:1$, the conductivity is increased by only about 3 per cent., which would correspond to an alteration of little more than a comma (§ 18) in the pitch of a resonator. There seems no reason to suppose that this approximate independence of shape is a property peculiar to the ellipse, and we may conclude with some confidence that in the case of any moderately elongated oval aperture, the conductivity may be calculated from the area alone with a considerable degree of accuracy.

If the area be given, there is no superior limit to c. For suppose the area σ to be distributed over n equal circles sufficiently far apart to act independently. The area of each circle is σ/n, and its conductivity is $2(n\pi)^{-\frac{1}{2}}\sigma^{\frac{1}{2}}$. The whole conductivity is n times as great, and therefore increases indefinitely with n. As a general rule, the more the opening is elongated or broken up, the greater will be the conductivity for a given area.

To find a superior limit to the conductivity of a given aperture we may avail ourselves of the principle that any addition to the aperture must be attended by an increase in the value of c. Thus in the case of a square, we may be sure that c is less than for the circumscribed circle, and we have already seen that it is greater than for the circle of equal area. If b be the side of the square,

$$\frac{2b}{\sqrt{\pi}} < c < \sqrt{2}\,b.$$

The tones of a resonator with a square aperture calculated from these two limits would differ by about a whole tone; the graver of them would doubtless be much the nearer to the truth. This example shews that even when analysis fails to give a solution in the mathematical sense, we need not be altogether in the dark as to the magnitudes of the quantities with which we are dealing.

In the case of similar orifices, or systems of orifices, c varies as the linear dimension.

307. Most resonators used in practice have necks of greater or less length, and even when there is nothing that would be called a neck, the thickness of the side of the reservoir cannot always be neglected. We shall therefore examine the conductivity of a channel formed by a cylindrical boring through an obstructing plate bounded by parallel planes, and, though we fail to solve the problem rigorously, we shall obtain information sufficient for most practical purposes. The thickness of the plate we shall call L, and the radius of the cylindrical channel R.

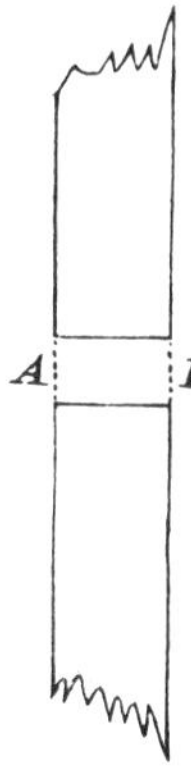

Whatever the resistance of the channel may be, it will be lessened by the introduction of infinitely thin discs of perfect conductivity at A and B, fig. 59. The effect of the discs is to produce constant potential over their areas, and the problem thus modified is susceptible of rigorous solution. Outside A and B the motion is the same as that previously investigated, when the obstructing plate is infinitely thin; between A and B the flow is uniform. The resistance is therefore on the whole

Fig. 59.

$$\frac{1}{2R} + \frac{L}{\pi R^2},$$

whence
$$c = \frac{\pi R^2}{L + \frac{1}{2}\pi R} \dots\dots\dots\dots\dots\dots(1).$$

If α denote the correction, which must be added to L on account of an open end,

$$\alpha = \tfrac{1}{4}\pi R \dots\dots\dots\dots\dots\dots (2).$$

This correction is in general under the mark, but, when L is very small in comparison with R, the assumed motion coincides more and more nearly with the actual motion, and thus the value of α in (2) tends to become correct.

A superior limit to the resistance may be calculated from a hypothetical motion of the fluid. For this purpose we will suppose infinitely thin pistons introduced at A and B, the effect of which will be to make the normal velocity constant at those places. Within the tube the flow will be uniform as before, but for the external space we have a new problem to consider:—To determine the motion of a fluid bounded by an infinite plane, the normal velocity over a circular area of the plane having a given constant value, and over the remainder of the plane being zero.

The potential may still be regarded as due to matter distributed over the disc, but it is no longer constant over the area; the *density* of the matter, however, being proportional to $d\phi/dn$ is constant.

The kinetic energy of the motion

$$= \tfrac{1}{2} \iint \phi \frac{d\phi}{dn} d\sigma = \tfrac{1}{2} \frac{d\phi}{dn} \iint \phi \, d\sigma,$$

the integration going over the area of the circle.

The total current through the plane

$$= \iint \frac{d\phi}{dn} \, d\sigma = \pi R^2 \frac{d\phi}{dn} .$$

Hence
$$\frac{2 \text{ kinetic energy}}{(\text{current})^2} = \frac{\iint \phi \, d\sigma}{\pi^2 R^4 \dfrac{d\phi}{dn}} .$$

If the density of the matter be taken as unity, $d\phi/dn = 2\pi$, and the required ratio is expressed by $P/\pi^3 R^4$, where P denotes the potential on itself of a circular layer of matter of unit density and of radius R.

The simplest method of calculating P depends upon the consideration that it represents the work required to break up the disc into infinitesimal elements and to remove them from each other's influence[1]. If we take polar co-ordinates (ρ, θ), the pole being at the edge of the disc whose radius is a, we have for the potential at the pole, $V = \iint d\theta d\rho$, the limits of ρ being 0 and $2a \cos \theta$, and those of θ being $-\frac{1}{2}\pi$ and $+\frac{1}{2}\pi$.

Thus
$$V = 4a \dots\dots\dots\dots\dots\dots\dots (3).$$

Now let us cut off a strip of breadth da from the edge of the disc. The work required to remove this to an infinite distance is $2\pi a\, da \cdot 4a$. If we gradually pare the disc down to nothing and carry all the parings to infinity[2], we find for the total work by integrating with respect to a from 0 to R,

$$P = \frac{8\pi R^3}{3} .$$

The limit to the resistance (for one side) is thus $8/3\pi^2 R$; we conclude that the resistance of the whole channel is less than

$$\frac{L}{\pi R^2} + \frac{16}{3\pi^2 R} \dots\dots\dots\dots\dots\dots(4).$$

Collecting our results, we see that

$$\frac{\pi}{4} R < \alpha < \frac{8}{3\pi} R \dots\dots\dots\dots\dots (5),$$

---

[1] A part of § 302 is repeated here for the sake of those who may wish to avoid the difficulties of the more complete investigation.

[2] This method of calculating P was suggested to the author by Professor Clerk Maxwell.

or in decimals,

$$\left. \begin{matrix} \alpha > \cdot785\ R \\ \alpha < \cdot849\ R \end{matrix} \right\} \quad \dots\dots\dots\dots\dots\dots\dots (6).$$

It must be observed that α here denotes the correction for one end. The whole resistance corresponds to a length $L + 2\alpha$ of tube having the section πR^2.

When L is very great in relation to R, we may take simply

$$c = \frac{\pi R^2}{L} \quad \dots\dots\dots\dots\dots\dots\dots (7).$$

In this case we have from (6) § 304

$$\lambda = \frac{2\sqrt{\pi} \cdot \sqrt{(SL)}}{R} \quad \dots\dots\dots\dots\dots\dots(8).$$

The correction for an open end (α) is a function of L, coinciding with the lower limit, viz. $\frac{1}{4}\pi R$, when L vanishes. As L increases, α increases with it; but does not, even when L is infinite, attain the superior limit $8R/3\pi$. For consider the motion going on in any middle piece of the tube. The kinetic energy is greater than corresponds merely to the length of the piece. If therefore the piece be removed, and the free ends brought together, the motion otherwise continuing as before, the kinetic energy will be diminished more than corresponds to the length of the piece subtracted. A *fortiori* will this be true of the real motion which would exist in the shortened tube. That, when $L = \infty$, α does not become $8R/3\pi$ is evident, because the normal velocity at the end, far from being constant, as was assumed in the calculation of this result, must increase from the centre outwards and become infinite at the edge.

A further approximation to the value of α may be obtained by assuming a variable velocity at the plane of the mouth. The calculation will be found in Appendix A. It appears that in the case of an infinitely long tube α cannot be so great as $\cdot82422\ R$. The real value of α is probably not far from $\cdot82\ R$.

308. Besides the cylinder there are very few forms of channel whose conductivity can be determined mathematically. When however the form is approximately cylindrical we may obtain limits, which are useful as allowing us to estimate the effect of such departures from mathematical accuracy as must occur in practice.

An inferior limit to the resistance of any elongated and approximately straight conductor may be obtained immediately by the imaginary introduction of an infinite number of plane perfectly conducting layers perpendicular to the axis. If σ denote the area of the section at any point x, the resistance between two layers distant dx will be $\sigma^{-1} dx$, and therefore the whole actual resistance is certainly greater than

$$\int \sigma^{-1} dx \dots\dots\dots\dots\dots (1),$$

unless indeed the conductor be truly cylindrical.

In order to find a superior limit we may calculate the kinetic energy of the current on the hypothesis that the velocity parallel to the axis is uniform over each section. The hypothetical motion is that which would follow from the introduction of an infinite number of rigid pistons moving freely, and the calculated result is necessarily in excess of the truth, unless the section be absolutely constant. We shall suppose for the sake of simplicity that the channel is symmetrical about an axis, in which case of course the motion of the fluid is symmetrical also.

If U denote the total current, we have *ex hypothesi* for the axial velocity at any point x

$$u = \sigma^{-1} U \dots\dots\dots\dots\dots (2),$$

from which the radial velocity v is determined by the equation of continuity (6 § 238),

$$\frac{d (ru)}{dx} + \frac{d (rv)}{dr} = 0.$$

Thus　　　　　　$rv = \text{const.} - \tfrac{1}{2} U r^2 \dfrac{d\sigma^{-1}}{dx},$

or, since there is no source of fluid on the axis,

$$v = - \tfrac{1}{2} U r \frac{d\sigma^{-1}}{dx} \dots\dots\dots\dots\dots (3).$$

The kinetic energy may now be calculated by simple integration :—

$$\int u^2 \sigma \, dx = U'^2 \int \sigma^{-1} dx,$$

$$\iint v^2 2\pi r \, dr \, dx = \frac{\pi U'^2}{8} \int y^4 \left(\frac{d\sigma^{-1}}{dx}\right)^2 dx,$$

if y be the radius of the channel at the point x, so that $\sigma = \pi y^2$.

Thus $\dfrac{2 \text{ kinetic energy}}{(\text{current})^2} = \int \dfrac{1}{\pi y^2} \left\{1 + \tfrac{1}{2} \left(\dfrac{dy}{dx}\right)^2\right\} dx\ldots\ldots\ldots(4).$

This is the quantity which gives a superior limit to the resistance. The first term, which corresponds to the component velocity u, is the same as that previously obtained for the lower limit, as might have been foreseen. The difference between the two, which gives the utmost error involved in taking either of them as the true value, is

$$\frac{1}{2\pi} \int \frac{1}{y^2} \left(\frac{dy}{dx}\right)^2 dx \ldots\ldots\ldots\ldots\ldots\ldots(5).$$

In a nearly cylindrical channel dy/dx is a small quantity and so the result found in this manner is closely approximate. It is not necessary that the section should be nearly constant, but only that it should vary slowly. The success of the approximation in this and similar cases depends upon the fact that the quantity to be estimated is at a minimum. Any reasonable approximation to the real motion will give a result very near the truth according to the principles of the differential calculus.

By means of the properties of the potential and stream functions (§ 238) the present problem admits of actual approximate solution. If ϕ and ψ denote the values of these functions at any point x, r; u, v denote the axial and transverse velocities,

$$u = \frac{d\phi}{dx} = \frac{1}{r}\frac{d\psi}{dr}, \qquad v = \frac{d\phi}{dr} = -\frac{1}{r}\frac{d\psi}{dx} \ldots\ldots\ldots\ldots(6),$$

whence by elimination

$$\frac{d^2\phi}{dr^2} + \frac{1}{r}\frac{d\phi}{dr} + \frac{d^2\phi}{dx^2} = 0 \ldots\ldots\ldots\ldots\ldots (7),$$

$$\frac{d^2\psi}{dr^2} - \frac{1}{r}\frac{d\psi}{dr} + \frac{d^2\psi}{dx^2} = 0 \ldots\ldots\ldots\ldots\ldots (8).$$

If F denote the value of ϕ as a function of x when $r = 0$, the general values of ϕ and ψ may be expressed in terms of F by means of (7) and (8) in the series

$$\left.\begin{array}{l} \phi = F - \dfrac{r^2 F''}{2^2} + \dfrac{r^4 F^{\mathrm{iv}}}{2^2 \cdot 4^2} - \dfrac{r^6 F^{\mathrm{vi}}}{2^2 \cdot 4^2 \cdot 6^2} + \cdots \\[2mm] \psi = \dfrac{r^2 F'}{2} - \dfrac{r^4 F'''}{2^2 \cdot 4} + \dfrac{r^6 F^{\mathrm{v}}}{2^2 \cdot 4^2 \cdot 6} - \dfrac{r^8 F^{\mathrm{vii}}}{2^2 \cdot 4^2 \cdot 6^2 \cdot 8} + \cdots \end{array}\right\} \quad \ldots\ldots\ldots (9),$$

where accents denote differentiation with respect to x. At the boundary of the channel where $r = y$, ψ is constant, say ψ_1. Then

$$\psi_1 = \frac{y^2 F'}{2} - \frac{y^4 F'''}{2^2 \cdot 4} + \frac{y^6 F^{\mathrm{v}}}{2^2 \cdot 4^2 \cdot 6} - \ldots\ldots\ldots\ldots (10)$$

is the equation connecting y and F. In the present problem y is given, and we have to express F by means of it. By successive approximation we obtain from (10)

$$F' = \frac{2\psi_1}{y^2} + \frac{y^2}{8} \left\{ \frac{d^2}{dx^2}\left(\frac{2\psi_1}{y^2}\right) + \frac{1}{8} \frac{d^2}{dx^2} y^2 \frac{d^2}{dx^2}\left(\frac{2\psi_1}{y^2}\right) \right\}$$

$$- \frac{y^4}{12 \cdot 4^2} \frac{d^4}{dx^4}\left(\frac{2\psi_1}{y^2}\right)\ldots\ldots\ldots\ldots\ldots (11).$$

The total stream is given by the integral

$$\int_0^y \frac{d\phi}{dx} 2\pi r \, dr = \int_0^y \frac{1}{r} \frac{d\psi}{dr} 2\pi r \, dr = 2\pi\psi_1;$$

and therefore the resistance between any two equipotential surfaces is represented by

$$\frac{1}{2\pi\psi_1} \int F' dx.$$

The expression for the resistance admits of considerable simplification by integration by parts in the case when the channel is truly cylindrical in the neighbourhood of the limits of integration. In this way we find for the final result,

$$\text{resistance} = \int \frac{dx}{\pi y^2} \left\{ 1 + \tfrac{1}{2} y'^2 - \frac{(3y'^2 - yy'')^2}{48} \right\}\ldots\ldots (12)[1],$$

y', y'' denoting the differential coefficients of y with respect to x.

It thus appears that the superior limit of the preceding investigation is in fact the correct result to the second order o

[1] *Proceedings of the London Mathematical Society*, Vol. VII. p. 70, 1876.

approximation. If we regard y as a function of ωx, where ω is a small quantity, (12) is correct as far as terms containing ω^4.

309. Our knowledge of the laws on which the pitch of resonators depends, is due to the labours of several experimenters and mathematicians.

The observation that for a given mouthpiece the pitch of a resonator depends mainly upon the volume S is due to Liscovius, who found that the pitch of a flask partly filled with water was not altered when the flask was inclined. This result was confirmed by Sondhauss[1]. The latter observer found further, that in the case of resonators without necks, the influence of the aperture depended mainly upon its area, although when the shape was very elongated, a certain rise of pitch ensued. He gave the formula

$$N = 52400 \frac{\sigma^{\frac{1}{2}}}{S^{\frac{1}{2}}} \dots\dots\dots\dots\dots\dots(1),$$

the unit of length being the millimetre.

The *theory* of this kind of resonator we owe to Helmholtz[2], whose formula is

$$N = \frac{a\sigma^{\frac{1}{2}}}{2^{\frac{1}{2}} \pi^{\frac{3}{4}} S^{\frac{1}{2}}} \dots\dots\dots\dots\dots\dots(2),$$

applicable to circular apertures.

For flasks with long necks, Sondhauss[3] found

$$N = 46705 \frac{\sigma^{\frac{1}{2}}}{L^{\frac{1}{2}} S^{\frac{1}{2}}} \dots\dots\dots\dots\dots\dots(3),$$

corresponding to the theoretical

$$N = \frac{a}{2\pi} \frac{\sigma^{\frac{1}{2}}}{L^{\frac{1}{2}} S^{\frac{1}{2}}} \dots\dots\dots\dots\dots\dots(4).$$

In practice it does not often happen either that the neck is so long that the correction for the open ends can be neglected, as (4) supposes, or, on the other hand, so short that it can itself be neglected, as supposed in (2). Wertheim[4] was the first

[1] Ueber den Brummkreisel und das Schwingungsgesetz der cubischen Pfeifen. *Pogg. Ann.* LXXXI. pp. 235, 347. 1850.

[2] Crelle, Bd. LVII. 1—72. 1860.

[3] Ueber die Schallschwingungen der Luft in erhitzten Glasröhren und in gedeckten Pfeifen von ungleicher Weite. *Pogg. Ann.* LXXIX. p. 1. 1850.

[4] Mémoire sur les vibrations sonores de l'air. *Ann. d. Chim.* (3) XXXI. p. 385. 1851.

to shew that the effect of an open end could be represented by an addition (α) to the length, independent, or nearly so, of L and λ.

The approximate theoretical determination of α is due to Helmholtz, who gave $\frac{1}{4}\pi R$ as the correction for an open end fitted with an infinite flange. His method consisted in inventing forms of tube for which the problem was soluble, and selecting that one which agreed most nearly with a cylinder. The correction $\frac{1}{4}\pi R$ is rigorously applicable to a tube whose radius at the open end and at a great distance from it is R, but which in the neighbourhood of the open end bulges slightly.

From the fact that the true cylinder may be derived by introducing an obstruction, we may infer that the result thus obtained is too small.

It is curious that the process followed in this work, which was first given in the memoir on resonance, leads to exactly the same result, though it would be difficult to conceive two methods more unlike each other.

The correction to the length will depend to some extent upon whether the flow of air from the open end is obstructed, or not. When the neck projects into open space, there will be less obstruction than when a backward flow is prevented by a flange as supposed in our approximate calculations. However, the uncertainty introduced in this way is not very important, and we may generally take $\alpha = \frac{1}{4}\pi R$ as a sufficient approximation. In practice, when the necks are short, the hypothesis of the flange agrees pretty well with fact, and when the necks are long, the correction is itself of subordinate importance.

The general formula will then run

$$N = \frac{a}{2\pi} \sqrt{\frac{\sigma}{S\{L + \frac{1}{2}\sqrt{(\pi\sigma)}\}}} \quad \ldots\ldots\ldots\ldots\ldots(5),$$

where σ is the area of the section of the neck, or in numbers

$$N = \frac{a}{6\cdot2832}\frac{\sigma^{\frac{1}{2}}}{S^{\frac{1}{2}}\sqrt{(L + \cdot8863\sqrt{\sigma})}} \quad \ldots\ldots\ldots\ldots\ldots(6).$$

A formula not differing much from this was given, as the embodiment of the results of his measurements, by Sondhauss[1] who

---

[1] *Pogg. Ann.* cxl. pp. 53, 219.   1870.

at the same time expressed a conviction that it was no mere empirical formula of interpolation, but the expression of a natural law. The theory of resonators with necks was given about the same time[1] in a memoir 'on Resonance' published in the *Philosophical Transactions* for 1871, from which most of the last few pages is derived.

310. The simple method of calculating the pitch of resonators with which we have been occupied is applicable to the gravest mode of vibration only, the character of which is quite distinct. The overtones of resonators with contracted necks are relatively very high, and the corresponding modes of vibration are by no means independent of the inertia of the air in the interior of the reservoir. The character of these modes will be more evident, when we come to consider the vibrations of air within a completely closed vessel, such as a sphere, but it will rarely happen that the pitch can be calculated theoretically.

There are, however, cases of multiple resonance to which our theory is applicable. These occur when two or more vessels communicate by channels with each other and with the external air; and are readily treated by Lagrange's method, provided of course that the wave-length of the vibration is sufficiently large in comparison with the dimensions of the vessels.

Suppose that there are two reservoirs, S, S', communicating with each other and with the external air by narrow passages or

Fig. 60.

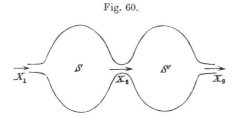

necks. If we were to consider SS' as a single reservoir and apply our previous formula, we should be led to an erroneous result; for that formula is founded on the assumption that within the reservoir the inertia of the air may be left out of account, whereas it is evident that the energy of the motion through the connecting passage may be as great as through the two others. However, an

---

[1] *Proceedings of the Royal Society*, Nov. 24, 1870.

investigation on the same general plan as before meets the case perfectly. Denoting by X_1, X_2, X_3 the total transfers of fluid through the three passages, we have as in (2) § 304 for the kinetic energy the expression

$$T = \tfrac{1}{2}\rho \left\{ \frac{\dot{X}_1{}^2}{c_1} + \frac{\dot{X}_2{}^2}{c_2} + \frac{\dot{X}_3{}^2}{c_3} \right\} \dots\dots\dots\dots\dots(1),$$

and for the potential energy,

$$V = \tfrac{1}{2}\rho a^2 \left\{ \frac{(X_2 - X_1)^2}{S} + \frac{(X_3 - X_2)^2}{S'} \right\} \dots\dots\dots\dots(2).$$

An application of Lagrange's method gives as the differential equations of motion,

$$\left. \begin{array}{c} \dfrac{\ddot{X}_1}{c_1} + a^2 \dfrac{X_1 - X_2}{S} = 0 \\[2mm] \dfrac{\ddot{X}_2}{c_2} + a^2 \left\{ \dfrac{X_2 - X_1}{S} + \dfrac{X_2 - X_3}{S'} \right\} = 0 \\[2mm] \dfrac{\ddot{X}_3}{c_3} + a^2 \dfrac{X_3 - X_1}{S'} = 0 \end{array} \right\} \dots\dots\dots\dots(3).$$

By addition and integration,

$$\frac{X_1}{c_1} + \frac{X_2}{c_2} + \frac{X_3}{c_3} = 0 \dots\dots\dots\dots\dots(4).$$

Hence on elimination of X_2,

$$\left. \begin{array}{c} \ddot{X}_1 + \dfrac{a^2}{S} \left\{ (c_1 + c_2) X_1 + \dfrac{c_1 c_2}{c_3} X_3 \right\} = 0 \\[2mm] \ddot{X}_3 + \dfrac{a^2}{S'} \left\{ (c_3 + c_2) X_3 + \dfrac{c_3 c_2}{c_1} X_1 \right\} = 0 \end{array} \right\} \dots\dots(5).$$

Assuming $X_1 = Ae^{pt}$, $X_3 = Be^{pt}$, we obtain on substitution and determination of $A : B$,

$$p^4 + p^2 a^2 \left\{ \frac{c_1 + c_2}{S} + \frac{c_3 + c_2}{S'} \right\} + \frac{a^4}{SS'} \left\{ c_1 c_3 + c_2 (c_1 + c_3) \right\} = 0 \dots(6),$$

as the equation to determine the natural tones. If N be the frequency of vibration, $N^2 = -p^2/4\pi^2$, the two values of p^2 being of course real and negative. The formula simplifies considerably if $c_3 = c_1$, $S' = S$; but it will be more instructive to work out this case from the beginning. Let $c_1 = c_3 = mc_2 = mc$.

The differential equations take the form

$$\ddot{X}_1 + \frac{a^2 c}{S}\{(1+m)X_1 + X_3\} = 0 \atop \ddot{X}_3 + \frac{a^2 c}{S}\{(1+m)X_3 + X_1\} = 0 \Bigg\} \quad \text{............} (7),$$

while from (4) $X_2 = -\dfrac{X_1 + X_3}{m}$.

Hence

$$(\ddot{X}_1 + \ddot{X}_3) + \frac{a^2 c}{S}(m+2)(X_1 + X_3) = 0 \atop (\ddot{X}_1 - \ddot{X}_3) + \frac{a^2 c}{S}m(X_1 - X_3) = 0 \Bigg\} \quad \text{.........} (8).$$

The whole motion may be divided into two parts. For the first of these

$$X_1 + X_3 = 0 \text{..........................}(9),$$

which requires that $X_2 = 0$. The motion is therefore the same as might take place were the communication between S and S' cut off, and has its frequency given by

$$N^2 = \frac{a^2 c_1}{4\pi^2 S} = \frac{a^2 m c}{4\pi^2 S}\text{....................}(10).$$

The density of the air is the same in both reservoirs.

For the other component part, $X_1 - X_3 = 0$, so that

$$X_2 = -\frac{2X_1}{m}; \qquad N'^2 = \frac{a^2(m+2)c}{4\pi^2 S}\text{..........}(11).$$

The vibrations are thus opposed in phase. The ratio of frequencies is given by $N'^2 : N^2 = m+2 : m$, shewing that the second mode has the shorter period. In this mode of vibration the connecting passage acts in some measure as a second opening to both vessels, and thus raises the pitch. If the passage be contracted, the interval of pitch between the two notes is small.

A particular case of the general formula worthy of notice is obtained by putting $c_3 = 0$, which amounts to suppressing one of the communications with the external air. We thus obtain

$$p^4 + a^2 p^2 \left(\frac{c_1 + c_2}{S} + \frac{c_2}{S'}\right) + \frac{a^4 c_1 c_2}{SS'} = 0\text{............}(12),$$

or, if $\qquad\qquad S = S', \qquad c_1 = mc_2 = mc,$

$$p^4 + a^2 p^2 \frac{(m+2)\,c}{S} + \frac{a^4 m\, c^2}{S^2} = 0 \dots\dots\dots\dots(13),$$

whence $\qquad N^2 = \dfrac{a^2 c}{8\pi^2 S} \{m + 2 \pm \sqrt{(m^2 + 4)}\} \dots\dots\dots\dots (14).$

If we further suppose $m = 1$, or $c_2 = c_1$,

$$N^2 = \frac{a^2 c}{8\pi^2 S} (3 \pm \sqrt{5}).$$

If N' be the frequency for a simple resonator (S, c),

$$N'^2 = \frac{a^2 c}{4\pi^2 S},$$

and thus $\qquad N_1{}^2 : N'^2 = \dfrac{3 + \sqrt{5}}{2} = 2{\cdot}618,$

$$N'^2 : N_2{}^2 = \frac{2}{3 - \sqrt{5}} = 2{\cdot}618.$$

It appears that the interval from N_1 to N' is the same as from N' to N_2, namely, $\sqrt{(2{\cdot}618)} = 1{\cdot}618$, or rather more than a fifth. It will be found that whatever the value of m may be, the interval between the two tones cannot be less than $2{\cdot}414$, which is about an octave and a minor third. The corresponding value of m is 2.

A similar method is applicable to any combination, however complicated, of reservoirs and connecting passages under the single restriction as to the comparative magnitudes of the reservoirs and wave-lengths; but the example just given is sufficient to illustrate the theory of multiple resonance. A few measurements of the pitch of double resonators are detailed in my memoir on resonance, already referred to.

311. The equations which we have employed hitherto take no account of the escape of energy from a resonator. If there were really no transfer of energy between a resonator and the external atmosphere, the motion would be isolated and of little practical interest: nevertheless the characteristic of a resonator consists in its vibrations being in great measure independent. Vibrations, once excited, will continue for a considerable number of periods without much loss of energy, and their frequency will be almost entirely independent of the rate of dissipation. The rate of dissipation is, however, an important feature in the character

of a resonator, on which its behaviour under certain circumstances
materially depends.   It will be understood that the dissipation
here spoken of means only the escape of energy from the vessel
and its neighbourhood, and its diffusion in the surrounding
medium, and not the transformation of ordinary energy into heat.
Of such transformation our equations take no account, unless
special terms be introduced for the purpose of representing the
effects of viscosity, and of the conduction and radiation of heat.

[The influence of the conduction of heat has been considered
by Koláček[1].]

Fig. 61.

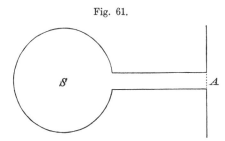

In a previous chapter (§ 278) we saw how to express the motion
on the right of the infinite flange (Fig. 61), in terms of the normal
velocity of the fluid over the disc A.   We found, § 278 (3),

$$\phi = -\frac{1}{2\pi}\iint\frac{d\phi}{dn}\frac{e^{-ikr}}{r}\,d\sigma,$$

where ϕ is proportional to e^{int}.

If r be the distance between any two points of the disc, kr is a
small quantity, and $e^{-ikr} = 1 - ikr$ approximately.

Thus $$\phi_A = -\frac{1}{2\pi}\left(\iint\frac{d\phi}{dn}\frac{d\sigma}{r} - ik\iint\frac{d\phi}{dn}\,d\sigma\right)\ldots\ldots\ldots\ldots(1).$$

The first term depends upon the distribution of the current.   If
we suppose that $d\phi/dn$ is constant, we obtain ultimately a term
representing an increase of inertia, or a correction to the length,
equal to $8R/3\pi$.   This we have already considered, under the
supposition of a piston at A.   The second term, on which the
dissipation depends, is independent of the distribution of current,

[1] *Wied. Ann.* t. 12, p. 353, 1881.

being a function of the total current (X) only. Confining our attention to this term, we have

$$\phi_A = \frac{ik\dot{X}}{2\pi} \quad\ldots\ldots\ldots\ldots\ldots\ldots (2).$$

Assuming now that $\phi \propto e^{int}$, we have for the part of the variation of pressure at A, on which dissipation depends,

$$\delta p = -\rho\dot{\phi}_A = -i\rho n \phi_A = \frac{\rho n k \dot{X}}{2\pi} = \frac{\rho n^2 \dot{X}}{2\pi a} \quad\ldots\ldots\ldots (3).$$

The corresponding work done during a transfer of fluid δX is $\dfrac{\rho n^2 \dot{X}}{2\pi a}\,\delta X$; and since, as in § 304, the expressions for the potential and kinetic energies are

$$V = \tfrac{1}{2}\rho a^2 \frac{X^2}{S}, \qquad T = \tfrac{1}{2}\rho \frac{\dot{X}^2}{c} \quad\ldots\ldots\ldots\ldots (4),$$

the equation of motion (§ 80) is

$$\ddot{X} + \frac{n^2 c}{2\pi a}\dot{X} + \frac{a^2 c}{S}X = 0 \quad\ldots\ldots\ldots\ldots\ldots (5)^1,$$

in place of (3) § 304. In the valuation of c an allowance must be included for the inertia of the fluid on the right-hand side of A, corresponding to the term omitted in the expression for δp.

Equation (5) is of the standard form for the free vibrations of dissipative systems of one degree of freedom (§ 45). The amplitude varies as $e^{-n^2 ct/4\pi a}$, being diminished in the ratio $e:1$ after a time equal to $4\pi a/n^2 c$. If the pitch (determined by n) be given, the vibrations have the greatest persistence when c is smallest, that is, when the neck is most contracted.

If S be given, we have on substituting for c its value in terms of S and n,

$$\frac{4\pi a}{n^2 c} = \frac{4\pi a^3}{n^4 S} \quad\ldots\ldots\ldots\ldots\ldots\ldots (6),$$

shewing that under these circumstances the duration of the motion increases rapidly as n diminishes.

In the case of similar resonators $c \propto n^{-1}$, and then

$$\frac{4\pi a}{n^2 c} \propto \frac{1}{n},$$

---

[1] Equation (5) is only approximate, inasmuch as the dissipative force is calculated on the supposition that the vibration is permanent; but this will lead to no material error when the dissipation is small.

which shews that in this case the same proportional loss of amplitude always occurs after the lapse of the same number of periods. This result may be obtained by the method of dimensions, as a consequence of the principle of dynamical similarity.

As an example of (5), I may refer to the case of a globe with a neck, intended for burning phosphorus in oxygen gas, whose capacity is ·251 cubic feet [7100 c.c.]. It was found by experiment that the note of maximum resonance made 120 vibrations per second, so that $n = 120 \times 2\pi$. Taking the velocity of sound (a) at 1120 feet [34200 cent.] per second, we find from these data

$$\frac{4\pi a^3}{n^4 S} = \frac{1}{5} \text{ of a second nearly.}$$

Judging from the sound produced when the globe is struck, I think that this estimate must be too low; but it should be observed that the absence of the infinite flange assumed in the theory must influence very materially the rate of dissipation.

We will now examine the forced vibrations due to a source of sound external to the resonator. If the pressure δp at the mouth of the resonator due to the source, i.e. calculated on the supposition that the mouth is closed, be $F e^{ikat}$, the equation of motion corresponding to (5), but applicable to the forced vibration only, is

$$\frac{\rho}{c} \ddot{X} + \frac{\rho k^2 a}{2\pi} \dot{X} + \frac{\rho a^2}{S} X = F e^{ikat} \quad \ldots\ldots\ldots\ldots(7).$$

If $X = X_0 e^{i(kat+\epsilon)}$, where X_0 is real,

$$\frac{F^2}{\rho^2 a^4 X_0^2} = \left(\frac{1}{S} - \frac{k^2}{c}\right)^2 + \left(\frac{k^3}{2\pi}\right)^2.$$

The maximum variation of pressure (G) inside the resonator is connected with X_0 by the equation

$$G = \frac{a^2 \rho X_0}{S} \quad \ldots\ldots\ldots\ldots\ldots\ldots\ldots\ldots\ldots(8),$$

since $X_0 \div S$ is the maximum condensation. Thus

$$\frac{F^2}{G^2} = \left(1 - \frac{k^2 S}{c}\right)^2 + \left(\frac{k^3 S}{2\pi}\right)^2 \quad \ldots\ldots\ldots\ldots\ldots(9),$$

which agrees with the equation obtained by Helmholtz for the case where the communication with the external air is by a simple aperture (§ 306). The present problem is nearly, but not quite, a case of that treated in § 46, the difference depending upon the fact that the coefficient of dissipation in (7) is itself a function of the period, and not an absolutely constant quantity. If the period, determined by k, and S be given, (9) shews that the internal variation of pressure (G) is a maximum when $c = k^2 S$, that is, when the natural note of the resonator (calculated without allowance for dissipation) is the same as that of the generating sound. The maximum vibration, when the coincidence of periods is perfect, varies inversely as S; but, if S be small, a very slight inequality in the periods is sufficient to cause a marked falling off in the intensity of the resonance (§ 49). In the practical use of resonators it is not advantageous to carry the reduction of S and c very far, probably because the arrangements necessary for connecting the interior with the ear or other sensitive apparatus involve a departure from the suppositions on which the calculations are founded, which becomes more and more important as the dimensions are reduced. When the sensitive apparatus is not in connection with the interior, as in the experiment of reinforcing the sound of a tuning-fork by means of a resonator, other elements enter into the question, and a distinct investigation is necessary (§ 319).

In virtue of the principle of reciprocity the investigation of the preceding paragraph may be applied to calculate the effect of a source of sound situated in the interior of a resonator.

312. We now pass on to the further discussion of the problem of the open pipe. We shall suppose that the open end of the pipe is provided with an infinite flange, and that its diameter is small in comparison with the wave-length of the vibration under consideration.

As an introduction to the question, we will further suppose that the mouth of the pipe is fitted with a freely moving piston without thickness and mass. The preceding problems, from which the present differs in reality but little, have already given us reason to think that the presence of the piston will cause no important modification. Within the tube we suppose (§ 255) that the velocity-potential is

$$\phi = (A \cos kx + B \sin kx)\, e^{int} \dots \dots \dots \dots \dots (1),$$

where, as usual, $k = 2\pi/\lambda = n/a$. At the mouth, where $x = 0$,

$$\phi_0 = A e^{int}; \qquad \left(\frac{d\phi}{dx}\right)_0 = kB\, e^{int} \dots \dots \dots \dots (2).$$

On the right of the piston the relation between ϕ_0 and $\left(\dfrac{d\phi}{dx}\right)_0$ is by § 302

$$\iint \phi_0 \, d\sigma \; : \; \left(\frac{d\phi}{dx}\right)_0 = i \cdot \frac{\pi R^2}{k}\left\{1 - \frac{J_1(2kR)}{kR}\right\} - \frac{\pi}{2k^3}K_1(2kR)\ldots\ldots(3),$$

R being the radius of the pipe. From this the solution of the problem may be obtained without any restriction as to the smallness of kR: since, however, it is only when kR is small that the presence of the piston would not materially modify the question, we may as well have the benefit of the simplification at once by taking as in (1) § 311

$$\iint \phi_0 \, d\sigma \; : \; \left(\frac{d\phi}{dx}\right)_0 = \frac{i\pi k R^4}{2} - \frac{8R^3}{3} \ldots\ldots\ldots\ldots(4).$$

Now, since the piston occupies no space, the values of $(d\phi/dx)_0$ must be the same on both sides of it; and since there is no mass, the like must be true of the values of $\iint \phi_0 d\sigma$. Thus

$$A\sigma = kB\left\{-\frac{8R^3}{3} + i\frac{k\pi R^4}{2}\right\}$$

or

$$A = B\left\{-\frac{8kR}{3\pi} + i\frac{k^2 R^2}{2}\right\} \ldots\ldots\ldots\ldots(5).$$

Substituting in (1), we find on rejecting the imaginary part, and putting for brevity $B = 1$,

$$\phi = \left\{\sin kx - \frac{8kR}{3\pi}\cos kx\right\}\cos nt - \tfrac{1}{2}k^2 R^2 \cos kx \; \sin nt\ldots\ldots(6).$$

In this expression the term containing $\sin nt$ depends upon the dissipation, and is the same as if there were no piston, while that involving $8kR/3\pi$ represents the effect of the inertia of the external air in the neighbourhood of the mouth. In order to compare with previous results, let α be such that

$$\sin kx - \frac{8kR}{3\pi}\cos kx = \sin k(x - \alpha);$$

then, the squares of small quantities being neglected,

$$\alpha = \frac{8R}{3\pi}\ldots\ldots\ldots\ldots\ldots\ldots(7),$$

and

$$\phi = \sin k(x - \alpha)\,\cos nt - \tfrac{1}{2}k^2 R^2 \cos kx \; \sin nt\ldots\ldots\ldots(8).$$

These formulæ shew that, if the dissipation be left out of account, the velocity-potential is the same as if the tube were lengthened by $8/3\pi$ of the radius, and the open end then behaved as a loop. The amount of the correction agrees with what previous investigations would have led us to expect as the result of the introduction of the piston. We have seen reason to know that the true value of α lies between $\frac{1}{4}\pi R$ and $8R/3\pi$, and that the presence of the piston does not affect the term representing the dissipation. But, before discussing our results, it will be advantageous to investigate them afresh by a rather different method, which besides being of somewhat greater generality, will help to throw light on the mechanics of the question.

313. For this purpose it will be convenient to shift the origin in the negative direction to such a distance from the mouth that the waves are there approximately plane, a displacement which according to our suppositions need not amount to more than a small fraction of the wave-length. The difficulty of the question consists in finding the connection between the waves in the pipe, which at a sufficient distance from the mouth are plane, and the diverging waves outside, which at a moderate distance may be treated as spherical. If the transition take place within a space small compared with the wave-length, which it must evidently do, if the diameter be small enough, the problem admits of solution, whatever may be the form of the pipe in the neighbourhood of the mouth.

Fig. 62.

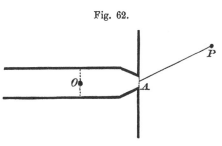

At a point P, whose distance from A is moderate, the velocity-potential is (§ 279)

$$\psi = \frac{A'}{r} e^{-ikr} e^{int} \quad\dots\dots\dots\dots\dots(1),$$

whence

$$\frac{d\psi}{dr} = -\frac{A'e^{i(nt-kr)}}{r^2} (1 + ikr)\dots\dots\dots\dots(2).$$

Let us consider the behaviour of the mass of air included between the plane section at O and a hemispherical surface whose centre is A, and radius r, r being large in comparison with the diameter of the pipe, but small in comparison with the wavelength. Within this space the air must move approximately as an incompressible fluid would do. Now the current across the hemispherical surface

$$= 2\pi r^2 \frac{d\psi}{dr} = -2\pi A'(1 + ikr)\, e^{i(nt-kr)} = -2\pi A'\, e^{int} \dots\dots(3),$$

if the square of kr be neglected.

If, as before, we take for the velocity-potential within the pipe

$$\phi = (A \cos kx + B \sin kx)\, e^{int}\dots\dots\dots\dots\dots(4),$$

we have for the current across the section at O,

$$\sigma \left(\frac{d\phi}{dx}\right)_0 = \sigma k B\, e^{int}\dots\dots\dots\dots\dots(5);$$

and thus

$$\sigma k B = -2\pi A' \dots\dots\dots\dots\dots(6).$$

This is the first condition; the second is to be found from the consideration that the total current (whose two values have just been equated) is proportional to the difference of potential at the terminals. Thus, if c denote the conductivity of the passage between the terminal surfaces,

$$\sigma \left(\frac{d\phi}{dx}\right)_0 = c\,(\psi_r - \phi_0),$$

or

$$\frac{\sigma k B}{c} = \frac{A'}{r}\, e^{-ikr} - A\dots\dots\dots\dots\dots(7).$$

On substituting for A' its value from (6), we have

$$-A = \sigma k B \left(\frac{1}{c} + \frac{e^{-ikr}}{2\pi r}\right) = \sigma k B \left\{\frac{1}{c} + \frac{1}{2\pi r} - \frac{ik}{2\pi}\right\}.$$

In this expression the second term is negligible in comparison with the first, for c is at most a quantity of the same order as the radius of the tube, and when the mouth is much contracted it is smaller still. Thus we may take

$$A = \sigma k B \left(-\frac{1}{c} + \frac{ik}{2\pi}\right)\dots\dots\dots\dots\dots(8).$$

Substituting this in (4), we have for the imaginary expression of the velocity-potential within the tube, if B be put equal to unity,

$$\phi = \left\{\sin kx + \sigma k \left(-\frac{1}{c} + \frac{ik}{2\pi}\right) \cos kx\right\} e^{int},$$

or, if only the real part be retained,

$$\phi = \left\{ \sin kx - \frac{\sigma k}{c} \cos kx \right\} \cos nt - \frac{k^2\sigma}{2\pi} \cos kx \sin nt \dots(9).$$

Following Helmholtz, we may simplify our results by introducing a quantity α defined by the equation

$$\tan k\alpha = \frac{k\sigma}{c} \dots(10).$$

Thus

$$\phi = \frac{\sin k (x - \alpha)}{\cos k\alpha} \cos nt - \frac{k^2\sigma}{2\pi} \cos kx \sin nt \dots(11),$$

and the corresponding potential outside the mouth is

$$\psi = - \frac{\sigma k}{2\pi r} \cos (nt - kr) \dots(12).$$

If R be the radius of the tube, we may replace σ by πR^2.

When the tube is a simple cylinder, and the origin lies at a distance ΔL from the mouth, we know that $\sigma c^{-1} = \Delta L + \mu R$, where μ is a number rather greater than $\frac{1}{4}\pi$. In such a case (the origin being taken sufficiently near the mouth) $k\alpha$ is a small quantity, and therefore from (10)

$$\alpha = \frac{\sigma}{c} = \Delta L + \mu R \dots(13).$$

At the same time $\cos k\alpha$ may be identified with unity. The principal term in ϕ, involving $\cos nt$, may then be calculated, as if the tube were prolonged, and there were a loop at a point situated at a distance μR beyond the actual position of the mouth, in accordance with what we found before. These results, approximate for ordinary tubes, become rigorous when the diameter is reduced without limit, friction being neglected.

If there be no flange at A, the value of c is slightly modified by the removal of what acts as an obstruction, but the principal effect is on the term representing the dissipation. If we suppose as an approximation that the waves diverging from A are spherical, we must take for the current $4\pi r^2 \, d\psi/dr$ instead of $2\pi r^2 \, d\psi/dr$. The ultimate effect of the alteration will be to halve the expression for the velocity-potential outside the mouth, as well as the corresponding second term in ϕ (involving $\sin nt$). The amount of dissipation is thus seen to depend materially on the degree in which the waves are free to diverge, and our analytical expressions must not be regarded as more than rough estimates.

The correct theory of the open organ-pipe, including equations (11) and (12), was discovered by Helmholtz[1], whose method, however, differs considerably from that here adopted. The earliest solutions of the problem by Lagrange, D. Bernoulli, and Euler, were founded on the assumption that at an open end the pressure could not vary from that of the surrounding atmosphere, a principle which may perhaps even now be considered applicable to an end whose openness is ideally perfect. The fact that in all ordinary cases energy escapes is a proof that there is not anywhere in the pipe an absolute loop, and it might have been expected that the inertia of the air just outside the mouth would have the effect of an increase in the length. The positions of the nodes in a sounding pipe were investigated experimentally by Savart[2] and Hopkins[3], with the result that the interval between the mouth and the nearest node is always less than the half of that separating consecutive nodes.

[The correction necessary for an open end is the origin of a departure from the simple law of octaves, which according to elementary theory would connect the notes of closed and open pipes of the same length. Thus in the application to an organ-pipe let αR denote the correction for the upper end when open, and l the length of the pipe including the correction for the mouth at the lower end. The whole effective length of the open pipe is then $l + \alpha R$, while the effective length of the pipe if closed at the upper end is l simply. The open pipe is practically the longer, and the interval between the notes is less than the octave of the simple theory[4].

It may be worthy of remark that the correction, assumed to be independent of wave-length, does not disturb the harmonic relations between the partial tones, whether a pipe be open or closed.]

314. Experimental determinations of the correction for an open end have generally been made without the use of a flange, and it therefore becomes important to form at any rate a rough estimate of its effect. No theoretical solution of the problem of an unflanged open end has hitherto been given, but it is easy to

[1] Crelle, Bd. 57, p. 1. 1860.

[2] Recherches sur les vibrations de l'air. *Ann. d. Chim.* t. xxiv. 1823.

[3] Aerial vibrations in cylindrical tubes. *Cambridge Transactions*, Vol. v. p. 231. 1833.

[4] Bosanquet, *Phil. Mag.* vi. p. 63, 1878.

see (§§ 79, 307) that the removal of the flange will reduce the correction materially below the value ·82 R (Appendix A). In the absence of theory I have attempted to determine the influence of a flange experimentally[1]. Two organ-pipes nearly enough in unison with one another to give countable beats were blown from an organ bellows; the effect of the flange was deduced from the difference in the frequencies of the beats according as one of the pipes was flanged or not. The correction due to the flange was about ·2R. A (probably more trustworthy) repetition of this experiment by Mr Bosanquet gave ·25R. If we subtract ·22R from ·82R, we obtain ·6R, which may be regarded as about the probable value of the correction for an unflanged open end, on the supposition that the wave-length is great in comparison with the diameter of the pipe.

Attempts to determine the correction entirely from experiment have not led hitherto to very precise results. Measurements by Wertheim[2] on doubly open pipes gave as a mean (for each end) ·663 R, while for pipes open at one end only the mean result was ·746 R. In two careful experiments by Bosanquet[3] on doubly open pipes the correction for one end was ·635 R, when $\lambda = 12\,R$, and ·543 R, when $\lambda = 30\,R$. Bosanquet lays it down as a general rule that the correction (expressed as a fraction of R) increases with the ratio of diameter to wave-length; part of this increase may however be due to the mutual reaction of the ends, which causes the plane of symmetry to behave like a rigid wall. When the pipe is only moderately long in proportion to its diameter, a state of things is approached which may be more nearly represented by the presence than by the absence of a flange. The comparison of theory and observation on this subject is a matter of some difficulty, because when the correction is small, its value, as calculated from observation, is affected by uncertainties as to absolute pitch and the velocity of sound, while for the case, when the correction is relatively larger, which experiment is more competent to deal with, there is at present no theory. Probably a more accurate value of the correction could be obtained from a resonator of the kind considered in § 306, where the communication with

[1] *Phil. Mag.* (5) III. 456. 1877. [The earliest experiments of the kind are those of Gripon (*Ann. d. Chim.* III. p. 384, 1874) who shewed that the effect of a large flange is proportional to the diameter of the pipe.]

[2] *Ann. d. Chim.* (3) t. XXXI. p. 394, 1851.

[3] *Phil. Mag.* (5) IV. p. 219. 1877.

the outside air is by a simple aperture; the "length" is in that case zero, and the "correction" is everything. Some measurements of this kind, in which, however, no great accuracy was attempted, will be found in my memoir on resonance[1].

[Careful experimental determinations of the correction for an unflanged open end have been made by Blaikley[2], who employed a vertical tube of thin brass 2·08 inches (5·3 cm.) in diameter. The lower part of the tube was immersed in water, the surface of which defined the "closed end," and the experiment consisted in varying the degree of immersion until the resonance to a fork of known pitch was a maximum. If the two shortest distances of the water surface from the open end thus found be l_1 and l_2, $(l_2 - l_1)$ represents the half wave-length, and the "correction for the open end" is $\frac{1}{2}(l_2 - l_1) - l_1$. The following are the results obtained by Blaikley, expressed as a fraction of the radius. They relate to the same tube resounding to forks of various pitch.

c'	253·68	·565
e'	317·46	·595
g'	380·81	·564
$b\flat'$	444·72	·587
c''	507·45	·568

The mean correction is thus ·576 R.]

Various methods have been used to determine the pitch of resonators experimentally. Most frequently, perhaps, the resonators have been made to *speak* after the manner of organ-pipes by a stream of air blown obliquely across their mouths. Although good results have been obtained in this way, our ignorance as to the mode of action of the wind renders the method unsatisfactory. In Bosanquet's experiments the pipes were not actually made to speak, but short discontinuous jets of air were blown across the open end, the pitch being estimated from the free vibrations as the sound died away. A method, similar in principle, that I have sometimes employed with advantage consists in exciting free vibrations by means of a blow. In order to obtain as well defined a note as possible, it is of importance to accommodate the hardness of the substance with which the resonator comes into contact to the pitch,

---

[1] *Phil. Trans.* 1871. See also Sondhauss, *Pogg. Ann.* t. 140, 53, 219 (1870), and some remarks thereupon by myself (*Phil. Mag.*, Sept. 1870).

[2] *Phil. Mag.* vol. 7, p. 339, 1879.

a low pitch requiring a soft blow. Thus the pitch of a test-tube may be determined in a moment by striking it against the bent knee.

In using this method we ought not entirely to overlook the fact that the natural pitch of a vibrating body is altered by a term depending upon the square of the dissipation. With the notation of § 45, the frequency is diminished from n to $n(1 - \frac{1}{8}k^2 n^{-2})$, or if x be the number of vibrations executed while the amplitude falls in the ratio $e : 1$, from n to

$$n\left(1 - \frac{1}{8\pi^2 x^2}\right).$$

The correction, however, would rarely be worth taking into account.

The measurements given in my memoir on resonance were conducted upon a different principle by estimating the note of maximum resonance. The ear was placed in communication with the interior of the cavity, while the chromatic scale was sounded. In this way it was found possible with a little practice to estimate the pitch of a good resonator to about a quarter of a semitone. In the case of small flasks with long necks, to which the above method would not be applicable, it was found sufficient merely to hold the flask near the vibrating wires of a pianoforte. The resonant note announced itself by a quivering of the body of the flask, easily perceptible by the fingers. In using this method it is important that the mind should be free from bias in subdividing the interval between two consecutive semitones. When the theoretical result is known, it is almost impossible to arrive at an independent opinion by experiment.

315. We will now, following Helmholtz, examine more closely the nature of the motion within the pipe, represented by the formula (11) § 313. We have

$$\phi = L \cos(nt - \theta) \quad\dots\dots\dots\dots\dots\dots\dots(1),$$

where

$$L^2 = \frac{\sin^2 k(x - \alpha)}{\cos^2 k\alpha} + \frac{k^4\sigma^2}{4\pi^2}\cos^2 kx\dots\dots\dots\dots\dots(2),$$

$$\tan\theta = -\frac{k^2\sigma\cos k\alpha\cos kx}{2\pi\sin k(x - \alpha)}\dots\dots\dots\dots\dots(3).$$

In the expression for L^2 the second term is very small, and therefore the maximum values of ϕ occur very nearly when

$$k(x-a) = (-m+\tfrac{1}{2})\pi,$$

or $$-x = \tfrac{1}{2}m\lambda - \tfrac{1}{4}\lambda - a \dots\dots\dots\dots\dots (4),$$

where m is a positive integer.

The distance between consecutive maxima is thus $\tfrac{1}{2}\lambda$, and the value of the maximum is $\sec^2 ka$. The minimum values of L^2 occur approximately when $k(x-a) = -m\pi$,

or $$-x = \tfrac{1}{2}m\lambda - a \dots\dots\dots\dots\dots\dots(5),$$

and their magnitude is given by

$$L^2 = \frac{k^4\sigma^2}{4\pi^2}\cos^2 kx = \frac{k^4\sigma^2}{4\pi^2}\cos^2 ka \dots\dots\dots\dots(6).$$

In like manner,

$$\frac{d\phi}{dx} = J\cos(nt-\chi)\dots\dots\dots\dots\dots(7),$$

where $$J^2 = k^2\frac{\cos^2 k(x-a)}{\cos^2 ka} + \frac{k^6\sigma^2}{4\pi^2}\sin^2 kx \dots\dots\dots\dots(8),$$

$$\tan\chi = \frac{k^2\sigma\cos ka\sin kx}{2\pi\cos k(x-a)}\dots\dots\dots\dots\dots(9).$$

The maximum values of J^2 occur when

$$-x = \tfrac{1}{2}m\lambda - a \dots\dots\dots\dots\dots (10),$$

and the minimum values, when

$$-x = \tfrac{1}{2}m\lambda - \tfrac{1}{4}\lambda - a \dots\dots\dots\dots\dots(11).$$

The approximate magnitude of the maximum is $k^2\sec^2 ka$, and that of the minimum $k^6\sigma^2\cos^2 ka \div 4\pi^2$. It appears that the maxima of velocity occur in the same parts of the tube as the minima of condensation (and rarefaction), and the minima of velocity in the same places as the maxima of condensation. The series of loops and nodes are arranged as if the first loop were at a distance a beyond the mouth.

With regard to the phases, we see that both θ and χ are in general small; and therefore with the exception of the places where L^2 and J^2 are near their minima the whole motion is synchronous, as if there were no dissipation.

Hitherto we have considered the problem of the passage of plane waves along the pipe and their gradual diffusion from the mouth, without regard to the origin of the plane waves themselves. All that we have assumed is that the origin of the motion is somewhere within the pipe. We will now suppose that the motion is due to the known vibration of a piston, situated at $x = -l$, the origin of co-ordinates being at the mouth. Thus, when $x = -l$,

$$\frac{d\phi}{dx} = G \cos nt \dots\dots\dots\dots\dots\dots(12),$$

and this must be made to correspond with the expression for the plane waves, generalized by the introduction of arbitrary amplitude and phase.

We may take

$$\frac{d\phi}{dx} = B J \cos(nt - \epsilon - \chi) \dots\dots\dots\dots(13),$$

where J and χ have the values given in (8), (9), while B and ϵ are arbitrary. Comparing (12) and (13) we conclude that

$$\tan \epsilon = \frac{k^2 \sigma \cos k\alpha \sin kl}{2\pi \cos k(l+\alpha)} \dots\dots\dots\dots(14),$$

$$G^2 = B^2 k^2 \left\{ \frac{\cos^2 k(l+\alpha)}{\cos^2 k\alpha} + \frac{k^4 \sigma^2}{4\pi^2} \sin^2 kl \right\} \dots\dots\dots (15),$$

by which B and ϵ are determined.

In accordance with (12) § 313, the corresponding divergent wave is represented by

$$\psi = -\frac{\sigma k B}{2\pi r} \cos(nt - \epsilon - kr) \dots\dots\dots\dots(16).$$

If G be given, B is greatest, when $\cos k(l+\alpha) = 0$, that is when the piston is situated at an approximate node. In that case

$$B = \frac{2\pi}{k^3 \sigma \cos k\alpha} G \dots\dots\dots\dots\dots(17),$$

shewing that the magnitude of the resulting vibration is very great, though not infinite, since $\cos k\alpha$ cannot vanish. When the mouth is much contracted, $\cos k\alpha$ may become small, but in this case it is necessary that the adjustment of periods be very exact in order that the first term of (15) may be negligible in comparison with the second. In ordinary pipes $\cos k\alpha$ is nearly equal to unity.

The minimum of vibration occurs when l is such that

$\cos k\,(l + \alpha) = \pm\,1$, that is, when the piston is situated at a loop. In that case

$$B = \frac{G \cos k\alpha}{k}\dots\dots\dots\dots\dots\dots(18).$$

The vibration outside the tube is then, according to the value of α, equal to or smaller than the vibration which there would be if there were no tube and the vibrating plate were made part of the yz plane.

316.   Our equations may also be applied to the investigation of the motion excited in a tube by external sources of sound. Let us suppose in the first place that the mouth of the tube is closed by a fixed plate forming part of the yz plane, and that the potential due to the external sources (approximately constant over the plate) is under these circumstances

$$\psi = H \cos nt\dots\dots\dots\dots\dots\dots(1),$$

where ψ is composed of the potential due to each source and its image in the yz plane, as explained in § 278.   Inside the tube let the potential be

$$\phi = H \cos kx \cos nt \dots\dots\dots\dots (2),$$

so that ϕ and its differential coefficient are continuous across the barrier.   The physical meaning of this is simple.   We imagine within the tube such a motion as is determined by the conditions that the velocity at the mouth is zero, and that the condensation at the mouth is the same as that due to the sources of sound when the mouth is closed.   It is obvious that under these circumstances the closing plate may be removed without any alteration in the motion.   Now, however, there is in general a finite velocity at $x = -l$, and therefore we cannot suppose the pipe to be there stopped.   But when there happens to be a node at $x = -l$, that is to say when l is such that $[\sin kl] = 0$, all the conditions are satisfied, and the actual motion within the pipe is that expressed by (2)[1].   This motion is evidently the same as might obtain if the pipe were closed at both ends; and in external space the potential is the same as if the mouth of the pipe were closed with the rigid plate.

In the general case in order to reduce the air at $x = -l$ to rest, we must superpose on the motion represented by (2) another of

[1] [An error, pointed out by Dr Burton, is here corrected.]

the kind investigated in § 313, so determined as to give at $x = -l$ a velocity equal and opposite to that of the first. Thus, if the second motion be given by

$$d\phi/dx = BJ \cos(nt - \epsilon - \chi),$$

we have $\epsilon + \chi = 0$, and

$$B^2 \left\{ \frac{\cos^2 k(l+\alpha)}{\cos^2 k\alpha} + \frac{k^4\sigma^2}{4\pi^2} \sin^2 kl \right\} = H^2 \sin^2 kl \ldots \ldots (3).$$

When $\sin kl = 0$, we have, as above explained, $B = 0$. The maximum value of B occurs when $\cos k(l + \alpha) = 0$, and then

$$B = \frac{2\pi H}{k^2\sigma} \ldots \ldots \ldots \ldots (4)^1.$$

It appears, as might have been expected, that the resonance is greatest when the reduced length is an odd multiple of $\frac{1}{4}\lambda$.

317. From the principle that in the neighbourhood of a node the inertia of the air does not come much into play, we see that in such places the form of a tube is of little consequence, and that only the capacity need be attended to. This consideration allows us to calculate the pitch of a pipe which is cylindrical through most of its length (l), but near the closed end expands into a bulb of small capacity (S). The reduced length is then evidently

$$l + \alpha + S\sigma^{-1} \ldots \ldots \ldots \ldots (1),$$

where α is the correction for the open end, and σ is the area of the transverse section of the cylindrical part. This formula is often useful, and may be applied also when the deviation from the cylindrical form does not take the shape of an enlargement.

When the enlargement represented by S is too large to allow of the above treatment, we may proceed as follows. The dissipation being neglected, the velocity-potential in the tube may be taken to be

$$\phi = \sin k(x - \alpha) \cos nt,$$

the origin being at the mouth, while $\alpha = \frac{1}{4}\pi R$ approximately. At $x = -l$, we have

$$\dot{\phi} = n \sin k(l + \alpha) \sin nt,$$

and

$$\frac{d\phi}{dx} = k \cos k(l + \alpha) \cos nt.$$

---

[1] Helmholtz, *Crelle*, Bd. 57, 1860.

Now the condensation is given by $s = -a^{-2}\dot{\phi}$, and the condition to be satisfied at $x = -l$ is

$$S\frac{ds}{dt} = -\sigma\frac{d\phi}{dx} \quad\dots\dots\dots\dots\dots\dots(2),$$

if it be assumed that the condensation within S is sensibly uniform. Thus

$$Sn^2 a^{-2} \sin k(l+\alpha) = \sigma k \cos k(l+\alpha),$$

or, since $n = ak$,

$$\tan k(l+\alpha) = \frac{\sigma}{kS} \quad\dots\dots\dots\dots\dots\dots(3)$$

is the equation determining the pitch. Numerical examples of the application of (3) are given in my memoir on resonance (*Phil. Trans.* 1871, p. 117).

Similar reasoning proves that in any case of stationary vibrations, for which the wave-length is several times as great as the diameter of the bulb, the end of the tube adjoining the bulb behaves approximately as an open end if kS be much greater than σ, and as a stopped end if kS be much less than σ.

318. The action of a resonator when under the influence of a source of sound in unison with itself is a point of considerable delicacy and importance, and one on which there has been a good deal of confusion among acoustical writers, the author not excepted.

There are cases where a resonator absorbs sound, as it were attracting the vibrations to itself and so diverting them from regions where otherwise they would be felt. For example, suppose that there is a simple source of sound B situated in a narrow tube at a distance $\frac{1}{4}\lambda$ (or any odd multiple thereof) from a closed end, and not too near the mouth: then at any distant external point A, its effect is nil. This is an immediate consequence of the principle of reciprocity, because if A were the source, there could be no variation of potential at B. The restriction, precluding too great a proximity to the mouth, may be dispensed with, if we suppose the source B to be diffused uniformly over the cross section, instead of concentrated in one point. Then, whatever may be the size and shape of the section, there is absolutely no disturbance on the further side. This is clear from the theory of vibrations in one dimension; the reciprocal form of the proposition—that whatever sources of disturbance may exist beyond the section, $\iint\psi d\sigma = 0$—may be proved from Helmholtz's formula (2) § 293, by taking for ϕ the velocity potential of the purely axial vibration of the same period.

It is scarcely necessary to say that, whenever no energy is emitted, the source does no work; and this requires, not that there shall be no variation of pressure at the source, for that in the case of a simple source is impossible, but that the variable part of the pressure shall have exactly the phase of the acceleration, and no component with the phase of the velocity.

Other examples of the absorption of sound by resonators are afforded by certain modifications of Herschel's[1] interference tube used by Quincke[2] to stop tones of definite pitch from reaching the ear.

In the combinations of pipes represented in Fig. 63, the sound enters freely at A; at B it finds itself at the mouth of a resonator of pitch identical with its own. Under these circumstances it is absorbed, and there is no vibration propagated along BD. It is clear that the cylindrical tube BC may be replaced by any other resonator of the same pitch (γ), without prejudice to the action of the apparatus. The ordinary explanation by interference (so called) of direct and reflected waves is then less applicable.

Fig. 63.

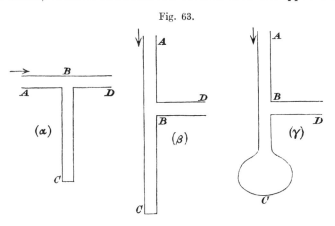

These cases where the source is at the mouth of a resonator must not be confused with others where the source is in the interior. If B be a source at the bottom of a stopped tube whose

[1] *Phil Mag.* 1833, Vol. III. p. 405.   [2] Pogg. *Ann.* CXXVIII. 177, 1866.

reduced length is $\frac{1}{4}\lambda$, the intensity at an external point A may be vastly greater than if there had been no tube. In fact the potential at A due to the source at B is the same as it would be at B were the source at A.

319. For a closer examination of the mechanics of resonance, we shall obtain the problem in a form disembarrassed of unnecessary difficulties by supposing the resonator to consist of a small circular plate, backed by a spring, and imbedded in an indefinite rigid plane. It was proved in a previous chapter, (30) § 302, that if M be the mass of the plate, ξ its displacement, $\mu\xi$ the force of restitution, R the radius, and σ the density of the air, the equation of vibration is

$$\left(M + \frac{8\sigma R^3}{3}\right)\ddot{\xi} + \frac{a\sigma\pi k^2 R^4}{2}\dot{\xi} + \mu\xi = F\ldots\ldots\ldots\ldots(1),$$

where F and ξ are proportional to e^{ikat}.

If the natural period of vibration (the reaction of external air included) coincide with that imposed, the equation reduces to

$$\tfrac{1}{2}a\sigma\pi k^2 R^4\dot{\xi} = F \ldots\ldots\ldots\ldots\ldots\ldots(2).$$

Let us now suppose that F is due to an external source of sound, giving when the plate is at rest a potential ψ_0, which will be nearly constant over the area of the plate. Thus

$$F = -\delta p \,.\, \pi R^2 = ika\sigma \,.\, \pi R^2 \,.\, \psi_0 \ldots\ldots\ldots\ldots\ldots(3);$$

so that $$\pi R^2\dot{\xi} = \dot{X} = 2i\pi k^{-1}\psi_0 = i\lambda\psi_0 \ldots\ldots\ldots\ldots\ldots(4),$$

and the potential ϕ due to the motion of the plate at a distance r will be

$$\phi = -\frac{\dot{X}}{2\pi}\frac{e^{-ikr}}{r} = -\frac{i\psi_0}{k}\frac{e^{-ikr}}{r} = \psi_0\frac{e^{-ikr}}{ikr} \ldots\ldots\ldots\ldots(5),$$

independent, it should be observed, of the area of the plate.

Leaving for the present the case of perfect isochronism, let us suppose that

$$-\left(M + \frac{8\sigma R^3}{3}\right)k'^2 a^2 + \mu = 0 \ldots\ldots\ldots\ldots\ldots(6),$$

so that $2\pi/k'$ is the wave-length of the natural note of the resonator. If M' be written for $M + \tfrac{8}{3}\sigma R^3$, the equation corresponding to (5) takes the form

$$\phi = \psi_0\frac{e^{-ikr}}{ikr} \div \left[1 - 2iM'\frac{k'^2 - k^2}{\pi\sigma k^3 R^4}\right]\ldots \ldots\ldots(7),$$

from which we may infer as before that if $k' = k$ the efficiency of the resonator as a source is independent of R. When the adjustment is imperfect, the law of falling off depends upon $M'R^{-4}$. Thus if M' be great and R small, although the maximum efficiency of the resonator is no less, a greater accuracy of adjustment is required in order to approach the maximum (§ 49). In the case of resonators with simple apertures $M' = \frac{16}{3}\sigma R^3$, so that $M'R^{-4}$ varies as R^{-1}. Accordingly resonators with small apertures require the greatest precision of tuning, but the difference is not important. From a comparison of the present investigation with that of § 311 it appears that the conditions of efficiency are different according as internal or external effects are considered.

We will now return to the case of isochronism and suppose further that the external source of sound to which the resonator A responds, is the motion of a similar plate B, whose distance c from A is a quantity large in comparison with the dimensions of the plates. The intensity of B may be supposed to be such that its potential is

$$\psi = \frac{e^{-ikr}}{r} \quad \dots\dots\dots\dots\dots\dots\dots(8).$$

Accordingly $\psi_0 = c^{-1}e^{-ikc}$, and therefore by (5)

$$\phi = \psi_0 \frac{e^{-ikr}}{ikr} = \frac{e^{-ikc}}{ikc} \cdot \frac{e^{-ikr}}{r} \quad \dots\dots\dots\dots\dots(9),$$

shewing that at equal distances from their sources

$$\phi : \psi = e^{-ikc} : ikc \quad \dots\dots\dots\dots\dots\dots(10).$$

The relation of phases may be represented by regarding the induced vibration ϕ as proceeding from B by way of A, and as being subject to an additional retardation of $\frac{1}{4}\lambda$, so that the whole retardation between B and A is $c + \frac{1}{4}\lambda$. In respect of amplitude ϕ is greater than ψ in the ratio of $1 : kc$.

Thus when kc is small, the induced vibration is much the greater, and the total sound is much louder than if A were not permitted to operate. In this case the phase is retarded by a quarter of a period.

It is important to have a clear idea of the cause of this augmentation of sound. In a previous chapter (§ 280) we saw that, when A is fixed, B gives out much less sound than might at first have been expected from the pressure developed. The explanation was that the *phase* of the pressure was unfavourable;

the larger part of it is concerned only in overcoming the inertia of the surrounding air, and is ineffective towards the performance of work. Now the pressure which sets A in motion is the whole pressure, and not merely the insignificant part that would of itself do work. The motion of A is determined by the condition that that component of the whole pressure upon it, which has the phase of the velocity, shall vanish. But of the pressure that is due to the motion of A, the larger part has the phase of the acceleration; and therefore the prescribed condition requires an equality between the small component of the pressure due to A's motion, and a pressure comparable with the large component of the pressure due to B's motion. The result is that A becomes a much more powerful source than B. Of course no work is done by the piston A; its effect is to augment the work done at B, by modifying the otherwise unfavourable relation between the phases of the pressure and of the velocity.

The infinite plane in the preceding discussion is only required in order that we may find room behind it for our machinery of springs. If we are content with still more highly idealized sources and resonators, we may dispense with it. To each piston must be added a duplicate, vibrating in a similar manner, but in the opposite direction, the effect of which will be to make the normal velocity of the fluid vanish over the plane AB. Under these circumstances the plane is without influence and may be removed. If the size of the plates be reduced without limit they become ultimately equivalent to simple sources of fluid; and we conclude that a simple source B will become more efficient than before in the ratio of $1 : kc$, when at a small distance c from it there is allowed to operate a simple resonator (as we may call it) of like pitch, that is, a source in which the inertia of the immediately surrounding fluid is compensated by some adequate machinery, and which is set in motion by external causes only.

In the present state of our knowledge of the mechanics of vibrating fluids, while the difficulties of deduction are for the most part still to be overcome, any simplification of conditions which allows progress to be made, without wholly destroying the practical character of the question, may be a step of great importance. Such, for example, was the introduction by Helmholtz of the idea of a source concentrated in one point, represented analytically by the violation at that point of the equation of

continuity. Perhaps in like manner the idea of a simple resonator may be useful, although the thing would be still more
impossible to construct than a simple source.

320. We have seen that there is a great augmentation of
sound, when a suitably tuned resonator is close to a simple
source. Much more is this the case, when the source of sound is
compound. The potential due to a double source is (§§ 294, 324)

$$r\psi = \mu e^{-ikr}\left(1 + \frac{1}{ikr}\right)\dots\dots\dots\dots\dots(1).$$

If the resonator be at a small distance c,

$$\psi_0 = \mu_0 \frac{e^{-ikc}}{ikc^2},$$

and therefore the potential due to the resonator at a distance r' is

$$\phi = \mu_0 \frac{e^{-ikc}}{ikc^2} \cdot \frac{e^{-ikr'}}{ikr'} = \mu_0 \frac{e^{-ikc}}{i^2k^2c^2} \cdot \frac{e^{-ikr'}}{r'}\dots\dots\dots(2).$$

If μ_0 vanish, the resonator is without effect; but when $\mu_0 = \pm 1$,
that is, when the resonator lies on the axis of the double source,
we have

$$\phi = \mp \frac{e^{-ikc}}{k^2c^2} \cdot \frac{e^{-ikr'}}{r'}\dots\dots\dots\dots(3).$$

At a distance from the double source its potential is

$$\psi = \mu \frac{e^{-ikr}}{r}\dots\dots\dots\dots\dots(4).$$

Thus we may consider that the potential due to the resonator
is greater than that due to the double source in the ratio $k^2c^2 : 1$,
the angular variation being disregarded.

A vibrating rigid sphere gives the same kind of motion to the
surrounding air as a double source situated at its centre; but the
substitution suggested by this fact is only permissible when the
radius of the sphere is small in comparison with c: otherwise
the presence of the sphere modifies the action of the resonator.
Nevertheless the preceding investigation shews how powerful
in general the action of a resonator is when placed in a suitable
position close to a compound source of sound, whose character
is such that it would of itself produce but little effect at a
distance.

One of the best examples of this use of a resonator is afforded by a vibrating bar of glass, or metal, held at the nodes. A strip of plate glass about a foot [30 cm.] long and an inch [2·5 cm.] broad, of medium thickness (say $\frac{1}{8}$ inch [·32 cm.]), supported at about 3 inches [7·6 cm.] from the ends by means of string twisted round it, answers the purpose very well. When struck by a hammer it gives but little sound except overtones; and even these may almost be got rid of by choosing a hammer of suitable softness. This deficiency of sound is a consequence of the small dimensions of the bar in comparison with the wave-length, which allows of the easy transference of air from one side to the other. If now the mouth of a resonator of the right pitch[1] be held over one of the free ends, a sound of considerable force and purity may be obtained by a well-managed blow. In this way an improved harmonicon may be constructed, with tones much lower than would be practicable without resonators. In the ordinary instrument the wave-lengths are sufficiently short to permit the bar to communicate vibrations to the air independently.

The reinforcement of the sound of a bell in a well-known experiment due to Savart[2] is an example of the same mode of action; but perhaps the most striking instance is in the arrangement adopted by Helmholtz in his experiments requiring pure tones, which are obtained by holding tuning-forks over the mouths of resonators.

321. When two simple resonators A_1, A_2, separately in tune with the source, are close together, the effect is less than if there were only one. If the potentials due respectively to A_1, A_2 be ϕ_1, ϕ_2, we may take

$$\phi_1 = A_1 \frac{e^{-ikr_1}}{r_1}, \qquad \phi_2 = A_2 \frac{e^{-ikr_2}}{r_2}.$$

Let R represent the distance A_1A_2, and ψ_1, ψ_2, the potentials that would exist at A_1, A_2, if there were no resonators; then the conditions to determine A_1, A_2 are by (5) § 319

$$\left.\begin{array}{l} \psi_1 + A_2/R = + ik A_1 \\ \psi_2 + A_1/R = + ik A_2 \end{array}\right\} \quad \ldots\ldots\ldots\ldots\ldots\ldots (1).$$

---

[1] To get the best effect, the mouth of the resonator ought to be pretty close to the bar; and then the pitch is decidedly lower than it would be in the open. The final adjustment may be made by varying the amount of obstruction. This use of resonators is of great antiquity.

[2] *Ann. d. Chim.* t. xxiv. 1823.

By hypothesis ψ_1 and ψ_2 are nearly equal, and therefore

$$A_1 = A_2 = \frac{R}{-1 + ikR} \psi \dots\dots\dots\dots (2).$$

Since ikR is small, the effect is much less than if there were only one resonator. It must be observed however that the diminished effectiveness is due to the resonators putting one another out of tune, and if this tendency be compensated by an alteration in the spring, any number of resonators near together have just the effect of one. This point is illustrated by § 302, where it will be seen (32) that though the resonance does not depend upon the size of the plate, still the inertia of the air, which has to be compensated by a spring, does depend upon it.

322. It will be proper to say a few words in this place on an objection, which has been brought forward by Bosanquet[1] as possibly invalidating the usual calculations of the pitch of resonators and of the correction to the length of organ-pipes. When fluid flows in a steady stream through a hole in a thin plate, the motion on the low pressure side is by no means of the character investigated in § 306. Instead of diverging after passing the hole so as to follow the surface of the plate, the fluid shapes itself into an approximately cylindrical jet, whose form for the case of two dimensions can be calculated[2] from formulæ given by Kirchhoff. On the high pressure side the motion does not deviate so widely from that determined by the electrical law. In like manner fluid passing outwards from a pipe continues to move in a cylindrical stream. If the external pressure be the greater, the character of the motion is different. In this case the stream lines converge from all directions to the mouth of the pipe, afterwards gathering themselves into a parallel bundle, whose section is considerably less than that of the pipe. It is clear that, if the formation of jets took place to any considerable extent during the passage of air through the mouths of resonators, our calculations of pitch would have to be seriously modified.

The precise conditions under which jets are formed is a subject of great delicacy. It may even be doubted whether they would occur at all in frictionless fluid moving with velocities so small that the corresponding pressures, which are proportional to the squares of

[1] *Phil. Mag.* Vol. IV. p. 125, 1877.

[2] *Phil. Mag.* Vol. II. p. 441, 1876.

the velocities, are inconsiderable.　But with air, as we actually have it, moving under the action of the pressures to be found in resonators, it must be admitted that jets may sometimes occur. While experimenting about two years ago with one of König's brass resonators of pitch c', I noticed that when the corresponding fork, strongly excited, was held to the mouth, a wind of considerable force issued from the nipple at the opposite side.　This effect may rise to such intensity as to blow out a candle upon whose wick the stream is directed.　It does not depend upon any peculiar motion of the air near the ends of the fork, as is proved by mounting the fork upon its resonance-box and presenting the open end of the box, instead of the fork itself, to the mouth of the resonator, when the effect is obtained with but slightly diminished intensity.　A similar result was obtained with a fork and resonator, of pitch an octave lower (c).　Closer examination revealed the fact that at the sides of the nipple the outward flowing stream was replaced by one in the opposite direction, so that a tongue of flame from a suitably placed candle appeared to enter the nipple at the same time that another candle situated immediately in front was blown away.　The two effects are of course in reality alternating, and only appear to be simultaneous in consequence of the inability of the eye to follow such rapid changes.　The formation of jets must make a serious draft on the energy of the motion, and this is no doubt the reason why it is necessary to close the nipple in order to obtain a powerful sound from a resonator of this form, when a suitably tuned fork is presented to it.

At the same time it does not appear probable that jet formation occurs to any appreciable extent at the mouths of resonators as ordinarily used.　The near agreement between the observed and the calculated pitch is almost a sufficient proof of this.　Another argument tending to the same conclusion may be drawn from the persistence of the free vibrations of resonators (§ 311), whose duration seems to exclude any important cause of dissipation beyond the communication of motion to the surrounding air.

In the case of organ-pipes, where the vibrations are very powerful, these arguments are less cogent, but I see no reason for thinking that the motion at the upper open end differs greatly from that supposed in Helmholtz's calculation.　No conclusion to the contrary can, I think, safely be drawn from the phenomena of steady motion. In the opposite extreme case of impulsive motion jets certainly cannot be formed, as follows from Thomson's principle of least energy (§ 79), and it is doubtful to which extreme the case of periodic motion may with greatest plausibility be assimilated. Observation by the method of intermittent illumination (§ 42) might lead to further information upon this subject.

322 a. As has already been mentioned, the free vibrations of the body of air contained in a resonator may be excited by a suitable blow delivered to the latter. The gas does not at first partake of the sudden movement imposed upon the walls, and the relative motion thus initiated is the origin of free vibrations of the kind considered in preceding sections. When corks are drawn from partially empty bottles, or when the lids are suddenly removed from tubular pasteboard pencil-cases, free vibrations of the resonating air columns are initiated in like manner.

If the vibrations are to be maintained with a view to the emission of a continued sound, the vibrating body must be in communication with a source of energy (§ 68 a), and the reaction between the two must be rightly accommodated with respect to phase. The question whether the source of energy or the resonator is to be regarded as the origin of the sound is of no particular significance and will be variously answered according to the point of view of the moment. In the organ the pipe, rather than the compressed air within the bellows or even the escaping wind, is regarded as the parent of the sound, but when a similar pipe is maintained in action by a flame the credit of the joint performance is usually given to the latter.

Up to this point the explanation of maintained vibrations is simple enough; but the complete theory in any particular case demands such an investigation of the reaction as will determine the phase relation. On this depends the whole question whether the reaction is favourable or unfavourable to the continuance of the vibrations, and the determination is often a matter of difficulty.

Before proceeding to discuss the action of the blast it will be desirable to say something further upon the organ-pipe considered simply as a resonator. We have seen (§ 314) how to take account of an upper open end, but according to the rule of Cavaillé-Coll the whole addition which must be made to the measured length of an open pipe in order to bring about agreement with the simple formula (8) § 255 amounts to as much as $3\frac{1}{3}R$, very much greater than the correction ($1\cdot2\,R$) necessary for a simple tube of circular section open at both ends. This discrepancy is sometimes attributed to the blast. But it must be remembered that the lower end is very much less open than the upper end, and that if a sensible correction on account of deficient openness is required for the latter, a much more important correction will probably be necessary for the former. Observations by the author[1] have shewn that this is the case. A pipe fitted with a sliding prolongation was tuned to maximum resonance with a given (256) fork as in Blaikley's experiment (§ 314). It was then blown from a well-regulated bellows with measured pressures of wind, and the pitch of the sounds so obtained was referred to that of the fork by the method of beats (§ 30). The results shewed that at practical pressures the pitch of the pipe as sounded by wind was *higher* than its natural note of maximum resonance; so that the considerable correction to the length found by Cavaillé-Coll is not attributable to the blast, but to the contracted character of the lower end treated as open in the elementary theory. In order to estimate the natural note an even larger " correction to the length " would be required.

The rise of pitch due to the wind increases with pressure. Thus in the case referred to above the pipe under a pressure of $1\cdot06$ inches ($2\cdot7$ cm.) of water gave a note about 2 vibrations per second sharper than that of the fork, but when the wind pressure was raised to $4\cdot2$ inches ($10\cdot7$ cm.) the excess was as much as 11 vibrations per second. When the pressure was raised much further, the pipe was " over blown " and gave the octave of its proper pitch. This, of course, corresponds to another mode of vibration of the aerial column.

It remains to consider the maintaining action of the blast. The vibrations of a column of air may be encouraged either by the introduction of fluid at a place where the density varies and at a moment of condensation (and by the similar abstraction of fluid at a moment of rarefaction), or by a suitable acceleration of the parts of the column situated near a loop. Since the blast of an organ acts at an open end of the pipe, it is clear that here we

[1] *Phil. Mag.* III. p. 462, 1877 ; XIII. p. 340, 1882.

have to do with the latter alternative. The sheet of wind directed across the lip of the pipe is easily deflected. When during the vibration the external air tends to enter the pipe, it carries the jet with it more or less completely. Half a period later when the natural flow is outwards, the jet is deflected in the corresponding direction. In either case the jet encourages the prevailing motion, and thus renders possible the maintenance of the vibration.

For ready speech it is necessary that the sheet of wind be accurately adjusted. But Schneebeli[1] has shewn that when the vibration is once started there is more latitude. In an experimental arrangement the jet was so adjusted as to pass entirely outside the pipe. Under these circumstances there was failure to speak until by a temporary strong blast directed upon it from outside the jet was bent inwards to the proper position. The pipe then spoke and continued in action until by a pressure in the reverse direction the jet was bent back. The motion of the jet may be made apparent with the aid of smoke or by means of a piece of tissue paper held so as to vibrate with it. Both Schneebeli and H. Smith[2] insist upon a comparison between the jet and the tongue of a reed organ-pipe, but the modes of action appear to be essentially different.

The above view of the matter, which is that adopted by v. Helmholtz in the fourth edition of his great work, appears to be satisfactory as a general explanation of the maintenance of a continued vibration, but it cannot be regarded as complete. In matters of this kind practice is usually in advance of theory; and many generations of practical men have brought the organ-pipe to a high degree of excellence.

Another view that has been favourably entertained by many good authorities regards the pipe as merely reinforcing by its resonance a sound primarily due to the friction of the jet playing against the lip, and there seems to be no doubt that sounds may thus originate[3]. Perhaps after all there is less difference than might at first appear between the two views, and the latter may be especially appropriate when the initiation of the sound rather than its maintenance is under consideration. A detailed discussion

[1] *Pogg. Ann.* Bd. 153, p. 301, 1874.

[2] *Nature*, 1873, 1874, 1875.

[3] See for example Melde's *Akustik*, p. 252; Sondhauss *Pogg. Ann.* xci. p. 126, 1854.

of the question will be found in an essay by Van Schaik[1]. For a fuller explanation we must probably await a better knowledge of the mechanics of jets.

322 b. The character of the sound emitted from a pipe depends upon the presence or absence of the various overtones, a matter which requires further consideration. When a system vibrates freely, the overtones may be harmonic or inharmonic according to the nature of the system, and the composition of the sound depends upon the initial circumstances. But in the case of a maintained vibration like that now before us the motion is strictly periodic, and the overtones must be harmonic if present at all. The frequency of the whole vibration will correspond approximately with that natural to the pipe in its gravest mode[2], but the agreement between the pitch of an audible overtone and that of any free vibration may be much less close. The strength of any overtone thus depends upon two things: first upon the extent to which the maintaining forces possess a component of the right kind, and secondly upon the degree of approximation between the overtone and some natural tone of the vibrating body. In organ-pipes the sharpness of the upper lip and the comparative thinness of the sheet of wind are favourable to the production of overtones; so that in narrow open pipes v. Helmholtz was able to hear plainly the first six partial tones. In wider open pipes, on the other hand, the agreement between the overtones and the natural tones is less close. In consequence, pipes of this class, especially if of wood, give a softer quality of sound, in which besides the fundamental only the octave and twelfth are to be detected[3].

When a bottle (§ 26), or a spherical resonator, is blown by wind after the manner of an organ-pipe, there are no natural tones in the neighbourhood of the harmonics, and the resulting sound is almost free from overtones.

322 c. When two organ-pipes of the same pitch stand side by side, complications ensue which not unfrequently give trouble in practice. In extreme cases the pipes may almost reduce one another to silence. Even when the mutual influence is more moderate, it may still go so far as to cause the pipes to speak

---

[1] *Ueber die Tonerregung in Labialpfeifen.* Rotterdam, 1891.

[2] We are not now speaking of "over blowing."

[3] *Tonempfindungen.* Fourth edition, p. 155, 1877.

in absolute unison, in spite of inevitable small natural differences. The simplest case that can be considered is that of a pipe, along the median plane of which a thin resisting wall is supposed to be introduced. If this wall occupy the whole plane, the original pipe is divided into two, independent of, and perfectly similar to one another. And the pitch of these segments is the same as that of the original pipe, fluid friction being neglected, since during the vibrations of the latter there is no motion across the median plane of symmetry. But the case is altered if the wall be limited to the part of the plane included within the pipe, for then the two vibrating columns are free to react upon one another. The system as a whole has two degrees of freedom— we are not now regarding overtones—and free vibrations are performed in two distinct periods. The first of these is characterised by synchronism of phase between the vibrations of the component columns, and the pitch is accordingly the same as before the separation into two parts. But in the second mode the phases of vibration of the component columns are opposed, so that the air which escapes from one open end is absorbed by the contiguous open end of the other part. In consequence the "correction for the open ends" is much diminished in amount, and the pitch in this mode is correspondingly raised. So long as the motion is free, temporary vibrations in both modes may co-exist, and would give rise to beats; but it does not follow that both can be maintained by the blast. This would indeed seem improbable beforehand, and experiment shews that after the first moment the vibrations are confined to the second mode. The contiguous open ends act as opposed sources, and but little sound escapes, although within the pipes, and indeed outside in the immediate neighbourhood of their mouths, the vigour of the vibrations is unimpaired. Effects of the same kind are produced when two distinct but similar pipes are mounted side by side, and under the influence of the blast the compound system may vibrate in one mode only, in spite of small differences of pitch between the notes of the pipes when sounding separately[1].

322 d. Direct observation of the state of things within a vibrating air column is of course a matter of great difficulty, but

---

[1] *Proceedings of the Musical Association*, Dec. 1878.

interesting results have been obtained by Töpler and Boltzmann[1], calling to their aid the method of optical interference to meet the difficulty arising out of the invisibility of air and the method of stroboscopic vision to meet that arising out of the rapidity of the changes. The upper end of an organ-pipe, closed by a thin plate of metal, was provided with sides of worked glass projecting above beyond the metal plate, and by suitable optical arrangements interference was produced between light which passed above and below. The space above being occupied by air at normal density and that below by air in a state of increased or diminished density according to the phase at the moment, the interference bands undergo displacements synchronous with the aerial vibration. Observed directly these displacements would escape the eye; but by the aid of a fork electrically maintained and provided with suitable slits (§ 42) the light may be rendered intermittent in a period nearly coincident with that of the vibration, and then the sequence of changes becomes apparent. From the observed movement of the bands it is possible to infer not merely the total change of density from maximum to minimum, but the law of the variation of density as a function of time.

When a pipe of large section was but moderately blown, the change of density at the node amounted to ·009 of an atmosphere, and the law was very nearly simple harmonic. Under a greater pressure of wind the simple harmonic law was widely departed from, the bands shifting themselves almost suddenly from one extreme position to the other. In this case the amplitude of the first overtone (the twelfth) was about one quarter of that of the fundamental tone. The whole variation of density was ·019 atmosphere.

322 *e.* In some experimental investigations a form of pipe more completely symmetrical with respect to the axis has been employed[2]. The lip is constituted by the entire circular edge of the pipe as defined by a plane perpendicular to the axis, and upon this an annular sheet of wind is brought to bear. A similar arrangement is adopted in the ordinary steam whistle.

Another way of applying wind to evoke the speech of small pipes has been experimented upon by Sondhauss[3], and the rationale

---

[1] *Pogg. Ann.* CXLI. p. 321, 1870.

[2] Gripon, *Ann. d. Chemie*, III. p. 384, 1874.

[3] *Pogg. Ann.* XCI. p. 126, 1854.

is even less understood. A tube entirely open at one end is partially closed at the other by a plate of wood or metal 2 or 3 mm. thick and pierced by a cylindrical aperture with sharp edges (Fig. 63 a).

Fig. 63 a.

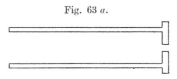

To set the pipe into action it is only necessary to insert the open end into a reservoir of wind. For rough purposes when it is not required to register the pressure nor to preserve a constant temperature, the mouth suffices, and the sound may be evoked either by pressure or by suction. The cylindrical aperture may be replaced by one of conical form, but in that case the wind must flow from the narrower towards the wider end. The sounds tabulated by Sondhauss vary from a' to f^6, corresponding in all cases to proper tones of the tube.

The whistling sounds of the unaided mouth are evidently of this class, the adjustment of pitch (from about c'' to c^5) being effected mainly by varying the internal capacity (§ 304). The formation of sound in whistling is sometimes said to be connected with a vibration of the lips, but this appears to be a mistake. I have found it possible to whistle through a suitable conical aperture in a piece of box-wood held tightly between the lips.

The occurrence of vibration may be taken as evidence that the steady flow of air through the passages in question is unstable. Contrary to what occurs in the organ-pipe and in sensitive flames, the deformations of the jet would seem here to be of the symmetrical sort. There is perhaps a tendency alternately to follow and to depart from the course marked out by the walls.

322 f. An important part of our present subject relates to the maintenance of vibrations by means of heat, and it will be possible to give at least a general account of the manner in which the effect takes place. In almost all cases where heat is communicated to a body expansion ensues, and this expansion may be made to do mechanical work. If the phases of the forces thus operative be favourable, a vibration may be maintained.

An instructive example is afforded by Trevelyan's rocker, consisting of a mass of iron or copper, so shaped that during vibration the weight is alternately carried on one or other of two adjacent and parallel ridges. When the instrument is heated and placed upon a block of cold lead, the vibrations persist so long as the heat remains sufficient. "Sir John Leslie first suggested that the cause of these vibrations is to be found in the expansion of the cold block by the heat which flows into it from the hot metal at the points of contact. Faraday [1], Seebeck [2], and Tyndall [3] have adopted this explanation; and they have shewn that most of the facts that they and others have ascertained respecting these vibrations are easily explained upon this view of their cause, supposing only that the expansion is sufficiently great to produce any sensible effect. Forbes [4], on the other hand, after an extensive series of experiments, was led to reject Sir John Leslie's explanation, one of his principal reasons for doing so being the impossibility, as it appeared to him, that the expansion occasioned by so slow a process as the conduction of heat could produce any sensible mechanical effect."

Davis, from whom [5] the above sentences are quoted, has examined the question mathematically, and has shewn that the explanation is adequate. It is evidently important that the lower body should possess a high rate of expansibility with temperature. In this respect lead stands high among the metals, and rock salt, which Tyndall found to answer well, is even more expansible.

The objection taken by Forbes may be met by the reply that the conduction of heat is not a slow process when small distances and masses are in question; and the special repulsion invoked by him as the basis of an alternative explanation would be of unsuitable character in respect of phase. It is essential that the phase of the force should be in arrear of the phase of the negative displacement.

In an experiment due to Page [6] the vibrations are made independent of an initial difference of temperature, the local heating at the points of contact being obtained with the aid of an

---

[1] *Proc. of Roy. Inst.* vol. ii. p. 119, 1831.

[2] *Pogg. Ann.* vol. li. p. 1, 1840. [3] *Phil. Mag.* vol. viii. p. 1, 1854.

[4] *Phil. Mag.* vol. iv. pp. 15, 182, 1834.

[5] *Phil. Mag.* vol. xlv. p. 296, 1873.

[6] Silliman's *Journal*, vol. ix. p. 105, 1850.

electric current caused to pass from one body to the other. In this arrangement there is no contraction in the upper body to be deducted from the expansion in the lower. On a similar principle Gore [1] has contrived a continuous motion of a copper ball which travels upon circular rails themselves connected with a powerful battery.

322 *g.* But the most interesting examples of vibrations maintained by heat are those which occur when the resonating body is gaseous. "If heat be periodically communicated to, and abstracted from, a mass of air vibrating (for example) in a cylinder bounded by a piston, the effect produced will depend upon the phase of the vibration at which the transfer of heat takes place. If heat be given to the air at the moment of greatest condensation, or be taken from it at the moment of greatest rarefaction, the vibration is encouraged. On the other hand if heat be given at the moment of greatest rarefaction, or abstracted at the moment of greatest condensation, the vibration is discouraged. The latter effect takes place of itself (§ 247) when the rapidity of alternation is neither very great nor very small in consequence of radiation; for when air is condensed it becomes hotter, and communicates heat to surrounding bodies. The two extreme cases are exceptional, though for different reasons. In the first, which corresponds to the suppositions of Laplace's theory of the propagation of sound, there is not sufficient time for a sensible transfer to be effected. In the second, the temperature remains nearly constant, and the loss of heat occurs during the *process* of condensation, and not when the condensation is effected. This case corresponds to Newton's theory of the velocity of sound. When the transfer of heat takes place at the moment of greatest condensation or of greatest rarefaction, the pitch is not affected.

If the air be at its normal density at the moment when the transfer of heat takes place, the vibration is neither encouraged nor discouraged, but the pitch is altered. Thus the pitch is *raised* if heat be communicated to the air a quarter period *before* the phase of greatest condensation; and the pitch is *lowered* if the heat be communicated a quarter period *after* the phase of greatest condensation.

[1] *Phil. Mag.* vol. xv. p. 519, 1858 ; vol. xviii. p. 94, 1859.

In general both kinds of effects are produced by a periodic transfer of heat. The pitch is altered, and the vibrations are either encouraged or discouraged. But there is no effect of the second kind if the air concerned be at a loop, i.e. a place where the density does not vary, nor if the communication of heat be the same at any stage of rarefaction as at the corresponding stage of condensation [1]."

Thus in any problem which may present itself of the maintenance of a vibration by heat, the principal question to be considered is the *phase* of the communication of heat relatively to that of the vibration.

322 *h*.   The sounds emitted by a jet of hydrogen burning in a pipe open at both ends, were noticed soon after the discovery of the gas, and have been the subject of several elaborate inquiries. The fact that the notes are substantially the same as those which may be elicited in other ways, e.g. by blowing, was announced by Chladni. Faraday[2] proved that other gases were competent to take the place of hydrogen, though not without disadvantage. But it is to Sondhauss[3] that we owe the most detailed examination of the circumstances under which the sound is produced. His experiments prove the importance of the part taken by the column of gas in the tube which supplies the jet. For example, sound cannot be got with a supply tube which is plugged with cotton in the neighbourhood of the jet, although no difference can be detected by the eye between the flame thus obtained and others which are competent to excite sound. When the supply tube is unobstructed, the sounds obtainable by varying the resonator are limited as to pitch, often dividing themselves into distinct groups. In the intervals between the groups no coaxing will induce a maintained sound; and it may be added that, for a part of the interval at any rate, the influence of the flame is inimical, so that a vibration started by a blow is damped more rapidly than if the jet were not ignited.

Forms of resonator other than the open pipe may be employed, and sometimes with advantage. Very low notes can be got from spherical resonators, such as the large globes employed for demon-

---

[1] *Proc. Roy. Inst.* vol. VIII. p. 536, 1878; *Nature*, vol. XVIII. p. 319, 1878.

[2] *Quart. Journ. Sci.* vol. V. p. 274, 1818.

[3] *Pogg. Ann.* vol. CIX. pp. 1, 426, 1860.

strating the combustion of phosphorus in oxygen gas. A globe of this kind gave in its natural condition a deep and pure tone of 64 vibrations per second. When it was fitted with a longer and narrower neck formed from a pasteboard tube, the calculated frequency fell to 25, and the vibrations, though vigorous enough to extinguish the flame, were hardly audible. When it is desired to excite very deep sounds, the supply tube should be made of considerable length, and the orifice must not be much contracted.

Singing flames may sometimes replace electrically maintained tuning-forks for the production of pure tones, when absolute constancy of pitch is not insisted upon. In order to avoid progressive deterioration of the air, it is advisable to use a resonator open above as well as below. A bulbous chimney, such as are often used with paraffin lamps, meets this requirement, and at the same time emits a pure tone. Or an otherwise cylindrical pipe may be blocked in the middle by a loosely fitting plug[1].

As Wheatstone shewed, the intermittence of a singing flame is easily made manifest by an oscillating, or a revolving, mirror. A more minute examination is best effected by the stroboscopic method, § 42. Drawings of the transformations thus observed have been given by Töpler[2], from which it appears that at one phase the flame may withdraw itself entirely within the supply tube.

Vibrations capable of being maintained are not always self-starting. The initial impulse may be given by a blow administered to the resonator, or by a gentle blast directed across the mouth. In the striking experiments of Schaffgotsch and Tyndall[3] a flame, previously silent, responds to a sound in unison with its own. In some cases the vibrations thus initiated rise to such intensity as to extinguish the flame.

The experiments of Sondhauss shew that a relationship of proportionality subsists between the lengths of the supply tubes and of the sounding columns. When the nature of the gas is varied, the same supply tube being retained, the mean lengths of

---

[1] *Phil. Mag.* vol. VII. p. 149, 1879.

[2] *Pogg. Ann.* vol. CXXVIII. p. 126, 1866.

[3] *Sound*, 3rd edition, p. 224, 1875.

the speaking columns are approximately as the square roots of the density of the gas. A connection is thus established between the natural note of a supply tube and the notes which can be sounded with its aid.

Partly in consequence of the peculiar and ill understood behaviour of flames, and partly for other reasons, the thorough explanation of the phenomena now under consideration is a matter of some difficulty; but there can be no doubt that they fall under the head of vibrations maintained by heat, the heat being communicated periodically to the mass of air confined in the sounding tube at a place where, in the course of a vibration, the pressure varies. Although some authors have shewn a tendency to lay stress upon the effects of the draught of air through the pipe, the sounds, as we have seen, can be readily produced, not only when there is no through draught, but even when the flame is so situated that there is no sensible periodic motion of the air in its neighbourhood.

In consequence of the variable pressure within the resonator, the issue of gas, and therefore the development of heat, varies during the vibration. The question is under what circumstances the variation is of the kind necessary for the maintenance of the vibration. If we were to suppose, as we might at first be inclined to do, that the issue of gas is greatest when the pressure in the resonator is least, and that the phase of greatest development of heat coincides with that of the greatest issue of gas, we should have the condition of things the most unfavourable of all to the persistence of the vibration. It is not difficult, however, to see that both suppositions are incorrect. In the supply tube (supposed to be unplugged, and of not too small bore) stationary, or approximately stationary, vibrations are excited, whose phase is either the same or the opposite of that of the vibration in the resonator. If the length of the supply tube from the burner to the open end in the gas-generating flask be less than a quarter of the wave-length in hydrogen of the actual vibration, the greatest issue of gas *precedes* by a quarter period the phase of greatest condensation; so that, if the development of heat is *retarded* somewhat in comparison with the issue of gas, a state of things exists *favourable* to the maintenance of the sound. Some such retardation is inevitable, because a jet of inflammable gas can burn only at the outside; but in many cases a still more potent cause may be found in the fact that during the retreat of the gas in the supply tube small quantities of air may enter from the interior of the resonator, whose expulsion must be effected before the inflammable gas can again begin to escape.

If the length of the supply tube amounts to exactly one quarter of the wave-length, the stationary vibration within it will be of such a character that a node is formed at the burner, the variable part of the pressure just inside the burner being the same as in the interior of the resonator. Under these circumstances there is nothing to make the flow of gas, or the development of heat, variable, and therefore the vibration cannot be maintained. This particular case is free from some of the difficulties which attach themselves to the general problem, and the conclusion is in accordance with Sondhauss' observations.

When the supply tube is somewhat longer than a quarter of the wave, the motion of the gas is materially different from that first described. Instead of preceding, the greatest outward flow of gas *follows* at a quarter period interval the phase of greatest condensation, and therefore if the development of heat be somewhat retarded, the whole effect is unfavourable. This state of things continues to prevail, as the supply tube is lengthened, until the length of half a wave is reached, after which the motion again changes sign, so as to restore the possibility of maintenance. Although the size of the flame and its position in the tube (or nec : of resonator) are not without influence, this sketch of the theory is sufficient to explain the fact, formulated by Sondhauss, that the principal element in the question is the length of the supply tube.

322 *i*. Another phenomenon of the class now under consideration occasionally obtrudes itself upon the notice of glass-blowers. When a bulb about 2 cm. in diameter is blown at the end of a somewhat narrow tube, 12 or 15 cm. long, a sound is sometimes heard proceeding from the heated glass. For experimental purposes it is well to use hard glass, which can afterwards be reheated at pleasure without losing its shape. As was found by De la Rive, the production of sound is facilitated by the presence of vapour, especially of alcohol and ether.

It was proved by Sondhauss[1] that a vibration of the glass

---

[1] *Pogg. Ann.* vol. LXXIX. p. 1, 1850.

self is no essential part of the phenomenon, and the same indefatigable observer was very successful in discovering the connection (§§ 303, 309) between the pitch of the note and the dimensions of the apparatus. But no adequate explanation of the production of sound was given.

For the sake of simplicity, a simple tube, hot at the closed end and getting gradually cooler towards the open end, may be considered. At a quarter of a period *before* the phase of greatest condensation (which occurs almost simultaneously at all parts of the column) the air is moving inwards, i.e. towards the closed end, and therefore is passing from colder to hotter parts of the tube; but the heat received at this moment (of normal density) has no effect either in encouraging or discouraging the vibration. The same would be true of the entire operation of the heat, if the adjustment of temperature were instantaneous, so that there was never any sensible difference between the temperatures of the air and of the neighbouring parts of the tube. But in fact the adjustment of temperature takes *time*, and thus the temperature of the air deviates from that of the neighbouring parts of the tube, inclining towards the temperature of that part of the tube from which the air has just come. From this it follows that at the phase of greatest condensation heat is received by the air, and at the phase of greatest rarefaction heat is given up from it, and thus there is a tendency to maintain the vibrations. It must not be forgotten, however, that apart from transfer of heat altogether, the condensed air is hotter than the rarefied air, and that in order that the whole effect of heat may be on the side of encouragement, it is necessary that previous to condensation the air should pass not merely towards a hotter part of the tube, but towards a part of the tube which is hotter than the air will be when it arrives there. On this account a great range of temperature is necessary for the maintenance of vibration, and even with a great range the influence of the transfer of heat is necessarily unfavourable at the closed end, where the motion is very small. This is probably the reason of the advantage of a bulb. It is obvious that if the *open* end of the tube were heated, the effect of the transfer of heat would be even more unfavourable than in the case of a temperature uniform throughout.

322 *j*. The last example of the production of sound by heat which we shall here consider is a very striking phenomenon discovered by Rijke[1]. When a piece of fine metallic gauze, stretching across the lower part of a tube open at both ends and held vertically, is heated by a gas flame placed under it, a sound of considerable power and lasting for several seconds is observed almost immediately *after* the removal of the flame. Differing in this respect from the case of sonorous flames (§ 322), the generation of sound was found by Rijke to be closely connected with the formation of a through draught impinging upon the heated gauze. In this form of the experiment the heat is soon abstracted, and then the sound ceases; but by keeping the gauze hot by the current from a powerful galvanic battery Rijke was able to obtain the prolongation of the sound for an indefinite period.

These notes may be obtained upon a large scale and form a very effective lecture experiment. For this purpose a cast iron pipe 5 feet (152 cm.) long and $4\frac{3}{4}$ inches (12 cm.) in diameter may be employed. The gauze (iron wire) is of about 32 meshes to the linear inch (2·54 cm.), and may advantageously be used in two thicknesses. It may be moulded with a hammer on a circular wooden block of somewhat smaller diameter than that of the pipe, and will then retain its position in the pipe by friction. When it is desired to produce the sound, the gauze caps are pushed up the pipe to a distance of about a foot (30·5 cm.), and a gas flame from a large rose burner is adjusted underneath, at such a level as to heat the gauze to bright redness. For this purpose the vertical tube of the lamp should be prolonged, if necessary, by an additional length of brass tubing. When a good red heat is attained, the flame is suddenly removed, either by withdrawing the lamp or by stopping the supply of gas. In about a second the sound begins, and presently rises to such intensity as to shake the room, after which it gradually dies away. The whole duration of the sound may be about 10 seconds[2].

In discussing the question of maintenance in accordance with the views already explained, we have to examine the character of the variable communication of heat from the gauze to the air. So far as the communication is affected directly by variations of pressure or density, the influence is unfavourable, inasmuch as the air will receive less heat from the gauze when its own temperature is raised by condensation. The maintenance depends

[1] *Pogg. Ann.* vol. cvii. p. 339, 1859; *Phil. Mag.* vol. xvii. p. 419, 1859.
[2] *Phil. Mag.* vol. vii. p. 155, 1879.

upon the variable transfer of heat due to the varying *motions* of the air through the gauze, this motion being compounded of a uniform motion upwards with a motion, alternately upwards and downwards, due to the vibration. In the lower half of the tube these motions conspire a quarter period *before* the phase of greatest condensation, and oppose one another a quarter period after that phase. The rate of transfer of heat will depend mainly upon the temperature of the air in contact with the gauze, being greatest when that temperature is lowest. Perhaps the easiest way to trace the mode of action is to begin with the case of a simple vibration without a steady current. Under these circumstances the whole of the air which comes in contact with the metal, in the course of a complete period, becomes heated; and after this state of things is established, there is comparatively little further transfer of heat. The effect of superposing a small steady upwards current is now easily recognized. At the limit of the inwards motion, i.e. at the phase of greatest condensation, a small quantity of air comes into contact with the metal, which has not done so before, and is accordingly cool; and the heat communicated to this quantity of air acts in the most favourable manner for the maintenance of the vibration.

A quite different result ensues if the gauze be placed in the *upper* half of the tube. In this case the fresh air will come into the field at the moment of greatest rarefaction, when the communication of heat has an unfavourable instead of a favourable effect. The principal note of the tube therefore cannot be sounded.

A complementary phenomenon discovered by Bosscha[1] and Riess[2] may be explained upon the same principles. If a current of *hot* air impinge upon *cold* gauze, sound is produced; but in order to obtain the principal note of the tube the gauze must be in the upper, and not as before in the lower, half of the tube. In an experiment due to Riess the sound is maintained indefinitely. The upper part of a brass tube is kept cool by water contained in a surrounding vessel, through the bottom of which the tube passes. In this way the gauze remains comparatively cool, although exposed to the heat of a gas flame situated an inch or two below it. The experiment sometimes succeeds better when the draught

[1] *Pogg. Ann.* vol. cvii. p. 342, 1859.
[2] *Pogg. Ann.* vol. cviii. p. 653, 1859; cix. p. 145, 1860.

is checked by a plate of wood placed somewhat closely over the top of the tube.

Both in Rijke's and Riess' experiments the variable transfer of heat depends upon the motion of vibration, while the effect of the transfer depends upon the variation of pressure. The gauze must therefore be placed where both effects are sensible, i.e. neither near a node nor near a loop. About a quarter of the length of the tube, from the lower or upper end, as the case may be, appears to be the most favourable position[1].

322 *k*. It remains to consider briefly another class of maintained aerial vibrations where the maintenance is provided for by the actual mechanical introduction of fluid, taking effect at a node and near the phase of maximum condensation. Well-known examples are afforded by such reed instruments as the clarinette and by the various wind instruments actuated directly by the lips. The notes obtained are determined in the main by the aerial columns, modified, it may be, to some extent by the maintaining appliances. The reeds of the harmonium and organ come under a different head. The pitch is there determined almost entirely by the tongues themselves vibrating under their own elasticity, resonating air columns being either altogether absent or playing but a subordinate part.

In the instruments now under discussion the air column and the wind-pipe are connected by a narrow aperture, which is opened and closed periodically by a vibrating tongue. Tongues are distinguished by v. Helmholtz as in-beating and out-beating. In the first case the passage is opened when the tongue moves inwards, i.e. against the wind, as happens in the clarinette. Lip instruments, such as the trombone, belong to the second class, the passage being open when the lips are moved outwards or with the wind.

Let us consider the case of a cylindrical pipe, open at the further end, in which the air vibrates at such a pitch as to make the quarter wave-length equal to the length of the pipe. The end of the column where the tongue is situated thus coincides with a approximate node, and the aerial vibration can be maintained the passage is open at the moment of greatest condensation,

[1] *Proc. Roy. Inst.* vol. VIII. p. 536, 1879; *Nature*, vol. XVIII. p. 319, 1878.

that air from the wind-pipe is then forcibly injected. The tongue is maintained in motion by the variable pressure within the pipe, and the phase of its motion will depend upon whether it is in-beating or out-beating. In the latter case its phase is nearly the opposite to that of the forces operative upon it, being open when the pressure tending to close it is greatest. This is the state of things in lip instruments, the lips being heavy in relation to the rapidity of the vibrations actually performed, § 46. When the tongue is light and stiff, it must be made in-beating, as in the clarinette, and its phase is then in approximate agreement with the phase of the forces. A slight departure in the proper direction from precise opposition or precise agreement of phase, as the case may be, will allow of the communication of sufficient energy to maintain the motion in spite of dissipative influences. A more complete analytical statement of the circumstances has been given by v. Helmholtz[1], to whom the whole theory is due.

The character of the sounds from the various wind instruments used in music differs greatly. Strongly contrasted qualities are obtained from the trombone and the euphonion, the former brilliant and piercing, and the latter mellow. Blaikley[2] has analysed the sounds from a number of instruments, and has called attention to various circumstances, such as the size of the bell-mouth, and the shape of the cup applied to the lips, upon which the differences probably depend. The pressures used in practice, rising to 40 inches (102 cm.) of water in the case of the euphonion, have been measured by Stone[3].

---

[1] *Tonempfindungen*, 4th edition, appendix VII.
[2] *Phil. Mag.* vol. VI. p. 119, 1878. [3] *Phil. Mag.* vol. XLVIII. p. 113, 1874.

CHAPTER XVII.

APPLICATIONS OF LAPLACE'S FUNCTIONS.

323. THE general equation of a velocity potential, when referred to polar co-ordinates, takes the form (§ 241)

$$r^2 \frac{d^2\psi}{dr^2} + 2r \frac{d\psi}{dr} + \frac{1}{\sin\theta} \frac{d}{d\theta}\left(\sin\theta \frac{d\psi}{d\theta}\right) + \frac{1}{\sin^2\theta} \frac{d^2\psi}{d\omega^2} + k^2 r^2 \psi = 0 \dots(1).$$

If k vanish, we have the equation of the ordinary potential, which, as we know, is satisfied, if $\psi = r^n S_n$, where S_n denotes the spherical surface harmonic[1] of order n. On substitution it appears that the equation satisfied by S_n is

$$\frac{1}{\sin\theta} \frac{d}{d\theta}\left(\sin\theta \frac{dS_n}{d\theta}\right) + \frac{1}{\sin^2\theta} \frac{d^2 S_n}{d\omega^2} + n(n+1) S_n = 0 \dots(2).$$

Now, whatever the form of ψ may be, it can be expanded in a series of spherical harmonics

$$\psi = \psi_0 + \psi_1 + \psi_2 + \dots + \psi_n + \dots(3),$$

where ψ_n will satisfy an equation such as (2).

Comparing (1) and (2) we see that to determine ψ_n as a function of r, we have

$$r^2 \frac{d^2\psi_n}{dr^2} + 2r \frac{d\psi_n}{dr} - n(n+1)\psi_n + k^2 r^2 \psi_n = 0;$$

or, as it may also be written,

$$\frac{d^2(r\psi_n)}{d(kr)^2} - \frac{n(n+1)}{(kr)^2}(r\psi_n) + r\psi_n = 0 \dots(4).$$

---

[1] On the theory of these functions the latest English works are Todhunter' *The Functions of Laplace, Lamé, and Bessel*, Ferrers' *Spherical Harmonics* and Gray and Mathews' *Bessel's Functions*, Macmillan, 1895.

In order to solve this equation, we may observe that when r is very great, the middle term is relatively negligible, and that then the solution is

$$r\psi_n = Ae^{ikr} + Be^{-ikr} \dots\dots\dots\dots\dots (5).$$

The same form may be assumed to hold good for the complete equation (4), if we look upon A and B no longer as constants, but as functions of r, whose nature is to be determined. Substituting in (4), we find for B,

$$-\frac{d^2B}{d(ikr)^2} + 2\frac{dB}{d(ikr)} + \frac{n(n+1)}{(ikr)^2}B = 0\dots\dots\dots(6).$$

Let us assume

$$B = B_0 + B_1(ikr)^{-1} + B_2(ikr)^{-2} + \dots + B_s(ikr)^{-s} + \dots (7),$$

and substitute in (6). Equating to zero the coefficient of $(ikr)^{-s-2}$, we obtain

$$B_{s+1} = B_s\frac{n(n+1) - s(s+1)}{2(s+1)} = B_s\frac{(n-s)(n+s+1)}{2(s+1)} \dots\dots(8).$$

Thus $\qquad\qquad B_1 = \frac{1}{2}n(n+1)B_0,$

$$B_2 = B_1\frac{(n-1)(n+2)}{2.2} = \frac{(n-1)n(n+1)(n+2)}{2.4}B_0, \quad \&\text{c.} ;$$

so that

$$B = B_0\left\{1 + \frac{n(n+1)}{2.ikr} + \frac{(n-1)\dots(n+2)}{2.4.(ikr)^2} + \frac{(n-2)\dots(n+3)}{2.4.6.(ikr)^3}\right.$$

$$\left. + \dots + \frac{1.2.3\dots 2n}{2.4.6\dots 2n.(ikr)^n}\right\} \dots\dots\dots\dots(9).$$

Denoting with Prof. Stokes[1] the series within brackets by $f_n(ikr)$, we have

$$B = B_0 f_n(ikr)\dots\dots\dots\dots\dots\dots(10).$$

In like manner by changing the sign of i, we get

$$A = A_0 f_n(-ikr)\dots\dots\dots\dots\dots(11).$$

The symbols A_0 and B_0, though independent of r, are functions of the angular co-ordinates: in the most general case, they are any two spherical surface harmonics of order n. Equation (5) may therefore be written

$$r\psi_n = S_n e^{-ikr}f_n(ikr) + S_n' e^{+ikr}f_n(-ikr)\dots\dots\dots(12).$$

---

[1] On the Communication of Vibrations from a Vibrating Body to a surrounding Gas. *Phil. Trans.* 1868.

By differentiation of (12)

$$\frac{d\psi_n}{dr} = -\frac{S_n}{r^2} e^{-ikr} F_n(ikr) - \frac{S_n{}'}{r^2} e^{+ikr} F_n(-ikr)\ldots(13),$$

where

$$F_n(ikr) = (1 + ikr) f_n(ikr) - ikr f_n{}'(ikr)\ldots\ldots(14).$$

The forms of the functions F, as far as $n = 7$, are exhibited in the accompanying table:

$F_0(y) = y + 1$
$F_1(y) = y + 2 + 2y^{-1}$
$F_2(y) = y + 4 + 9y^{-1} + 9y^{-2}$
$F_3(y) = y + 7 + 27y^{-1} + 60y^{-2} + 60y^{-3}$
$F_4(y) = y + 11 + 65y^{-1} + 240y^{-2} + 525y^{-3} + 525y^{-4}$
$F_5(y) = y + 16 + 135y^{-1} + 735y^{-2} + 2625y^{-3} + 5670y^{-4} + 5670y^{-5}$
$F_6(y) = y + 22 + 252y^{-1} + 1890y^{-2} + 9765y^{-3} + 34020y^{-4} + 72765y^{-5} + 72765y^{-6}$
$F_7(y) = y + 29 + 434y^{-1} + 4284y^{-2} + 29925y^{-3} + 148995y^{-4} + 509355y^{-5} + 1081080y^{-6}$
    $+ 1081080y^{-7}$

In order to find the leading terms in $F_n(ikr)$ when ikr is small, we have on reversing the series in (9)

$$f_n(ikr) = 1.3.5\ldots(2n-1)(ikr)^{-n}\left\{1 + ikr + \frac{n-1}{2n-1}(ikr)^2 + \ldots\right\}$$
$$\ldots\ldots\ldots\ldots\ldots\ldots\ldots\ldots(15),$$

whence by (14) we find

$$F_n(ikr) = 1.3.5\ldots(2n-1)(n+1)(ikr)^{-n}$$
$$\times\left\{1 + ikr + \frac{n^2(ikr)^2}{(n+1)(2n-1)} + \ldots\right\}\ldots\ldots(16).$$

324. An important case of our general formulæ occurs when ψ represents a disturbance which is propagated wholly *outwards*. At a great distance from the origin, $f_n(ikr) = f_n(-ikr) = 1$, and thus, if we restore the time factor (e^{ikat}), we have

$$r\psi_n = S_n e^{ik(at-r)} + S_n{}' e^{ik(at+r)}\ldots\ldots\ldots\ldots(1),$$

of which the second part represents a disturbance travelling inwards. Under the circumstances contemplated we are therefore to take $S_n{}' = 0$, and thus

$$r\psi_n = S_n f_n(ikr) e^{ik(at-r)}\ldots\ldots\ldots\ldots\ldots(2),$$

which represents in the most general manner the nth harmonic component of a disturbance of the given period diffusing itself outwards into infinite space.

The origin of the disturbance may be in a prescribed normal motion of the surface of a sphere of radius c. Let us suppose that at any point on the sphere the outward velocity is represented by $U e^{ikat}$, U being in general a function of the position of the point considered.

If U be expanded in the spherical harmonic series

$$U = U_0 + U_1 + U_2 + \dots + U_n + \dots\dots\dots\dots(3),$$

we must have by (13) § 323

$$U_n = - \frac{S_n}{c^2} e^{-ikc} F_n (ikc)\dots\dots\dots\dots\dots(4).$$

The complete value of ψ is thus

$$\psi = - \frac{c^2}{r} e^{ik(at-r+c)} \Sigma \frac{U_n}{F_n (ikc)} f_n (ikr)\dots\dots\dots(5),$$

where the summation is to be extended to all (integral) values of n. The real part of this equation will give the velocity potential due to the normal velocity $U \cos kat$[1] at the surface of the sphere $r = c$.

Prof. Stokes has applied this solution to the explanation of a remarkable experiment by Leslie, according to which it appeared that the sound of a bell vibrating in a partially exhausted receiver is diminished by the introduction of hydrogen. This paradoxical phenomenon has its origin in the augmented wave-length due to the addition of hydrogen, in consequence of which the bell loses its hold (so to speak) on the surrounding gas. The general explanation cannot be better given than in the words of Prof. Stokes:

"Suppose a person to move his hand to and fro through a small space. The motion which is occasioned in the air is almost exactly the same as it would have been if the air had been an incompressible fluid. There is a mere local reciprocating motion, in which the air immediately in front is pushed forward, and that immediately behind impelled after the moving body, while in the interior space generally the air recedes from the encroachment of the moving body, and in the posterior space generally flows in from all sides to supply the vacuum which tends to be created; so that in lateral directions the flow of the fluid is backwards, a

---

[1] The assumption of a real value for U is equivalent to limiting the normal velocity to be in the same phase all over the sphere $r=c$. To include the most general aerial motion U would have to be treated as complex.

portion of the excess of fluid in front going to supply the deficiency behind. Now conceive the periodic time of the motion to be continually diminished. Gradually the alternation of movement becomes too rapid to permit of the full establishment of the merely local reciprocating flow; the air is sensibly compressed and rarefied, and a sensible sound wave (or wave of the same nature, in case the periodic time be beyond the limits suitable to hearing) is propagated to a distance. The same takes place in any gas; and the more rapid be the propagation of condensations and rarefactions in the gas, the more nearly will it approach, in relation to the motions we have under consideration, to the condition of an incompressible fluid; the more nearly will the conditions of the displacement of the gas at the surface of the solid be satisfied by a merely local reciprocating flow."

In discussing the solution (5), Prof. Stokes goes on to say,

"At a great distance from the sphere the function $f_n(ikr)$[1] becomes ultimately equal to 1, and we have

$$\psi = -\frac{c^2}{r} e^{ik(at-r+c)} \Sigma \frac{U_n}{F_n(ikc)} \quad \dots\dots\dots\dots\dots (6).$$

"It appears (from the value of $d\psi/dr$) that the component of the velocity along the radius vector is of the order r^{-1}, and that in any direction perpendicular to the radius vector of the order r^{-2} so that the lateral motion may be disregarded except in the neighbourhood of the sphere.

"In order to examine the influence of the lateral motion in the neighbourhood of the sphere, let us compare the actual disturbance at a great distance with what it would have been if all lateral motion had been prevented, suppose by infinitely thin conical partitions dividing the fluid into elementary canals, each bounded by a conical surface having its vertex at the centre.

"On this supposition the motion in any canal would evidently be the same as it would be in all directions if the sphere vibrated by contraction and expansion of the surface, the same all round, and such that the normal velocity of the surface was the same as it is at the particular point at which the canal in question abuts on the surface. Now if U were constant the expansion of U would

---

[1] I have made some slight changes in Prof. Stokes' notation.

be reduced to its first term U_0, and seeing that $f_0(ikr) = 1$, we should have from (5),

$$\psi = -\frac{c^2}{r} e^{ik(at-r+c)} \frac{U_0}{F_0(ikc)}.$$

This expression will apply to any particular canal if we take U_0 to denote the normal velocity at the sphere's surface for that particular canal; and therefore to obtain an expression applicable at once to all the canals, we have merely to write U for U_0. To facilitate a comparison with (5) and (6), I shall, however, write ΣU_n for U. We have then,

$$\psi = -\frac{c^2}{r} e^{ik(at-r+c)} \frac{\Sigma U_n}{F_0(ikc)} \quad \dots\dots\dots\dots\dots (7).$$

It must be remembered that this is merely an expression applicable at once to all the canals, the motion in each of which takes place wholly along the radius vector, and accordingly the expression is not to be differentiated with respect to θ or ω with the view of finding the transverse velocities.

"On comparing (7) with the expression for the function ψ in the actual motion at a great distance from the sphere (6), we see that the two are identical with the exception that U_n is divided by two different constants, namely $F_0(ikc)$ in the former case and $F_n(ikc)$ in the latter. The same will be true of the leading terms (or those of the order r^{-1}) in the expressions for the condensation and velocity. Hence if the mode of vibration of the sphere be such that the normal velocity of its surface is expressed by a Laplace's function of any one order, the disturbance at a great distance from the sphere will vary from one direction to another according to the same law as if lateral motions had been prevented, the amplitude of excursion at a given distance from the centre varying in both cases as the amplitude of excursion, in a normal direction, of the surface of the sphere itself. The only difference is that expressed by the symbolic ratio $F_n(ikc) : F_0(ikc)$. If we suppose $F_n(ikc)$ reduced to the form $\mu_n(\cos\alpha_n + i\sin\alpha_n)$, the amplitude of vibration in the actual case will be to that in the supposed case as μ_0 to μ_n, and the phases in the two cases will differ by $\alpha_0 - \alpha_n$.

"If the normal velocity of the surface of the sphere be not expressible by a single Laplace's Function, but only by a series, finite or infinite, of such functions, the disturbance at a given great distance from the centre will no longer vary from one direction to another according to the same law as the normal velocity of the surface of the sphere, since the modulus μ_n and likewise the amplitude α_n of the imaginary quantity $F_n(ikc)$ vary with the order of the function.

" Let us now suppose the disturbance expressed by a Laplace's function of some one order, and seek the numerical value of the alteration of intensity at a distance, produced by the lateral motion which actually exists.

"The intensity will be measured by the *vis viva* produced in a given time, and consequently will vary as the density multiplied by the velocity of propagation multiplied by the square of the amplitude of vibration. It is the last factor alone that is different from what it would have been if there had been no lateral motion. The amplitude is altered in the proportion of μ_0 to μ_n, so that if $\mu_n{}^2 : \mu_0{}^2 = I_n$, I_n is the quantity by which the intensity that would have existed if the fluid had been hindered from lateral motion has to be divided.

"If λ be the length of the sound-wave corresponding to the period of the vibration, $k = 2\pi/\lambda$, so that kc is the ratio of the circumference of the sphere to the length of a wave. If we suppose the gas to be air and λ to be 2 feet, which would correspond to about 550 vibrations in a second, and the circumference $2\pi c$ to be 1 foot (a size and pitch which would correspond with the case of a common house-bell), we shall have $kc = \frac{1}{2}$. The following table gives the values of the squares of the modulus and of the

kc	$n=0$	$n=1$	$n=2$	$n=3$	$n=4$	
4	17	16·25	14·879	13·848	20·177	Values of $\mu_n{}^2$.
2	5	5	9·3125	80	1495·8	
1	2	5	89	3965	300137	
0·5	1·25	16·25	1330·2	236191	72086371	
0·25	1·0625	64·062	20878	14837899	18160×10^6	
4	1	0·95588	0·87523	0·81459	1·1869	Values of I_n.
2	1	1	1·8625	16	299·16	
1	1	2·5	44·5	1982·5	150068	
0·5	1	13	1064·2	188953	57669097	
0·25	1	60·294	19650	13965×10^3	17092×10^6	

ratio I_n for the functions $F_n(ikc)$ of the first five orders, for each of the values 4, 2, 1, $\frac{1}{2}$, and $\frac{1}{4}$ of kc. It will presently appear why the table has been extended further in the direction of values greater than $\frac{1}{2}$ than it has in the opposite direction. Five significant figures at least are retained.

" When $kc = \infty$ we get from the analytical expressions $I_n = 1$. We see from the table that when kc is somewhat large I_n is liable to be a *little* less than 1, and consequently the sound to be *a little* more intense than if lateral motion had been prevented. The possibility of that is explained by considering that the waves of condensation spreading from those compartments of the sphere which at a given moment are vibrating positively, *i.e.* outwards, after the lapse of a half period may have spread over the neighbouring compartments, which are now in their turn vibrating positively, so that these latter compartments in their outward motion work against a somewhat greater pressure than if such compartment had opposite to it only the vibration of the gas which it had itself occasioned; and the same explanation applies *mutatis mutandis* to the waves of rarefaction. However, the increase of sound thus occasioned by the existence of lateral motion is but small in any case, whereas when kc is somewhat small I_n increases enormously, and the sound becomes a mere nothing compared with what it would have been had lateral motion been prevented.

"The higher be the order of the function, the greater will be the number of compartments, alternately positive and negative as to their mode of vibration at a given moment, into which the surface of the sphere will be divided. We see from the table that for a given periodic time as well as radius the value of I_n becomes considerable when n is somewhat high. However practically vibrations of this kind are produced when the elastic sphere executes, not its principal, but one of its subordinate vibrations, the pitch corresponding to which rises with the order of vibration, so that k increases with that order. It was for this reason that the table was extended from $kc = 0.5$ further in the direction of high pitch than low pitch, namely, to three octaves higher and only one octave lower.

" When the sphere vibrates symmetrically about the centre, *i.e.* so that any two opposite points of the surface are at a given moment moving with equal velocities in opposite directions, or more generally when the mode of vibration is such that there is no change of position of the centre of gravity of the volume, there is no term of order 1. For a sphere vibrating in the manner of a bell the principal vibration is that expressed by a term of the order 2, to which I shall now more particularly attend.

" Putting, for shortness, $k^2c^2 = q$, we have

$$\mu_0{}^2 = q + 1, \quad \mu_2{}^2 = (q^{\frac{1}{2}} + 9q^{-\frac{1}{2}})^2 + (4 - 9q^{-1})^2 = q - 2 + 9q^{-1} + 81q^{-2},$$

$$I_2 = \frac{q^3 - 2q^2 + 9q + 81}{q^2 (q + 1)}.$$

" The minimum value of I_2 is determined by

$$q^3 - 6q^2 - 84q - 54 = 0,$$

giving approximately,

$$q = 12\cdot859, \qquad kc = 3\cdot586, \qquad \mu_0{}^2 = 13\cdot859, \qquad \mu_2{}^2 = 12\cdot049,$$

$$I_2 = \cdot86941 ;$$

so that the utmost increase of sound produced by lateral motion amounts to about 15 per cent.

"I now come more particularly to Leslie's experiments. Nothing is stated as to the form, size, or pitch of his bell; and even if these had been accurately described, there would have been a good deal of guess-work in fixing on the size of the sphere which should be considered the best representative of the bell. Hence all we can do is to choose such values for k and c as are comparable with the probable conditions of the experiment.

" I possess a bell, belonging to an old bell-in-air apparatus, which may probably be somewhat similar to that used by Leslie. It is nearly hemispherical, the diameter is $1\cdot96$ inch, and the pitch an octave above the middle c of a piano. Taking the number of vibrations 1056 per second, and the velocity of sound in air 1100 feet per second, we have $\lambda = 12\cdot5$ inches. To represent the bell by a sphere of the same radius would be very greatly to underrate the influence of local circulation, since near the mouth the gas has but a little way to get round from the outside to the inside or the reverse. To represent it by a sphere of half the radius would still apparently be to underrate the effect. Nevertheless for the sake of rather under-estimating than exaggerating the influence of the cause here investigated, I will make these two suppositions successively, giving respectively $c = \cdot98$ and $c = \cdot49$, $kc = \cdot4926$, and $kc = \cdot2463$ for air.

" If it were not for lateral motion the intensity would vary from gas to gas in the proportion of the density into the velocity of propagation, and therefore as the pressure into the square root of the density under a standard pressure, if we take the factor depending on the development of heat as sensibly the same for the gases and gaseous mixtures with which we have to deal. In the following Table the first column gives the gas, the second the

Gas.	P	D	Q_r	$c = \cdot98$			$c = \cdot49$		
				q	I_2	Q	q	I_2	Q
Air..................	1	1	1	·2427	1136	1	·06067	20890	1
Hydrogen..........	1	·0690	·2627	·01674	284700	·001048	·004186	4604000	·001191
Air rarefied	·01	·01	·01	·2427	1136	·01	·06067	20890	·01
The same filled with H...	1	·0783	·2798	·01900	220600	·001440	·004751	3572000	·001637
Air of same density	·0783	·0783	·0783	·2427	1136	·0783	·06067	20890	·0783
Air rarefied ½	·5	·5	·5	·2427	1136	·5	·06067	20890	·5
The same filled with H....	1	·5345	·7311	·1297	4322	·1921	·0324	74890	·2039

pressure p, in atmospheres, the third the density D under the pressure p, referred to the density of the air at the atmospheric pressure as unity, the fourth, Q_r, what would have been the intensity had the motion been wholly radial, referred to the intensity in air at atmospheric pressure as unity, or, in other words, a quantity varying as $p \times$ (the density at pressure 1)$^{\frac{1}{2}}$. Then follow the values of q, I_2, and Q, the last being the actual intensity referred to air as before.

"An inspection of the numbers contained in the columns headed Q will shew that the cause here investigated is amply sufficient to account for the facts mentioned by Leslie."

The importance of the subject, and the masterly manner in which it has been treated by Prof. Stokes, will probably be thought sufficient to justify this long quotation. The simplicity of the true explanation contrasts remarkably with conjectures that had previously been advanced. Sir J. Herschel, for example, thought that the mixture of two gases tending to propagate sound with different velocities might produce a confusion resulting in a rapid stifling of the sound.

[The subject now under consideration may be still more simply illustrated by the problems of §§ 268, 301. The former, for instance, may be regarded as the extreme case of the present, in which the spherical surface is reduced to a plane vibrating in rectangular segments. If we suppose the size of these segments, determined by p and q, to be given, and trace the effect of gradually increasing frequency, we see that it is only when the frequency attains a certain value that sensible vibrations are propagated to infinity, the law of diminution with distance being exponential in its form. On the other hand vibrations whose frequency exceeds the critical value are propagated without loss, escaping the attenuation to which spherical waves must of necessity submit.]

325. The term of zero order

$$\psi_0 = \frac{S_0}{r} e^{ik(at-r)} \quad \dots \dots \dots \dots \dots (1).$$

where S_0 is a complex constant, corresponds to the potential of a *simple source* of arbitrary intensity and phase, situated at the centre of the sphere (§ 279). If, as often happens in practice, the source of sound be a solid body vibrating without much change of volume, this term is relatively deficient. In the case of a rigid sphere vibrating about a position of equilibrium, the deficiency is absolute[1], inasmuch as the whole motion will then be represented by a term of order 1; and whenever the body is very small in comparison with the wave-length, the term of zero order must be insignificant. For if we integrate the equation of motion, $\nabla^2 \psi + k^2 \psi = 0$, over the small volume included between the body and a sphere closely surrounding it, we see that the whole quantity of fluid which enters and leaves this space is small, and that therefore there is but little total flow across the surface of the sphere.

Putting $n = 1$, we get for the term of the first order

$$r\psi_1 = S_1 e^{ik(at-r)} \left\{ 1 + \frac{1}{ikr} \right\} \dots\dots\dots\dots(2),$$

and S_1 is proportional to the cosine of the angle between the direction considered and some fixed axis. This expression is of the same form as the potential of a *double* source (§ 294), situated at the centre, and composed of two equal and opposite simple sources lying on the axis in question, whose distance apart is infinitely small, and intensities such that the product of the intensities and distance is finite. For, if x be the axis, and the cosine of the angle between x and r be μ, it is evident that the potential of the double source is proportional to

$$\frac{d}{dx}\left(\frac{e^{-ikr}}{r}\right) = \mu \frac{d}{dr}\left(\frac{e^{-ikr}}{r}\right) = -ik\frac{\mu e^{-ikr}}{r}\left\{1 + \frac{1}{ikr}\right\}.$$

It appears then that the disturbance due to the vibration of a sphere as a rigid body is the same as that corresponding to a double source at the centre whose axis coincides with the line of the sphere's vibration.

The reaction of the air on a small sphere vibrating as a rigid body with a harmonic motion, may be readily calculated from preceding formulæ. If $\dot{\xi}$ denote the velocity of the sphere at time t,

$$U_1 e^{ikat} = \dot{\xi}\mu \dots\dots\dots\dots\dots(3),$$

and therefore for the value of ψ at the surface of the sphere, we have from (5) § 324,

$$\dot{\psi} = -ikac\dot{\xi}\mu \frac{f_1(ikc)}{F_1(ikc)} \dots\dots\dots\dots(4).$$

---

[1] The centre of the sphere being the origin of coordinates.

The force Ξ due to aerial pressures accelerating the motion is given by

$$\Xi = - \iint \mu \, \delta p \, dS = \rho \iint \mu \dot{\psi} \, dS$$

$$= - ika\rho c^3 \dot{\xi} \frac{f_1(ikc)}{F_1(ikc)} \int 2\pi \mu^2 d\mu = - ika \frac{4\pi c^3}{3} \rho \dot{\xi} \frac{f_1(ikc)}{F_1(ikc)}.$$

If we write

$$\frac{f_1(ikc)}{F_1(ikc)} = p - i q \dots\dots\dots\dots\dots\dots(5),$$

then

$$\Xi = - p \cdot \tfrac{4}{3}\pi\rho c^3 \cdot \ddot{\xi} - qka \cdot \tfrac{4}{3}\pi\rho c^3 \cdot \dot{\xi} \dots\dots\dots\dots(6),$$

inasmuch as

$$\ddot{\xi} = ika \dot{\xi}.$$

The operation of the air is therefore to increase the effective inertia of the sphere by p times the inertia of the air displaced, and to retard the motion by a force proportional to the velocity, and equal to $\tfrac{4}{3}\pi\rho c^3 \cdot qka\dot{\xi}$, these effects being in general functions of the frequency of vibration. By introduction of the values of f_1 and F_1 we find

$$\frac{f_1(ikc)}{F_1(ikc)} = \frac{2 + k^2c^2 - i k^3 c^3}{4 + k^4 c^4} \dots\dots\dots\dots(7);$$

so that,

$$p = \frac{2 + k^2 c^2}{4 + k^4 c^4}, \qquad q = \frac{k^3 c^3}{4 + k^4 c^4} \dots\dots\dots\dots(8).$$

When kc is small, we have approximately $p = \tfrac{1}{2}$, $q = \tfrac{1}{4} k^3 c^3$. Hence the effective inertia of a small sphere is increased by one-half of that of the air displaced—a quantity independent of the frequency and the same as if the fluid were incompressible. The dissipative term, which corresponds to the energy emitted, is of high order in kc, and therefore (the effects of viscosity being disregarded) the vibrations of a small sphere are but slowly damped.

The motion of an ellipsoid through an incompressible fluid has been investigated by Green[1], and his result is applicable to the calculation of the increase of effective inertia due to a compressible fluid, provided the dimensions of the body be small in comparison with the wave-length of the vibration. For a small circular disc vibrating at right angles to its plane, the increase of effective inertia is to the mass of a sphere of fluid, whose radius is equal to

[1] *Edinburgh Transactions*, Dec. 16, 1833. Also Green's *Mathematical Papers*, edited by Ferrers. Macmillan & Co., 1871.

that of the disc, as 2 to π. The result for the case of a sphere given above was obtained by Poisson[1], a short time before the publication of Green's paper.

It has been proved by Maxwell[2] that the various terms of the harmonic expansion of the common potential may be regarded as due to *multiple points* of corresponding degrees of complexity. Thus V_i is proportional to $\dfrac{d^i}{dh_1 dh_2 \dots dh_i}\left(\dfrac{1}{r}\right)$, where there are i differentiations of r^{-1} with respect to the axes h_1, h_2, &c., any number of which may in particular cases coincide. It might perhaps have been expected that a similar law would hold for the velocity potential with the substitution of $r^{-1}e^{-ikr}$ for r^{-1}. This however is not the case; it may be shewn that the potential of a quadruple source, denoted by $\dfrac{d^2}{dh_1 dh_2}\cdot\dfrac{e^{-ikr}}{r}$, corresponds in general not to the term of the second order simply, viz., $S_2\dfrac{e^{-ikr}}{r}f_2(ikr)$, but to a combination of this with a term of zero order. The analogy therefore holds only in the single instance of the *double* point or source, though of course the function $r^{-1}e^{-ikr}$ after any number of differentiations continues to satisfy the fundamental equation

$$(\nabla^2 + k^2)\,\psi = 0.$$

It is perhaps worth notice that the disturbance outside any imaginary sphere which completely encloses the origin of sound may be represented as due to the normal motion of the surface of any smaller concentric sphere, or, as a particular case when the radius of the sphere is infinitely small, as due to a source concentrated in one point at the centre. This source will in general be composed of a combination of multiple sources of all orders of complexity.

326. When the origin of the disturbance is the vibration of a rigid body parallel to its axis of revolution, the various spherical harmonics S_n reduce to simple multiples of the zonal harmonic $P_n(\mu)$, which may be defined as the coefficient of e^n in the expansion of $\{1 - 2e\mu + e^2\}^{-\frac{1}{2}}$ in rising powers of e. [For the forms of these functions see § 334.] And whenever the solid, besides being

---

[1] *Mémoires de l'Académie des Sciences*, Tom. XI. p. 521.

[2] Maxwell's *Electricity and Magnetism*, Ch. IX.

symmetrical about an axis, is also symmetrical with respect to an equatorial plane (whose intersection with the axis is taken as origin of co-ordinates), the expansion of the resulting disturbance in spherical harmonics will contain terms of odd order only. For example, if the vibrating body were a circular disc moving perpendicularly to its plane, the expansion of ψ would contain terms proportional to $P_1(\mu)$, $P_3(\mu)$, $P_5(\mu)$, &c. In the case of the sphere, as we have seen, the series reduces absolutely to its first term, and this term will generally be preponderant.

On the other hand we may have a vibrating system symmetrical about an axis and with respect to an equatorial plane, but in such a manner that the motions of the parts on the two sides of the plane are opposed. Under this head comes the ideal tuning-fork, composed of equal spheres or parallel circular discs, whose distance apart varies periodically. Symmetry shews that the velocity-potential, being the same at any point and at its image in the plane of symmetry, must be an even function of μ, and therefore expressible by a series containing only the even functions $P_0(\mu)$, $P_2(\mu)$, &c. The second function $P_2(\mu)$ would usually preponderate, though in particular cases, as for example if the body were composed of two discs very close together in comparison with their diameter, the symmetrical term of zero order might become important. A comparison with the known solution for the sphere whose surface vibrates according to any law, will in most cases furnish material for an estimate as to the relative importance of the various terms.

[The accompanying table, p. 251, giving P_n as a function of θ, or $\cos^{-1}\mu$, is abbreviated from that of Perry[1].]

327. The total emission of energy by a vibrating sphere is found by multiplying the variable part of the pressure (proportional to $\dot\psi$) by the normal velocity and integrating over the surface (§ 245). In virtue of the conjugate property the various spherical harmonic terms may be taken separately without loss of generality. We have (§ 323)

$$\left.\begin{aligned} \dot\psi_n &= ika\, \frac{S_n\, e^{ik(at-r)}}{r}\, f_n(ikr) \\ \frac{d\psi_n}{dr} &= -\, \frac{S_n\, e^{ik(at-r)}}{r^2}\, F_n(ikr) \end{aligned}\right\} \quad \dots\dots\dots\dots\dots(1),$$

[1] Phil. Mag. vol. xxxii., p. 516, 1891.

Table of Zonal Spherical Harmonics.

θ	P_1	P_2	P_3	P_4	P_5	P_6	P_7
0	1·0000	1·0000	1·0000	1·0000	1·0000	1·0000	1·0000
2	·9994	·9982	·9963	·9939	·9909	·9872	·9829
4	·9976	·9927	·9854	·9758	·9638	·9495	·9329
6	·9945	·9836	·9674	·9459	·9194	·8881	·8522
8	·9903	·9709	·9423	·9048	·8589	·8053	·7448
10	·9848	·9548	·9106	·8532	·7840	·7045	·6164
12	·9781	·9352	·8724	·7920	·6966	·5892	·4732
14	·9703	·9122	·8283	·7224	·5990	·4635	·3219
16	·9613	·8860	·7787	·6454	·4937	·3322	·1699
18	·9511	·8568	·7240	·5624	·3836	·2002	·0289
20	·9397	·8245	·6649	·4750	·2715	·0719	− ·1072
22	·9272	·7895	·6019	·3845	·1602	− ·0481	− ·2201
24	·9135	·7518	·5357	·2926	·0525	− ·1559	− ·3095
26	·8988	·7117	·4670	·2007	− ·0489	− ·2478	− ·3717
28	·8829	·6694	·3964	·1105	− ·1415	− ·3211	− ·4052
30	·8660	·6250	·3248	·0234	− ·2233	− ·3740	− ·4101
32	·8480	·5788	·2527	− ·0591	− ·2923	− ·4052	− ·3876
34	·8290	·5310	·1809	− ·1357	− ·3473	− ·4148	− ·3409
36	·8090	·4818	·1102	− ·2052	− ·3871	− ·4031	− ·2738
38	·7880	·4314	·0413	− ·2666	− ·4112	− ·3719	− ·1918
40	·7660	·3802	− ·0252	− ·3190	− ·4197	− ·3234	− ·1003
42	·7431	·3284	− ·0887	− ·3616	− ·4128	− ·2611	− ·0065
44	·7193	·2762	− ·1485	− ·3940	− ·3914	− ·1878	·0846
46	·6947	·2238	− ·2040	− ·4158	− ·3568	− ·1079	·1666
48	·6691	·1716	− ·2547	− ·4270	− ·3105	− ·0251	·2349
50	·6428	·1198	− ·3002	− ·4275	− ·2545	+ ·0563	·2854
52	·6157	·0686	− ·3401	− ·4178	− ·1910	+ ·1326	·3153
54	·5878	·0182	− ·3740	− ·3984	− ·1223	+ ·2002	·3234
56	·5592	− ·0310	− ·4016	− ·3698	− ·0510	+ ·2559	·3095
58	·5299	− ·0788	− ·4229	− ·3331	·0206	+ ·2976	·2752
60	·5000	− ·1250	− ·4375	− ·2891	·0898	+ ·3232	·2231
62	·4695	− ·1694	− ·4455	− ·2390	·1545	+ ·3321	·1571
64	·4384	− ·2117	− ·4470	− ·1841	·2123	+ ·3240	·0818
66	·4067	− ·2518	− ·4419	− ·1256	·2615	+ ·2996	·0021
68	·3746	− ·2896	− ·4305	− ·0650	·3005	+ ·2605	− ·0763
70	·3420	− ·3245	− ·4130	− ·0038	·3281	+ ·2089	− ·1485
72	·3090	− ·3568	− ·3898	·0568	·3434	+ ·1472	− ·2099
74	·2756	− ·3860	− ·3611	·1153	·3461	+ ·0795	− ·2559
76	·2419	− ·4132	− ·3275	·1705	·3362	+ ·0076	− ·2848
78	·2079	− ·4352	− ·2894	·2211	·3143	− ·0644	− ·2943
80	·1736	− ·4548	− ·2474	·2659	·2810	− ·1321	− ·2835
82	·1392	− ·4709	− ·2020	·3040	·2378	− ·1926	− ·2536
84	·1045	− ·4836	− ·1539	·3345	·1861	− ·2431	− ·2067
86	·0698	− ·4927	− ·1038	·3569	·1278	− ·2811	− ·1460
88	·0349	− ·4982	− ·0522	·3704	·0651	− ·3045	− ·0735
90	·0000	− ·5000	− ·0000	·3750	·0000	− ·3125	·0000

or on rejecting the imaginary part

$$\left.\begin{aligned}
\dot{\psi}_n &= -\frac{kaS_n}{r}\{\beta'\cos k\,(at-r)+\alpha'\sin k\,(at-r)\}\\
\frac{d\psi_n}{dr} &= -\frac{S_n}{r^2}\{\ \alpha\cos k\,(at-r)-\beta\sin k\,(at-r)\}
\end{aligned}\right\} \quad \ldots\ldots(2),$$

where $$F=\alpha+i\beta, \qquad f=\alpha'+i\beta' \ldots\ldots\ldots\ldots\ldots(3).$$

Thus $$\iint \dot{\psi}_n \frac{d\psi_n}{dr}\,dS = \iint \dot{\psi}_n \frac{d\psi_n}{dr}\,.\,r^2 d\sigma$$

$$= \frac{ka}{r}\iint S_n^2 d\sigma\,\{\alpha\beta'\cos^2 k\,(at-r)-\alpha'\beta\sin^2 k\,(at-r)$$
$$+ (\alpha\alpha'-\beta\beta')\sin k\,(at-r)\cos k\,(at-r)\}.$$

When this is integrated over a long range of time, the periodic terms may be omitted, and thus

$$\int . \iint \dot{\psi}_n \frac{d\psi_n}{dr}\,dS\,.\,dt = \frac{kat}{2r}\,(\alpha\beta'-\alpha'\beta)\iint S_n^2 d\sigma \ \ldots\ldots(4).$$

Now, since there can be on the whole no accumulation of energy in the space included between two concentric spherical surfaces, the rates of transmission of energy across these surfaces must be the same, that is to say $r^{-1}\,(\alpha'\beta-\beta'\alpha)$ must be independent of r. In order to determine the constant value, we may take the particular case of r indefinitely great, when

$$F_n\,(ikr)=ikr, \qquad \alpha=0, \qquad \beta=kr,$$
$$f_n\,(ikr)=1, \qquad \alpha'=1, \qquad \beta'=0.$$

Thus $$\alpha'\beta-\beta'\alpha=kr, \quad \text{identically} \ldots\ldots\ldots\ldots\ldots(5).$$

It may be observed that the left-hand member of (5) when multiplied by i is the imaginary part of $(\alpha+i\beta)\,(\alpha'-i\beta')$ or of $F_n\,(ikr)f_n\,(-ikr)$, so that our result may be expressed by saying that the imaginary part of $F_n\,(ikr)\,f_n\,(-ikr)$ is ikr, or

$$F_n\,(ikr)\,f_n\,(-ikr)-F_n\,(-ikr)\,f_n\,(ikr)=2ikr \ \ldots\ldots(6).$$

In this form we shall have occasion presently to make use of it.

The same conclusion may be arrived at somewhat more directly by an application of Helmholtz's theorem (§ 294), i.e. that if two functions u and v satisfy through a closed space S the equation $(\nabla^2 + k^2)\,u = 0$, then

$$\iint\left(u\frac{dv}{dn}-v\frac{du}{dn}\right)dS=0\ldots\ldots\ldots\ldots\ldots(7).$$

If we take for S the space between two concentric spheres, making

$$u = \frac{S_n e^{-ikr} f_n(ikr)}{r}, \qquad v = \frac{S_n e^{+ikr} f_n(-ikr)}{r},$$

we find that $r^{-1}\{F_n(ikr) f_n(-ikr) - F_n(-ikr) f_n(ikr)\}$ must be independent of r.

We have therefore

$$\int . \iint \dot{\psi}_n \frac{d\psi_n}{dr} \, dS \, . \, dt = -\tfrac{1}{2} k^2 a t \iint S_n^2 \, d\sigma \, ;$$

so that the expression for the energy emitted in time t is (since $\delta p = -\rho \dot{\psi}$)

$$W = \tfrac{1}{2} k^2 \rho \, at \iint S_n^2 \, d\sigma \, \dotfill (8).$$

It will be more instructive to exhibit W as a function of the normal motion at the surface of a sphere of radius c. From (2)

$$\frac{d\psi_n}{dr} = -\frac{S_n}{c^2} [\cos kat \, (\alpha \cos kc + \beta \sin kc)$$
$$+ \sin kat \, (\alpha \sin kc - \beta \cos kc)],$$

so that, if the amplitude of $d\psi_n/dr$ be U_n, we have as the relation between S_n and U_n

$$c^4 U_n^2 = (\alpha^2 + \beta^2) \, S_n^2 \, \dotfill (9).$$

Thus
$$W = \frac{k^2 c^4 \rho at}{2 \, (\alpha^2 + \beta^2)} \iint U_n^2 \, d\sigma \dotfill (10).$$

This formula may be verified for the particular cases $n = 0$ and $n = 1$, treated in §§ 280, 325 respectively.

328. If the source of disturbance be a normal motion of a small part of the surface of the sphere ($r = c$) in the immediate neighbourhood of the point $\mu = 1$, we must take in the general solution applicable to divergent waves, viz.

$$\psi = -\frac{c^2}{r} \, e^{ik(at-r+c)} \, \Sigma \, \frac{U_n}{F_n(ikc)} \, f_n(ikr) \, \dotfill (1),$$

$$U_n = \tfrac{1}{2}(2n+1) \, P_n(\mu) . \int_{-1}^{+1} U P_n(\mu) \, d\mu$$

$$= \tfrac{1}{2}(2n+1) \, P_n(\mu) \int_{-1}^{+1} U d\mu = \frac{2n+1}{4\pi c^2} \, P_n(\mu) \iint U \, dS \dotfill (2) \, ;$$

for where U is sensible, $P_n(\mu) = 1$. Thus

$$\psi = -\frac{e^{ik(at-r+c)}}{4\pi r} \cdot \iint U dS \cdot \Sigma (2n+1) P_n(\mu) \frac{f_n(ikr)}{F_n(ikc)} \quad \ldots\ldots(3).$$

In this formula $\iint U dS$ measures the intensity of the source.

If ikc be very small,

$$\frac{f_0(ikr)}{F_0(ikc)} = 1 - ikc + \ldots, \qquad \frac{f_1(ikr)}{F_1(ikc)} = \tfrac{1}{2}ikc\left(1 + \frac{1}{ikr}\right) + \ldots \&c.;$$

so that ultimately

$$\psi = -\frac{e^{ik(at-r)}}{4\pi r} \iint U dS \ldots\ldots\ldots\ldots\ldots \ldots(4),$$

and the waves diverge as from a simple source of equal magnitude.

We will now examine the problem when kc is not very small, taking for simplicity the case where ψ is required at a great distance only, so that $f_n(ikr) = 1$. The factor on which the relative intensities in various directions depend is

$$\Sigma \frac{(2n+1)}{2} \frac{P_n(\mu)}{F_n(ikc)} \ldots\ldots\ldots\ldots\ldots\ldots (5),$$

and a complete solution of the question would involve a discussion of this series as a function of μ and kc.

Thus, if

$$\Sigma \frac{(2n+1)}{2} \frac{P_n(\mu)}{F_n(ikc)} = F + iG \ldots\ldots\ldots\ldots (6),$$

$$\psi = -\frac{1}{2\pi r} \iint U dS \cdot \{F^2 + G^2\}^{\frac{1}{2}} \cdot e^{ik(at-r+c)+i\theta} \ldots\ldots\ldots(7),$$

where

$$\tan\theta = G : F \ldots\ldots \ldots\ldots\ldots\ldots (8).$$

The intensity of the vibrations in the various directions is thus measured by $F^2 + G^2$. If, as before, $F_n = \alpha + i\beta$,

$$\left.\begin{aligned} F &= \Sigma \frac{2n+1}{2} \frac{\alpha P_n(\mu)}{\alpha^2 + \beta^2} \\ -G &= \Sigma \frac{2n+1}{2} \frac{\beta P_n(\mu)}{\alpha^2 + \beta^2} \end{aligned}\right\} \ldots\ldots\ldots\ldots (9).$$

The following table gives the means of calculating F and G for any value of μ, when $kc = \tfrac{1}{2}$, 1, or 2. In the last case it is necessary to go as far as $n = 7$ to get a tolerably accurate result, and for larger values of kc the calculation would soon become very laborious. In all problems of this sort the harmonic analysis seems to lose its power when the waves are very small in comparison with the dimensions of bodies.

$$kc = \tfrac{1}{2}.$$

n	$2a$	2β	$(n+\tfrac{1}{2})a \div (a^2+\beta^2)$	$(n+\tfrac{1}{2})\beta \div (a^2+\beta^2)$
0	$+\quad 2$	$+\quad 1$	$+ \cdot 4$	$+ \cdot 2$
1	$+\quad 4$	$-\quad 7$	$+ \cdot 1846153$	$- \cdot 3230768$
2	$-\quad 64$	$-\quad 35$	$- \cdot 0601391$	$- \cdot 0328885$
3	$-\quad 466$	$+\quad 853$	$- \cdot 0034527$	$+ \cdot 0063201$
4	$+\ 14902$	$+\ 8141$	$+ \cdot 0004653$	$+ \cdot 0002542$
5	$+175592$	-321419	$+ \cdot 0000144$	$- \cdot 0000264$

$$kc = 1.$$

n	a	β	$(n+\tfrac{1}{2})a \div (a^2+\beta^2)$	$(n+\tfrac{1}{2})\beta \div (a^2+\beta^2)$
0	$+\quad 1$	$+\quad 1$	$+ \cdot 25$	$+ \cdot 25$
1	$+\quad 2$	$-\quad 1$	$+ \cdot 6$	$- \cdot 3$
2	$-\quad 5$	$-\quad 8$	$- \cdot 140449$	$- \cdot 224719$
3	$-\quad 53$	$+\quad 34$	$- \cdot 046784$	$+ \cdot 030013$
4	$+\quad 296$	$+\quad 461$	$+ \cdot 004438$	$+ \cdot 006912$
5	$+\ 4951$	$-\ 3179$	$+ \cdot 000787$	$- \cdot 000505$
6	$-\ 40613$	$-\ 63251$	$- \cdot 000047$	$- \cdot 000073$
7	-936340	$+601217$	$- \cdot 000006$	$+ \cdot 000004$

$$kc = 2.$$

n	a	β	$(n+\tfrac{1}{2})a \div (a^2+\beta^2)$	$(n+\tfrac{1}{2})\beta \div (a^2+\beta^2)$
0	$+\quad 1$	$+\quad 2$	$+ \cdot 1$	$+ \cdot 2$
1	$+\quad 2$	$+\quad 1$	$+ \cdot 6$	$+ \cdot 3$
2	$+\quad 1\cdot 75$	$-\quad 2\cdot 5$	$+ \cdot 46980$	$- \cdot 67114$
3	$-\quad 8$	$-\quad 4$	$- \cdot 35$	$- \cdot 175$
4	$-\ 16\cdot 1875$	$+\ 35\cdot 125$	$- \cdot 04870$	$+ \cdot 10567$
5	$+\ 186\cdot 625$	$+\ 85\cdot 4375$	$+ \cdot 02436$	$+ \cdot 01115$
6	$+\ 538\cdot 80$	$-1177\cdot 3$	$+ \cdot 00209$	$- \cdot 00456$
7	$-8621\cdot 7$	$-3945\cdot 8$	$- \cdot 00072$	$- \cdot 00033$

The most interesting question on which this analysis informs us is the influence which a rigid sphere, situated close to the source, has on the intensity of sound in different directions. By the principle of reciprocity (§ 294) the source and the place of observation may be interchanged. When therefore we know the relative intensities at two distant points B, B', due to a source A on the surface of the sphere, we have also the relative intensities (measured by potential) at the point A, due to distant sources at B and B'. On this account the problem has a double interest.

As a numerical example I have calculated the values of $F + iG$ and $F^2 + G^2$ for the above values of kc, when $\mu = 1$, $\mu = -1$, $\mu = 0$, that is, looking from the centre of the sphere, in the direction of the source, in the opposite direction, and laterally.

When kc is zero, the value of $F^2 + G^2$ is ·25, which therefore represents on the same scale as in the table the intensity due to an unobstructed source of equal magnitude. We may interpret kc as the ratio of the circumference of the sphere to the wave-length of the sound.

kc	μ	$F + iG$	$F^2 + G^2$
$\frac{1}{2}$	1	·521503 + ·149417i	·294291
	−1	·159149 − ·484149i	·259729
	0	·430244 − ·216539i	·231999
1	1	·667938 + ·238369i	·502961
	−1	− ·440055 − ·302609i	·285220
	0	+ ·321903 − ·364974i	·236828
2	1	·79683 + ·23421i	·6898
	−1	·24954 + ·50586i	·3182
	0	− ·15381 − ·57662i	·3562

In looking at these figures the first point which attracts attention is the comparatively slight deviation from uniformity in the intensities in different directions. Even when the circumference of the sphere amounts to twice the wave-length, there is scarcely anything to be called a sound shadow. But what is perhaps still more unexpected is that in the first two cases the intensity behind the sphere exceeds that in a transverse direction. This result depends mainly on the preponderance of the term of the first order, which vanishes with μ. The order of the more important terms increases with kc; when kc is 2, the principal term is that of the second order.

Up to a certain point the augmentation of the sphere will increase the total energy emitted, because a simple source emits twice as much energy when close to a rigid plane as when entirely in the open. Within the limits of the table this effect masks the obstruction due to an increasing sphere, so that when $\mu = -1$, the intensity is greater when the circumference is twice the wave-length than when it is half the wave-length, the source itself remaining constant.

If the source be not simple harmonic with respect to time, the relative proportions of the various constituents will vary to some extent both with the size of the sphere and with the direction of the point of observation, illustrating the fundamental character of the analysis into simple harmonics.

When kc is decidedly less than one-half, the calculation may be conducted with sufficient approximation algebraically. The result is

$$F^2 + G^2 = \tfrac{1}{4} + \tfrac{1}{12} k^2 c^2 \left(\tfrac{7}{4}\mu^2 - \tfrac{4}{3}\right)$$
$$+ \tfrac{1}{4} k^4 c^4 \left(1 + \tfrac{3}{2}\mu + \tfrac{50}{81}P_2 + \tfrac{25}{81}P_2{}^2 - \tfrac{7}{20}\mu P_3 + \tfrac{6}{175}P_4\right)$$
$$+ \text{ terms in } k^6 c^6 \quad\ldots\ldots\ldots\ldots\ldots\ldots \quad (10).$$

It appears that so far as the term in $k^2 c^2$, the intensity is an even function of μ, viz. the same at any two points diametrically opposed. For the principal directions $\mu = \pm 1$, or 0, the numerical calculation of the coefficient of $k^4 c^4$ is easy on account of the simple values then assumed by the functions P. Thus

$$(\mu = 1), \qquad F^2 + G^2 = \tfrac{1}{4} + \tfrac{5}{144} k^2 c^2 + \cdot 77755\, k^4 c^4 + \ldots\ldots$$
$$(\mu = -1), \quad F^2 + G^2 = \tfrac{1}{4} + \tfrac{5}{144} k^2 c^2 + \cdot 02755\, k^4 c^4 + \ldots\ldots$$
$$(\mu = 0), \qquad F^2 + G^2 = \tfrac{1}{4} - \tfrac{1}{9}\, k^2 c^2 + \cdot 19534\, k^4 c^4 + \ldots\ldots$$

When $k^4 c^4$ can be neglected, the intensity is less in a lateral direction than immediately in front of or behind the sphere. Or, by the reciprocal property, a source at a distance will give a greater intensity on the surface of a small sphere at the point furthest from the source than in a lateral position.

If we apply these formulæ to the case of $kc = \tfrac{1}{2}$, we get

$$(\mu = 1), \qquad F^2 + G^2 = \cdot 3073,$$
$$(\mu = -1), \quad F^2 + G^2 = \cdot 2604,$$
$$(\mu = 0), \qquad F^2 + G^2 = \cdot 2344,$$

which agree pretty closely with the results of the more complete calculation.

For other values of μ, the coefficient of k^4c^4 in (10) might be calculated with the aid of tables of Legendre's functions, or from the following algebraic expression in terms of μ[1],

$$1 + \tfrac{3}{2}\mu + \tfrac{50}{81}P_2 + \tfrac{25}{81}P_2{}^2 - \tfrac{7}{20}\mu P_3 + \tfrac{6}{175}P_4$$
$$= \cdot 78138 + 1\cdot5\,\mu + \cdot85938\,\mu^2 - \cdot03056\,\mu^4.$$

The *difference* of intensities in the directions $\mu = +1$ and $\mu = -1$ may be very simply expressed. Thus

$$(F^2 + G^2)_{\mu=1} - (F^2 + G^2)_{\mu=-1} = \tfrac{3}{4}k^4c^4.$$

If $kc = \tfrac{3}{8}$, $\tfrac{3}{4}k^4c^4 = \cdot0148.$

If $kc = \tfrac{2}{8}$, $\tfrac{3}{4}k^4c^4 = \cdot0029.$

If $kc = \tfrac{1}{8}$, $\tfrac{3}{4}k^4c^4 = \cdot0002.$

At the same time the total value of $F^2 + G^2$ approximates to $\cdot25$, when kc is small.

These numbers have an interesting bearing on the explanation of the part played by the two ears in the perception of the quarter from which a sound proceeds.

It should be observed that the variations of intensity in different directions about which we have been speaking are due to the presence of the sphere as an obstacle, and not to the fact that the source is on the circumference of the sphere instead of at the centre. At a great distance a small displacement of a source of sound will affect the *phase* but not the *intensity* in any direction.

In order to find the alteration of phase we have for a small sphere

$$F = \tfrac{1}{2} + k^2c^2\left(-\tfrac{1}{2} + \tfrac{3}{4}\mu - \tfrac{5}{18}P_2\right), \quad G = kc\left(-\tfrac{1}{2} + \tfrac{3}{4}\mu\right),$$

$$\tan\theta = G : F = kc\left(-1 + \tfrac{3}{2}\mu\right), \quad \text{or} \quad \theta = kc\left(-1 + \tfrac{3}{2}\mu\right) \text{ nearly.}$$

Thus in (7) $e^{ik\,(at-r+c)+i\theta} = e^{ik\,(at-r+\frac{3}{2}\mu c)},$

from which we may infer that the phase at a distance is the same as if the source had been situated at the point $\mu = 1$, $r = \tfrac{3}{2}c$ (instead of $r = c$), and there had been no obstacle.

329. The functional symbols f and F may be expressed in terms of P. It is known[2] that

$$P_n(\mu) = 1 - \frac{n}{1}\cdot\frac{n+1}{1}\cdot\frac{1-\mu}{2} + \frac{n(n-1)}{1\cdot2}\frac{(n+1)(n+2)}{1\cdot2}\frac{(1-\mu)^2}{2^2} - \cdots$$

[1] For the forms of the functions P, see § 334.

[2] Thomson and Tait's *Nat. Phil.* § 782 (quoted from Murphy).

or, on changing μ into $1 - \mu$,

$$P_n(1-\mu) = 1 - \frac{n}{1} \cdot \frac{n+1}{1} \cdot \frac{\mu}{2} + \frac{n(n-1)}{1 \cdot 2} \cdot \frac{(n+1)(n+2)}{1 \cdot 2} \cdot \frac{\mu^2}{2^2} - \ldots (1).$$

Consider now the symbolic operator $P_n\left(1 - \dfrac{d}{dy}\right)$, and let it operate on y^{-1}.

Since $\left(\dfrac{d}{dy}\right)^s \cdot \dfrac{1}{y} = (-1)(-2) \ldots \ldots (-s) y^{-s-1}$,

$$P_n\left(1 - \frac{d}{dy}\right) \cdot \frac{1}{y} = y^{-1} + \frac{n(n+1)}{1 \cdot 2} y^{-2} + \frac{(n-1) \ldots \ldots (n+2)}{2 \cdot 4} y^{-3} + \ldots \ldots$$

A comparison with (9) § 323 now shews that

$$f_n(y) = y P_n\left(1 - \frac{d}{dy}\right) \cdot \frac{1}{y} \ldots \ldots \ldots \ldots \ldots (2),$$

from which we deduce by a known formula,

$$\frac{e^{-y}}{y} f_n(y) = e^{-y} P_n\left(1 - \frac{d}{dy}\right) \frac{1}{y} = (-1)^n P_n\left(\frac{d}{dy}\right) \cdot \frac{e^{-y}}{y} \ldots \ldots (3).$$

In like manner,

$$\frac{e^{+y}}{y} f_n(-y) = P_n\left(\frac{d}{dy}\right) \cdot \frac{e^{+y}}{y}.$$

If we now identify y with ikr, we see that the general solution, (12) § 323, may be written

$$\psi_n = (-1)^n ik S_n P_n\left(\frac{d}{d \cdot ikr}\right) \cdot \frac{e^{-ikr}}{ikr} + ik S_n' P_n\left(\frac{d}{d \cdot ikr}\right) \cdot \frac{e^{+ikr}}{ikr} \ldots (4),$$

from which the second term is to be omitted, if no part of the disturbance be propagated inwards.

Again from (14) § 323 we see that

$$\frac{F_n(y)}{y^2} = \left(1 - \frac{d}{dy}\right) \cdot \frac{f_n(y)}{y},$$

whence $\quad F_n(y) = y^2 P_n\left(1 - \dfrac{d}{dy}\right)\left(1 - \dfrac{d}{dy}\right) \cdot \dfrac{1}{y} \ldots \ldots \ldots (5),$

and $\quad \dfrac{F_n(y) e^{-y}}{y^2} = -(-)^n P_n\left(\dfrac{d}{dy}\right) \dfrac{d}{dy} \cdot \dfrac{e^{-y}}{y} \ldots \ldots \ldots \ldots (6).$

Similarly, $\quad \dfrac{F_n(-y) e^{+y}}{y^2} = -P_n\left(\dfrac{d}{dy}\right) \dfrac{d}{dy} \cdot \dfrac{e^{+y}}{y} \ldots \ldots \ldots \ldots (7).$

Using these expressions in (13) § 323, we get

$$\frac{d\psi_n}{dr} = (-)^{n+1} k^2 S_n P_n \left(\frac{d}{d.ikr}\right) \frac{d}{d.ikr} \cdot \frac{e^{-ikr}}{ikr}$$

$$- k^2 S_n' P_n \left(\frac{d}{d.ikr}\right) \frac{d}{d.ikr} \cdot \frac{e^{+ikr}}{ikr} \ldots\ldots\ldots (8).$$

330. We have already considered in some detail the form assumed by our general expressions when there is no source at infinity. An equally important class of cases is defined by the condition that there be no source at the origin. We shall now investigate what restriction is thereby imposed on our general expressions.

Reversing the series for f_n, we have

$$r\psi_n = \frac{1 \cdot 3 \cdot 5 \ldots (2n-1)}{(ikr)^n} \{S_n e^{-ikr} (1 + ikr + \ldots) + (-1)^n S_n' e^{+ikr} (1 - ikr + \ldots)\},$$

shewing that, as r diminishes without limit, $r\psi_n$ approximates to

$$r\psi_n = \frac{1 \cdot 3 \cdot 5 \ldots (2n-1)}{(ikr)^n} \{S_n + (-1)^n S_n'\}.$$

In order therefore that ψ_n may be finite at the origin,

$$S_n + (-1)^n S_n' = 0 \ldots\ldots\ldots\ldots\ldots\ldots(1)$$

is a necessary condition; that it is sufficient we shall see later.

Accordingly (12) § 323 becomes

$$r\psi_n = S_n \{e^{-ikr} f_n (ikr) - (-1)^n e^{+ikr} f_n (-ikr)\} \ldots\ldots(2).$$

If, separating the real and imaginary parts of f_n, we write (as before)

$$f_n = \alpha' + i\beta' \ldots\ldots\ldots\ldots\ldots\ldots\ldots (3),$$

(2) may be put into the form

$$r\psi_n = - 2i^{n+1} S_n \{\alpha' \sin (kr + \tfrac{1}{2} n\pi) - \beta' \cos (kr + \tfrac{1}{2} n\pi)\} \ldots\ldots(4).$$

Another form may be derived from (4) § 329. We have

$$\psi_n = - 2ik (-1)^n S_n P_n \left(\frac{d}{d.ikr}\right) \cdot \frac{e^{+ikr} - e^{-ikr}}{2ikr}$$

$$= - 2ik (-1)^n S_n P_n \left(\frac{d}{d.ikr}\right) \cdot \frac{\sin kr}{kr} \ldots\ldots\ldots(5).$$

Since the function P_n is either wholly odd or wholly even, the expression for ψ_n is wholly real or wholly imaginary.

In order to prove that the value of ψ_n in (5) remains finite when r vanishes, we begin by observing that

$$\frac{2\sin kr}{kr} = \int_{-1}^{+1} e^{-ikr\mu}\, d\mu \quad \text{.....................(6)},$$

so that

$$2P_n\left(\frac{d}{d\,.\,ikr}\right)\frac{\sin kr}{kr} = \int_{-1}^{+1} P_n\left(\frac{d}{d\,.\,ikr}\right) e^{ikr\mu}\, d\mu$$

$$= \int_{-1}^{+1} P_n(\mu)\, e^{ikr\mu}\, d\mu \text{................(7)},$$

as is obvious when it is considered that the effect of differentiating $e^{ikr\mu}$ any number of times with respect to ikr is to multiply it by the corresponding power of μ. It remains to expand the expression on the right in ascending powers of r. We have

$$\int_{-1}^{+1} P_n(\mu)\, e^{ikr\mu}\, d\mu = \int_{-1}^{+1} d\mu\, P_n(\mu)\left\{1 + ikr\,.\,\mu + \frac{(ikr)^2}{1\,.\,2}\,.\,\mu^2 + \dots \right.$$
$$\left. + \frac{(ikr)^n}{1\,.\,2\dots n}\,.\,\mu^n + \dots\right\}.$$

Now any positive integral power of μ, such as μ^p, can be expanded in a terminating series of the functions P, the function of highest order being P_p. It follows that, if $p < n$,

$$\int_{-1}^{+1} \mu^p\, P_n(\mu)\, d\mu = 0,$$

by known properties of these functions; so that the lowest power of ikr in $\int_{-1}^{+1} P_n(\mu)\, e^{ikr\mu}\, d\mu$ is $(ikr)^n$. Retaining only the leading term, we may write

$$\int_{-1}^{+1} P_n(\mu)\, e^{ikr\mu}\, d\mu = \frac{(ikr)^n}{1\,.\,2\dots n}\int_{-1}^{+1} \mu^n\, P_n(\mu)\, d\mu.$$

From the expression for $P_n(\mu)$ in terms of μ, viz.

$$P_n(\mu) = \frac{1\,.\,3\,.\,5\dots(2n-1)}{1\,.\,2\,.\,3\dots n}\left\{\mu^n - \frac{n(n-1)}{2(2n-1)}\,\mu^{n-2} \right.$$
$$\left. + \frac{n(n-1)(n-2)(n-3)}{2\,.\,4\,.\,(2n-1)(2n-3)}\,\mu^{n-4} - \dots\right\} \text{...........(8)},$$

we see that

$$\mu^n = \frac{1\,.\,2\,.\,3\dots n}{1\,.\,3\,.\,5\dots(2n-1)}\, P_n(\mu) + \text{terms in } \mu \text{ of lower order than } \mu^n\,;$$

and therefore

$$\int_{-1}^{+1} \mu^n P_n(\mu)\, d\mu = \frac{1.2.3\ldots n}{1.3.5\ldots(2n-1)} \int_{-1}^{+1} [P_n(\mu)]^2\, d\mu$$

$$= \frac{1.2.3\ldots n}{1.3.5\ldots(2n-1)} \cdot \frac{2}{2n+1} \ldots\ldots\ldots\ldots (9).$$

Accordingly, by (5) and (7)

$$\psi_n = -2ik(-1)^n S_n \frac{(ikr)^n}{1.3.5\ldots(2n+1)} + \ldots \ldots\ldots\ldots(10),$$

which shews that ψ_n vanishes with r, except when $n = 0$.

The complete series for ψ_n, when there is no source at the pole, is more conveniently obtained by the aid of the theory of Bessel's functions. The differential equations (4) § 200, satisfied by these functions, viz.

$$\frac{d^2y}{dz^2} + \frac{1}{z}\frac{dy}{dz} + \left(1 - \frac{m^2}{z^2}\right) y = 0 \ldots\ldots\ldots\ldots\ldots (11),$$

may also be written in the form

$$\frac{d^2(yz^{\frac{1}{2}})}{dz^2} + \left(1 - \frac{4m^2-1}{4z^2}\right) yz^{\frac{1}{2}} = 0 \ldots\ldots\ldots\ldots (12).$$

It is known (§ 200) that the solution of (11) subject to the condition of finiteness when $z = 0$, is $y = A J_m(z)$, where

$$J_m(z) = \frac{z^m}{2^m \Gamma(m+1)} \left\{ 1 - \frac{z^2}{2.(2m+2)} \right.$$
$$\left. + \frac{z^4}{2.4.(2m+2)(2m+4)} - \ldots \right\} \ldots\ldots\ldots(13),$$

is the Bessel's function of order m.

When m is integral, $\Gamma(m+1) = 1.2.3\ldots m$; but here we have to do with m fractional and of the form $n + \frac{1}{2}$, n being integral. In this case

$$\Gamma(m+1) = \frac{1.3.5\ldots(2n+1)}{2^{n+1}} \cdot \sqrt{\pi} \ldots\ldots\ldots\ldots(14).$$

Referring now to (12), we see that the solution of

$$\frac{d^2\theta}{dz^2} + \left(1 - \frac{4m^2-1}{4z^2}\right) \theta = 0 \ldots\ldots\ldots\ldots\ldots (15),$$

under the same condition of finiteness when $z = 0$, is

$$\theta = A z^{\frac{1}{2}} J_m(z) \ldots\ldots\ldots\ldots\ldots\ldots(16).$$

Now the function ψ_n, with which we are at present concerned, satisfies (4) § 323, viz.

$$\frac{d^2(r\psi_n)}{d(kr)^2} + \left(1 - \frac{n(n+1)}{(kr)^2}\right) r\psi_n = 0 \dots\dots\dots (17),$$

which is of the same form as (15), if $m = n + \frac{1}{2}$; so that the solution is

$$\psi_n = A(kr)^{-\frac{1}{2}} J_{n+\frac{1}{2}}(kr)$$

$$= A \frac{(kr)^n \sqrt{2}}{1.3\dots(2n+1)\sqrt{\pi}} \left\{1 - \frac{(kr)^2}{2.(2n+3)}\right.$$

$$\left. + \frac{(kr)^4}{2.4.(2n+3)(2n+5)} - \dots \right\} \dots\dots\dots (18).$$

Determining the constant by a comparison with (10), we find

$$\psi_n = -2(-1)^n i^{n+1} k S_n \left(\frac{\pi}{2kr}\right)^{\frac{1}{2}} J_{n+\frac{1}{2}}(kr)$$

$$= -2ik(-1)^n S_n \frac{(ikr)^n}{1.3.5\dots(2n+1)} \left\{1 - \frac{k^2 r^2}{2(2n+3)}\right.$$

$$\left. + \frac{k^4 r^4}{2.4.(2n+3)(2n+5)} - \frac{k^6 r^6}{2.4.6.(2n+3)(2n+5)(2n+7)} + \dots \right\}$$

$$\dots\dots\dots\dots (19),$$

as the complete expression for ψ_n in rising powers of r.

Comparing the different expressions (5) and (19) for ψ_n, we obtain

$$P_n\left(\frac{d}{d.ikr}\right) . \frac{\sin kr}{kr} = i^n \left(\frac{\pi}{2kr}\right)^{\frac{1}{2}} J_{n+\frac{1}{2}}(kr) \dots\dots\dots\dots (20).$$

If $F = \alpha + i\beta$, the corresponding expressions for $d\psi_n/dr$, are

$$\frac{d\psi_n}{dr} = -\frac{S_n}{r^2} \{e^{-ikr} F_n(ikr) - (-1)^n e^{+ikr} F_n(-ikr)\}$$

$$= \frac{2i^{n+1} S_n}{r^2} \{\alpha \sin(kr + \tfrac{1}{2}n\pi) - \beta \cos(kr + \tfrac{1}{2}n\pi)\}$$

$$= -2ik^2(-1)^n S_n P_n\left(\frac{d}{d.ikr}\right) \frac{d}{d.kr} . \frac{\sin kr}{kr}$$

$$= \frac{2n(-1)^n k^2 S_n(ikr)^{n-1}}{1.3.5\dots(2n+1)} \left\{1 - \frac{n+2}{2n(2n+3)} k^2 r^2 + \dots \right\} \dots\dots (21).$$

It will be convenient to write down for reference the forms of ψ and $d\psi/dr$ for the first three orders.

$$
n=0
\begin{cases}
\psi_0 = -2ikS_0 \dfrac{\sin kr}{kr}, \\[2mm]
\dfrac{d\psi_0}{dr} = \dfrac{2ikS_0}{r}\left\{\dfrac{\sin kr}{kr} - \cos kr\right\}.
\end{cases}
$$

$$
n=1
\begin{cases}
\psi_1 = \dfrac{2S_1}{r}\left\{\cos kr - \dfrac{\sin kr}{kr}\right\}, \\[2mm]
\dfrac{d\psi_1}{dr} = -\dfrac{2S_1}{r^2}\left\{2\cos kr + \left(kr - \dfrac{2}{kr}\right)\sin kr\right\}.
\end{cases}
$$

$$
n=2
\begin{cases}
\psi_2 = -\dfrac{2iS_2}{r}\left\{\left(1 - \dfrac{3}{k^2 r^2}\right)\sin kr + \dfrac{3}{kr}\cos kr\right\}, \\[2mm]
\dfrac{d\psi_2}{dr} = \dfrac{2iS_2}{r^2}\left\{\left(4 - \dfrac{9}{k^2 r^2}\right)\sin kr - \left(kr - \dfrac{9}{kr}\right)\cos kr\right\}.
\end{cases}
$$

331. One of the most interesting applications of these results is to the investigation of the motion of a gas within a rigid spherical envelope. To determine the free periods we have only to suppose that $d\psi/dr$ vanishes, when r is equal to the radius of the envelope. Thus in the case of the symmetrical vibrations, we have to determine k,

$$\tan kr = kr \quad\dotfill(1),$$

an equation which we have already considered in the chapter on membranes, § 207. The first finite root ($kr = 1\cdot4303\,\pi$) corresponds to the symmetrical vibration of lowest pitch. In the case of a higher root, the vibration in question has spherical nodes, whose radii correspond to the inferior roots.

Any cone, whose vertex is at the origin, may be made rigid without affecting the conditions of the question.

The loops, or places of no pressure variation, are given by $(kr)^{-1}\sin kr = 0$, or $kr = m\pi$, where m is any integer, except zero.

The case of $n=1$, when the vibrations may be called diametral, is perhaps the most interesting. S_1, being a harmonic of order 1, is proportional to $\cos\theta$ where θ is the angle between r and some fixed direction of reference. Since $d\psi_1/d\theta$ vanishes only at the poles, there are no conical nodes[1] with vertex at the centre. Any meridianal plane, however, is nodal, and may be supposed rigid. Along any specified radius vector, ψ_1 and $d\psi_1/d\theta$ vanish, and change sign, with $\cos kr - (kr)^{-1} \sin kr$, viz. when $\tan kr = kr$.

To find the spherical nodes, we have

$$\tan kr = \frac{2kr}{2 - k^2 r^2} \quad\text{.....................(2)}.$$

The first root is $kr = 0$. Calculating from Trigonometrical Tables by trial and error, I find for the next root, which corresponds to the vibration of most importance within a sphere, $kr = 119\cdot26 \times \pi/180$; so that $r : \lambda = \cdot3313$.

The air sways from side to side in much the same manner as in a doubly closed pipe. Without analysis we might anticipate that the pitch would be higher for the sphere than for a closed pipe of equal length, because the sphere may be derived from the cylinder with closed ends, by filling up part of the latter with obstructing material, the effect of which must be to sharpen the *spring*, while the mass to be moved remains but little changed. In fact, for a closed pipe of length $2r$,

$$r : \lambda = \cdot25.$$

The sphere is thus higher in pitch than the cylinder by about a Fourth.

The vibration now under consideration is the gravest of which the sphere is capable; it is more than an octave graver than the gravest radial vibration. The next vibration of this type is such that $kr = 340\cdot35 \, \pi/180$, or

$$r : \lambda = \cdot9454,$$

and is therefore higher than the first radial.

When kr is great, the roots of (2) may be conveniently calculated by means of a series. If $kr = \sigma\pi - y$, [where σ is an integer,] then

$$\tan y = \frac{2(\sigma\pi - y)}{(\sigma\pi - y)^2 - 2},$$

from which we find

$$kr = \sigma\pi - \frac{2}{\sigma\pi} - \frac{16}{3\sigma^3\pi^3} + \quad\text{.....................} (3).$$

---

[1] A node is a surface which might be supposed rigid, viz. one across which there is no motion.

When $n = 2$, the general expression for S_n is

$$S_2 = A_0 \left(\cos^2\theta - \tfrac{1}{3} \right) + (A_1 \cos\omega + B_1 \sin\omega) \sin\theta \cos\theta$$
$$+ (A_2 \cos 2\omega + B_2 \sin 2\omega) \sin^2\theta \dots (4),$$

from which we may select for special consideration the following notable cases:

(α)   the zonal harmonic,

$$S_2 = A_0 \left(\cos^2\theta - \tfrac{1}{3} \right) \dots (4a).$$

Here $d\psi_2/d\theta$ is proportional to $\sin 2\theta$, and therefore vanishes when $\theta = \tfrac{1}{2}\pi$. This shews that the equatorial plane is a nodal surface, so that the same motion might take place within a closed hemisphere. Also since S_2 does not involve ω, any meridianal plane may be regarded as rigid.

(β)   the sectorial harmonic

$$S_2 = A_2 \cos 2\omega \sin^2\theta \dots (5).$$

Here again $d\psi_2/d\theta$ varies as $\sin 2\theta$, and the equatorial plane is nodal. But $d\psi_2/d\omega$ varies as $\sin 2\omega$, and therefore does not vanish independently of θ, except when $\sin 2\omega = 0$. It appears accordingly that two, and but two, meridianal planes are nodal, and that these are at right angles to one another.

(γ)   the tesseral harmonic,

$$S_2 = A_1 \cos\omega \sin\theta \cos\theta \dots (6).$$

In this case $d\psi_2/d\theta$ vanishes independently of ω with $\cos 2\theta$, that is, when $\theta = \tfrac{1}{4}\pi$, or $\tfrac{3}{4}\pi$, which gives a nodal cone of revolution whose vertical angle is a right angle. $d\psi_2/d\omega$ varies as $\sin\omega$, and thus there is one meridianal nodal plane, and but one [1].

The spherical nodes are given by

$$\tan kr = \frac{k^3 r^3 - 9kr}{4k^2 r^2 - 9} \dots (7),$$

of which the first finite solution is

$$kr = 3\cdot 3422,$$

giving a tone graver than any of the radial group.

In the case of the general harmonic, the equation giving the

---

[1] [I owe to Prof. Lamb the remark that the difference between (β) and (γ) is only in relation to the axes of reference.]

tones possible within a sphere of radius r may be written (21) § 330

$$\tan (kr + \tfrac{1}{2} n\pi) = \beta : \alpha \dots\dots\dots\dots\dots(8),$$

or

$$P_n \left(\frac{d}{d \cdot ikr} \right) \frac{d}{d.kr} \cdot \frac{\sin kr}{kr} = 0 \dots\dots\dots\dots (9),$$

or again,

$$2\,kr\,J'_{n+\frac{1}{2}}(kr) = J_{n+\frac{1}{2}}(kr)\dots\dots\dots\dots(10).$$

[For the roots of

$$\frac{d}{dz} \{ z^{-\frac{1}{2}} J_\nu (z) \} = 0 \dots\dots\dots\dots\dots (11),$$

equivalent to (10), Prof. McMahon gives [1]

$$z_\nu{}^{(s)} = \beta' - \frac{m+7}{8\beta'} - \frac{4\,(7m^2 + 154m + 95)}{3\,(8\beta')^3}$$
$$- \frac{32\,(83m^3 + 3535m^2 + 3561m + 6133)}{15\,(8\beta')^5} - \dots\dots\dots(12),$$

where $m = 4\nu^2$, and

$$\beta' = \tfrac{1}{4}\,(2\nu + 4s + 1)\dots\dots\dots\dots\dots(13).$$

If $n = 1$, so that $\nu = \tfrac{3}{2}$,

$$m = 9, \qquad \beta' = s + 1,$$

and (12) gives a result in harmony with (3).]

Table A shews the values of λ for a sphere of radius unity, corresponding to the more important modes of vibration. In B is exhibited the frequency of the various vibrations referred to the gravest of the whole system. The Table is extended far enough to include two octaves.

TABLE A,

Giving the values of λ for a sphere of unit radius.

Order of Harmonic.

		0	1	2	3	4	5	6
	0	1·3983	3·0186	1·8800	1·392	1·113	·9300	·8002
	1	·81334	1·0577	·86195	·7320	·6385		
Number of internal spherical nodes.	2	·57622	·68251	·59208	·5248			
	3	·44670	·50653	·45380				
	4	·36485	·40330					
	5	·30833	·33523					

[1] *Annals of Mathematics*, vol. IX. no. 1.

TABLE B.

Pitch of each tone, referred to gravest.	Order of Harmonic.	Number of internal spherical nodes.	Pitch of each tone, referred to gravest.	Order of Harmonic.	Number of internal spherical nodes.
1·0000	1	0	2 8540	1	1
1·6056	2 .	0	3·2458	5	0
2·1588	0	0	3·5021	2	1
2·169	3	0	3·7114	0	1
2·712	4	0	3·772	6	0

332. If we drop unnecessary constants, the particular solution for the vibrations of gas within a spherical case of radius unity is represented by

$$\psi_n = S_n (kr)^{-\frac{1}{2}} J_{n+\frac{1}{2}} (kr) \cos (kat - \theta) \dots \dots (1),$$

where k is a root of

$$2k J'_{n+\frac{1}{2}} (k) = J_{n+\frac{1}{2}} (k) \dots \dots \dots (2).$$

In generalising this, we must remember that S_n may be composed of several terms, corresponding to each of which there may exist a vibration of arbitrary amplitude and phase. Further, each term in S_n may be associated with any, or all, of the values of k determined by (2). For example, under the head of $n = 2$, we might have

$$\psi_2 = A (\cos^2\theta - \tfrac{1}{3}) (k_1 r)^{-\frac{1}{2}} J_{n+\frac{1}{2}} (k_1 r) \cos (k_1 at + \theta_1)$$
$$+ B \cos 2\omega \sin^2\theta (k_2 r)^{-\frac{1}{2}} J_{n+\frac{1}{2}} (k_2 r) \cos (k_2 at + \theta_2),$$

k_1 and k_2 being different roots of

$$2k J'_{\frac{5}{2}} (k) = J_{\frac{5}{2}} (k).$$

Any two of the constituents of ψ are conjugate, *i.e.* will vanish when multiplied together and integrated over the volume of the sphere. This follows from the property of the spherical harmonics wherever the two terms considered correspond to different values of n, or to two different constituents of S_n[1]. The only case remaining for consideration requires us to shew that

$$\int_0^1 r^2 dr \cdot (k_1 r)^{-\frac{1}{2}} J_{n+\frac{1}{2}} (k_1 r) \cdot (k_2 r)^{-\frac{1}{2}} J_{n+\frac{1}{2}} (k_2 r) = 0 \dots \dots (3),$$

[1] Thomson and Tait's *Nat. Phil.* p. 151.

where k_1 and k_2 are *different* roots of

$$2k\,J'_{n+\frac{1}{2}}(k) = J_{n+\frac{1}{2}}(k)\dots\dots\dots\dots\dots(4),$$

and this is an immediate consequence of a fundamental property of these functions (§ 203). There is therefore no difficulty in adapting the general solution to prescribed initial circumstances.

In order to illustrate this subject we will take the case where initially the gas is in its position of equilibrium but is moving with constant velocity parallel to x. This condition of things would be approximately realised, if the case, having been previously in uniform motion, were suddenly stopped.

Since there is no initial condensation or rarefaction, all the quantities θ_n vanish. If $d\psi/dx$ be initially unity, we have $\psi = x = r\mu$, which shews that the solution contains only terms of the first order in spherical harmonics. The solution is therefore of the form

$$\psi = A_1 (k_1 r)^{-\frac{1}{2}} J_{\frac{3}{2}}(k_1 r)\,\mu \cos k_1 at$$
$$+ A_2 (k_2 r)^{-\frac{1}{2}} J_{\frac{3}{2}}(k_2 r)\,\mu \cos k_2 at + \dots\dots\dots\dots(5),$$

where k_1, k_2, &c. are roots of

$$2k J_{\frac{3}{2}}'(k) = J_{\frac{3}{2}}(k)\dots\dots\dots\dots\dots\dots(6).$$

To determine the coefficients, we have initially for values of r from 0 to 1,

$$r = A_1 (k_1 r)^{-\frac{1}{2}} J_{\frac{3}{2}}(k_1 r) + A_2 (k_2 r)^{-\frac{1}{2}} J_{\frac{3}{2}}(k_2 r) + \dots\dots\dots(7).$$

Multiplying by $r^{\frac{3}{2}} J_{\frac{3}{2}}(kr)$ and integrating with respect to r from 0 to 1, we find

$$\int_0^1 r^{\frac{5}{2}} J_{\frac{3}{2}}(kr)\,dr = A k^{-\frac{1}{2}} \int_0^1 [J_{\frac{3}{2}}(kr)]^2\,r\,dr\dots\dots\dots\dots(8),$$

the other terms on the right vanishing in virtue of the conjugate property. Now by (16), § 203,

$$2\int_0^1 [J_{\frac{3}{2}}(kr)]^2\,r\,dr = [J_{\frac{3}{2}}'(k)]^2 + \left(1 - \frac{9}{4k^2}\right)[J_{\frac{3}{2}}(k)]^2$$
$$= \left(1 - \frac{2}{k^2}\right)[J_{\frac{3}{2}}(k)]^2 \dots\dots\dots\dots\dots(9),$$

by (6).

The evaluation of $\int_0^1 r^{\frac{5}{2}} J_{\frac{3}{2}}(kr)\,dr$ may be effected by the aid of

a general theorem relating to these functions. By the fundamental
differential equation

$$\int_0^r r^{n+1} \left[\frac{1}{r} \frac{d}{dr} \left(r \frac{dJ_n(kr)}{dr} \right) + \left(k^2 - \frac{n^2}{r^2} \right) J_n(kr) \right] dr = 0,$$

whence by integration by parts we obtain,

$$k^2 \int_0^r r^{n+1} J_n(kr) \, dr = n r^n J_n(kr) - r^{n+1} \frac{dJ_n(kr)}{dr} \quad \dots\dots\dots(10),$$

or, if we make $r = 1$,

$$k^2 \int_0^1 r^{n+1} J_n(kr) \, dr = n J_n(k) - k J_n'(k) \dots\dots\dots\dots(11).$$

Thus in the case, with which we are here concerned,

$$k^2 \int_0^1 r^{\frac{5}{2}} J_{\frac{3}{2}}(kr) \, dr = \tfrac{3}{2} J_{\frac{3}{2}}(k) - k J_{\frac{3}{2}}'(k) = J_{\frac{1}{2}}(k) \quad \text{by (6)}.$$

Equation (8) therefore takes the form

$$A = \frac{2 k^{\frac{1}{2}}}{(k^2 - 2) J_{\frac{3}{2}}(k)} \quad \dots\dots\dots\dots\dots\dots(12),$$

and the final solution is

$$\psi = \Sigma \frac{2 r^{-\frac{1}{2}} \mu}{k^2 - 2} \frac{J_{\frac{3}{2}}(kr)}{J_{\frac{3}{2}}(k)} \cos kat \dots\dots\dots\dots\dots(13),$$

where the summation is to be extended to all the admissible
values of k.

When $t = 0$, and $r = 1$, we must have $\psi = \mu$, and accordingly

$$\Sigma \frac{2}{k^2 - 2} = 1 \quad \dots\dots\dots\dots\dots\dots\dots\dots(14).$$

It will be remembered that the higher values of k are approximately, (3) § 331,

$$k = \sigma \pi - \frac{2}{\sigma \pi} \quad \dots\dots\dots\dots\dots\dots\dots\dots(15).$$

The first value of k is 2·0815, and the second 5·9402, whence

$$\frac{2}{k_1^2 - 2} = ·85742, \qquad \frac{2}{k_2^2 - 2} = ·06009,$$

shewing that the first term in the series for ψ is by far the most
important.

It may be well to recall here that

$$J_{\frac{3}{2}}(z) = \sqrt{\frac{2}{\pi z}} \left(\frac{\sin z}{z} - \cos z \right) \quad \ldots\ldots\ldots\ldots\ldots(16).$$

Equation (14) may be verified thus: the quantities k are the roots of

$$\frac{d}{dz} \left\{ z^{-\frac{1}{2}} J_{\frac{3}{2}}(z) \right\} = 0,$$

or, if $\phi = z^{-\frac{1}{2}} J_{\frac{3}{2}}(z)$, the roots of $\phi' = 0$, where ϕ satisfies

$$\phi'' + \frac{2}{z}\phi' + \left(1 - \frac{2}{z^2} \right)\phi = 0 \ldots\ldots\ldots\ldots\ldots(17).$$

Now, since the leading term in the expansion of ϕ' in ascending powers of z is independent of z, we may write

$$\phi' = \text{const.} \left\{ 1 - \frac{z^2}{k_1^2} \right\} \left\{ 1 - \frac{z^2}{k_2^2} \right\} \ldots\ldots$$

whence, by taking the logarithms and differentiating,

$$-\frac{\phi''}{\phi'} = \frac{2z}{k_1^2 - z^2} + \frac{2z}{k_2^2 - z^2} + \ldots$$

If we now put $z^2 = 2$, we get by (17),

$$\Sigma \frac{2}{k^2 - 2} = -\frac{\phi''}{z\phi'} \quad (z^2 = 2) = 1.$$

333. In a similar manner we may treat the problem of the vibrations of air included between rigid concentric spherical surfaces, whose radii are r_1 and r_2. For by (13) § 323, if $d\psi_n/dr$ vanish for these values of r,

$$e^{2ikr_1} \frac{F_n(-ikr_1)}{F_n(+ikr_1)} = e^{2ikr_2} \frac{F_n(-ikr_2)}{F_n(+ikr_2)},$$

whence

$$\tan k(r_1 - r_2) = \frac{(\beta/\alpha)_1 - (\beta/\alpha)_2}{1 + (\beta/\alpha)_1 (\beta/\alpha)_2} \ldots\ldots\ldots\ldots(1),$$

where as before

$$F_n(+ikr) = \alpha + i\beta \ldots\ldots\ldots\ldots\ldots\ldots(2).$$

When the difference between r_1 and r_2 is very small compared with either, the problem identifies itself with that of the vibration of a spherical sheet of air, and is best solved independently. In (1)

§ 323, if ψ be independent of r, as it is evident that it must approximately be in the case supposed, we have

$$\frac{1}{\sin \theta} \frac{d}{d\theta} \left(\sin \theta \frac{d\psi}{d\theta} \right) + \frac{1}{\sin^2 \theta} \frac{d^2\psi}{d\omega^2} + k^2 r^2 \psi = 0 \quad \ldots\ldots (3),$$

whose solution is simply

$$\psi_n = S_n \ldots\ldots\ldots\ldots\ldots\ldots\ldots\ldots\ldots(4),$$

while the admissible values of k^2 are given by

$$k^2 r^2 = n (n + 1) \ldots\ldots\ldots\ldots\ldots\ldots\ldots(5).$$

The interval between the gravest tone ($n = 1$) and the next is such that two of them would make a twelfth (octave + fifth). The problem of the spherical sheet of gas will be further considered in the following chapter. [For a derivation of (5) from the fundamental determinant, equivalent to (1), the reader may be referred to a short paper[1] by Mr Chree.]

334. The next application that we shall make of the spherical harmonic analysis is to investigate the disturbance which ensues when plane waves of sound impinge on an obstructing sphere. Taking the centre of the sphere as origin of polar co-ordinates, and the direction from which the waves come as the axis of μ, let ϕ be the potential of the unobstructed plane waves. Then, leaving out an unnecessary complex coefficient, we have

$$\phi = e^{ik\,(at+x)} = e^{ikat} \cdot e^{ikr\mu} \ldots\ldots\ldots\ldots\ldots\ldots(1),$$

and the solution of the problem requires the expansion of $e^{ikr\mu}$ in spherical harmonics. On account of the symmetry the harmonics reduce themselves to Legendre's functions $P_n(\mu)$, so that we may take

$$e^{ikr\mu} = A_0 + A_1 P_1 + \ldots + A_n P_n + \ldots\ldots\ldots(2),$$

where $A_0 \ldots$ are functions of r, but not of μ. From what has been already proved we may anticipate that A_n, considered as a function of r, must vary as

$$P_n \left(\frac{d}{d \cdot ikr} \right) \frac{\sin kr}{kr}, \quad \text{or as} \quad r^{-\frac{1}{2}} J_{n+\frac{1}{2}} (kr),$$

but the same result may easily be obtained directly. Multiplying

[1] *Messenger of Mathematics*, vol. xv. p. 20, 1886.

(2) by $P_n(\mu)$, and integrating with respect to μ from $\mu = -1$ to $\mu = +1$, we find

$$\int_{-1}^{+1} P_n(\mu)\, e^{ikr\mu}\, d\mu = A_n \int_{-1}^{+1} (P_n)^2\, d\mu = \frac{2A_n}{2n+1} \quad\ldots\ldots(3);$$

and, as in § 330,

$$\int_{-1}^{+1} P_n(\mu)\, e^{ikr\mu}\, d\mu = 2P_n\left(\frac{d\,.}{d\,.\,ikr}\right).\frac{\sin kr}{kr},$$

so that finally

$$\frac{A_n}{2n+1} = P_n\left(\frac{d}{d\,.\,ikr}\right).\frac{\sin kr}{kr} = i^n \sqrt{\frac{\pi}{2kr}}.\, J_{n+\frac{1}{2}}(kr) \quad\ldots\ldots(4).$$

In the problem in hand the whole motion outside the sphere may be divided into two parts; the first, that represented by ϕ and corresponding to undisturbed plane waves, and the second a disturbance due to the presence of the sphere, and radiating outwards from it. If the potential of the latter part be ψ, we have (2) § 324 on replacing the general harmonic S_n by $a_n P_n(\mu)$,

$$\left.\begin{array}{l} r\psi_n = a_n P_n(\mu)\,.\, e^{-ikr} f_n(ikr) \\[2mm] r^2\dfrac{d\psi_n}{dr} = -a_n P_n(\mu)\,.\, e^{-ikr} F_n(ikr) \end{array}\right\} \quad\ldots\ldots\ldots(5).$$

The velocity-potential of the whole motion is found by addition of ϕ and ψ, the constants a_n being determined by the boundary conditions, whose form depends upon the character of the obstruction presented by the sphere. The simplest case is that of a rigid and fixed sphere, and then the condition to be satisfied when $r = c$ is that

$$\frac{d\phi}{dr} + \frac{d\psi}{dr} = 0 \quad\ldots\ldots\ldots\ldots\ldots\ldots(6),$$

a relation which must of course hold good for each harmonic element separately. For the element of order n, we get

$$a_n = (2n+1)\frac{kc^2 e^{ikc}}{F_n(ikc)} P_n\left(\frac{d}{d\,.\,ikc}\right)\frac{d}{d\,.\,kc}.\frac{\sin kc}{kc}\quad\ldots\ldots(7).$$

Corresponding to the plane waves $\phi = e^{ik(at+x)}$, the disturbance due to the presence of the sphere is expressed by

$$\psi = \frac{kc^2}{r} e^{ik(at-r+c)}$$

$$\times \sum_{n=0}^{n=\infty} \frac{2n+1}{F_n(ikc)} P_n\left(\frac{d}{d\,.\,ikc}\right)\frac{d}{d\,.\,kc}.\frac{\sin kc}{kc}.\, P_n(\mu)\,.\, f_n(ikr)\ldots(8).$$

At a sufficient distance from the source of disturbance we may take $f_n(ikr) = 1$. In order to pass to the solution of a real problem, we may separate the real and imaginary parts, and throw away the latter. On this supposition the plane waves are represented by

$$[\phi] = \cos k\,(at + x)\dots\dots\dots\dots\dots\dots(9).$$

Confining ourselves for simplicity's sake to parts of space at a great distance from the sphere, where $f_n(ikr) = 1$, we proceed to extract the real part of (8). Since the functions P are wholly even or wholly odd,

$$P_n\left(\frac{d}{d\,.\,ikc}\right)\frac{d}{d\,.\,kc}\cdot\frac{\sin kc}{kc}$$

is wholly real or wholly imaginary, so that this factor presents no difficulty. $\{F_n(ikc)\}^{-1}$, however, is complex, and since $F_n(ikc) = \alpha + i\beta$,

$$\{F_n(ikc)\}^{-1} = \frac{\alpha - i\beta}{\alpha^2 + \beta^2} = \frac{e^{i\gamma}}{\sqrt{(\alpha^2 + \beta^2)}},$$

where $\tan\gamma = -\beta/\alpha$. [If the positive value of $\sqrt{(\alpha^2 + \beta^2)}$ be taken in all cases, γ must be so chosen that $\cos\gamma$ has the same sign as α.]

Thus

$$\psi = \Sigma\,(2n+1)\,\frac{kc^2}{r}\,e^{i[k(at-r+c)+\gamma]}$$
$$\times\,\{\alpha^2 + \beta^2\}^{-\frac{1}{2}}\,P_n\left(\frac{d}{d\,.\,ikc}\right)\frac{d}{d\,.\,kc}\cdot\frac{\sin kc}{kc}\cdot P_n(\mu)\dots\dots(10).$$

When therefore n is even,

$$[\psi] = (2n+1)\,\frac{kc^2}{r}\cos\{k\,(at - r + c) + \gamma\}$$
$$\times\,\{\alpha^2 + \beta^2\}^{-\frac{1}{2}}\,P_n\left(\frac{d}{d\,.\,ikc}\right)\frac{d}{d\,.\,kc}\cdot\frac{\sin kc}{kc}\cdot P_n(\mu)\dots(11),$$

while, if n be odd,

$$[\psi] = (2n+1)\,\frac{kc^2}{r}\,i\sin\{k\,(at - r + c) + \gamma\}$$
$$\times\,\{\alpha^2 + \beta^2\}^{-\frac{1}{2}}\,P_n\left(\frac{d}{d\,.\,ikc}\right)\frac{d}{d\,.\,kc}\cdot\frac{\sin kc}{kc}\cdot P_n(\mu)\dots(12).$$

As examples we may write down the terms in $[\psi]$, involving harmonics of orders 0, 1, 2. The following table of the functions $P_n(\mu)$ will be useful.

$$P_0 = 1, \qquad\qquad P_1 = \mu,$$
$$P_2 = \tfrac{3}{2}(\mu^2 - \tfrac{1}{3}), \qquad\qquad P_3 = \tfrac{5}{2}(\mu^3 - \tfrac{3}{5}\mu),$$
$$P_4 = \tfrac{35}{8}(\mu^4 - \tfrac{6}{7}\mu^2 + \tfrac{3}{35}), \qquad P_5 = \tfrac{63}{8}(\mu^5 - \tfrac{10}{9}\mu^3 + \tfrac{5}{21}\mu).$$

We have,

$$n = 0, \quad \alpha^2 + \beta^2 = 1 + k^2 c^2, \quad \tan\gamma_0 = -kc,$$

$$[\psi_0] = \frac{kc^2}{r}\{1 + k^2 c^2\}^{-\frac{1}{2}} \frac{d}{d\,.\,kc} \frac{\sin kc}{kc} . \cos\{k(at - r + c) + \gamma_0\}\ldots(13);$$

$$n = 1, \quad \alpha^2 + \beta^2 = k^2 c^2 + \frac{4}{k^2 c^2}, \quad \tan\gamma_1 = -\frac{k^2 c^2 - 2}{2kc},$$

$$[\psi_1] = \frac{3kc^2}{r}\left\{k^2 c^2 + \frac{4}{k^2 c^2}\right\}^{-\frac{1}{2}} \frac{d^2}{d(kc)^2} . \frac{\sin kc}{kc} . \mu . \sin\{k(at - r + c) + \gamma_1\}$$
$$\ldots\ldots\ldots\ldots\ldots\ldots\ldots\ldots(14);$$

$$n = 2, \quad \alpha^2 + \beta^2 = k^2 c^2 - 2 + \frac{9}{k^2 c^2} + \frac{81}{k^4 c^4}, \quad \tan\gamma_2 = -\frac{kc(k^2 c^2 - 9)}{4k^2 c^2 - 9},$$

$$[\psi_2] = -\frac{45 kc^2}{4r}\left\{k^2 c^2 - 2 + \frac{9}{k^2 c^2} + \frac{81}{k^4 c^4}\right\}^{-\frac{1}{2}}$$
$$\times \left\{\frac{d^3}{d(kc)^3} + \tfrac{1}{3}\frac{d}{d(kc)}\right\} \frac{\sin kc}{kc} . (\mu^2 - \tfrac{1}{3}) \cos\{k(at - r + c) + \gamma_2\}\ldots(15).$$

The solution of the problem here obtained, though analytically quite general, is hardly of practical use except when kc is a small quantity. In this case we may advantageously expand our results in rising powers of kc.

$$[\psi_0] = -\frac{k^2 c^3}{3r}(1 - \tfrac{3}{5}k^2 c^2 + \tfrac{3}{7}k^4 c^4 - \tfrac{19}{54}k^6 c^6 + \ldots)$$
$$\times \cos\{k(at - r + c) + \gamma_0\}\ldots\ldots\ldots\ldots\ldots(16).$$

$$[\psi_1] = -\frac{k^2 c^3}{2r}(1 - \tfrac{3}{10}k^2 c^2 - \tfrac{3}{28}k^4 c^4 + \tfrac{1}{27}k^6 c^6 + \ldots)$$
$$\times \mu . \sin\{k(at - r + c) + \gamma_1\}\ldots\ldots\ldots\ldots(17),$$

$$[\psi_2] = -\frac{k^4 c^5}{9r}(1 - \tfrac{25}{126}k^2 c^2 + \tfrac{13}{567}k^4 c^4 + \ldots)$$
$$\times (\mu^2 - \tfrac{1}{3}) \cos\{k(at - r + c) + \gamma_2\}\ldots\ldots(18).$$

It appears that while $[\psi_0]$ and $[\psi_1]$ are of the same order in the small quantity kc, $[\psi_2]$ is two orders higher. We shall find presently that the higher harmonic components in $[\psi]$ depend upon still more elevated powers of kc. For a first approximation, then, we may confine ourselves to the elements of order 0 and 1.

Although $[\psi_0]$ contains a cosine, and $[\psi_1]$ a sine, they nevertheless differ in phase by a small quantity only. Comparing two of the values of $d\psi_n/dr$ in (21) § 330 we see that

$$\alpha \sin\left(kc + \tfrac{1}{2}n\pi\right) - \beta \cos\left(kc + \tfrac{1}{2}n\pi\right)$$
$$= -(-1)^n \frac{n(kc)^{n+1}}{1.3.5\ldots(2n+1)} + \text{higher powers of } kc$$

identically. Dividing by $\alpha \cos\left(kc + \tfrac{1}{2}n\pi\right)$, we get ultimately

$$\tan\left(kc + \tfrac{1}{2}n\pi\right) - \frac{\beta}{\alpha} = -\frac{(-1)^n}{\alpha\cos\left(kc + \tfrac{1}{2}n\pi\right)} \cdot \frac{n(kc)^{n+1}}{1.3.5\ldots(2n+1)}.$$

When n is *even*, this equation becomes on substitution for α of its leading term from (16) § 323,

$$\tan kc - \frac{\beta}{\alpha} = -\frac{n}{(n+1)(2n+1)} \frac{(kc)^{2n+1}}{\{1.3.5\ldots(2n-1)\}^2} \ldots(19).$$

For example, if $n = 2$,

$$\tan kc - \left(\frac{\beta}{\alpha}\right)_2 = -\frac{2(kc)^5}{3^3.5} + \ldots\ldots$$

When n is at all high, the expressions $\tan kc$ and β/α become very nearly identical for moderate values of kc.

When n is *odd*, we get in a nearly similar manner,

$$\cot kc + \frac{\beta}{\alpha} = \frac{n(kc)^{2n-1}}{(n+1)(2n+1)\{1.3.5\ldots(2n-1)\}^2} + \ldots\ldots(20).$$

[From (19) we see that when n is even $\tan\gamma$, or $-\beta/\alpha$, is approximately equal to $-\tan kc$, and from (20) when n is odd that $\cot\gamma = \tan kc$. In the first case, by (16) § 323, α has the sign of i^{-n} or of $(-1)^{\frac{1}{2}n}$; and in the second case α has the sign of i^{-n+1} or of $(-1)^{\frac{1}{2}(n-1)}$. In both cases the approximate solution may be expressed

$$\gamma = -kc + \tfrac{1}{2}n\pi \ldots\ldots\ldots\ldots\ldots\ldots(20').[1]]$$

The velocity-potential of the disturbance due to a small rigid and fixed sphere is therefore approximately,

$$[\psi_0] + [\psi_1] = -\frac{k^2 c^3}{3r}\left(1 + \tfrac{3}{2}\mu\right)\cos k(at - r)$$
$$= -\frac{\pi T}{r\lambda^2}\left(1 + \tfrac{3}{2}\mu\right)\cos k(at - r)\ldots\ldots(21),$$

---

[1] This emendation and others consequential to it are due to Dr Burton.

if T denote the volume of the obstacle, the corresponding direct wave being

$$[\phi] = \cos k\,(at + x)\dots\dots\dots\dots\dots\dots(22).$$

For a given obstacle and a given distance the ratio of the amplitudes of the scattered and the direct waves is in general proportional to the inverse square of the wave-length, and the ratio of intensities is proportional to the inverse fourth power (§ 296).

In order to compare the intensities of the primary and scattered sounds, we may suppose the former to originate in a simple source, provided it be sufficiently distant (R) from T. Thus, if

$$[\phi] = \frac{\cos k\,(at - R)}{R}\dots\dots\dots\dots\dots(23),$$

$$[\psi] = -\frac{\pi T}{r\,R\lambda^2}(1 + \tfrac{3}{2}\mu)\cos k\,(at - r)\dots\dots\dots(24);$$

so that at equal distances from their sources the secondary and the primary waves are in the ratio

$$-\frac{\pi T}{R\lambda^2}(1 + \tfrac{3}{2}\mu)\dots\dots\dots\dots\dots\dots(25).$$

The intensities are therefore in the ratio

$$\frac{\pi^2 T^2}{R^2\lambda^4}(1 + \tfrac{3}{2}\mu)^2\dots\dots\dots\dots\dots\dots(26),$$

which, in the case of $\mu = +1$, gives approximately

$$\frac{61{\cdot}72\,T^2}{R^2\lambda^4}\dots\dots\dots\dots\dots\dots\dots(27).$$

It must be well understood that in order that this result may apply, λ must be great compared with the linear dimension of T, and R must be great compared with λ.

To find the leading term in the expression for ψ_n, when kc is small, we have in the first place,

$$(2n + 1) P_n\left(\frac{d}{d\,.\,ikc}\right)\frac{d}{d\,.\,kc}\cdot\frac{\sin kc}{kc}$$

$$= \frac{n\,i^n\,(kc)^{n-1}}{1\,.\,3\,.\,5\dots(2n - 1)}\left\{1 - \frac{(n + 2)\,k^2 c^2}{2\,.\,n\,.\,(2n + 3)} + \dots\right\}\dots\dots(28).$$

Again,

$$\alpha^2 + \beta^2 = F_n(ikc) \times F_n(-ikc)$$

$$= \{1 \cdot 3 \cdot 5 \ldots (2n-1)(n+1)(kc)^{-n}\}^2 \left\{1 + \frac{(n-1)k^2c^2}{(n+1)(2n-1)} + \ldots\right\}$$

$$\ldots\ldots\ldots\ldots\ldots\ldots(29);$$

so that

$$\{\alpha^2 + \beta^2\}^{-\frac{1}{2}} = \frac{(kc)^n}{1 \cdot 3 \ldots (2n-1)(n+1)} \left\{1 - \frac{(n-1)k^2c^2}{2 \cdot (n+1)(2n-1)} + \ldots\right\}$$

$$\ldots\ldots\ldots\ldots\ldots\ldots(30).$$

Hence, from (10),

$$\psi_n = \frac{c(kc)^{2n} n i^n P_n(\mu)}{r \{1 \cdot 3 \cdot 5 \ldots (2n-1)\}^2 (n+1)} e^{i[k(at-r+c)+\gamma_n]}$$

$$\times \left\{1 - k^2c^2 \left(\frac{n-1}{(2n+2)(2n-1)} + \frac{n+2}{2n(2n+3)}\right) + \ldots\right\} \ldots(31).$$

When n is even, [since $\gamma = -kc + \frac{1}{2}n\pi$ approximately,]

$$[\psi_n] = \frac{c(kc)^{2n} n i^n P_n(\mu)}{r\{1 \cdot 3 \ldots\ldots (2n-1)\}^2 (n+1)} \cos\{k(at-r) + \frac{1}{2}n\pi\}$$

$$\times \left\{1 - k^2c^2\left(\frac{n-1}{(2n+2)(2n-1)} + \frac{n+2}{2n(2n+3)}\right) + \ldots\ldots\right\} \ldots\ldots(32);$$

while if n be odd, we have merely to replace i^n by i^{n+1} [and cos by sin], the result being then still real.

By means of (31) we may verify the first two terms in the expressions for $[\psi_1]$, $[\psi_2]$, in (17), (18). To the case of $n = 0$, (31) does not apply.

Again, by (31),

$$[\psi_3] = \frac{k^6 c^7}{120\,r}\left\{1 - \tfrac{77}{540}k^2c^2\right\}\{\mu^3 - \tfrac{3}{5}\mu\}\sin\{k(at-r+c)+\gamma_3\} \ldots(33),$$

$$[\psi_4] = \frac{k^8 c^9}{3150\,r}\{\mu^4 - \tfrac{6}{7}\mu^2 + \tfrac{3}{35}\}\cos\{k(at-r+c)+\gamma_4\}\ldots\ldots\ldots\ldots(34).$$

Combining (17), (18), (33), (34), we have the value of $[\psi]$ complete as far as the terms which are of the order k^6c^6 compared with the two leading terms given in (21). In compounding the partial expressions, it is as necessary to be exact with respect to the phases of the components as with respect to their amplitudes; but for purposes requiring only one harmonic element at a time, the phase is often of subordinate importance. In such cases we may take

$$\gamma = \mp kc + \tfrac{1}{2} n\pi.$$

From (31) or (32) it appears that the leading term in ψ_n rises two orders in kc with each step in the order of the harmonic; and that ψ_n is itself expressed by a series containing only even, or only odd, powers of kc. But besides being of higher order in kc, the leading term becomes rapidly smaller as n increases, on account of the other factors which it contains. This is evident, because for all values of n and μ, $P_n(\mu) < 1$; the same is true of $n/(n+1)$; while i^n only affects the phase.

In particular cases any one of the harmonic elements of $[\psi]$ may vanish. From (11), (12), since $(\alpha^2 + \beta^2)^{-\frac{1}{2}}$ cannot vanish, we have in such a case

$$P_n\left(\frac{d}{d \cdot ikc}\right) \frac{d}{d \cdot kc} \frac{\sin kc}{kc} = 0,$$

the same equation as that which gives the periods of the vibrations of order n in a closed sphere of radius c. A little consideration will shew that this result might have been expected. The table of § 331 is applicable to this question and shews, among other things, that when kc is small, no harmonic element in $[\psi]$ can vanish.

In consequence of the aerial pressures the sphere is acted on by a force parallel to the axis of μ, whose tendency is to set the sphere into vibration. The magnitude of this force, if σ be the density of the fluid, is given by

$$2\pi c^2 \sigma \int_{-1}^{+1} (\dot{\phi} + \dot{\psi}) \, \mu \, d\mu,$$

in which, by the conjugate property of Legendre's functions, only the term of the first order affects the result of the integration. Now, when $r = c$,

$$\phi_1 = 3 \, e^{ikat} \frac{d}{d \cdot ikc} \frac{kc}{kc} \cdot \mu,$$

$$\psi_1 = 3kc \, e^{ikat} \frac{f_1(ikc)}{F_1(ikc)} \frac{d}{d \cdot ikc} \cdot \frac{d}{d \cdot kc} \cdot \frac{\sin kc}{kc} \cdot \mu,$$

where

$$f_1(ikc) = 1 + \frac{1}{ikc}, \qquad F_1(ikc) = ikc + 2 + \frac{2}{ikc}.$$

In order that the force may vanish, it would be necessary that

$$\frac{d}{d \cdot kc} \cdot \frac{\sin kc}{kc} + kc \frac{f_1(ikc)}{F_1(ikc)} \frac{d^2}{(d \cdot kc)^2} \frac{\sin kc}{kc} = 0,$$

which cannot be satisfied by any real value of kc. We conclude that, if the sphere be free to move, it will always be set into vibration.

If instead of being absolutely plane, the primary waves have their origin in a unit source at a great, though finite, distance R from the centre of the sphere, we have

$$\phi = -\frac{1}{4\pi R} e^{ik(at-R)} \Sigma (2n+1) P_n(\mu) \times P_n\left(\frac{d}{d \cdot ikr}\right) \frac{\sin kr}{kr} \dots(35),$$

$$\psi = -\frac{kc^2}{4\pi r R} e^{ik(at-R-r+c)} \Sigma (2n+1) \frac{P_n(\mu) f_n(ikr)}{F_n(ikc)}$$

$$\times P_n\left(\frac{d}{d \cdot ikc}\right) \frac{d}{d \cdot kc} \frac{\sin kc}{kc} \dots\dots\dots\dots\dots(36).$$

On the sphere itself $r = c$, so that the value of the total potential at any point at the surface is

$$\phi + \psi = -\frac{e^{ik(at-R)}}{4\pi R} \Sigma (2n+1) P_n(\mu)$$

$$\times \left[P_n\left(\frac{d}{d \cdot ikc}\right) \frac{\sin kc}{kc} + kc \frac{f_n(ikc)}{F_n(ikc)} P_n\left(\frac{d}{d \cdot ikc}\right) \frac{d}{d \cdot kc} \frac{\sin kc}{kc} \right].$$

This expression may be simplified. We have

$$P_n\left(\frac{d}{d \cdot ikc}\right) \frac{\sin kc}{kc} = \frac{1}{2ikc} \{-(-1)^n e^{-ikc} f_n(ikc) + e^{+ikc} f_n(-ikc)\},$$

$$\frac{d}{d \cdot kc} \cdot P_n\left(\frac{d}{d \cdot ikc}\right) \frac{\sin kc}{kc} = \frac{1}{2ik^2c^2} \{(-1)^n e^{-ikc} F_n(ikc) - e^{+ikc} F_n(-ikc)\},$$

and thus the quantity within square brackets may be written

$$\frac{e^{ikc}}{2ikc} \frac{F_n(ikc) f_n(-ikc) - F_n(-ikc) f_n(ikc)}{F_n(ikc)},$$

which by (6) § 327 is identical with $e^{ikc} [F_n(ikc)]^{-1}$. Thus

$$\phi + \psi = -\frac{e^{ik(at-R+c)}}{4\pi R} \Sigma (2n+1) \frac{P_n(\mu)}{F_n(ikc)} \dots\dots\dots(37),$$

which is the same as if the source had been on the sphere, and the point at which the potential is required at a great distance (§ 328), and is an example of the general Principle of Reciprocity.

By assuming the principle, and making use of the result (3) of § 328, we see that if the source of the primary waves be at a finite distance R, the value of the total potential at any point on the sphere is

$$\phi + \psi = -\frac{1}{4\pi R} e^{ik(at-R+c)} \Sigma (2n+1) P_n(\mu) \frac{f_n(ikR)}{F_n(ikc)} \dots\dots(38).$$

If A and B be any two points external to the sphere, a unit source at A will give the same total potential at B, as a unit source at B would give at A. In either case the total potential is made up of two parts, of which the first is the same as if there were no obstacle to the free propagation of the waves, and the second represents the disturbance due to the obstacle. Of these two parts the first is obviously the same, whichever of the two points be regarded as source, and therefore the other parts must also be equal, that is the value of ψ at B when A is a source is equal to the value of ψ at A when B is an equal source. Now when the source A is at a great distance R, the value of ψ at a point B whose angular distance from A is $\cos^{-1}\mu$, and linear distance from the centre is r, is (36)

$$\psi = -\frac{kc^2}{4\pi rR} e^{ik(at-R-r+c)} \Sigma (2n+1) \frac{P_n(\mu) f_n(ikr)}{F_n(ikc)}$$

$$\times P_n\left(\frac{d}{d.ikc}\right) \frac{d}{d.kc} \cdot \frac{\sin kc}{kc},$$

and accordingly this is also the value of ψ at a great distance R, when the source is at B. But since ψ is a disturbance radiating outwards from the sphere, its value at any finite distance R may be inferred from that at an infinite distance by introducing into each harmonic term the factor $f_n(ikR)$. We thus obtain the following symmetrical expression

$$\psi = -\frac{kc^2}{4\pi rR} e^{ik(at-R-r+c)} \Sigma (2n+1) \frac{P_n(\mu)}{F_n(ikc)}$$

$$\times f_n(ikR) . f_n(ikr) P_n\left(\frac{d}{d.ikc}\right) \frac{d}{d.kc} \frac{\sin kc}{kc} \dots\dots\dots(39),$$

which gives this part of the potential at either point, when the other is a unit source.

It should be observed that the general part of the argument does not depend upon the obstacle being either spherical or rigid.

From the expansion of $e^{ikr\mu}$ in spherical harmonics, we may deduce that of the potential of waves issuing from a unit simple source A finitely distant (r) from the origin of co-ordinates. The potential at a point B at an infinite distance R from the origin, and in a direction making an angle $\cos^{-1}\mu$ with r, will be

$$\phi = \frac{e^{-ik(R-\mu r)}}{4\pi R},$$

the time factor being omitted.

Hence by the expansion of $e^{ikr\mu}$

$$\phi = \frac{e^{-ikR}}{4\pi R} \Sigma (2n+1) P_n\left(\frac{d}{d.ikr}\right) \frac{\sin kr}{kr} . P_n(\mu);$$

from which we pass to the case of a finite R by the simple introduction of the factor $f_n(ikR)$.

Thus the potential at a finitely distant point B of a unit source at A is

$$\phi = \frac{e^{ik(at-R)}}{4\pi R} \Sigma (2n+1) P_n\left(\frac{d}{d.ikr}\right) \frac{\sin kr}{kr} f_n(ikR) . P_n(\mu)\ldots(40).$$

335. Having considered at some length the case of a rigid spherical obstacle, we will now sketch briefly the course of the investigation when the obstacle is gaseous. Although in all natural gases the compressibility is nearly the same, we will suppose for the sake of generality that the matter occupying the sphere differs in compressibility, as well as in density, from the medium in which the plane waves advance.

Exterior to the sphere, ϕ is the same exactly, and ψ is of the same form as before. For the motion inside the sphere, if $k' = 2\pi\lambda'$ be the internal wave-length, (2) § 330,

$$\psi_n = \frac{a_n' P_n}{r} \{e^{-ik'r} f_n(ik'r) - (-1)^n e^{+ik'r} f_n(-ik'r)\},$$

$$\frac{d\psi_n}{dr} = \frac{2a_n' P_n}{r^2} . i^{n+1} \{\alpha \sin(k'r + \tfrac{1}{2}n\pi) - \beta \cos(k'r + \tfrac{1}{2}n\pi)\},$$

satisfying the condition of continuity through the centre.

If σ, σ' be the natural densities, m, m' the compressibilities,

$$k'^2/k^2 = \sigma'/\sigma . m/m'\ldots\ldots\ldots\ldots\ldots\ldots(1);$$

and the conditions, to be satisfied by each harmonic element separately, are

$$d\phi/dr + d\psi/dr \text{ (outside)} = d\psi/dr \text{ (inside)} \ldots\ldots\ldots(2),$$

$$\sigma \{\phi + \psi \text{ (outside)}\} = \sigma'\psi \text{ (inside)} \ldots\ldots\ldots\ldots(3),$$

expressing respectively the equalities of the normal motions and of the pressures on the two sides of the bounding surface. From these equations the complete solution may be worked out; but we will here confine ourselves to finding the value of the leading terms, when kc, $k'c$ are very small.

In this case, when $r = c$,

$$\left. \begin{aligned} \psi_0 \text{ (inside)} &= -2ik'a_0' \\ d\psi_0/dr \text{ (inside)} &= \tfrac{2}{3}ik'^3 ca_0' \end{aligned} \right\} \ldots\ldots\ldots\ldots(4),$$

$$\left. \begin{aligned} \phi_0 &= 1 \\ d\phi_0/dr &= -\tfrac{1}{3}k^2 c \end{aligned} \right\} \ldots\ldots\ldots\ldots(5),$$

$$\left. \begin{aligned} \psi_0 \text{ (outside)} &= a_0/c \\ d\psi_0/dr \text{ (outside)} &= -a_0/c^2 \end{aligned} \right\} \ldots\ldots\ldots\ldots(6).$$

Using these in (2), (3), and eliminating a_0', retaining only the principal term, we find

$$a_0 = -\frac{k^2 c^3}{3} \cdot \frac{m' - m}{m'} \ldots\ldots\ldots\ldots(7).$$

In like manner for the term of first order,

$$\left. \begin{aligned} \psi_1 \text{ (inside)} &= -\tfrac{2}{3}a_1' k'^2 c\mu \\ d\psi_1/dr \text{ (inside)} &= -\tfrac{2}{3}a_1' k'^2 \mu \end{aligned} \right\} \ldots\ldots\ldots\ldots(8),$$

$$\left. \begin{aligned} \phi_1 &= ikc\mu \\ d\phi_1/dr &= ik\mu \end{aligned} \right\} \ldots\ldots\ldots\ldots(9),$$

$$\left. \begin{aligned} \psi_1 \text{ (outside)} &= a_1/ikc^2 \cdot \mu \\ d\psi_1/dr \text{ (outside)} &= -2a_1/ikc^3 \cdot \mu \end{aligned} \right\} \ldots\ldots\ldots\ldots(10),$$

which give

$$a_1 = \frac{k^2 c^3 (\sigma - \sigma')}{\sigma + 2\sigma'} \ldots\ldots\ldots\ldots(11).$$

At a distance from the sphere the disturbance due to it is expressed by

$$\psi = \frac{1}{r} e^{ik(at-r)} \{a_0 + a_1\mu\}$$

$$= -\frac{k^2 c^3}{3r} e^{ik(at-r)} \left\{ \frac{m' - m}{m'} + 3\frac{\sigma' - \sigma}{\sigma + 2\sigma'}\mu \right\} \ldots\ldots\ldots\ldots(12).$$

If we introduce the relations

$$T = \tfrac{4}{3}\pi c^3, \qquad k = 2\pi/\lambda,$$

and throw away the imaginary part, we obtain

$$\psi = -\frac{\pi T}{\lambda^2 r}\left\{\frac{m'-m}{m'} + 3\frac{\sigma'-\sigma}{\sigma+2\sigma'}\mu\right\}\cos k\,(at-r)\ldots\ldots(13),$$

as the expression for the most important part of the disturbance, corresponding to (21) § 334 for a fixed rigid sphere. It appears, as might have been expected, that the term of zero order is due to the variation of compressibility, and that of order one to the variation of density.

From (13) we may fall back on the case of a rigid fixed sphere, by making both σ' and m' infinite. It is not sufficient to make σ' by itself infinite, apparently because, if m' at the same time remained finite, $k'c$ would not be small, as the investigation has assumed.

When $m'-m$, $\sigma'-\sigma$ are small, (13) becomes equivalent to

$$\psi = -\frac{\pi T}{\lambda^2 r}\left\{\frac{m'-m}{m} + \frac{\sigma'-\sigma}{\sigma}\mu\right\}\cos k\,(at-r),$$

corresponding to $\phi = \cos kat$ at the centre of the sphere. This agrees with the result (13) of § 296, in which the obstacle may be of any form.

In actual gases $m' = m$, and the term of zero order disappears. If the gas occupying the spherical space be incomparably lighter than the other gas, $\sigma' = 0$, and

$$\psi = 3\frac{\pi T}{\lambda^2 r}\mu\cos k\,(at-r)\ldots\ldots\ldots\ldots\ldots(14),$$

so that in the term of order one, the effect is twice that of a rigid body, and has the reverse sign.

The greater part of this chapter is taken from two papers by the author "On the vibrations of a gas contained within a rigid spherical envelope," and an "Investigation of the disturbance produced by a spherical obstacle on the waves of sound[1]," and from the paper by Professor Stokes already referred to.

---

[1] *Math. Society's Proceedings*, March 14, 1872; Nov. 14, 1872.

335 a.   An interesting function, which has been considered by Prof. Lamb,[1] relates to the maximum disturbance that can be produced by an infinitesimal resonator exposed to plane waves.

The value of ϕ for the expansion of the primary waves is given by (1), (2), (4), § 334.   By (5) § 334 and (3) § 329 the value of ψ for the secondary waves may be taken to be

$$\psi = ika_n(-1)^n P_n(\mu) \cdot P_n\left(\frac{d}{d \cdot ikr}\right)\left(\frac{e^{-ikr}}{ikr}\right),$$

in which

$$P_n\left(\frac{d}{d \cdot ikr}\right)\left(\frac{e^{-ikr}}{ikr}\right) = \frac{1}{i} P_n\left(\frac{d}{d \cdot ikr}\right)\left(\frac{\cos kr}{kr} - i\frac{\sin kr}{kr}\right).$$

If we omit the common factor $P_n(\mu)$, we have

$$\phi + \psi = \{2n+1-(-1)^n ika_n\} Pn\left(\frac{d}{d \cdot ikr}\right)\frac{\sin kr}{kr}$$

$$+ (-1)^n ka_n P_n\left(\frac{d}{d \cdot ikr}\right)\frac{\cos kr}{kr} \quad \ldots\ldots\ldots\ldots(1).$$

Now the only condition imposed upon the appliances introduced at r is that they shall do no work.   This requires that $\phi + \psi$ be in the same phase as $d\phi/dr + d\psi/dr$, viz. that the ratio of (1) and of its derivative with respect to r shall be *real*. Since P_n is a wholly odd or wholly even function, this requires that

$$\frac{2n+1-(-1)^n ika_n}{(-1)^n ka_n} \text{ be real.}$$

If a_n, which may be complex, be written $Ae^{i\alpha}$, we get

$$kA = -(-1)^n(2n+1)\sin \alpha \quad \ldots\ldots\ldots\ldots\ldots(2).$$

Thus A is a maximum when

$$\sin \alpha = -(-1)^n \ldots\ldots\ldots\ldots\ldots\ldots(3),$$

and the maximum value is

$$A = \frac{2n+1}{k} \quad \ldots\ldots\ldots\ldots\ldots\ldots(4).$$

---

[1] *London Math. Soc. Proc.* Vol. XXXII. p. 11, 1900.

By (3) and (4), $a_n = -(-1)^n i \dfrac{2n+1}{k}$ ....(5),

so that in (1), $2n+1 - (-1)^n ika_n = 0$....................(6),

but $\phi + \psi$ does not itself vanish.

If the incident plane waves are regarded as due to a source at a great distance R, we have, in order to secure the value unity at the resonator as supposed,

$$\phi = \frac{Re^{-ikR}}{R} \quad \dots\dots\dots\dots\dots\dots(7),$$

with which we may compare

$$\psi = \frac{a_n e^{-ikr}}{r} P_n(\mu)\dots\dots\dots\dots\dots\dots(8).$$

The work emitted by the *primary* source being represented by

$$R^2 \int_{-1}^{+1} d\mu,$$

that emitted, or rather diverted, by the resonator will be

$$\mathrm{Mod}^2 a_n \int_{-1}^{+1} P_n^2(\mu)\, d\mu.$$

Now $$\int_{-1}^{+1} P_n^2(\mu)\, d\mu = \frac{2}{2n+1}$$

and $$\int_{-1}^{+1} d\mu = 2.$$

Also $$\mathrm{Mod}^2 a_n = \frac{(2n+1)^2}{k^2} \quad \dots\dots\dots\dots\dots(9);$$

so that the ratio of works is

$$\frac{2n+1}{k^2 R^2} \quad \dots\dots\dots\dots\dots\dots(10).$$

This agrees with the result of § 319 for a symmetrical resonator

$$(n = 0).$$

Prof. Lamb expresses his conclusion in terms of the energy transmitted in the primary waves across unit of area. This being taken as unity, the work emitted by the resonator is given by multiplying (10) by the area of the sphere of radius R, viz. $4\pi R^2$. We get accordingly

$$\frac{(2n+1)\lambda^2}{\pi} \dots\dots\dots\dots\dots\dots(11),$$

$2\pi/\lambda$ being substituted for k. This formula, given by Prof. Lamb, expresses the whole energy emitted by the resonator in terms of the energy of the primary waves per unit of area.

It is worthy of remark that we have nowhere assumed that r at the surface of the resonator is small. The results therefore apply to resonators of finite size, provided that the symmetrical constitution implied in the harmonic analysis is maintained. And the maximum energy emitted is the same whatever be the size of the resonator.

The case of $n = 1$ is in some respects the simplest, inasmuch as the resonator may then consist of a rigid sphere held to a fixed point by elastic attachments. As a particular case of (1) we have

$$i\,(\phi_1 + \psi_1) = \mu(3 + ika_1)\frac{d}{dkr}\frac{\sin kr}{kr} - ka_1\mu\frac{d}{dkr}\frac{\cos kr}{kr} \quad (12).$$

This may be considered to represent the force acting upon the sphere due to the pressures. Its derivative with respect to r will represent in like manner the acceleration of the sphere, and by suitable choice of mass and spring all the conditions may be satisfied, provided that the ratio of these quantities is real. The maximum a_1 is, as in (5),

$$a_1 = 3i/k \dots\dots\dots\dots\dots\dots(13);$$

and, as in (11), the energy emitted by the resonator is, on the scale there adopted, $3\lambda^2/\pi$.

It may occasion surprise that the energy emissible in the present case is 3 times that emissible from a symmetrical resonator; but the 3 may be got rid of by another presentation of the matter. We have supposed hitherto that the sphere is capable of vibration in the line of symmetry defined by the direction of propagation of the primary waves. If the sphere, considered as infinitesimal, be capable of vibration along one line only, its efficiency as a resonator is proportional to the cosine squared of the angle between this direction and that of the primary waves. This limitation to a single direction of vibration is really the standard case, while that previously considered involves three degrees of freedom. If now we inquire what is the *average* efficiency of the resonator for primary waves reaching it in one direction, we see that the above specially favoured efficiencies must be reduced, in the ratio

$$\int_0^1 \mu^2 d\mu : \int_0^1 d\mu \; ;$$

that is, in the ratio of $3:1$. The *average* efficiency of the resonator when $n=1$ is then the same as when $n=0$, and a like result applies whatever n may be. The increased efficiency represented by the factors $2n+1$ must be regarded as due to the cooperation of $2n+1$ degrees of freedom.

CHAPTER XVIII.

336. In a former chapter (§ 135), we saw that a proof of Fourier's theorem might be obtained by considering the mechanics of a vibrating string. A similar treatment of the problem of a spherical sheet of air will lead us to a proof of Laplace's expansion for a function which is arbitrary at every point of a spherical surface.

As in § 333, if ψ is the velocity-potential, the equation of continuity, referred to the ordinary polar co-ordinates θ, ω, takes the form,

$$c^2 \frac{d^2\psi}{dt^2} = a^2 \left\{ \frac{1}{\sin\theta} \frac{d}{d\theta}\left(\sin\theta \frac{d\psi}{d\theta}\right) + \frac{1}{\sin^2\theta} \frac{d^2\psi}{d\omega^2} \right\}.$$

Whatever may be the character of the free motion, it can be analysed into a series of simple harmonic vibrations, the nature of which is determined by the corresponding functions ψ, considered as dependent on space. Thus, if $\psi \propto e^{ikat}$, the equation to determine ψ as a function of θ and ω is

$$\frac{1}{\sin\theta} \frac{d}{d\theta}\left(\sin\theta \frac{d\psi}{d\theta}\right) + \frac{1}{\sin^2\theta} \frac{d^2\psi}{d\omega^2} + k^2 c^2 \psi = 0 \ldots\ldots\ldots(1).$$

Again, whatever function ψ may be, it can be expanded by Fourier's theorem[1] in a series of sines and cosines of the multiples of ω. Thus

$$\psi = \psi_0 + \psi_1 \cos\omega + \psi_1' \sin\omega + \psi_2 \cos 2\omega + \psi_2' \sin 2\omega$$
$$+ \ldots\ldots + \psi_s \cos s\omega + \psi_s' \sin s\omega + \ldots\ldots\ldots(2),$$

---

[1] We here introduce the condition that ψ recurs after one revolution round the sphere.

where the coefficients ψ_0, $\psi_1 \ldots \psi_1'$, $\psi_2' \ldots$ are functions of θ only; and by the conjugate property of the circular functions, each term of the series must satisfy the equation independently. Accordingly,

$$\frac{1}{\sin\theta}\frac{d}{d\theta}\left(\sin\theta\frac{d\psi_s}{d\theta}\right) - \frac{s^2\psi_s}{\sin^2\theta} + k^2 c^2 \psi_s = 0 \ \ldots\ldots\ldots (3)$$

is the equation from which the character of ψ_s or ψ_s' is to be determined. This equation may be written in various ways.

In terms of μ ($= \cos\theta$),

$$\frac{d}{d\mu}\left\{(1-\mu^2)\frac{d\psi_s}{d\mu}\right\} + h^2\psi_s - \frac{s^2}{1-\mu^2}\psi_s = 0 \ \ldots\ldots\ldots (4);$$

or, if $\nu = \sin\theta$,

$$\nu^2(1-\nu^2)\frac{d^2\psi_s}{d\nu^2} + \nu(1-2\nu^2)\frac{d\psi_s}{d\nu} + \nu^2 h^2 \psi_s - s^2 \psi_s = 0 \ldots(5),$$

where h^2 is written for $k^2 c^2$.

When the original function ψ is symmetrical with respect to the pole, that is, depends upon latitude only, s vanishes, and the equations simplify. This case we may conveniently take first. In terms of μ,

$$(1-\mu^2)\frac{d^2\psi_0}{d\mu^2} - 2\mu\frac{d\psi_0}{d\mu} + h^2\psi_0 = 0 \ldots\ldots\ldots\ldots(6).$$

The solution of this equation involves two arbitrary constants, multiplying two definite functions of μ, and may be obtained in the ordinary way by assuming an ascending series and determining the exponents and coefficients by substitution. Thus

$$\psi_0 = A\left\{1 - \frac{h^2}{1\,.\,2}\mu^2 + \frac{h^2(h^2-2\,.\,3)}{1\,.\,2\,.\,3\,.\,4}\mu^4\right.$$

$$\left. - \frac{h^2(h^2-2\,.\,3)(h^2-4\,.\,5)}{1\,.\,2\,.\,3\,.\,4\,.\,5\,.\,6}\mu^6 + \&\text{c.}\right\}$$

$$+ B\left\{\mu - \frac{h^2-1\,.\,2}{1\,.\,2\,.\,3}\mu^3 + \frac{(h^2-1\,.\,2)(h^2-3\,.\,4)}{1\,.\,2\,.\,3\,.\,4\,.\,5}\mu^5 - \&\text{c.}\right\} \ \ldots\ldots (7),$$

in which A and B are arbitrary constants.

Let us now further suppose that ψ besides being symmetrical round the pole is also symmetrical with respect to the equator (which is accordingly *nodal*), or in other words that ψ is an *even* function of the sine of the latitude (μ). Under these circumstances it is clear that B must vanish, and the value of ψ be expressed simply by the first series, multiplied by the arbitrary constant A. This value of the velocity-potential is the logical consequence of the original differential equation and of the two restrictions as to symmetry. The value of h^2 might appear to be arbitrary, but from what we know of the mechanics of the problem, it is certain beforehand that h^2 is really limited to a series of particular values. The condition, which yet remains to be introduced and by which h is determined, is that the original equation is satisfied at the pole itself, or in other words that the pole is not a source; and this requires us to consider the value of the series when $\mu = 1$. Since the series is an even function of μ, if the pole $\mu = +1$ be not a source, neither will be the pole $\mu = -1$. It is evident at once that if h^2 be of the form $n(n+1)$, where n is an even integer, the series terminates, and therefore remains finite when $\mu = 1$; but what we now want to prove is that, if the series remain finite for $\mu = 1$, h^2 is necessarily of the above-mentioned form. By the ordinary rule it appears at once that, whatever be the value of h^2, the ratio of successive terms tends to the limit μ^2, and therefore the series is convergent for all values of μ less than unity. But for the extreme value $\mu = 1$, a higher method of discrimination is necessary.

It is known[1] that the infinite hypergeometrical series

$$1 + \frac{ab}{cd} + \frac{a(a+1)b(b+1)}{c(c+1)d(d+1)} + \frac{a(a+1)(a+2)b(b+1)(b+2)}{c(c+1)(c+2)d(d+1)(d+2)} + \ldots (8)$$

is convergent, if $c + d - a - b$ be greater than 1, and divergent if $c + d - a - b$ be equal to, or less than 1. In the latter case the value of $c + d - a - b$ affords a criterion of the degree of divergency. Of two divergent series of the above form, for which the values of $c + d - a - b$ are different, that one is *relatively* infinite for which the value of $c + d - a - b$ is the smaller.

Our present series (7) may be reduced to the standard form by taking $h^2 = n(n+1)$, where n is not assumed to be integral. Thus

$$1 - \frac{h^2}{1 \cdot 2} \mu^2 + \frac{h^2 (h^2 - 2 \cdot 3)}{1 \cdot 2 \cdot 3 \cdot 4} \mu^4 - \dots$$

$$= 1 - \frac{n(n+1)}{1 \cdot 2} \mu^2 + \frac{n(n+1)(n-2)(n+3)}{1 \cdot 2 \cdot 3 \cdot 4} \mu^4 - \dots$$

$$= 1 + \frac{(-\frac{1}{2}n)(\frac{1}{2}n + \frac{1}{2})}{1 \cdot \frac{1}{2}} \mu^2 + \frac{(-\frac{1}{2}n)(-\frac{1}{2}n + 1)(\frac{1}{2}n + \frac{1}{2})(\frac{1}{2}n + \frac{1}{2} + 1)}{1 \cdot 2 \cdot \frac{1}{2} \cdot \frac{3}{2}} \mu^4$$
$$+ \dots \dots \dots (9),$$

which is of the standard form, if

$$a = -\tfrac{1}{2}n, \qquad b = \tfrac{1}{2}n + \tfrac{1}{2}, \qquad c = \tfrac{1}{2}, \qquad d = 1.$$

Accordingly, since $c + d - a - b = 1$, the series is divergent for $\mu = 1$, *unless it terminate;* and it terminates only when n is an even integer. We are thus led to the conclusion that when the pole is not a source, and ψ_0 is an even function of μ, h^2 must be of the form $n(n+1)$, where n is an even integer.

In like manner, we may prove that when ψ_0 is an odd function of μ, and the poles are not sources, $A = 0$, and h^2 must be of the form $n(n+1)$, n being an *odd* integer.

If n be fractional, both series are divergent for $\mu = \pm 1$, and although a combination of them may be found which remains finite at one or other pole, there can be no combination which remains finite at *both* poles. If therefore it be a condition that no point on the surface of the sphere is a source, we have no alternative but to make n integral, and even then we do not secure finiteness at the poles unless we further suppose $A = 0$, when n is odd, and $B = 0$, when n is even. We conclude that for a complete spherical layer, the only admissible values of ψ, which are functions of latitude only, and proportional to harmonic functions of the time, are included under

$$\psi = C P_n(\mu),$$

where $P_n(\mu)$ is Legendre's function, and n is any odd or even integer. The possibility of expanding an arbitrary function of latitude in a series of Legendre's functions is a necessary consequence of what has now been proved. Any possible motion of the layer of gas is represented by the series

$$\psi = A_0 + P_1(\mu) \left(A_1 \cos \frac{\sqrt{(1 \cdot 2)} \cdot at}{c} + B_1 \sin \frac{\sqrt{(1 \cdot 2)} \cdot at}{c} \right) + \dots$$
$$+ P_n(\mu) \left(A_n \cos \frac{\sqrt{\{n(n+1)\}} \cdot at}{c} + B_n \sin \frac{\sqrt{\{n(n+1)\}} \cdot at}{c} \right) + \dots (10).$$

When $t = 0$,

$$\psi = A_0 + A_1 P_1(\mu) + \ldots + A_n P_n(\mu) + \ldots\ldots\ldots(11),$$

and the value of ψ when $t = 0$ is an *arbitrary* function of latitude.

The method that we have here followed has also the advantage of proving the conjugate property,

$$\int_{-1}^{+1} P_n(\mu) P_m(\mu)\, d\mu = 0 \ldots\ldots\ldots\ldots\ldots(12),$$

where n and m are different integers. For the functions $P(\mu)$ are the *normal* functions (§ 94) for the vibrating system under consideration, and accordingly the expression for the kinetic energy can only involve the *squares* of the generalized velocities. If (12) do not hold good, the *products* also of the velocities must enter.

The value of ψ appropriate to a *plane* layer of vibrating gas can of course be deduced as a particular case of the general solution applicable to a spherical layer. Confining ourselves to the case where there is no source at the pole ($\mu = 1$), we have to investigate the limiting form of $\psi = C P_n(\mu)$, where $n(n+1) = k^2 c^2$, when c^2 and n^2 are infinite. At the same time $\mu - 1$ and ν are infinitesimal, and $c\nu$ passes into the plane polar radius (r), so that $n\nu = kr$. For this purpose the most convenient form of $P_n(\mu)$ is that of Murphy[1]:

$$P_n(\cos\theta) = 1 - \frac{n(n+1)}{1^2}\sin^2\frac{\theta}{2} + \frac{(n-1)n(n+1)(n+2)}{1^2 \cdot 2^2}\sin^4\frac{\theta}{2}$$
$$- \ldots\ldots\ldots\ldots(13).$$

The limit is evidently

$$\psi = C\left\{1 - \frac{k^2 r^2}{2^2} + \frac{k^4 r^4}{2^2 \cdot 4^2} - \frac{k^6 r^6}{2^2 \cdot 4^2 \cdot 6^2} + \ldots\right\} = C J_0(kr)\ldots\ldots(14),$$

shewing that the Bessel's function of zero order is an extreme case of Legendre's functions.

When the spherical layer is not complete, the problem requires a different treatment. Thus, if the gas be bounded by walls stretching along two parallels of latitude, the complete integral involving two arbitrary constants will in general be necessary.

---

[1] Thomson and Tait's *Nat. Phil.* § 782. [$t = \sin^2\frac{1}{2}\theta$, not $4\sin^2\frac{1}{2}\theta$.] Todhunter's *Laplace's Functions*, § 19.

The ratio of the constants and the admissible values of h^2 are to be determined by the two boundary conditions expressing that at the parallels in question the motion is wholly in longitude. The value of μ being throughout numerically less than unity, the series are always convergent.

If the portion of the surface occupied by gas be that included between two parallels of latitude at equal distances from the equator, the question becomes simpler, since then one or other of the constants A and B in (7) vanishes in the case of each normal function.

337. When the spherical area contemplated includes a pole, we have, as in the case of the complete sphere, to introduce the condition that the pole is not a source. For this purpose the solution in terms of ν, i.e. $\sin \theta$, will be more convenient.

If we restrict ourselves for the present to the case of symmetry, we have, putting $s = 0$ in (5) § 336,

$$\nu (1 - \nu^2) \frac{d^2\psi_0}{d\nu^2} + (1 - 2\nu^2) \frac{d\psi_0}{d\nu} + h^2 \nu \, \psi_0 = 0 \ldots\ldots\ldots (1).$$

One solution of this equation is readily obtained in the ordinary way by assuming an ascending series and substituting in the differential equation to determine the exponents and coefficients. We get[1]

$$\psi_0 = A \left\{ 1 + \frac{0 \cdot 1 - h^2}{2^2} \nu^2 + \frac{(0 \cdot 1 - h^2)(2 \cdot 3 - h^2)}{2^2 \cdot 4^2} \nu^4 \right.$$
$$\left. + \frac{(0 \cdot 1 - h^2)(2 \cdot 3 - h^2)(4 \cdot 5 - h^2)}{2^2 \cdot 4^2 \cdot 6^2} \nu^6 + \ldots \right\} \ldots\ldots (2).$$

This value of ψ_0 is the most general solution of (1), subject to the condition of finiteness when $\nu = 0$. The complete solution involving two arbitrary constants provides for a source of arbitrary intensity at the pole, in which case the value of ψ_0 is infinite when $\nu = 0$. Any solution which remains finite when $\nu = 0$ and involves one arbitrary constant, is therefore the most general possible under the restriction that the pole be not a source. Accordingly it is unnecessary for our purpose to complete the solution. The nature of the second function (involving a logarithm of ν) will be illustrated in the particular case of a plane layer to be considered presently.

[1] Heine's *Kugelfunctionen*, § 28.

By writing $n(n+1)$ for h^2 the series within brackets becomes

$$1 - \frac{n(n+1)}{2^2}\nu^2 + \frac{(n-2)n(n+1)(n+3)}{2^2 \cdot 4^2}\nu^2 - \ldots\ldots \quad (3),$$

or, when reduced to the standard hypergeometrical form,

$$1 + \frac{(-\tfrac{1}{2}n)(\tfrac{1}{2}n+\tfrac{1}{2})}{1 \cdot 1}\nu^2 + \frac{(-\tfrac{1}{2}n)(-\tfrac{1}{2}n+1)(\tfrac{1}{2}n+\tfrac{1}{2})(\tfrac{1}{2}n+\tfrac{1}{2}+1)}{1 \cdot 2 \cdot 1 \cdot 2}\nu^4 + \ldots,$$

corresponding to

$$a = -\tfrac{1}{2}n, \quad b = \tfrac{1}{2}n + \tfrac{1}{2}, \quad c = 1, \quad d = 1.$$

Since $c + d - a - b = \tfrac{3}{2}$, the series converges for all values of ν from 0 to 1 inclusive. To values of $\theta (= \sin^{-1}\nu)$ greater than $\tfrac{1}{2}\pi$ the solution is inapplicable.

When n is an integer, the series becomes identical with Legendre's function $P_n(\mu)$. If the integer be even, the series terminates, but otherwise remains infinite. Thus, when $n = 1$, the series is identical with the expansion of μ, viz. $\sqrt{(1-\nu^2)}$, in powers of ν.

The expression for ψ in terms of ν may be conveniently applied to the investigation of the free symmetrical vibrations of a spherical layer of air, bounded by a small circle, whose radius is less than the quadrant. The condition to be satisfied is simply $d\psi/d\nu = 0$, an equation by which the possible values of h^2, or k^2c^2, are connected with the given boundary value of ν.

Certain particular cases of this problem may be treated by means of Legendre's functions. Suppose, for example, that $n = 6$, so that $h^2 = k^2c^2 = 42$. The corresponding solution is $\psi = AP_6(\mu)$. The greatest value of μ for which $d\psi/d\mu = 0$ is $\mu = \cdot8302$, corresponding to $\theta = 33° 53' = \cdot59137$ radians[1].

If we take $c\theta = r$, so that r is the radius of the small circle measured along the sphere, we get

$$kr = \sqrt{(42)} \times \cdot59137 = 3\cdot8325,$$

which is the equation connecting the value of $k (= 2\pi/\lambda)$ with the curved radius r, in the case of a small circle, whose angular radius is $33° 53'$. If the layer were plane ($\S 339$), the value of kr would be $3\cdot8317$; so that it makes no perceptible difference in the pitch of the gravest tone whether the radius (r) of given length be

---

[1] The radian is the unit of circular measure.

straight, or be curved to an arc of $33°$. The result of the comparison would, however, be materially different, if we were to take the length of the circumference as the same in the two cases, that is, replace $c\theta = r$ by $cv = r$.

In order to deduce the symmetrical solution for a plane layer, it is only necessary to make c infinite, while cv remains finite. On account of the infinite value of h^2, the solution assumes the simple form

$$\psi = A \left\{ 1 - \frac{h^2 v^2}{2^2} + \frac{h^4 v^4}{2^2 \cdot 4^2} - \frac{h^6 v^6}{2^2 \cdot 4^2 \cdot 6^2} + \ldots \right\} \ldots\ldots\ldots(4),$$

or, if we write $cv = r$, where r is the polar radius in two dimensions,

$$\psi = A \left\{ 1 - \frac{k^2 r^2}{2^2} + \frac{k^4 r^4}{2^2 \cdot 4^2} - \ldots\ldots\ldots \right\} = A\, J_0\,(kr)\ldots\ldots\ldots(5),$$

as in (14) § 336.

The differential equation for ψ in terms of v, when c is infinite and $cv = r$, becomes

$$\frac{d^2\psi}{dr^2} + \frac{1}{r}\frac{d\psi}{dr} + k^2\psi = 0 \ldots\ldots\ldots\ldots\ldots\ldots(6).$$

An independent investigation and solution for the plane problem will be given presently.

338. When s is different from zero, the differential equation satisfied by the coefficients of $\sin s\omega$, $\cos s\omega$, is

$$v^2(1 - v^2)\frac{d^2\psi_s}{dv^2} + v(1 - 2v^2)\frac{d\psi_s}{dv} + v^2 h^2 \psi_s - s^2 \psi_s = 0 \ldots\ldots(1),$$

and the solution, subject to the condition of finiteness when $v = 0$[1], is easily found to be

$$\psi_s = A v^s \left\{ 1 + \frac{s(s+1) - h^2}{2(2s+2)} v^2 \right.$$
$$\left. + \frac{s(s+1) - h^2}{2(2s+2)} \cdot \frac{(s+2)(s+3) - h^2}{4(2s+4)} v^4 + \ldots \right\};$$

or, if we put $h^2 = n(n+1)$,

$$\psi_s = A v^s \left\{ 1 + \frac{(s-n)(s+n+1)}{2 \cdot (2s+2)} v^2 \right.$$
$$\left. + \frac{(s-n)(s-n+2)(s+n+1)(s+n+3)}{2 \cdot 4 \cdot (2s+2)(2s+4)} v^4 + \ldots \right\} \ldots\ldots(2).$$

---

[1] The solution may be completed by the addition of a second function derived from (2) by changing the sign of s, which occurs in (1) only as s^2, but a modification is necessary, when s is a positive integer. The method of procedure will be exemplified presently in the case of the plane layer.

We have here the complete solution of the problem of the vibrations of a spherical layer of gas bounded by a small circle whose radius is less than the quadrant. For each value of s, there are a series of possible values of n, determined by the condition $d\psi_s/d\nu = 0$; with any of these values of n the function on the right-hand side of (2), when multiplied by $\cos s\omega$ or $\sin s\omega$, is a normal function of the system. The aggregate of all the normal functions corresponding to every admissible value of s and n, with an arbitrary coefficient prefixed to each, gives an expression capable of being identified with the initial value of ψ, i.e. with a function given arbitrarily over the area of the small circle.

When the radius of the sphere c is infinitely great, h^2 is infinite. If $c\nu = r$, $h^2\nu^2 = k^2 r^2$, and (2) becomes

$$\psi_s = A'r^s \left\{ 1 - \frac{k^2 r^2}{2 \cdot (2s+2)} + \frac{k^4 r^4}{2 \cdot 4 \cdot (2s+2)(2s+4)} - \dots \right\} \dots\dots(3),$$

a function of r proportional to $J_s(kr)$.

In terms of μ, the differential equation satisfied by the coefficient of $\cos s\omega$, or $\sin s\omega$, is

$$\frac{d}{d\mu} \left\{ (1-\mu^2) \frac{d\psi_s}{d\mu} \right\} + h^2\psi_s - \frac{s^2}{1-\mu^2} \psi_s = 0 \dots\dots\dots(4).$$

Assuming $\psi_s = (1-\mu^2)^{\frac{1}{2}s}\phi_s$, we find as the equation for ϕ_s

$$(1-\mu^2)\frac{d^2\phi_s}{d\mu^2} - 2(s+1)\mu\frac{d\phi_s}{d\mu} + \{h^2 - s(s+1)\}\phi_s = 0 \dots(5),$$

which will be more easily dealt with.

To solve it, let

$$\phi_s = \mu^\alpha + a_2\mu^{\alpha+2} + a_4\mu^{\alpha+4} + \dots + a_{2m}\mu^{\alpha+2m} + \dots,$$

and substitute in (5). The coefficient of the lowest power of μ is $\alpha(\alpha-1)$; so that $\alpha = 0$, or $\alpha = 1$. The relation between a_{2m+2}, and a_{2m}, found by equating to zero the coefficient of $\mu^{\alpha+2m}$, is

$$a_{2m+2} = a_{2m} \frac{(\alpha+2m+s-n)(\alpha+2m+s+n+1)}{(\alpha+2m+1)(\alpha+2m+2)},$$

where $n(n+1) = h^2$.

The complete value of ϕ_s is accordingly given by

$$\phi_s = A \left\{ 1 + \frac{(s-n)(s+n+1)}{1 \cdot 2} \mu^2 + \frac{(s-n)(s-n+2)(s+n+1)(s+n+3)}{1 \cdot 2 \cdot 3 \cdot 4} \mu^4 \right.$$

$$\left. + \frac{(s-n)(s-n+2)(s-n+4)(s+n+1)(s+n+3)(s+n+5)}{1 \cdot 2 \cdot 3 \cdot 4 \cdot 5 \cdot 6} \mu^6 + \dots \right\}$$

$$+ B \left\{ \mu + \frac{(s-n+1)(s+n+2)}{2 \cdot 3} \mu^3 \right.$$

$$\left. + \frac{(s-n+1)(s-n+3)(s+n+2)(s+n+4)}{2 \cdot 3 \cdot 4 \cdot 5} \mu^5 + \dots \right\} \dots (6),$$

where A and B are arbitrary constants;

and
$$\psi_s = (1 - \mu^2)^{\frac{1}{2}s} \phi_s \dots \dots \dots (7).$$

We have now to prove that the condition that neither pole is a source requires that $n - s$ be a positive integer, in which case one or other of the series in the expression for ϕ_s terminates. For this purpose it will not be enough to shew that the series (unless terminating) are infinite when $\mu = \pm 1$; it will be necessary to prove that they remain divergent after multiplication by $(1 - \mu^2)^{\frac{1}{2}s}$, or as we may put it more conveniently, that they are infinite when $\mu = \pm 1$ *in comparison with* $(1 - \mu^2)^{-\frac{1}{2}s}$. It will be sufficient to consider in detail the case of the first series.

We have

$$1 + \frac{(s-n)(s+n+1)}{1 \cdot 2} + \frac{(s-n)(s-n+2)(s+n+1)(s+n+3)}{1 \cdot 2 \cdot 3 \cdot 4} + \dots \dots$$

$$= 1 + \frac{(\frac{1}{2}s - \frac{1}{2}n)(\frac{1}{2}s + \frac{1}{2}n + \frac{1}{2})}{1 \cdot \frac{1}{2}}$$

$$+ \frac{(\frac{1}{2}s - \frac{1}{2}n)(\frac{1}{2}s - \frac{1}{2}n + 1)(\frac{1}{2}s + \frac{1}{2}n + \frac{1}{2})(\frac{1}{2}s + \frac{1}{2}n + \frac{1}{2} + 1)}{1 \cdot 2 \cdot \frac{1}{2} \cdot \frac{3}{2}} + \dots;$$

which is of the standard form (8) § 336

$$1 + \frac{ab}{cd} + \frac{a(a+1)b(b+1)}{c(c+1)d(d+1)} + \dots,$$

if
$$a = \tfrac{1}{2}s - \tfrac{1}{2}n, \quad b = \tfrac{1}{2}s + \tfrac{1}{2}n + \tfrac{1}{2}, \quad c = 1, \quad d = \tfrac{1}{2}.$$

The degree of divergency is determined by the value of $a + b - c - d$, which is here equal to $s - 1$.

On the other hand, the binomial theorem gives for the expansion of $(1 - \mu^2)^{-\frac{1}{2}s}$

$$1 + \frac{\frac{1}{2}s}{1} \mu^2 + \frac{\frac{1}{2}s\left(\frac{1}{2}s + 1\right)}{1 \cdot 2} \mu^4 + \ldots\ldots,$$

which is of the standard form, if

$$a = \tfrac{1}{2}s, \quad c = 1, \quad b = d, \quad \text{and makes} \quad a + b - c - d = \tfrac{1}{2}s - 1.$$

Since $s - 1 > \tfrac{1}{2}s - 1$, it appears that the series in the expression for ϕ_s are infinities of a higher order than $(1 - \mu^2)^{-\frac{1}{2}s}$, and therefore remain infinite after multiplication by $(1 - \mu^2)^{\frac{1}{2}s}$. Accordingly ψ_s cannot be finite at both poles unless one or other of the series terminate, which can only happen when $n - s$ is zero, or a positive integer. If the integer be even, we have still to suppose $B = 0$; and if the integer be odd, $A = 0$, in order to secure finiteness at the poles.

In either case the value of ϕ_s for the complete sphere may be put into the form

$$\phi_s = \frac{d^{n+s}}{d\mu^{n+s}} (1 - \mu^2)^n = \frac{d^s P_n(\mu)}{d\mu^s} \ \ldots\ldots\ldots\ldots (8),$$

where the constant multiplier is omitted. The complete expression for that part of ψ which contains $\cos s\omega$ or $\sin s\omega$ as a factor is therefore

$$\psi = \frac{\cos s\omega}{\sin s\omega} \sum_{n=s}^{n=\infty} A_n \nu^s \frac{d^s}{d\mu^s} P_n(\mu) \ \ldots\ldots\ldots\ldots\ldots(9),$$

where A_n is constant with respect to μ and ω, but as a function of the time will vary as

$$\cos\left(\frac{\sqrt{(n \cdot n + 1)} \, at}{c} + \epsilon\right)\ldots\ldots\ldots\ldots\ldots(10).$$

For most purposes, however, it is more convenient to group the terms for which n is the same, rather than those for which s is the same. Thus for any value of n

$$\psi = \sum_{s=0}^{s=n} \nu^s \frac{d^s P_n(\mu)}{d\mu^s} (A_s \cos s\omega + B_s \sin s\omega) \ \ldots\ldots\ldots (11),$$

where every coefficient A_s, B_s may be regarded as containing a time factor of the form (10).

Initially ψ is an arbitrary function of μ and ω, and therefore any such function is capable of being represented in the form

$$\psi = \sum_{n=0}^{n=\infty} \sum_{s=0}^{s=n} \nu^s \frac{d^s P_n(\mu)}{d\mu^s}(A_s{}^n \cos s\omega + B_s{}^n \sin s\omega)\ldots(12),$$

which is Laplace's expansion in spherical surface harmonics.

From the differential equation (5), or from its general solution (6), it is easy to prove that ϕ_s is of the same form as $d\phi_{s-1}/d\mu$, so that we may write

$$\phi_s = \left(\frac{d}{d\mu}\right)^s \phi_0 \ldots\ldots\ldots\ldots\ldots (13),$$

(in which no connection between the arbitrary constants is asserted), or in terms of ψ by (7),

$$\psi_s = (1 - \mu^2)^{\frac{1}{2}s} \left(\frac{d}{d\mu}\right)^s \psi_0 \ldots\ldots\ldots\ldots (14).$$

Equation (13) is a generalization of the property of Laplace's functions used in (8).

The corresponding relations for the plane problem may be deduced, as before, by attaching an infinite value to n, which in (13), (14) is arbitrary, and writing $n\nu = kr$. Since $\mu^2 + \nu^2 = 1$,

$$\psi_s = \nu^s \left(-\frac{\mu}{\nu} \frac{d}{d\nu} \right)^s \psi_0,$$

ψ_0 being regarded as a function of ν. In the limit μ (even though subject to differentiation) may be identified with unity, and thus we may take

$$\psi_s = (-2kr)^s \left(\frac{d}{d.(kr)^2}\right)^s \psi_0 \ldots\ldots\ldots\ldots(15).$$

When the pole is not a source, ψ_s is proportional to $J_s(kr)$. The constant coefficient, left undetermined by (15), may be readily found by a comparison of the leading terms. It thus appears that

$$J_s(kr) = (-2kr)^s \left(\frac{d}{d.(kr)^2}\right)^s J_0(kr)\ldots\ldots\ldots(16),$$

a well-known property of Bessel's functions[1].

The vibrations of a plane layer of gas are of course more easily dealt with, than those of a layer of finite curvature, but I have preferred to exhibit the indirect as well as the direct method of investigation, both for the sake of the spherical problem

---

[1] Todhunter's *Laplace's Functions*, § 390.

itself with the corresponding Laplace's expansion[1], and because the connection between Bessel's and Laplace's functions appears not to be generally understood. We may now, however, proceed to the independent treatment of the plane problem.

339.   If in the general equation of simple aerial vibrations

$$\nabla^2 \psi + k^2 \psi = 0,$$

we assume that ψ is independent of z, and introduce plane polar coordinates, we get (§ 241)

$$\frac{d^2\psi}{dr^2} + \frac{1}{r}\frac{d\psi}{dr} + \frac{1}{r^2}\frac{d^2\psi}{d\theta^2} + k^2\psi = 0 \ldots \ldots \ldots \ldots (1);$$

or, if ψ be expanded in Fourier's series

$$\psi = \psi_0 + \psi_1 + \ldots + \psi_n + \ldots \ldots \ldots \ldots \ldots (2),$$

where ψ_n is of the form $A_n \cos n\theta + B_n \sin n\theta$,

$$\frac{d^2\psi_n}{dr^2} + \frac{1}{r}\frac{d\psi_n}{dr} + \left(k^2 - \frac{n^2}{r^2}\right)\psi_n = 0 \ldots \ldots \ldots \ldots (3)[2].$$

This equation is of the same form as that with which we had to deal in treating of circular membranes (§ 200); the principal mathematical difference between the two questions lies in the fact that while in the case of membranes the condition to be satisfied at the boundary is $\psi = 0$, in the present case interest attaches itself rather to the boundary condition $d\psi/dr = 0$, corresponding to the confinement of the gas by a rigid cylindrical envelope[3].

The pole not being a source, the solution of (3) is

$$\psi_n = A\, J_n(kr) \ldots \ldots \ldots \ldots \ldots \ldots (4),$$

and the equation giving the possible periods of vibration within a cylinder of radius r, is

$$J_n'(kr) = 0 \ldots \ldots \ldots \ldots \ldots \ldots \ldots (5).$$

The lower values of kr satisfying (5) are given in the following table[4], which was calculated from Hansen's tables of the functions

---

[1] I have been much assisted by Heine's *Handbuch der Kugelfunctionen*, Berlin, 1861, and by Sir W. Thomson's papers on Laplace's Theory of the Tides, *Phil. Mag.* Vol. L. 1875.

[2] I here recur to the usual notation, but the reader will understand that n corresponds to the s of preceding sections. The n of Laplace's functions is now infinite.

[3] [The symmetrical vibrations within a cylindrical boundary, corresponding to $n = 0$, were considered by Duhamel (*Liouville Journ. Math.* Vol. 14, p. 69, 1849).]

[4] Notes on Bessel's Functions. *Phil. Mag.* Nov. 1872.

J by means of the relations allowing J_n to be expressed in terms of J_0 and J_1.

Number of internal circular nodes.	$n = 0$	$n = 1$	$n = 2$	$n = 3$
0	3·832	1·841	3·054	4·201
1	7·015	5·332	6·705	8·015
2	10·174	8·536	9·965	11·344
3	13·324	11·706		
4	16·471	14·864		
5	19·616	18·016		

[For the roots of the equation $J_n'(z) = 0$, Prof. McMahon[1] finds

$$z_n^{(s)} = \beta' - \frac{m+3}{8\beta'} - \frac{4(7m^2 + 82m - 9)}{3(8\beta')^3}$$
$$- \frac{32(83m^3 + 2075m^2 - 3039m + 3527)}{15(8\beta')^5} \ldots\ldots(6\,a),$$

where $m = 4n^2$, and $\beta' = \frac{1}{4}\pi(2n + 4s + 1)$. It will be found that $n = 0$ in $(6\,a)$ gives the same result as $n = 1$ in (4) § 206, in accordance with the identity $J_0'(z) = -J_1(z)$.]

The particular solution may be written

$$\psi_n = (A \cos n\theta + B \sin n\theta) J_n(kr) \cos kat$$
$$+ (C \cos n\theta + D \sin n\theta) J_n(kr) \sin kat \ldots\ldots\ldots (6),$$

where A, B, C, D are arbitrary for every admissible value of n and k. As in the corresponding problems for the sphere and circular membrane, the sum of all the particular solutions must be general enough to represent, when $t = 0$, arbitrary values of ψ and $\dot{\psi}$.

As an example of compound vibrations we may suppose, as in § 332, that the initial condition of the gas is that defined by

$$\dot{\psi} = 0, \quad \psi = x = r \cos \theta.$$

Under these circumstances (6) reduces to

$$\psi = A_1 \cos \theta \, J_1(k_1 r) \cos k_1 at + A_2 \cos \theta \, J_1(k_2 r) \cos k_2 at + \ldots (7),$$

and, if we suppose the radius of the cylinder to be unity, the admissible values of k are the roots of

$$J_1'(k) = 0 \ldots\ldots\ldots\ldots\ldots\ldots\ldots (8).$$

The condition to determine the coefficients A is that for all values of r from $r = 0$ to $r = 1$,

$$r = A_1 J_1 (k_1 r) + A_2 J_1 (k_2 r) + \dots\dots\dots\dots (9),$$

whence, as in § 332,

$$A = \frac{2}{(k^2 - 1) J_1 (k)} \dots\dots\dots\dots\dots (10).$$

The complete solution is therefore

$$\psi = \Sigma \frac{2 \cos \theta \, J_1 (kr)}{(k^2 - 1) J_1 (k)} \cos kat \dots\dots\dots (11),$$

where the summation extends to all the values of k determined by (8).

If we put $t = 0$ and $r = 1$, we get from (9) and (10)

$$\Sigma \frac{2}{k^2 - 1} = 1 \dots\dots\dots\dots\dots\dots (12),$$

an equation which may be verified numerically, or by an analytical process similar to that applied in the case of (14) § 332. We may prove that

$$\log J_1' (z) = \text{constant} + \Sigma \log \left(1 - \frac{z^2}{k^2} \right),$$

whence by differentiation

$$\frac{J_1'' (z)}{J_1' (z)} = - \Sigma \frac{2z}{k^2 - z^2}.$$

From this (12) is derived by putting $z = 1$, and having regard to the fundamental differential equation satisfied by J_1, which shews that

$$J_1'' (1) : J_1' (1) = - 1.$$

[More generally, if $J_n' (k) = 0$,

$$\Sigma \frac{2n}{k^2 - n^2} = 1. \;]$$

Hitherto we have supposed the cylinder complete, so that ψ recurs after each revolution, which requires that n be integral; but if instead of the complete cylinder we take the sector included between $\theta = 0$ and $\theta = \beta$, fractional values of n will in general present themselves. Since $d\psi/d\theta$ vanishes at both limits of θ, ψ must be of the form

$$\psi = A \cos (kat + \epsilon) \cos n\theta \, J_n (kr) \dots\dots\dots (13),$$

where $n = \nu\pi/\beta$, ν being integral. If β be an aliquot part of π (or π itself), the complete solution involves only integral values of n, as might have been foreseen; but, in general, functions of fractional order must be introduced.

An interesting example occurs when $\beta = 2\pi$, which corresponds to the case of a cylinder, traversed by a rigid wall stretching from the centre to the circumference (compare § 207). The effect of the wall is to render possible a difference of pressure on its two sides; but when no such difference occurs, the wall may be removed, and the vibrations are included under the theory of a complete cylinder. This state of things occurs when ν is even. But when ν is odd, n is of the form (integer $+ \frac{1}{2}$), and the pressures on the two sides of the wall are different. In the latter case J_n is expressible in finite terms. The gravest tone is obtained by taking $\nu = 1$, or $n = \frac{1}{2}$, when

$$\psi = A \cos(kat + \epsilon) . \cos \tfrac{1}{2}\theta . \frac{\sin kr}{\sqrt{(kr)}} \dots\dots\dots\dots(14),$$

and the admissible values of k are the roots of $\tan k = 2k$. The first root (after $k = 0$) is $k = 1\cdot 1655$, corresponding to a tone decidedly graver than any of which the complete cylinder is capable.

The preceding analysis has an interesting application to the mathematically analogous problem of the vibrations of water in a cylindrical vessel of uniform depth. The reader may consult a paper on waves by the author in the *Philosophical Magazine* for April, 1876, and papers by Prof. Guthrie to which reference is there made. The observation of the periodic time is very easy, and in this way may be obtained an experimental solution of problems, whose theoretical treatment is far beyond the power of known methods.

340. Returning to the complete cylinder, let us suppose it closed by rigid transverse walls at $z = 0$, and $z = l$, and remove the restriction that the motion is to be the same in all transverse sections. The general differential equation (§ 241) is

$$\frac{d^2\psi}{dr^2} + \frac{1}{r}\frac{d\psi}{dr} + \frac{1}{r^2}\frac{d^2\psi}{d\theta^2} + \frac{d^2\psi}{dz^2} + k^2\psi = 0 \dots\dots\dots\dots(1).$$

Let ψ be expanded by Fourier's theorem in the series

$$\psi = H_0 + H_1 \cos\frac{\pi z}{l} + H_2 \cos\frac{2\pi z}{l} + \dots + H_p \cos\left(p\,\frac{\pi z}{l}\right) + \dots(2),$$

where the coefficients H_p may be functions of r and θ. This form secures the fulfilment of the boundary conditions, when $z = 0$, $z = l$, and each term must satisfy the differential equation separately. Thus

$$\frac{d^2H_p}{dr^2} + \frac{1}{r}\frac{dH_p}{dr} + \frac{1}{r^2}\frac{d^2H_p}{d\theta^2} + \left(k^2 - p^2\frac{\pi^2}{l^2}\right)H_p = 0 \ \ldots\ldots (3),$$

which is of the same form as when the motion is independent of z, k^2 being replaced by $k^2 - p^2\pi^2 l^{-2}$. The particular solution may therefore be written

$$\psi = (A_n\cos n\theta + B_n\sin n\theta).\cos p\,\frac{\pi z}{l}.J_n(\sqrt{k^2 - p^2\pi^2 l^{-2}}.r)\cos kat$$

$$+ (C_n\cos n\theta + D_n\sin n\theta)\cos p\,\frac{\pi z}{l}.J_n(\sqrt{k^2 - p^2\pi^2 l^{-2}}.r)\sin kat\ldots(4),$$

which must be generalized by a triple summation, with respect to all integral values of p and n, and also with respect to all the values of k, determined by the equation,

$$J_n'(\sqrt{k^2 - p^2\pi^2 l^{-2}}.r) = 0\ldots\ldots\ldots\ldots\ldots\ldots\ldots(5).$$

If $r = 1$, and K denote the values of k given in the table (§ 339), corresponding to purely transverse vibrations, we have

$$k^2 = K^2 + p^2\pi^2/l^2 \ \ldots\ldots\ldots\ldots\ldots\ldots\ldots(6).$$

The purely axial vibrations correspond to a zero value of K, not included in the table.

341. The complete integral of the equation

$$\frac{d^2\psi_n}{dr^2} + \frac{1}{r}\frac{d\psi_n}{dr} + \left(k^2 - \frac{n^2}{r^2}\right)\psi_n = 0\ldots\ldots\ldots\ldots(1),$$

when there is no limitation as to the absence of a source at the pole, involves a second function of r, which may be denoted by $J_{-n}(kr)$. Thus, omitting unnecessary constant multipliers, we may take (§ 200)

$$\psi_n = Ar^{+n}\left\{1 - \frac{k^2r^2}{2.2 + 2n} + \frac{k^4r^4}{2.4.2 + 2n.4 + 2n} - \ldots\right\}$$

$$+ Br^{-n}\left\{1 - \frac{k^2r^2}{2.2 - 2n} + \frac{k^4r^4}{2.4.2 - 2n.4 - 2n} - \ldots\right\}\ldots\ldots\ldots(2),$$

but the second series requires modification, if n be integral. When $n = 0$, the two series become identical, and thus the immediate result of supposing $n = 0$ in (2) lacks the necessary generality. The required solution may, however, be obtained by the ordinary rule applicable to such cases. Denoting the coefficients of A and B in (2) by $f(n)$, $f(-n)$, we have

$$\psi = A f(n) + B f(-n)$$
$$= (A + B) f(0) + (A - B) f'(0) n + (A + B) f''(0) \frac{n^2}{1 \cdot 2} + \dots.$$

by Maclaurin's theorem. Hence, taking new arbitrary constants, we may write as the limiting form of (2),

$$\psi_0 = A f(0) + B f'(0).$$

In this equation $f(0)$ is $J_0(kr)$; to find $f'(0)$ we have

$$f'(n) = r^n \log r \left\{ 1 - \frac{k^2 r^2}{2 \cdot 2 + 2n} + \frac{k^4 r^4}{2 \cdot 4 \cdot 2 + 2n \cdot 4 + 2n} - \dots \right\}$$
$$+ r^n \frac{d}{dn} \left\{ 1 - \frac{k^2 r^2}{2 \cdot 2 + 2n} + \frac{k^4 r^4}{2 \cdot 4 \cdot 2 + 2n \cdot 4 + 2n} - \dots \right\}.$$

If u denote the general term (involving r^{2m}) of the series within brackets, taken without regard to sign,

$$\frac{1}{u} \frac{du}{dn} = \frac{d \log u}{dn} = -\frac{2}{2 + 2n} - \frac{2}{4 + 2n} - \dots - \frac{2}{2m + 2n},$$

so that $$\left(\frac{du}{dn} \right)_{n=0} = -u_{n=0} S_m,$$

if $$S_m = \frac{1}{1} + \frac{1}{2} + \frac{1}{3} + \dots + \frac{1}{m} \dots \dots \dots \dots (3).$$

Thus $$f'(0) = \log r \left\{ 1 - \frac{k^2 r^2}{2^2} + \frac{k^4 r^4}{2^2 \cdot 4^2} - \frac{k^6 r^6}{2^2 \cdot 4^2 \cdot 6^2} + \dots \right\}$$
$$+ \left\{ \frac{k^2 r^2}{2^2} - \frac{k^4 r^4}{2^2 \cdot 4^2} S_2 + \frac{k^6 r^6}{2^2 \cdot 4^2 \cdot 6^2} S_3 - \dots \dots \right\},$$

and the complete integral for the case $n = 0$ is

$$\psi_0 = (A + B \log r) \left\{ 1 - \frac{k^2 r^2}{2^2} + \frac{k^4 r^4}{2^2 \cdot 4^2} - \dots \right\}$$
$$+ B \left\{ \frac{k^2 r^2}{2^2} - \frac{k^4 r^4}{2^2 \cdot 4^2} S_2 + \frac{k^6 r^6}{2^2 \cdot 4^2 \cdot 6^2} S_3 - \dots \right\} \dots \dots \dots (4).$$

For the general integral value of n the corresponding expression may be derived by means of (15) § 338

$$\psi_n = (-2kr)^n \left(\frac{d}{d(kr)^2} \right)^n \psi_0 \dots \dots \dots \dots (5).$$

The formula of derivation (5) may be obtained directly from the differential equation (1). Writing z for kr and putting

$$\psi_n = z^n \phi_n \dots\dots\dots\dots\dots(6),$$

we find in place of (1)

$$\frac{d^2\phi_n}{dz^2} + \frac{2n+1}{z}\frac{d\phi_n}{dz} + \phi_n = 0 \dots\dots\dots(7).$$

Again (7) may be put into the form

$$z^2 \frac{d^2\phi_n}{d(z^2)^2} + (n+1)\frac{d\phi_n}{d.z^2} + \tfrac{1}{4}\phi_n = 0\dots\dots\dots(8),$$

from which it follows at once that

$$\phi_n = \frac{d}{d.z^2}\phi_{n-1}\dots\dots\dots\dots(9);$$

so that

$$\phi_n = \left(\frac{d}{d.z^2}\right)^n \phi_0 \dots\dots\dots\dots(10),$$

or by (6)

$$\psi_n = z^n \left(\frac{d}{d.z^2}\right)^n \psi_0\dots\dots\dots\dots(11),$$

which is equivalent to (5), since the constants in ψ_0 are arbitrary in both equations.

The serial expressions for ψ_n thus obtained are convergent for all values of the argument, but are practically useless when the argument is great. In such cases we must have recourse to semi-convergent series corresponding to that of (10) § 200.

Equation (1) may be put into the form

$$\frac{d^2(z^{\frac{1}{2}}\psi_n)}{dz^2} - \frac{(n-\frac{1}{2})(n+\frac{1}{2})}{z^2}(z^{\frac{1}{2}}\psi_n) + z^{\frac{1}{2}}\psi_n = 0\dots\dots\dots(12),$$

whence by § 323 (4), (12), we find as the general solution of (1)

$$\psi_n = C(ikr)^{-\frac{1}{2}}e^{-ikr}\left\{1 - \frac{1^2-4n^2}{1.8ikr} + \frac{(1^2-4n^2)(3^2-4n^2)}{1.2.(8ikr)^2}\right.$$
$$\left. - \frac{(1^2-4n^2)(3^2-4n^2)(5^2-4n^2)}{1.2.3.(8ikr)^3} + \dots\right\}$$
$$+ D(ikr)^{-\frac{1}{2}}e^{+ikr}\left\{1 + \frac{1^2-4n^2}{1.8ikr} + \frac{(1^2-4n^2)(3^2-4n^2)}{1.2.(8ikr)^2}\right.$$
$$\left. + \frac{(1^2-4n^2)(3^2-4n^2)(5^2-4n^2)}{1.2.3.(8ikr)^3} + \dots\right\}\dots\dots(13).$$

When n is integral, these series are infinite and ultimately divergent, but (§§ 200, 302) this circumstance does not interfere with their practical utility.

The most important application of the complete integral of (1) is to represent a disturbance diverging from the pole, a problem which has been treated by Stokes in his memoir on the communication of vibrations to a gas. The condition that the disturbance represented by (13) shall be exclusively divergent is simply $D = 0$, as appears immediately on introduction of the time factor e^{ikat} by supposing r to be very great; the principal difficulty of the question consists in discovering what relation between the coefficients of the ascending series corresponds to this condition, for which purpose Stokes employs the solution of (1) in the form of a definite integral. We shall attain the same object, perhaps more simply, by using the results of § 302.

By (22), (24) § 302

$$-\left(\frac{\pi}{2iz}\right)^{\frac{1}{2}} e^{-iz} \left\{1 - \frac{1^2}{1 \cdot 8iz} + \frac{1^2 \cdot 3^2}{1 \cdot 2 \cdot (8iz)^2} + \ldots\right\}$$

$$= \tfrac{1}{2}\pi \left\{K(z) + i J_0(z)\right\} - \int_0^\infty \frac{e^{-\beta} d\beta}{\sqrt{(\beta^2 + z^2)}} \ldots \ldots (14),$$

and thus the question reduces itself to the determination of the form of the right-hand member of (14) when z is small. By (5) § 302 and (5) § 200 we have

$$\tfrac{1}{2}\pi \left\{K(z) + i J_0(z)\right\} = z + \tfrac{1}{2} i\pi + \text{higher terms in } z \ldots\ldots(15),$$

so that all that remains is to find the form of the definite integral in (14), when z is small. Putting $\sqrt{(\beta^2 + z^2)} = y - \beta$, we have

$$\int_0^\infty \frac{e^{-\beta} d\beta}{\sqrt{(\beta^2 + z^2)}} = \int_z^\infty e^{-\frac{y^2 - z^2}{2y}} \frac{dy}{y} = \int_z^\infty e^{\frac{1}{2}z^2/y} e^{-\frac{1}{2}y} \frac{dy}{y}.$$

When z is small, $z^2/2y$ is also small throughout the range of integration, and thus we may write

$$\int_0^\infty \frac{e^{-\beta} d\beta}{\sqrt{(\beta^2 + z^2)}} = \int_z^\infty \left\{1 + \frac{z^2}{2y} + \tfrac{1}{2} \frac{z^4}{4y^2} + \ldots\right\} \frac{e^{-\frac{1}{2}y}}{y} dy.$$

The first integral on the right is

$$\int_z^\infty \frac{e^{-\frac{1}{2}y}}{y} dy = \int_{\frac{1}{2}z}^\infty \frac{e^{-v} dv}{v} = -\gamma - \log\left(\tfrac{1}{2}z\right) + \tfrac{1}{2}z + \ldots\ldots (16)^1,$$

---

[1] De Morgan's *Differential and Integral Calculus*, p. 653.

where γ is Euler's constant ($\cdot 5772\ldots$); and, as we may easily satisfy ourselves by integration by parts, the other integrals do not contribute anything to the leading terms. Thus, when z is very small,

$$-\left(\frac{\pi}{2iz}\right)^{\frac{1}{2}} e^{-iz} \left\{1 - \frac{1^2}{1\cdot 8iz} + \frac{1^2\cdot 3^2}{1\cdot 2\cdot(8iz)} - \frac{1^2\cdot 3^2\cdot 5^2}{1\cdot 2\cdot 3\cdot(8iz)^3} + \cdots\right\}$$

$$= \gamma + \log(\tfrac{1}{2}z) + \tfrac{1}{2}i\pi + \ldots\ldots\ldots (17).$$

Replacing z by kr, and comparing with the form assumed by (4) when r is small, we see that in order to make the series identical we must take

$$A = \gamma + \log\tfrac{1}{2} + \log k + \tfrac{1}{2}i\pi, \qquad B = 1\,;$$

so that a series of waves diverging from the pole, whose expression in descending series is

$$\psi_0 = -\left(\frac{\pi}{2ikr}\right)^{\frac{1}{2}} e^{-ikr}\left\{1 - \frac{1^2}{1\cdot 8ikr} + \frac{1^2\cdot 3^2}{1\cdot 2\cdot(8ikr)^2} - \cdots\right\} \ldots\ldots(18),$$

is represented also by the ascending series

$$\psi_0 = \left(\gamma + \log\frac{ikr}{2}\right)\left\{1 - \frac{k^2r^2}{2^2} + \frac{k^4r^4}{2^2\cdot 4^2} - \cdots\right\}$$

$$+ \frac{k^2r^2}{2^2}S_1 - \frac{k^4r^4}{2^2\cdot 4^2}\cdot S_2 + \frac{k^6r^6}{2^2\cdot 4^2\cdot 6^2}S_3 - \ldots\ldots(19).$$

In applying the formula of derivation (11) to the descending series, the parts containing e^{-ikr} and e^{+ikr} as factors will evidently remain distinct, and the complete integral for the general value of n, subject to the condition that the part containing e^{+ikr} shall not appear, will be got by differentiation from the complete integral for $n = 0$ subject to the same condition. Thus, since by (5) $\psi_1 = d\psi_0/dr$,

$$\psi_1 = \left(\frac{\pi i}{2kr}\right)^{\frac{1}{2}} e^{-ikr}\left\{1 - \frac{-1\cdot 3}{1\cdot 8ikr} + \frac{-1\cdot 1\cdot 3\cdot 5}{1\cdot 2\cdot(8ikr)^2}\right.$$

$$\left. - \frac{-1\cdot 1\cdot 3^2\cdot 5\cdot 7}{1\cdot 2\cdot 3\cdot(8ikr)^3} + \cdots\right\} \ldots\ldots(20),$$

or, in terms of the ascending series,

$$\psi_1 = \frac{1}{kr}\left\{1 - \frac{k^2 r^2}{2^2} + \frac{k^4 r^4}{2^2 . 4^2} - \ldots\right\}$$

$$- \left(\gamma + \log\frac{ikr}{2}\right)\left\{\frac{kr}{2} - \frac{k^3 r^3}{2^2 . 4} + \frac{k^5 r^5}{2^2 . 4^2 . 6} - \ldots\right\}$$

$$+ \frac{kr}{2} S_1 - \frac{k^3 r^3}{2^2 . 4} S_2 + \frac{k^5 r^5}{2^2 . 4^2 . 6} S_3 - \ldots\ldots\ldots(21).$$

These expressions are applied by Prof. Stokes to shew how feebly the vibrations of a string, (corresponding to the term of order one), are communicated to the surrounding gas. For this purpose he makes a comparison between the actual sound, and what would have been emitted in the same direction, were the lateral motion of the gas in the neighbourhood of the string prevented. For a piano string corresponding to the middle C, the radius of the wire may be about ·02 inch, and λ is about 25 inches; and it appears that the sound is nearly 40,000 times weaker than it would have been if the motion of the particles of air had taken place in planes passing through the axis of the string. "This shews the vital importance of sounding-boards in stringed instruments. Although the amplitude of vibration of the particles of the sounding-board is extremely small compared with that of the particles of the string, yet as it presents a broad surface to the air it is able to excite loud sonorous vibrations, whereas were the string supported in an absolutely rigid manner, the vibrations which it could excite directly in the air would be so small as to be almost or altogether inaudible."

Fig. 64.

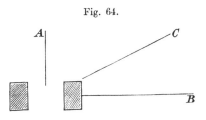

"The increase of sound produced by the stoppage of lateral motion may be prettily exhibited by a very simple experiment. Take a tuning-fork, and holding it in the fingers after it has been made to vibrate, place a sheet of paper, or the blade of a broad knife, with its edge parallel to the axis of the fork, and as near to the fork as conveniently may be without touching. If the plane of the obstacle coincide with either of the planes of symmetry of the fork, as represented in section at A or B, no effect is produced; but if it be placed in an intermediate position, such as C, the sound becomes much stronger[1]."

342. The real expression for the velocity-potential of symmetrical waves diverging in two dimensions is obtained from (18) § 341 after introduction of the time factor e^{ikat} by rejecting the imaginary part; it is

$$\psi_0 = -\left(\frac{\pi}{2kr}\right)^{\frac{1}{2}} \cos k\,(at - r - \tfrac{1}{8}\lambda) \left\{ 1 - \frac{1^2 \cdot 3^2}{1 \cdot 2 \cdot (8kr)^2} + \dots \right\}$$

$$+ \left(\frac{\pi}{2kr}\right)^{\frac{1}{2}} \sin k\,(at - r - \tfrac{1}{8}\lambda) \left\{ \frac{1^2}{1 \cdot 8kr} - \frac{1^2 \cdot 3^2 \cdot 5^2}{1 \cdot 2 \cdot 3 \cdot (8kr)^3} + \dots \right\} \dots\dots(1),$$

in which, as usual, two arbitrary constants may be inserted, one as a multiplier of the whole expression and the other as an addition to the time.

The problem of a linear source of uniform intensity may also be treated by the general method applicable in three dimensions. Thus by (3) § 277, if ρ be the distance of any element dx from O, the point at which the potential is to be estimated, and r be the smallest value of ρ, so that $\rho^2 = r^2 + x^2$, we may take

$$\phi = 2 \int_0^\infty \frac{e^{-ik\rho}\,dx}{\rho} = 2 \int_r^\infty \frac{e^{-ik\rho}\,d\rho}{\sqrt{(\rho^2 - r^2)}} \dots\dots\dots\dots(2),$$

which must be of the same form as (1). Taking $y = \rho - r$, we may write in place of (2)

$$\phi = 2 \int_0^\infty \frac{e^{-ikr}e^{-iky}\,dy}{\sqrt{y} \cdot \sqrt{(2r + y)}} \dots\dots\dots\dots(3),$$

from which the various expressions follow as in (14) § 341. When kr is great, an approximate value of the integral may be obtained by neglecting the variation of $\sqrt{(2r + y)}$, since on account of the rapid fluctuation of sign caused by the factor e^{-iky} we need attend

[1] *Phil. Trans.* vol. 158, p. 447, 1868.

only to small values of y. Now

$$\int_0^\infty \frac{\cos x\, dx}{\sqrt{x}} = \int_0^\infty \frac{\sin x\, dx}{\sqrt{x}} = \sqrt{\left(\frac{\pi}{2}\right)} \quad \ldots\ldots\ldots\ldots(4),$$

so that $\quad \phi = \sqrt{\left(\frac{\pi}{kr}\right)} e^{-ikr}(1-i) = \sqrt{\left(\frac{2\pi}{kr}\right)} e^{-ik(r+\frac{1}{8}\lambda)} \ldots\ldots\ldots(5).$

Introducing the factor e^{ikat}, and rejecting the imaginary part of the expression, we have finally

$$\phi = \sqrt{\left(\frac{2\pi}{kr}\right)} \cos k\,(at - r - \tfrac{1}{8}\lambda) \ldots\ldots\ldots\ldots(6),$$

as the value of the velocity-potential at a great distance. A similar argument is applicable to shew that (1) is also the expression for the velocity-potential on one side of an infinite plane (§ 278) due to the uniform normal motion of an infinitesimal strip bounded by parallel lines.

In like manner we may regard the term of the first order (20) § 341 as the expression of the velocity-potential due to double sources uniformly distributed along an infinite straight line.

From the point of view of the present section we see the significance of the retardation of $\tfrac{1}{8}\lambda$, which appears in (1) and in the results of the following section (16), (17). In the ordinary integration for surface distributions by Fresnel's zones (§ 283) the whole effect is the half of that of the first zone, and the phase of the effect of the first zone is midway between the phases due to its extreme parts, i.e. $\tfrac{1}{4}\lambda$ behind the phase due to the central point. In the present case the retardation of the resultant relatively to the central element is less, on account of the preponderance of the central parts.

[From the formulæ of the present section for the velocity-potential of a linear source we may obtain by integration a corresponding expression for a source which is uniformly distributed over a *plane*. The waves issuing from this latter are necessarily plane waves, of which the velocity-potential can at once be written down, and the comparison of results leads to the evaluation of certain definite integrals relating to Bessel's and allied functions[1].

[1] On Point-, Line-, and Plane-Sources of Sound. *Proc. London Math. Soc.* Vol. xix. p. 504, 1888.

343. In illustration of the formulæ of § 341 we may take the problem of the disturbance of plane waves of sound by a cylindrical obstacle, whose radius is small in comparison with the length of the waves, and whose axis is parallel to their plane. (Compare § 335.)

Let the plane waves be represented by

$$\phi = e^{ik(at+x)} = e^{ikat} \cdot e^{ikr\cos\theta} \quad \dots\dots\dots\dots\dots(1).$$

The general expansion of ϕ in Fourier's series may be readily effected, the coefficients of the various terms being, as might be anticipated, simply the Bessel's functions of corresponding orders. [Thus, as in (12) § 272 a,

$$e^{ikr\cos\theta} = J_0(kr) + 2i J_1(kr)\cos\theta + \dots + 2i^n J_n(kr)\cos n\theta + \dots.]$$

But, as we confine ourselves here to the case where c the radius of the cylinder is small, we will at once expand in powers of r.

Thus, when $r = c$, if e^{ikat} be omitted,

$$\phi = 1 - \tfrac{1}{4}k^2c^2 + ikc \cdot \cos\theta + \dots\dots\dots\dots\dots(2),$$

$$\frac{d\phi}{dr} = -\tfrac{1}{2}k^2c + ik \cdot \cos\theta + \dots\dots\dots\dots\dots(3).$$

The amount and even the law of the disturbance depends upon the character of the obstacle. We will begin by supposing the material of the cylinder to be a gas of density σ' and compressibility m'; the solution of the problem for a rigid obstacle may finally be derived by suitable suppositions with respect to σ', m'. If k' be the internal value of k, we have inside the cylinder by the condition that the axis is not a source (§ 339),

$$\psi = A_0\left\{1 - \frac{k'^2r^2}{2^2} + \frac{k'^4r^4}{2^2 \cdot 4^2} - \dots\right\} + A_1 r\left\{1 - \frac{k'^2r^2}{2 \cdot 4} + \frac{k'^4r^4}{2 \cdot 4^2 \cdot 6} - \dots\right\}\cos\theta\,;$$

so that, when $r = c$,

$$\psi\,(\text{inside}) = A_0\left(1 - \tfrac{1}{4}k'^2c^2\right) + A_1 c\left(1 - \tfrac{1}{8}k'^2c^2\right) \cdot \cos\theta \dots(4),$$

$$\frac{d\psi}{dr}\,(\text{inside}) = -\tfrac{1}{2}A_0 k'^2c + A_1\left(1 - \tfrac{3}{8}k'^2c^2\right)\cos\theta \dots\dots\dots(5).$$

Outside the cylinder, when $r = c$, we have by (19), (21) § 341,

$$\psi = B_0\left(\gamma + \log\frac{ikc}{2}\right) + \frac{B_1\cos\theta}{kc} \quad \dots\dots\dots\dots\dots(6),$$

$$\frac{d\psi}{dr} = \frac{B_0}{c} - \frac{B_1\cos\theta}{kc^2} \dots\dots\dots\dots\dots\dots(7).$$

The conditions to be satisfied at the surface of separation are thus

$$- A_0 k'^2 c^2 = - k^2 c^2 + 2B_0 \dots\dots\dots\dots\dots(8),$$

$$\frac{\sigma'}{\sigma} A_0 \left(1 - \tfrac{1}{4} k'^2 c^2\right) = 1 - \tfrac{1}{4} k^2 c^2 + B_0 \left(\gamma + \log \frac{ikc}{2}\right) \dots\dots(9),$$

$$A_1 \left(1 - \frac{3 k'^2 c^2}{8}\right) = ik - \frac{B_1}{kc^2} \dots\dots\dots\dots(10),$$

$$\frac{\sigma'}{\sigma} A_1 c \left(1 - \frac{k'^2 c^2}{8}\right) = ikc + \frac{B_1}{kc} \dots\dots\dots\dots(11),$$

from which by eliminating A_0, A_1 we get approximately

$$B_0 = \tfrac{1}{2} k^2 c^2 \left(1 - \frac{k'^2}{k^2} \cdot \frac{\sigma}{\sigma'}\right) = \tfrac{1}{2} k^2 c^2 \frac{m' - m}{m'} \dots\dots\dots(12),$$

$$B_1 = i k^2 c^2 \frac{\sigma' - \sigma}{\sigma' + \sigma} \dots\dots\dots\dots\dots(13).$$

Thus at a distance from the cylinder we have by (18) and (20) § 341,

$$\psi = - B_0 \left(\frac{\pi}{2ikr}\right)^{\frac{1}{2}} e^{-ikr} + B_1 \left(\frac{\pi i}{2kr}\right)^{\frac{1}{2}} e^{-ikr} \cdot \cos\theta$$

$$= - k^2 c^2 e^{-ikr} \left(\frac{\pi}{2ikr}\right)^{\frac{1}{2}} \left\{\frac{m' - m}{2m'} + \frac{\sigma' - \sigma}{\sigma' + \sigma} \cos\theta\right\}$$

$$= - \frac{2\pi \cdot \pi c^2}{r^{\frac{1}{2}} \lambda^{\frac{3}{2}}} e^{-i(kr + \frac{1}{4}\pi)} \left\{\frac{m' - m}{2m'} + \frac{\sigma' - \sigma}{\sigma' + \sigma} \cos\theta\right\} \dots(14).$$

Hence, corresponding to the primary wave

$$\phi = \cos \frac{2\pi}{\lambda} (at + x) \dots\dots\dots\dots\dots(15),$$

the scattered wave is approximately

$$\psi = - \frac{2\pi \cdot \pi c^2}{r^{\frac{1}{2}} \lambda^{\frac{3}{2}}} \left\{\frac{m' - m}{2m'} + \frac{\sigma' - \sigma}{\sigma' + \sigma} \cos\theta\right\} \cos \frac{2\pi}{\lambda} (at - r - \tfrac{1}{8}\lambda) \dots(16).$$

The fact that ψ varies inversely as $\lambda^{\frac{3}{2}}$ might have been anticipated by the method of dimensions, as in the corresponding problem for the sphere (§ 296). As in that case, the symmetrical part of the divergent wave depends upon the variation of compressibility, and would disappear in the application to an actual gas; and the term of the first order depends upon the variation of density.

By supposing σ' and m' to become infinite, in such a manner that their ratio remains finite, we obtain the solution corresponding to a rigid and immoveable obstacle,

$$\psi = -\frac{2\pi \cdot \pi c^2}{r^{\frac{1}{2}} \lambda^{\frac{3}{2}}} (\tfrac{1}{2} + \cos \theta) \cos \frac{2\pi}{\lambda} (at - r - \tfrac{1}{8}\lambda)\ldots\ldots(17).$$

The exceeding smallness of the obstruction offered by fine wires or fibres to the passage of sound is strikingly illustrated in some of Tyndall's experiments. A piece of stiff felt half an inch in thickness allows much more sound to pass than a *wetted* pocket-handkerchief, which in consequence of the closing of its pores behaves rather as a thin lamina. For the same reason fogs, and even rain and snow, interfere but little with the free propagation of sounds of moderate wave-length. In the case of a hiss, or other very acute sound, the effect would perhaps be apparent.

[The partial reflections from sheets of muslin may be utilized to illustrate an important principle. If a pure tone of high (inaudible) pitch be reflected from a single sheet so as to impinge upon a sensitive flame, the intensity will probably be insufficient to produce a visible effect. If, however, a moderate number of such sheets be placed parallel to one another and at such equal distances apart that the partial reflections agree in phase, then the flame may be powerfully affected. The parallelism and equidistance of the sheets may be maintained mechanically by a lazy-tongs arrangement, which nevertheless allows the common distance to be varied. It is then easy to trace the dependence of the action upon the accommodation of the interval to the wave length of the sound. Thus, if the incidence were perpendicular, the flame would be most powerfully influenced when the interval between adjacent sheets was equal to the *half* wave length; and although the exigencies of experiment make it necessary to introduce obliquity, allowance for this is readily made[1].]

---

[1] Iridescent Crystals, *Proc. Roy. Inst.* April 1889. See also *Phil. Mag.* vol. xxiv. p. 145, 1887; vol. xxvi. p. 256, 1888.

CHAPTER XIX.

FRICTION AND HEAT CONDUCTION.

344. THE equations of Chapter XI. and the consequences that we have deduced from them are based upon the assumption (§ 236), that the mutual action between any two portions of fluid separated by an imaginary surface is normal to that surface. Actual fluids however do not come up to this ideal; in many phenomena the defect of fluidity, usually called viscosity or fluid friction, plays an important and even a preponderating part. It will therefore be proper to inquire whether the laws of aerial vibrations are sensibly influenced by the viscosity of air, and if so in what manner.

In order to understand clearly the nature of viscosity, let us conceive a fluid divided into parallel strata in such a manner that while each stratum moves in its own plane with uniform velocity, a change of velocity occurs in passing from one stratum to another. The simplest supposition which we can make is that the velocities of all the strata are in the same direction, but increase uniformly in magnitude as we pass along a line perpendicular to the planes of stratification. Under these circumstances a tangential force between contiguous strata is called into play, in the direction of the relative motion, and of magnitude proportional to the rate at which the velocity changes, and to a coefficient of viscosity, commonly denoted by the letter μ. Thus, if the strata be parallel to xy and the direction of their motion be parallel to y, the tangential force, reckoned (like a pressure) per unit of area, is

$$\mu \frac{dv}{dz} \dots\dots\dots\dots\dots\dots\dots\dots (1).$$

The dimensions of μ are $[ML^{-1}T^{-1}]$.

The examination of the origin of the tangential force belongs to molecular science. It has been explained by Maxwell in ac-

cordance with the kinetic theory of gases as resulting from interchange of molecules between the strata, giving rise to diffusion of
momentum. Both by theory and experiment the remarkable
conclusion has been established that within wide limits the force
is independent of the density of the gas. For air at θ^0 Centigrade
Maxwell[1] found

$$\mu = \cdot0001878\,(1 + \cdot00366\,\theta)\dots\dots\dots\dots\dots\dots(2),$$

the centimetre, gramme, and second being units.

345. The investigation of the equations of fluid motion in
which regard is paid to viscous forces can scarcely be considered
to belong to the subject of this work, but it may be of service
to some readers to point out its close connection with the more
generally known theory of solid elasticity.

The potential energy of unit of volume of uniformly strained
isotropic matter may be expressed[2]

$$V = \tfrac{1}{2}m\delta^2 + \tfrac{1}{2}n\,(e^2+f^2+g^2-2fg-2ge-2ef+a^2+b^2+c^2)$$
$$= \tfrac{1}{2}\kappa\delta^2 + \tfrac{1}{2}n\,(2e^2+2f^2+2g^2-\tfrac{2}{3}\delta^2+a^2+b^2+c^2)\;\dots\dots\dots(1),$$

in which $\delta\,(=e+f+g)$ is the dilatation, e, f, g, a, b, c are the six
components of strain, connected with the actual displacements
α, β, γ by the equations

$$e = \frac{d\alpha}{dx}, \qquad f = \frac{d\beta}{dy}, \qquad g = \frac{d\gamma}{dz} \quad\dots\dots\dots\dots\dots(2),$$

$$a = \frac{d\beta}{dz} + \frac{d\gamma}{dy}, \quad b = \frac{d\gamma}{dx} + \frac{d\alpha}{dz}, \quad c = \frac{d\alpha}{dy} + \frac{d\beta}{dx} \quad\dots\dots\dots(3),$$

and m, n, κ are constants of elasticity, connected by the equation

$$\kappa = m - \tfrac{1}{3}n \quad\dots\dots\dots\dots\dots\dots\dots\dots\dots(4),$$

of which n measures the *rigidity*, or resistance to *shearing*, and κ
measures the resistance to change of *volume*. The components of
stress P, Q, R, S, T, U, corresponding respectively to e, f, g, a, b, c,
are found from V by simple differentiation with respect to those
quantities; thus

$$P = \kappa\delta + 2n\,(e - \tfrac{1}{3}\delta), \text{ \&c.}\dots\dots\dots\dots\dots(5),$$

$$S = na, \text{ \&c.}\dots\dots\dots\dots\dots\dots\dots\dots(6).$$

---

[1] On the Viscosity or Internal Friction of Air and other Gases. *Phil. Trans.*
vol. 156, p. 249, 1866.

[2] Thomson and Tait's *Natural Philosophy*. Appendix C.

If X, Y, Z be the components of the applied force reckoned per unit of volume, the equations of equilibrium are of the form

$$\frac{dP}{dx} + \frac{dU}{dy} + \frac{dT}{dz} + X = 0, \&c. \ldots\ldots\ldots\ldots (7),$$

from which the equations of motion are immediately obtainable by means of D'Alembert's principle. In terms of the displacements α, β, γ, these equations become

$$\kappa \frac{d\delta}{dx} + \tfrac{1}{3} n \frac{d\delta}{dx} + n \nabla^2 \alpha + X = 0, \&c. \ldots\ldots\ldots\ldots (8),$$

where
$$\delta = \frac{d\alpha}{dx} + \frac{d\beta}{dy} + \frac{d\gamma}{dz} \ldots\ldots\ldots\ldots\ldots (9).$$

In the ordinary theory of fluid friction no forces of restitution are included, but on the other hand we have to consider viscous forces whose relation to the velocities (u, v, w) of the fluid elements is of precisely the same character as that of the forces of restitution to the displacements (α, β, γ) of an isotropic solid. Thus if δ' be the velocity of dilatation, so that

$$\delta' = \frac{du}{dx} + \frac{dv}{dy} + \frac{dw}{dz} \ldots\ldots\ldots\ldots\ldots (10),$$

the force parallel to x due to viscosity is, as in (8),

$$\kappa \frac{d\delta'}{dx} + \tfrac{1}{3} n \frac{d\delta'}{dx} + n \nabla^2 u \ldots\ldots\ldots\ldots\ldots (11).$$

So far κ and n are arbitrary constants; but it has been argued with great force by Prof. Stokes, that there is no reason why a motion of dilatation uniform in all directions should give rise to viscous force, or cause the pressure to differ from the statical pressure corresponding to the actual density. In accordance with this argument we are to put $\kappa = 0$; and, as appears from (6), n coincides with the quantity previously denoted by μ. The frictional terms are therefore

$$\mu \left\{ \nabla^2 u + \tfrac{1}{3} \frac{d}{dx} \left(\frac{du}{dx} + \frac{dv}{dy} + \frac{dw}{dz} \right) \right\}, \ \&c. ;$$

and (§ 237) the equations of motion take the form

$$\rho \left(\frac{Du}{Dt} - X \right) + \frac{dp}{dx} - \mu \nabla^2 u - \tfrac{1}{3} \mu \frac{d}{dx} \left(\frac{du}{dx} + \frac{dv}{dy} + \frac{dw}{dz} \right) = 0 \ \ldots\ldots (12);$$

or, if there be no applied forces and the square of the motion be neglected,

$$\rho_0 \frac{du}{dt} + \frac{dp}{dx} - \mu \nabla^2 u - \tfrac{1}{3}\mu \frac{d}{dx}\left(\frac{du}{dx} + \frac{dv}{dy} + \frac{dw}{dz}\right) = 0 \dots\dots\dots(13).$$

We may observe that the dissipative forces here considered correspond to a dissipation function, whose form is the same with respect to u, v, w as that of V with respect to α, β, γ, in the theory of isotropic solids. Thus putting $\kappa = 0$, we have from (1)

$$F = \tfrac{1}{2}\mu \iiint \left[2\left(\frac{du}{dx}\right)^2 + 2\left(\frac{dv}{dy}\right)^2 + 2\left(\frac{dw}{dz}\right)^2 - \tfrac{2}{3}\left(\frac{du}{dx} + \frac{dv}{dy} + \frac{dw}{dz}\right)^2 \right.$$
$$\left. + \left(\frac{dv}{dz} + \frac{dw}{dy}\right)^2 + \left(\frac{dw}{dx} + \frac{du}{dz}\right)^2 + \left(\frac{du}{dy} + \frac{dv}{dx}\right)^2 \right] dx\,dy\,dz \dots\dots(14),$$

in agreement with Prof. Stokes' calculation[1]. The theory of friction for the case of a compressible fluid was first given by Poisson[2].

346. We will now apply the differential equations to the investigation of plane waves of sound. Supposing that v and w are zero and that u, p, &c. are functions of x only, we obtain from (13) § 345

$$\rho_0 \frac{du}{dt} + \frac{dp}{dx} - \frac{4\mu}{3}\frac{d^2u}{dx^2} = 0 \dots\dots\dots\dots\dots(1).$$

The equation of continuity (3) § 238 is in this case

$$\frac{ds}{dt} + \frac{du}{dx} = 0 \dots\dots\dots\dots\dots\dots(2),$$

and the relation between the variable part of the pressure δp and the condensation s is as usual (§ 244)

$$\delta p = a^2 \rho_0 s \dots\dots\dots\dots\dots\dots(3).$$

Thus, eliminating δp and s between (1), (2), (3), we obtain

$$\frac{d^2u}{dt^2} - a^2 \frac{d^2u}{dx^2} - \frac{4\mu}{3\rho_0}\frac{d^3u}{dx^2 dt} = 0 \dots\dots\dots\dots(4),$$

which is the equation given by Stokes[3].

Let us now inquire how a train of harmonic waves of wavelength λ, which are maintained at the origin $(x = 0)$, fade away

---

[1] *Cambridge Transactions*, vol. IX. § 49, 1851.

[2] *Journal de l'Ecole Polytechnique*, t. XIII. cah. 20, p. 139.

[3] *Cambridge Transactions*, vol. VIII. p. 287, 1845.

as x increases. Assuming that u varies as e^{int}, we find as in § 148,

$$u = A e^{-\alpha x} \cos{(nt - \beta x)} \ldots\ldots\ldots\ldots\ldots\ldots(5),$$

where $\quad \beta^2 - \alpha^2 = \dfrac{n^2 a^2}{a^4 + \dfrac{16\mu^2 n^2}{9\rho_0^2}}, \quad 2\alpha\beta = \dfrac{4\mu n^3/3\rho_0}{a^4 + \dfrac{16\mu^2 n^2}{9\rho_0^2}} \ldots\ldots\ldots\ldots(6).$

In the application to air at ordinary pressures μ may be considered to be a very small quantity and its square may be neglected. Thus

$$\beta = \frac{n}{a}, \qquad \alpha = \frac{2\mu n^2}{3\rho_0 a^3} \ldots\ldots\ldots\ldots\ldots\ldots\ldots(7).$$

It appears that to this order of approximation the velocity of sound is unaffected by fluid friction. If we replace n by $2\pi a \lambda^{-1}$, the expression for the coefficient of decay becomes

$$\alpha = \frac{8\pi^2 \mu}{3\lambda^2 \rho_0 a} \ldots\ldots\ldots\ldots\ldots\ldots\ldots(8),$$

shewing that the influence of viscosity is greatest on the waves of short wave-length. The amplitude is diminished in the ratio $e : 1$, when $x = \alpha^{-1}$. In C. G. S. measure we may take

$$\rho_0 = \cdot0013, \quad \mu = \cdot00019, \quad a = 33200 ;$$

whence $\qquad\qquad x = 8800\lambda^2 \ldots\ldots\ldots\ldots\ldots\ldots\ldots(9).$

Thus the amplitude of waves of one centimetre wave-length is diminished in the ratio $e : 1$ after travelling a distance of 88 metres. A wave-length of 10 centimetres would correspond nearly to g^{iv}; for this case $x = 8800$ metres. It appears therefore that at atmospheric pressures the influence of friction is not likely to be sensible to ordinary observation, except near the upper limit of the musical scale. The mellowing of sounds by distance, as observed in mountainous countries, is perhaps to be attributed to friction, by the operation of which the higher and harsher components are gradually eliminated. It must often have been noticed that the sound s is scarcely, if at all, returned by echos, and I have found [1] that at a distance of 200 metres a powerful hiss loses its character, even when there is no reflection. Probably this effect also is due to viscosity.

[1] Acoustical Observations, *Phil. Mag.* vol. III. p. 456, 1877.

In highly rarefied air the value of α as given in (8) is much increased, μ being constant. Sounds even of grave pitch may then be affected within moderate distances.

From the observations of Colladon in the lake of Geneva it would appear that in water grave sounds are more rapidly damped than acute sounds. At a moderate distance from a bell, struck under water, he found the sound short and sharp, without musical character.

347. The effect of viscosity in modifying the motion of air in contact with vibrating solids will be best understood from the solution of the problem for a very simple case given by Stokes. Let us suppose that an infinite plane (yz) executes harmonic vibrations in a direction (y) parallel to itself. The motion being in parallel strata, u and w vanish, and the variable quantities are functions of x only. The first of equations (13) § 345 shews that the pressure is constant; the corresponding equation in v takes the form

$$\frac{dv}{dt} = \frac{\mu}{\rho} \frac{d^2v}{dx^2} \cdots\cdots\cdots\cdots\cdots\cdots\cdots(1),$$

similar to the equation for the linear conduction of heat. If we now suppose that v is proportional to e^{int}, the resulting equation in x is

$$\frac{d^2v}{dx^2} = i\frac{n\rho}{\mu}v \cdots\cdots\cdots\cdots\cdots\cdots(2),$$

and its general solution

$$v = Ae^{-mx} + Be^{+mx} \cdots\cdots\cdots\cdots\cdots(3),$$

where

$$m = \sqrt{\left(\frac{n\rho}{2\mu}\right)}(1 + i) \cdots\cdots\cdots\cdots\cdots(4),$$

If the gas be on the positive side of the vibrating plane the motion is to vanish when $x = +\infty$. Hence $B = 0$, and the value of v becomes on rejection of the imaginary part

$$v = Ae^{-\sqrt{\left(\frac{n\rho}{2\mu}\right)}x}\cos\left\{nt - \sqrt{\left(\frac{n\rho}{2\mu}\right)}x\right\} \cdots\cdots\cdots(5),$$

corresponding to the motion

$$V = A\cos nt \cdots\cdots\cdots\cdots\cdots\cdots(6)$$

at $x = 0$. The velocity of the fluid in contact with the plane is usually assumed to be the same as that of the plane itself on the apparently sufficient ground that the contrary would imply an infinitely greater smoothness of the fluid with respect to the solid than with respect to itself. On this supposition (5) expresses the motion of the fluid on the positive side due to a motion of the plane given by (6).

The tangential force per unit area acting on the plane is

$$\mu \, dv/dx_{(x=0)},$$

or $\quad \mu \sqrt{\left(\dfrac{n\rho}{2\mu}\right)} \{- \cos nt + \sin nt\} = - \sqrt{(\tfrac{1}{2}n\rho\mu)} \left(V + \dfrac{1}{n}\dfrac{dV}{dt}\right)\dots(7),$

if $A = 1$. The first term represents a dissipative force tending to stop the motion; the second represents a force equivalent to an increase in the inertia of the vibrating body. The magnitude of both forces depends upon the frequency of the vibration.

We will apply this result to calculate approximately the velocity of sound in tubes so narrow that the viscosity of air exercises a sensible influence. As in § 265, let X denote the total transfer of fluid across the section of the tube at the point x. The force, due to hydrostatic pressure, acting on the slice between x and $x + \delta x$ is, as usual,

$$- S \frac{dp}{dx} \delta x = a^2 \rho \, \delta x \, \frac{d^2 X}{dx^2} \dots \dots \dots \dots \dots \dots (8).$$

The force due to viscosity may be inferred from the investigation for a vibrating plane, provided that the thickness of the layer of air adhering to the walls of the tube be small in comparison with the diameter. Thus, if P be the perimeter of the tube, and V be the velocity of the current at a distance from the walls of the tube, the tangential force on the slice, whose volume is $S\delta x$, is by (7)

$$- P\delta x\sqrt{(\tfrac{1}{2}n\rho\mu)} \left(V + \frac{1}{n}\frac{dV}{dt}\right),$$

or on replacing V by $\dfrac{dX}{dt} \div S$

$$- P\,\delta x\sqrt{(\tfrac{1}{2}n\rho\mu)}\left(\frac{dX}{dt} + \frac{1}{n}\frac{d^2X}{dt^2}\right) \div S \dots \dots \dots \dots (9).$$

The equation of motion *for this period* is therefore

$$\rho\delta x \frac{d^2 X}{dt^2} + \sqrt{(\tfrac{1}{2}n\rho\mu)}\frac{P\delta x}{S}\left(\frac{dX}{dt} + \frac{1}{n}\frac{d^2X}{dt^2}\right) = a^2\rho\,\delta x\,\frac{d^2X}{dx^2},$$

or
$$\frac{d^2 X}{dt^2}\left\{1 + \frac{P}{S}\sqrt{\left(\frac{\mu}{2n\rho}\right)}\right\} + \frac{P}{S}\sqrt{\left(\frac{n\mu}{\rho}\right)}\frac{dX}{dt} = a^2\frac{d^2 X}{dx^2}\dots(10).$$

The velocity of sound is approximately

$$a\left\{1 - \tfrac{1}{2}\frac{P}{S}\sqrt{\left(\frac{\mu}{2n\rho}\right)}\right\}\dots(11),$$

or in the case of a circular tube of radius r,

$$a\left\{1 - \frac{1}{r}\sqrt{\left(\frac{\mu}{2n\rho}\right)}\right\}\dots(12).$$

The result expressed in (12) was first obtained by Helmholtz.

348[1]. In the investigation of Kirchhoff[2], to which we now proceed, account is taken not only of viscosity but of the equally important effects arising from the generation of heat and its communication by conduction to and from the solid walls of a narrow tube.

The square of the motion being neglected, the "equation of continuity" (3) § 237 is

$$\frac{ds}{dt} + \frac{du}{dx} + \frac{dv}{dy} + \frac{dw}{dz} = 0 \dots (1);$$

so that the dynamical equations (13) § 345 may be written in the form

$$\frac{du}{dt} + \frac{1}{\rho_0}\frac{dp}{dx} = \frac{\mu}{\rho_0}\nabla^2 u + \frac{\mu}{3\rho_0}\frac{d^2 s}{dx\,dt}\dots(2).$$

The thermal questions involved have already been considered in § 247. By equation (4)

$$\frac{d\theta}{dt} = \beta\frac{ds}{dt} + \nu\nabla^2\theta \dots (3),$$

where ν is a constant representing the thermometric conductivity.

By (3) § 247
$$p/\rho_0 = b^2(1 + s + \alpha\theta)\dots(4),$$

in which b denotes Newton's value of the velocity of sound, viz. $\sqrt{(p_0/\rho_0)}$. If we denote Laplace's value for the velocity by a,

$$a^2/b^2 = \gamma = 1 + \alpha\beta\dots(5),$$
so that
$$\beta = (a^2 - b^2)/b^2\alpha \dots (6).$$

---

[1] This and the following §§ appear for the first time in the second edition. The first edition closed with § 348, there devoted to the question of dynamical similarity.

[2] *Pogg. Ann.* vol. cxxxiv., p. 177, 1868.

It will simplify the equations if we introduce a new symbol θ' in place of θ, connected with it by the relation $\theta' = \theta/\beta$. Thus (3) becomes

$$\frac{d\theta'}{dt} - \frac{ds}{dt} = \nu \nabla^2 \theta \dots\dots\dots\dots\dots(7),$$

and the typical equation (2) may be written

$$\frac{du}{dt} + b^2 \frac{ds}{dx} + (a^2 - b^2) \frac{d\theta'}{dx} = \mu' \nabla^2 u - \mu'' \frac{d^2 s}{dx\,dt} \dots\dots(8),$$

where μ' is equal to μ/ρ_0. μ'' represents a second constant, whose value according to Stokes' theory is $\frac{1}{3}\mu'$. This relation is in accordance with Maxwell's kinetic theory, which on the introduction of more special suppositions further gives

$$\nu = \tfrac{5}{2}\mu' \dots\dots\dots\dots\dots\dots(9).$$

In any case μ', μ'', ν may be regarded as being of the same order of magnitude.

We will now, following Kirchhoff closely, introduce the supposition that the variables u, v, w, s, θ' are functions of the time on account only of the factor e^{ht}, where h is a constant to be afterwards taken as imaginary. Differentiations with respect to t are then represented by the insertion of the factor h, and the equations become

$$du/dx + dv/dy + dw/dz + hs = 0 \dots\dots\dots\dots (10),$$

$$\left.\begin{array}{l} hu - \mu'\nabla^2 u = - dP/dx \\ hv - \mu'\nabla^2 v = - dP/dy \\ hw - \mu'\nabla^2 w = - dP/dz \end{array}\right\} \dots\dots\dots\dots(11),$$

$$P = (b^2 + h\mu'') s + (a^2 - b^2) \theta' \dots\dots\dots\dots(12),$$

$$s = \theta' - (\nu/h) \nabla^2 \theta' \dots\dots\dots\dots\dots(13).$$

By (13), if s be eliminated, (12) and (10) become

$$P = (a^2 + h\mu'') \theta' - \frac{\nu}{h} (b^2 + h\mu'') \nabla^2 \theta' \dots\dots\dots (14),$$

$$\frac{du}{dx} + \frac{dv}{dy} + \frac{dw}{dz} + h\theta' - \nu\nabla^2 \theta' = 0 \dots\dots\dots\dots (15).$$

By differentiation of equations (11) with respect to x, y, z, with subsequent addition and use of (14), (15), we find as the equation in θ'

$$h^2 \theta' - \{a^2 + h(\mu' + \mu'' + \nu)\} \nabla^2 \theta' + \frac{\nu}{h} \{b^2 + h(\mu' + \mu'')\} \nabla^4 \theta' = 0 \dots(16).$$

A solution of (16) may be obtained in the form
$$\theta' = A_1 Q_1 + A_2 Q_2 \quad\ldots\ldots\ldots\ldots\ldots\ldots (17),$$
where Q_1, Q_2 are functions satisfying respectively
$$\nabla^2 Q_1 = \lambda_1 Q_1, \qquad \nabla^2 Q_2 = \lambda_2 Q_2 \ldots\ldots\ldots\ldots (18),$$
λ_1, λ_2 being the roots of
$$h^2 - \{a^2 + h(\mu' + \mu'' + \nu)\}\lambda + \frac{\nu}{h}\{b^2 + h(\mu' + \mu'')\}\lambda^2 = 0\ldots(19),$$
while A_1, A_2 denote arbitrary constants.

In correspondence with this value of θ', particular solutions of equations (11) are obtained by equating u, v, w to the differential coefficients of
$$B_1 Q_1 + B_2 Q_2,$$
taken with respect to x, y, z. The relation of the constants B_1, B_2 to A_1, A_2 appears at once from (15), which gives
$$\nabla^2 (B_1 Q_1 + B_2 Q_2) + (h - \nu\nabla^2)(A_1 Q_1 + A_2 Q_2) = 0,$$
so that by (18)
$$B_1 = A_1\left(\nu - \frac{h}{\lambda_1}\right), \qquad B_2 = A_2\left(\nu - \frac{h}{\lambda_2}\right)\ldots\ldots\ldots(20).$$

More general solutions may be obtained by addition to u, v, w respectively of u', v', w', where u', v', w' satisfy
$$\nabla^2 u' = \frac{h}{\mu'} u', \qquad \nabla^2 v' = \frac{h}{\mu'} v', \qquad \nabla^2 w' = \frac{h}{\mu'} w' \ldots (21).$$
Thus
$$\left.\begin{array}{l} u = u' + B_1 dQ_1/dx + B_2 dQ_2/dx \\ v = v' + B_1 dQ_1/dy + B_2 dQ_2/dy \\ w = w' + B_1 dQ_1/dz + B_2 dQ_2/dz \end{array}\right\} \ldots\ldots\ldots\ldots (22),$$
where B_1, B_2 have the values above given.

By substitution in (15) of the values of u, v, w specified in (22) it appears that
$$\frac{du'}{dx} + \frac{dv'}{dy} + \frac{dw'}{dz} = 0 \ldots\ldots\ldots\ldots\ldots\ldots (23).$$

349. These results are first applied by Kirchhoff to the case of plane waves, supposed to be propagated in infinite space in the direction of $+x$. Thus v' and w' vanish, while u', Q_1, Q_2 are independent of y and z. It follows from (23) § 348 that u' also vanishes. The equations for Q_1 and Q_2 are
$$d^2 Q_1/dx^2 = \lambda_1 Q_1, \qquad d^2 Q_2/dx^2 = \lambda_2 Q_2 \ldots\ldots\ldots (1),$$

so that we may take

$$Q_1 = e^{-x\sqrt{\lambda_1}}, \qquad Q_2 = e^{-x\sqrt{\lambda_2}} \dots\dots\dots\dots (2),$$

where the signs of the square roots are to be so chosen that the real parts are positive. Accordingly

$$u = A_1\lambda_1^{\frac{1}{2}}\left(\frac{h}{\lambda_1} - \nu\right)e^{-x\sqrt{\lambda_1}} + A_2\lambda_2^{\frac{1}{2}}\left(\frac{h}{\lambda_2} - \nu\right)e^{-x\sqrt{\lambda_2}}\dots\dots(3),$$

$$\theta' = A_1 e^{-x\sqrt{\lambda_1}} + A_2 e^{-x\sqrt{\lambda_2}} \dots\dots\dots\dots\dots (4),$$

in which the constants A_1, A_2 may be regarded as determined by the values of u and θ' when $x = 0$.

The solution, as expressed by (3), (4), is too general for our present purpose, providing as it does for arbitrary communication of heat at $x = 0$. From the quadratic in λ, (19) § 348, we see that if μ', μ'', ν be regarded as small quantities, one of the values of λ, say λ_1, is approximately equal to h^2/a^2, while the other λ_2 is very great. The solution which we require is that corresponding to λ_1 simply. The second approximation to λ_1 is by (19) § 348

$$\lambda_1 = \frac{h^2}{a^2 + h(\mu' + \mu'' + \nu)} + \frac{\nu b^2 \lambda_1^2}{ha^2} = \frac{h^2}{a^2}\left\{1 - \frac{h(\mu' + \mu'' + \nu)}{a^2}\right\} + \frac{\nu b^2 h^3}{a^6};$$

so that

$$\pm\sqrt{\lambda_1} = \frac{h}{a} - \frac{h^2}{2a^3}\left\{\mu' + \mu'' + \nu(1 - b^2/a^2)\right\}\dots\dots\dots(5).$$

If we now write in for h, we see that the typical solution is

$$u = e^{-m'x} e^{in(t - x/a)} \dots\dots\dots\dots\dots\dots\dots(6),$$

where

$$m' = \frac{n^2}{2a^3}\left\{\mu' + \mu'' + \nu\left(1 - \frac{b^2}{a^2}\right)\right\}\dots\dots\dots\dots(7).$$

In (6) an arbitrary multiplier and an arbitrary addition to t may, as usual, be introduced; and, if desired, the solution may be realized by omitting the imaginary part.

These results are in harmony with those already obtained for particular cases. Thus, if $\nu = 0$, (7) gives

$$m' = \frac{n^2}{2a^3}(\mu' + \mu''),$$

in agreement with (7) § 346, where

$$\mu'' = \tfrac{1}{3}\mu' = \tfrac{1}{3}\mu/\rho.$$

On the other hand if viscosity be left out of account, so that $\mu' = \mu'' = 0$, we fall back upon (18) § 247. It is unnecessary to add anything to the discussions already given.

In the case of spherical waves, propagated in the direction of $+r$, Kirchhoff finds in like manner as the expression for the radial velocity

$$\frac{d}{dr}\frac{e^{-m'r}}{r} \cdot e^{in(t-r/a)} \dots\dots\dots\dots\dots(8),$$

where m' has the same value (7) as before.

350. We will now pass on to the more important problem and suppose that the air is contained in a cylindrical tube of circular section, and that the motion is symmetrical with respect to the axis of x. If $y^2 + z^2 = r^2$, and

$$v = q \cdot y/r, \qquad w = q \cdot z/r,$$
$$v' = q' \cdot y/r, \qquad w' = q' \cdot z/r,$$

then u, u', q, q', Q_1, Q_2 are to be regarded as functions of x and r. We suppose further that as functions of x these quantities are proportional to e^{mx}, where m is a complex constant to be determined. The equations (18) § 348 for Q_1, Q_2 become

$$\frac{d^2Q_1}{dr^2} + \frac{1}{r}\frac{dQ_1}{dr} = (\lambda_1 - m^2)\,Q_1 \dots\dots\dots\dots(1),$$

$$\frac{d^2Q_2}{dr^2} + \frac{1}{r}\frac{dQ_2}{dr} = (\lambda_2 - m^2)\,Q_2 \dots\dots\dots\dots(2).$$

For u', q' equations (21), (23) give

$$\frac{d^2u'}{dr^2} + \frac{1}{r}\frac{du'}{dr} = \left(\frac{h}{\mu'} - m^2\right)u' \dots\dots\dots\dots(3),$$

$$\frac{d^2q'}{dr^2} + \frac{1}{r}\frac{dq'}{dr} - \frac{q'}{r^2} = \left(\frac{h}{\mu'} - m^2\right)q' \dots\dots\dots\dots(4),$$

$$m\,u' + \frac{dq'}{dr} + \frac{q'}{r} = 0 \dots\dots\dots\dots\dots(5).$$

These three equations are satisfied if u' be determined by means of the first, and q' is chosen so that

$$q' = -\frac{m}{h/\mu' - m^2}\frac{du'}{dr} \dots\dots\dots\dots(6),$$

a relation obtained by subtracting from (4) the result of differentiating (5) with respect to r. The solution of (3) may be written $u' = AQ$, in which A is a constant, and Q a function of r satisfying

$$\frac{d^2Q}{dr^2} + \frac{1}{r}\frac{dQ}{dr} = \left(\frac{h}{\mu'} - m^2\right)Q \dots\dots\dots\dots(7).$$

Thus, by (20), (22) § 348,

$$u = A Q - A_1 m \left(\frac{h}{\lambda_1} - \nu\right) Q_1 - A_2 m \left(\frac{h}{\lambda_2} - \nu\right) Q_2 \ldots\ldots (8),$$

$$q = - A \frac{m}{h/\mu' - m^2} \frac{dQ}{dr} - A_1 \left(\frac{h}{\lambda_1} - \nu\right) \frac{dQ_1}{dr} - A_2 \left(\frac{h}{\lambda_2} - \nu\right) \frac{dQ_2}{dr} \ldots(9),$$

$$\theta' = A_1 Q_1 + A_2 Q_2 \ldots\ldots\ldots\ldots\ldots\ldots (10).$$

On the walls of the tube u, q, θ' must satisfy certain conditions. It will here be supposed that there is neither motion of the gas nor change of temperature; so that when r has a value equal to the radius of the tube, u, q, θ' vanish. The condition of which we are in search is thus expressed by the evanescence of the determinant of (8), (9), (10), viz.:

$$\frac{m^2 h}{h/\mu' - m^2} \left(\frac{1}{\lambda_1} - \frac{1}{\lambda_2}\right) \frac{d \log Q}{dr} + \left(\frac{h}{\lambda_1} - \nu\right) \frac{d \log Q_1}{dr}$$
$$- \left(\frac{h}{\lambda_2} - \nu\right) \frac{d \log Q_2}{dr} = 0 \ldots\ldots (11).$$

The three functions Q, Q_1, Q_2, which are required to remain finite when $r = 0$, are Bessel's functions of order zero (§ 200), so that we may write in the usual notation

$$\left. \begin{array}{l} Q = J_0 \{r \sqrt{(m^2 - h/\mu')}\} \\ Q_1 = J_0 \{r \sqrt{(m^2 - \lambda_1)}\} \\ Q_2 = J_0 \{r \sqrt{(m^2 - \lambda_2)}\} \end{array} \right\} \ldots\ldots\ldots\ldots\ldots(12).$$

In equation (11) the values of λ_1, λ_2 are independent of r, being determined by (19) § 348. In the application to air under normal conditions μ', μ'', ν may be regarded as small, and we have approximately

$$\lambda_1 = h^2/a^2, \qquad \lambda_2 = ha^2/\nu b^2 \ldots\ldots\ldots\ldots(13)$$

A second approximation to the value of λ_1 has already been given in (5) § 349. It is here assumed that the velocity of propagation of viscous effects of the pitch in question, viz. $\sqrt{(2\mu' n)}$, § 347, is small compared with that of sound, so that $in\mu'/a^2$, or $h\mu'/a^2$, is a small quantity.

In interpreting the solution there are two extreme cases worthy of special notice. The first of these, which is that considered by Kirchhoff, arises when μ', μ'', ν are treated as very small, so small that the layer of gas immediately affected by the walls of the tube is but an insignificant fraction of the whole contents. When μ' &c. vanish, we have

$$\lambda_1 = h^2/a^2, \qquad m^2 = h^2/a^2,$$

so that $r\sqrt{(m^2 - \lambda_1)}$ is here to be regarded as small. On the other hand $r\sqrt{(m^2 - h/\mu')}$, $r\sqrt{(m^2 - \lambda_2)}$ are large.

The value of $J_0(z)$, when z is small, is given by the ascending series (5) § 200; from which it follows at once that

$$d \log J_0(z)/dz = -\tfrac{1}{2}z.$$

When z is very large and such that its imaginary part is positive, (10) § 200 gives

$$d \log J_0(z)/dz = -\tan(z - \tfrac{1}{4}\pi) = -i.$$

Thus, if we retain only the terms of highest order,

$$\left.\begin{aligned}
d \log Q/dr &= \sqrt{(h/\mu')} \\
d \log Q_1/dr &= \tfrac{1}{2}r(\lambda_1 - m^2) \\
d \log Q_2/dr &= \sqrt{(ha^2/\nu b^2)}
\end{aligned}\right\} \dots\dots\dots\dots (14).$$

Using these in (11) with the approximate values of λ_1, λ_2 from (13), we find

$$m^2 = \frac{h^2}{a^2}\left(1 + \frac{2\gamma'}{r\sqrt{h}}\right) \dots\dots\dots\dots\dots (15),$$

where

$$\gamma' = \sqrt{\mu'} + (a/b - b/a)\sqrt{\nu} \dots\dots\dots\dots (16),$$

and the sign of $\sqrt{h}$ is to be so chosen that the real part is positive.

We now write

$$h = ni \dots\dots\dots\dots\dots\dots (17),$$

so that the frequency is $n/2\pi$. Thus

$$\sqrt{h} = \sqrt{(\tfrac{1}{2}n)} \cdot (1 + i) \dots\dots\dots\dots (18);$$

and

$$m = \pm(m' + im'') \dots\dots\dots\dots\dots (19),$$

where by (15)

$$m' = \frac{\sqrt{n} \cdot \gamma'}{\sqrt{2} \cdot ar}, \qquad m'' = \frac{n}{a} + \frac{\sqrt{n} \cdot \gamma'}{\sqrt{2} \cdot ar} \dots\dots\dots (20).$$

If we restore the hitherto suppressed factors dependent upon x and t, we have

$$u = BRe^{ht+mx}, \qquad q = BR'e^{ht+mx}, \qquad \theta' = BR''e^{ht+mx},$$

where B is an arbitrary constant, and R, R', R'' are certain functions of r, which vanish when r is equated to the radius of the tube, and which for points lying at a finite distance from the walls assume the values

$$R = 1, \qquad R' = 0, \qquad R'' = -1/a.$$

The realized solution for u, applicable at points which lie at a finite distance from the walls, may be written

$$u = C_1 e^{m'x} \sin (nt + m''x + \delta_1) + C_2 e^{-m'x} \sin (nt - m''x + \delta_2)\ldots(21),$$

where C_1, C_2, δ_1, δ_2 denote four real arbitrary constants. Accordingly m' determines the attenuation which the waves suffer in their progress, and m'' determines the velocity of propagation. This velocity is

$$n/m'' = a \left\{ 1 - \frac{\gamma'}{\sqrt{(2n) . r}} \right\} \ldots\ldots\ldots\ldots\ldots(22),$$

in harmony with (12) § 347.

The diminution of the velocity of sound in narrow tubes, as indicated by the wave-length of stationary vibrations, was observed by Kundt (§ 260), and has been specially investigated by Schneebeli[1] and A. Seebeck[2]. From their experiments it appears that the diminution of velocity varies as r^{-1}, in accordance with (22), but that, when n varies, it is proportional rather to $n^{-\frac{3}{4}}$ than to $n^{-\frac{1}{2}}$. Since μ is independent of the density (ρ), the effect would be increased in rarefied air.

We will now turn to the consideration of another extreme case of equation (11). This arises when the tube is such that the layer immediately affected by the friction, instead of merely forming a thin coating to the walls, extends itself over the whole section, as must inevitably happen if the diameter be sufficiently reduced. Under these circumstances hr^2/μ' is a small, and not, as in the case treated by Kirchhoff, a large quantity, and the arguments of *all* the three functions in (12) are to be regarded as small.

One result of the investigation may be foreseen. When the diameter of the tube is very much reduced, the conduction of heat from the centre to the circumference of the column of air becomes more and more free. In the limit the temperature of the solid walls controls that of the included gas, and the expansions and rarefactions take place isothermally. Under these circumstances there is no dissipation due to conduction, and everything is the same as if no heat were developed at all. Consequently the coefficient of heat-conduction will not appear in the result, which

[1] *Pogg. Ann.* vol. cxxxvi. p. 296, 1869.

[2] *Pogg. Ann.* vol. cxxxix. p. 104, 1870.

will involve, moreover, the Newtonian value (b) of the velocity of sound, and not that of Laplace (a).

When z is small,

$$J_0(z) = 1 - \frac{z^2}{2^2} + \frac{z^4}{2^2 \cdot 4^2},$$

so that approximately

$$d \log J_0(z)/dz = - \tfrac{1}{2} z \left(1 + \tfrac{1}{8} z^2\right) \dots \dots \dots \dots (23).$$

When the results of the application of (23) to Q, Q_1, Q_2 are introduced into (11), the equation may be divided by $\tfrac{1}{2} r$, and the left-hand member will then consist of two parts, of which the first is independent of r and the second is proportional to r^2. The first part reduces itself without further approximations to $\nu\left(\lambda_2 - \lambda_1\right)$. For the second part the leading terms only need be retained. Thus with use of (13)

$$\nu\left(\lambda_2 - \lambda_1\right) - \frac{a^2 r^2}{8} \left\{ \frac{m^2}{\mu'} + \frac{h^2\left(a^2 - b^2\right)}{\nu b^4} \right\} = 0,$$

whence

$$m^2 = \frac{8\,\mu' h}{b^2 r^2} - \frac{h^2 \mu'\left(a^2 - b^2\right)}{\nu b^4}.$$

The ratio of the second term to the first is of the order $h r^2/\nu$, by supposition a small quantity, so that we are to take simply

$$m^2 = \frac{8\mu' h}{b^2 r^2} = \frac{8 i \mu' n}{b^2 r^2} \dots \dots \dots \dots \dots \dots (24),$$

as the solution applicable under the supposed conditions.

Before leaving this question it may be worth while to consider briefly the corresponding problem in two dimensions, although it is of less importance than that of the circular tube treated by Kirchhoff. The analysis is a little simpler; but, as it follows practically the same course, we may content ourselves with a mere indication of the necessary changes. The motion is supposed to be independent of z and to take place between parallel walls at $y = \pm y_1$.

The equations (1) to (11) of the preceding investigation may be regarded as still applicable in the present problem, if we write v for q and y for r, with omission of the terms where r occurs in the denominator. The general solution of the equations corresponding to (1), (2), (7) contains two functions whose form is that of sines and cosines of multiples of y. But from (8), (9), (10) it is evident that the conditions of the problem at $y = 0$ require the absence of the sine function, so that in (12) we are simply to replace the function J_0 by the cosine.

In the case where μ' &c. are regarded as infinitely small we have as in (14), when $y = y_1$,

$$\left.\begin{array}{l} d \log Q/dy = \sqrt{(h/\mu')} \\ d \log Q_2/dy = \sqrt{(ha^2/\nu b^2)} \end{array}\right\} \quad \ldots\ldots\ldots\ldots\ldots (25),$$

but in place of the second of equations (14)

$$d \log Q_1/dy = y_1 (\lambda_1 - m^2) \ldots\ldots\ldots\ldots\ldots (26).$$

When these values are substituted in (11), the resulting equation is unchanged, except that r is replaced by $2y_1$. The same substitution is to be made in (15), (20), (22). The latter gives for the velocity of sound

$$a \left\{ 1 - \frac{1}{2y_1} \frac{\gamma'}{\sqrt{(2n)}} \right\} \ldots\ldots\ldots\ldots\ldots (27).$$

It is worth notice that (27) is what (11) § 347 becomes for this case when we replace $\sqrt{\mu'}$ by γ'; and we may perhaps infer that the same change is sufficient to render that equation applicable to a section of any form when thermal effects are to be taken into account.

In the second extreme case where the distance between the walls $(2y_1)$ is so small that $hy_1{}^2/\nu$ is to be neglected, we have in place of (23)

$$d \log \cos z/dz = -z (1 + \tfrac{1}{3} z^2) \ldots\ldots\ldots\ldots\ldots (28).$$

The equations following are thus adapted to our present purpose if we replace $\tfrac{1}{8} r^2$ by $\tfrac{1}{3} y_1{}^2$. The analogue of (24) is accordingly

$$m^2 = \frac{3\mu' h}{b^2 y_1{}^2} = \frac{3i\mu' n}{b^2 y_1{}^2} \ldots\ldots\ldots\ldots\ldots (29).$$

351. The results of § 350 have an important bearing upon the explanation of the behaviour of porous bodies in relation to sound. Tyndall has shewn that in many cases sound penetrates such bodies more freely than would have been expected, although it is reflected from thin layers of continuous solid matter. On the other hand a hay-stack seems to form a very perfect obstacle. It is probable that porous walls give a diminished reflection, so that within a building so bounded resonance is less prolonged than if the walls were formed of continuous matter.

When we inquire into the mechanical question, it is evident that sound is not destroyed by obstacles as such. In the absence of dissipative forces, what is not transmitted must be reflected. Destruction depends upon viscosity and upon conduction of heat; but the influence of these agencies is enormously augmented by the contact of solid matter exposing a large surface. At such a surface the tangential as well as the normal motion is hindered, and a passage of heat to and fro takes place, as the neighbouring air is heated and cooled during its condensations and rarefactions. With such rapidity of alternation as we are concerned with in the case of audible sounds, these influences extend to only a very thin layer of the air and of the solid, and are thus greatly favoured by a fine state of division.

Let us conceive an otherwise continuous wall, presenting a flat face, to be perforated by a great number of similar narrow channels, uniformly distributed, and bounded by surfaces everywhere perpendicular to the face of the wall. If the channels be sufficiently numerous, the transition, when sound impinges, from simple plane waves on the outside to the state on the inside of aerial vibration corresponding to the interior of a channel of unlimited length, occupies a space which is small relatively to the wave-length of the vibration, and then the connection between the condition of things inside and outside admits of simple expression.

Considering first the interior of one of the channels, and taking the axis of x parallel to the axis of the channel, we suppose that as functions of x the velocity components u, v, w and the condensation s are proportional to e^{ikx}, while as functions of t everything is proportional to e^{int}, n being real. The relationship between k and n depends upon the nature of the gas and upon the size and form of the channel, and has been determined for certain important cases in § 350, ik being there denoted by m. Supposing it to be known, we will go on to shew how the problem of reflection is to be dealt with.

For this purpose consider the equation of continuity as integrated over the cross-section σ of the channel. Since the walls of the channel are impenetrable,

$$\frac{d}{dt}\iint s\,d\sigma + \frac{d}{dx}\iint u\,d\sigma = 0,$$

so that　　　　　　$n\iint s\,d\sigma + k\iint u\,d\sigma = 0 \dots\dots\dots\dots\dots\dots(1).$

This equation is applicable at points distant from the open end more than several diameters of the channel.

Taking now the origin of x at the face of the wall, we have to form corresponding expressions for the waves outside; and we may there neglect the effects of viscosity of conduction of heat. If a be the velocity of sound in the open, and $k_0 = n/a$, we may write for waves incident and reflected perpendicularly

$$s = (e^{ik_0 x} + Be^{-ik_0 x})\, e^{int} \dots\dots\dots\dots\dots (2),$$

$$u = a\,(- e^{ik_0 x} + Be^{-ik_0 x})\, e^{int} \dots\dots\dots\dots (3);$$

so that the incident wave is

$$s = e^{i(nt + k_0 x)} \dots\dots\dots\dots\dots\dots (4),$$

or, on throwing away the imaginary part,

$$s = \cos{(nt + k_0 x)} \dots\dots\dots\dots\dots\dots (5).$$

These expressions are applicable when x exceeds a moderate multiple of the distance between the channels. Close up to the face the motion will be more complicated; but we have no need to investigate it in detail. The ratio of u and s at a place near the wall is given with sufficient accuracy by putting $x = 0$ in (2) and (3),

$$\frac{u}{as} = \frac{B - 1}{B + 1} \dots\dots\dots\dots\dots\dots (6).$$

We now assume that a space, defined by parallel planes one on either side of $x = 0$, may be taken so thin relatively to the wave-length that the mean pressures are sensibly the same at the two boundaries, and that the flow into the space at one boundary is sensibly equal to the flow out of the space at the other boundary, and yet broad enough relatively to the transverse dimensions of the channels to allow the application of (6) at one bounding plane and of (1) at the other bounding plane. The equality of flow does not imply an equality of mean velocities, since the areas concerned are different. The mean velocities will be inversely proportional to the corresponding areas—that is in the ratio $\sigma : \sigma + \sigma'$, if σ' denote the area of the unperforated part of the wall corresponding to each channel. By (1) and (6) the connection between the inside and outside motion is expressed by

$$- \frac{n}{k}\sigma = \frac{a\,(B - 1)}{B + 1}\,(\sigma + \sigma').$$

We will denote the ratio of the unperforated to the perforated parts of the wall by g, so that $g = \sigma'/\sigma$. Thus

$$\frac{1-B}{1+B} = \frac{k_0}{k\,(1+g)} \dots\dots\dots\dots\dots(7).$$

If $g = 0$, $k = k_0$, that is, if the wall be abolished, or if it be reduced to infinitely thin partitions between the channels while at the same time the dissipative effects are neglected, there is no reflection. If there are no perforations ($g = \infty$), then $B = 1$, signifying total reflection. Generally in place of (7) we may write

$$B = \frac{k\,(1+g) - k_0}{k\,(1+g) + k_0} \dots\dots\dots\dots\dots(8),$$

which is the solution of the problem proposed. It is understood that waves which have once entered the wall do not return. When dissipative forces act, this condition may always be satisfied by supposing the channels to be long enough. The necessary length of channel, or thickness of wall, will depend upon the properties of the gas and upon the size and shape of the channels. Even in the absence of dissipative forces there must be reflection, except in the extreme case $g = 0$. Putting $k = k_0$ in (8), we have

$$B = \frac{g}{2+g} \dots\dots\dots\dots\dots (9).$$

If $g = 1$, that is if half the wall be cut away, $B = \frac{1}{3}$, $B^2 = \frac{1}{9}$, so that the reflection is but small. If the channels be circular and arranged in square order as close as possible to one another, $g = (4 - \pi)/\pi$, whence $B = \cdot121$, $B^2 = \cdot015$, nearly all the motion being transmitted.

If the channels be circular in section and so small that nr^2/ν may be neglected, we have, (24) § 350,

$$-k^2 = m^2 = \frac{8i\mu' n}{b^2 r^2} \dots \dots\dots\dots\dots(10);$$

so that (21) the wave propagated into a channel is proportional to

$$e^{m'x} \sin\,(nt + m''x + \delta_1) \dots\dots\dots\dots\dots (11),$$

where

$$m' = m'' = \frac{2\,\sqrt{(\mu'n)}}{br} = \frac{2\sqrt{(\mu'\gamma n)}}{ar} \dots\dots\dots\dots (12),$$

γ being the ratio of specific heats § 246.

To take a numerical example, suppose that the pitch is 256, so that $n = 2\pi \times 256$. The value of μ' for air is ·16 C.G.S., and that of ν is ·256. If we take $r = \frac{1}{1000}$ cm., we find $nr^2/8\nu$ equal to about $\frac{1}{1000}$. If r were ten times as great, the approximation in (10) would perhaps still be sufficient.

From (12), if $n = 2\pi \times 256$,

$$m' = m'' = ·00115/r \dots\dots\dots\dots\dots(13);$$

so that, if $r = \frac{1}{1000}$ cm., $m' = 1·15$. In this case the amplitude is reduced in the ratio $e : 1$ in passing over the distance $1/m'$, that is about one centimetre. The distance penetrated is proportional to the radius of the channel.

The amplitude of the reflected wave is by (8)

$$B = \frac{m'(1+g)(1-i) - k_0}{m'(1+g)(1-i) + k_0},$$

or, as we may write it,

$$B = \frac{M - 1 - iM}{M + 1 - iM} \dots\dots\dots\dots\dots (14),$$

where

$$M = (1 + g)\, m'/k_0 \dots\dots\dots\dots\dots(15).$$

If I be the intensity of the reflected sound, that of the incident sound being unity,

$$I = \frac{2M^2 - 2M + 1}{2M^2 + 2M + 1} \dots\dots\dots\dots\dots(16).$$

The intensity of the intromitted sound is given by

$$1 - I = \frac{4M}{2M^2 + 2M + 1} \dots\dots\dots\dots (17).$$

By (12), (15)

$$M = \frac{2(1+g)\sqrt{(\mu'\gamma)}}{r\sqrt{n}} \dots\dots\dots\dots\dots(18).$$

If we suppose $r = \frac{1}{1000}$ cm., and $g = 1$, we shall have a wall of pretty close texture. In this case by (18), $M = 47·4$ and $1 - I = ·0412$. A loss of 4 per cent. may not appear to be important; but we must remember that in prolonged resonance we are concerned with the accumulated effect of a large number of reflections, so that a comparatively small loss in a single reflection may well be material. The thickness of the porous layer necessary to produce this effect is less than one centimetre.

Again, suppose $r = \frac{1}{100}$ cm., $g = 1$. We find $M = 4·74$, $1 - I = ·342$; and the necessary thickness would be less than 10 centimetres.

If r be much greater than $\frac{1}{100}$ cm., the exchange of heat between the air and the sides of the channel is no longer sufficiently free to allow of the use of (24) § 350. When the diameter is so great that the thermal and viscous effects extend through only a small fraction of it, we have the case discussed by Kirchhoff (15) § 350. Here

$$k = \frac{n}{a}\left\{1 + \frac{\gamma'(1-i)}{r\sqrt{(2n)}}\right\} \quad \dots\dots\dots\dots\dots (19),$$

which value is to be substituted in (8). If for simplicity we put $g = 0$, we find

$$B = \frac{\gamma'(1-i)}{2r\sqrt{(2n)}} \quad \dots\dots\dots\dots\dots\dots (20),$$

$$I = \gamma'^2/4r^2 n \quad \dots\dots\dots\dots\dots\dots(21).$$

The supposition that $g = 0$ is, however, inconsistent with the circular section; and it is therefore preferable to use the solution corresponding to (27) § 350, applicable when the channels assume the form of narrow crevasses[1]. We have merely to replace r in (19), (20), (21) by $2y_1$, $2y_1$ being the width of a crevasse. The incident sound is absorbed more and more completely as the width of the channels increases; but at the same time a greater length of channel, or thickness of wall, becomes necessary in order to prevent a return from the further side. If $g = 0$, there is no theoretical limit to the absorption; and, as we have seen, a moderate value of g does not of itself entail more than a comparatively small reflection. A loosely compacted hay-stack would seem to be as effective an absorbent of sound as anything likely to be met with.

In large spaces bounded by non-porous walls, roof, and floor, and with few windows, a prolonged resonance seems inevitable. The mitigating influence of thick carpets in such cases is well known. The application of similar material to the walls and to the roof appears to offer the best chance of further improvement.

352. One of the most curious consequences of viscosity is the generation in certain cases of regular vortices. Of this an example, discovered by Dvořák, has already been mentioned in § 260. In

---

[1] It may be remarked that even in the two-dimensional problem the supposition $g = 0$ involves an infinite capacity for heat in the material composing the partitions.

a theoretically inviscid fluid no such effect could occur, § 240; and, even when viscosity enters, the phenomenon is one of the second order, dependent, that is, upon the *square* of the motion. Three problems of this kind have been treated by the author[1] on a former occasion, but here we must limit ourselves to Dvořák's phenomenon, further simplifying the question by taking the case of two dimensions and by neglecting the terms dependent upon the development and conduction of heat.

If we suppose that $p = a^2\rho$, and write s for $\log(\rho/\rho_0)$, the fundamental equations (12) § 345 are

$$a^2\frac{ds}{dx} = -\frac{du}{dt} - u\frac{du}{dx} - v\frac{du}{dy} + \mu'\nabla^2 u + \mu''\frac{d}{dx}\left(\frac{du}{dx} + \frac{dv}{dy}\right)\dots(1),$$

with a corresponding equation for v, and the equation of continuity § 238

$$\frac{du}{dx} + \frac{dv}{dy} + \frac{ds}{dt} + u\frac{ds}{dx} + v\frac{ds}{dy} = 0\dots\dots\dots\dots(2).$$

Whatever may be the actual values of u and v, we may write

$$u = \frac{d\phi}{dx} + \frac{d\psi}{dy}, \qquad v = \frac{d\phi}{dy} - \frac{d\psi}{dx} \dots\dots\dots(3),$$

in which

$$\nabla^2\phi = \frac{du}{dx} + \frac{dv}{dy}, \qquad \nabla^2\psi = \frac{du}{dy} - \frac{dv}{dx} \dots\dots\dots(4).$$

From (1), (2)

$$\left(a^2 + \mu''\frac{d}{dt}\right)\frac{ds}{dx} = -\frac{du}{dt} + \mu'\nabla^2 u$$
$$- u\frac{du}{dx} - v\frac{du}{dy} - \mu''\frac{d}{dx}\left(u\frac{ds}{dx} + v\frac{ds}{dy}\right)\dots\dots(5),$$

$$\left(a^2 + \mu''\frac{d}{dt}\right)\frac{ds}{dy} = -\frac{dv}{dt} + \mu'\nabla^2 v$$
$$- u\frac{dv}{dx} - v\frac{dv}{dy} - \mu''\frac{d}{dy}\left(u\frac{ds}{dx} + v\frac{ds}{dy}\right)\dots\dots(6).$$

Again, from (5), (6),

$$\left(a^2 + \mu'\frac{d}{dt} + \mu''\frac{d}{dt}\right)\nabla^2 s - \frac{d^2 s}{dt^2} = \frac{d}{dt}\left(u\frac{ds}{dx} + v\frac{ds}{dy}\right)$$
$$- (\mu' + \mu'')\nabla^2\left(u\frac{ds}{dx} + v\frac{ds}{dy}\right) - \frac{d}{dx}\left(u\frac{du}{dx} + v\frac{du}{dy}\right) - \frac{d}{dy}\left(u\frac{dv}{dx} + v\frac{dv}{dy}\right)$$
$$\dots\dots\dots(7).$$

[1] On the Circulation of Air observed in Kundt's Tubes, and on some allied Acoustical Problems. *Phil. Trans.* vol. 175, p. 1, 1883.

For the first approximation the terms of the second order in u, v, s are to be omitted. If we assume that as functions of t all the periodic quantities are proportional to e^{int}, and write q for $a^2 + in\mu' + in\mu''$, (7) becomes

$$q\nabla^2 s + n^2 s = 0 \quad \dots\dots\dots\dots\dots\dots (8).$$

Now by (2), (4) $\nabla^2\phi = -ins = i(q/n)\nabla^2 s$,

so that $$\phi = iqs/n \dots\dots\dots\dots\dots\dots\dots (9)^1,$$

and $$u = \frac{iq}{n}\frac{ds}{dx} + \frac{d\psi}{dy}, \qquad v = \frac{iq}{n}\frac{ds}{dy} - \frac{d\psi}{dx} \dots\dots\dots (10).$$

Substituting in (5), (6), with omission of the terms of the second order, we get in view of (8),

$$(\mu'\nabla^2 - in)\frac{d\psi}{dy} = 0, \qquad (\mu'\nabla^2 - in)\frac{d\psi}{dx} = 0,$$

whence $$(\mu'\nabla^2 - in)\psi = 0 \dots\dots\dots\dots\dots (11).$$

If we eliminate s directly from equations (1), we get

$$\left(\mu'\nabla^4 - \frac{d}{dt}\nabla^2\right)\psi = \frac{d}{dy}\left(u\frac{du}{dx} + v\frac{du}{dy}\right) - \frac{d}{dx}\left(u\frac{dv}{dx} + v\frac{dv}{dy}\right)$$

$$= \frac{d}{dy}(v\nabla^2\psi) + \frac{d}{dx}(u\nabla^2\psi)$$

$$= \left(\frac{du}{dx} + \frac{dv}{dy}\right)\nabla^2\psi + u\frac{d\nabla^2\psi}{dx} + v\frac{d\nabla^2\psi}{dy} \quad \dots\dots (12).$$

If we now assume that as functions of x the quantities s, ψ, &c. are proportional to e^{ikx}, equations (8), (11) may be written

$$(d^2/dy^2 - k''^2)s = 0 \dots\dots\dots\dots\dots(13),$$

where $k''^2 = k^2 - n^2/q$,

$$(d^2/dy^2 - k'^2)\psi = 0 \dots\dots\dots\dots\dots(14),$$

where $k'^2 = k^2 + in/\mu'$.

If the origin for y be in the middle between the two parallel bounding planes, s must be an even function of y, and ψ must be an odd function. Thus we may write

$$s = A\cosh k''y . e^{int} . e^{ikx}, \qquad \psi = B\sinh k'y . e^{int} . e^{ikx}\dots(15),$$

$$\left. \begin{array}{l} u = (-kq/n . A\cosh k''y + k'B\cosh k'y)e^{int} . e^{ikx} \\ v = (iqk''/n . A\sinh k''y - ikB\sinh k'y)e^{int} . e^{ikx} \end{array} \right\}\dots(16).$$

---

[1] It is unnecessary to add a complementary function ϕ' satisfying $\nabla^2\phi' = 0$, for the motion corresponding thereto may be regarded as covered by ψ.

If the fixed walls are situated at $y = \pm y_1$, u and v must vanish for these values of y. Eliminating from (16) the ratio of A to B, we get as the equation for determining k,

$$k^2 \tanh k'y_1 = k'k'' \tanh k''y_1 \dots\dots\dots\dots (17),$$

where k', k'' are the functions of k above defined. Equation (17) may be regarded as a modified and simplified form of (11) § 350, modified on account of the change from symmetry about an axis to two dimensions, and simplified by the omission of the thermal terms represented by ν. The comparison is readily made. Since $\lambda_2 = \infty$, the third term in (11), involving Q_2, disappears altogether, and then λ_1^{-1} divides out. In (11), (12) r is to be replaced by y, and J_0 by cosine, as has already been explained. Further,

$$m^2 = -k^2, \quad h = in.$$

We now introduce further approximations dependent upon the assumption that the direct influence of viscosity extends through a layer whose thickness is a small fraction only of y_1. In this case $k^2 = n^2/a^2$ nearly, so that $k''y_1$ is a small quantity and $k'y_1$ is a large quantity, and we may take

$$\tanh k'y_1 = \pm 1, \qquad \tanh k''y_1 = \pm k''y_1.$$

Equation (17) then becomes

$$k^2 = k'k''^2 y_1 \dots\dots\dots\dots\dots\dots(18),$$

or, if we introduce the values of k', k'' from (13), (14),

$$k^2 = y_1 (k^2 - n^2/q)(k^2 + in/\mu')^{\frac{1}{2}}.$$

Thus approximately

$$k = \pm \frac{n}{a} \left\{ 1 + \frac{1-i}{2y_1 \sqrt{(2n/\mu')}} \right\} \dots\dots\dots (19),$$

in agreement with the result already indicated in § 350.

In taking approximate forms for (16) we must specify which half of the symmetrical motion we contemplate. If we choose that for which y is *negative*, we replace $\cosh k'y$ and $\sinh k'y$ by $\frac{1}{2}e^{-k'y}$. For $\cosh k''y$ we may write unity, and for $\sinh k''y$ simply $k''y$. If we change the arbitrary multiplier so that the maximum value of u is u_0 and for the present take u_0 equal to unity, we have

$$\left.\begin{array}{l} u = (-1 + e^{-k'(y+y_1)})\, e^{ikx}\, e^{int} \\ v = ik/k' \cdot (y/y_1 + e^{-k'(y+y_1)})\, e^{ikx}\, e^{int} \end{array}\right\} \dots\dots\dots(20),$$

in which, of course, u and v vanish when $y = -y_1$.

If in (20) we change k into $-k$ and then take the mean, we obtain

$$\left.\begin{array}{l} u = (-1 + e^{-k'(y+y_1)}) \cos kx\, e^{int} \\ v = -k/k' \cdot (y/y_1 + e^{-k'(y+y_1)}) \sin kx\, e^{int} \end{array}\right\} \dots \dots (21).$$

Although k is not absolutely a real quantity, we may consider it to be so with sufficient approximation for our purpose. We may also take in (14)

$$k' = \sqrt{(in/\mu')} = \beta\,(1 + i) \dots \dots \dots \dots (22),$$

if $\beta = \sqrt{(n/2\mu')}$. Using this approximation in (21), we get in terms of real quantities,

$$\left.\begin{array}{l} u = \cos kx\, [-\cos nt + e^{-\beta(y+y_1)} \cos \{nt - \beta\,(y + y_1)\}] \\[2mm] v = -\dfrac{k \sin kx}{\beta\,\sqrt{2}} \left[\dfrac{y}{y_1} \cos (nt - \tfrac{1}{4}\pi) \right. \\[2mm] \qquad\qquad \left. + e^{-\beta(y+y_1)} \cos \{nt - \tfrac{1}{4}\pi - \beta\,(y + y_1)\} \right] \end{array}\right\} \dots (23).$$

It will shorten the expressions with which we have to deal if we measure y from the wall (on the negative side) instead of, as hitherto, from the plane of symmetry, for which purpose we must write y for $y + y_1$. Thus

$$\left.\begin{array}{l} u_1 = \cos kx\, [-\cos nt + e^{-\beta y} \cos (nt - \beta y)] \\[2mm] v_1 = \dfrac{k \sin kx}{\beta\,\sqrt{2}} \left[\dfrac{y_1 - y}{y_1} \cos (nt - \tfrac{1}{4}\pi) - e^{-\beta y} \cos (nt - \tfrac{1}{4}\pi - \beta y) \right] \end{array}\right\} \dots (24),$$

the subscripts indicating the order of the terms.

These are the values of the velocities when the square of the motion is neglected. In proceeding to a second approximation we require to form expressions for the right-hand members of (7) and (12), which for the purposes of the first approximation were neglected altogether. The additional terms dependent upon the square of the motion are partly independent of the time and partly of double frequency involving $2nt$. The latter are not of much interest, so that we shall confine ourselves to the non-periodic part. Further simplifications are admissible in virtue of the small thickness of the retarded layer in proportion to the width of the channel ($2y_1$) and still more in proportion to the wave-length (λ). Thus k/β is a small quantity and may usually be neglected.

From (24)

$$\nabla^2 \psi_1 = \beta \sqrt{2} . \cos kx \, e^{-\beta y} \sin (nt - \tfrac{1}{4}\pi - \beta y) \dots\dots(25),$$

$$du_1/dx + dv_1/dy = k \sin kx \cos nt \dots\dots\dots (26),$$

$$u_1 \frac{d\nabla^2\psi_1}{dx} + v_1 \frac{d\nabla^2\psi_1}{dy} = \tfrac{1}{2}k\beta \sin 2kx \, e^{-\beta y} (-\cos \beta y + e^{-\beta y})$$
$$+ \text{ terms in } 2nt \dots\dots\dots\dots(27),$$

$$\left(\frac{du_1}{dx} + \frac{dv_1}{dy}\right)\nabla^2\psi_1 = -\tfrac{1}{4}k\beta \sin 2kx \, e^{-\beta y}(\sin \beta y + \cos \beta y)$$
$$+ \text{ terms in } 2nt \dots\dots\dots\dots(28).$$

Thus for the non-periodic part of ψ of the second order, we have from (12)

$$\nabla^4\psi_2 = -\frac{k\beta}{4\mu'} \sin 2kx \, e^{-\beta y} \{\sin \beta y + 3 \cos \beta y - 2e^{-\beta y}\}\dots(29).$$

In this we identify ∇^4 with $(d/dy)^4$, so that

$$\psi_2 = \frac{k \sin 2kx \, e^{-\beta y}}{16\mu'\beta^3} \{\sin \beta y + 3 \cos \beta y + \tfrac{1}{2}e^{-\beta y}\} \dots.(30),$$

to which may be added a complementary function, satisfying $\nabla^4\psi_2 = 0$, of the form

$$\psi_2 = \frac{\sin 2kx}{16\mu'\beta^3}\{A \sinh 2k\,(y_1 - y) + B\,(y_1 - y)\cosh 2k\,(y_1 - y)\}\dots(31),$$

or, as we may take it approximately, if y_1 be small compared with λ,

$$\psi_2 = \frac{k \sin 2kx}{16\mu'\beta^3}\{A'\,(y_1 - y) + B'\,(y_1 - y)^3\} \dots\dots(32).$$

Equations (30), (32) give the non-periodic part of ψ of the second order.

The value of s to a second approximation would have to be investigated by means of (7). It will be composed of two parts, the first independent of t, the second a harmonic function of $2nt$. In calculating the part of $d\phi/dx$ independent of t from

$$\nabla^2\phi = -ds/dt - u\,ds/dx - v\,ds/dy,$$

we shall obtain nothing from ds/dt. In the remaining terms on the right-hand side it will be sufficient to employ the values u, v, s of the first approximation. From

$$ds/dt = -du/dx - dv/dy,$$

in conjunction with (26), we get

$$s = -u_0/a \cdot \sin kx \, \sin nt,$$

whence $d^2\phi_2/d\,(\beta y)^2 = ku_0^2/2a\beta^2 \cdot \cos^2 kx \, e^{-\beta y} \sin \beta y.$

From this it is easily seen that the part of u_2 resulting from $d\phi/dx$ in (3) is of order k^2/β^2 in comparison with the part (33) resulting from ψ_2, and may be omitted. Accordingly by (30), with introduction of the value of β and (in order to restore homogeneity) of u_0^2,

$$u_2 = -\frac{u_0^2 \sin 2kx \, e^{-\beta y}}{8a} \{4 \sin \beta y + 2 \cos \beta y + e^{-\beta y}\} \ldots (33),$$

$$v_2 = -\frac{2ku_0^2 \cos 2kx \, e^{-\beta y}}{8\beta a} \{\sin \beta y + 3 \cos \beta y + \tfrac{1}{2} e^{-\beta y}\} \ldots (34);$$

and from (32)

$$u_2 = -\frac{u_0^2 \sin 2kx}{8\beta a} \{A' + 3B'\,(y_1 - y)^2\} \ldots \ldots \ldots \ldots (35),$$

$$v_2 = -\frac{2ku_0^2 \cos 2kx}{8\beta a} \{A'\,(y_1 - y) + B'\,(y_1 - y)^3\} \ldots \ldots (36).$$

The complete value of the terms of the second order in u, v are given by addition of (33), (35) and of (34), (36). The constants A', B' are to be determined by the condition that these values vanish when $y = 0$. We thus obtain as the complete expression of the terms of the second order

$$u_2 = -\frac{u_0^2 \sin 2kx}{8a} \left\{ e^{-\beta y}\,(4 \sin \beta y + 2 \cos \beta y + e^{-\beta y}) + \tfrac{3}{2} - \tfrac{9}{2}\frac{(y_1 - y)^2}{y_1^2} \right\}$$
$$\ldots \ldots \ldots \ldots (37),$$

$$v_2 = -\frac{2ku_0^2 \cos 2kx}{8\beta a} \left\{ e^{-\beta y}\,(\sin \beta y + 3 \cos \beta y + \tfrac{1}{2} e^{-\beta y}) \right.$$
$$\left. + \tfrac{3}{2}\beta\,(y_1 - y) - \tfrac{3}{2}\beta\frac{(y_1 - y)^3}{y_1^2} \right\} \ldots (38).$$

Outside the thin film of air immediately influenced by the friction we may put $e^{-\beta y} = 0$, and then

$$u_2 = -\frac{3u_0^2 \sin 2kx}{16a} \left\{ 1 - \frac{3\,(y_1 - y)^2}{y_1^2} \right\} \ldots \ldots \ldots \ldots (39),$$

$$v_2 = -\frac{3u_0^2 \cdot 2k \cos 2kx}{16a} \left\{ y_1 - y - \frac{(y_1 - y)^3}{y_1^2} \right\} \ldots \ldots (40).$$

From (39) we see that u_2 changes sign as we pass from the boundary $y = 0$ to the plane of symmetry $y = y_1$, the critical value of y being $y_1 (1 - \sqrt{\frac{1}{3}})$, or $\cdot 423 y_1$.

The value of u_1 from (24) corresponding to (39) is

$$u_1 = - u_0 \cos kx \cos nt \dots\dots\dots\dots\dots(41),$$

so that the loops correspond to $kx = 0$, π, 2π, ..., and the nodes correspond to $kx = \frac{1}{2}\pi$, $\frac{3}{2}\pi$, ...

The steady motion represented by (39), (40) is of a very simple character. It consists of a series of vortices periodic with respect to x in the distance $\frac{1}{2}\lambda$. From (40) it appears that v is positive at the nodes and negative at the loops, vanishing of course in each case both at the wall $y = 0$ and at the plane of symmetry $y = y_1$.

Fig. 65.

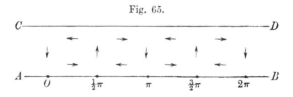

In the figure AB represents the wall, CD the plane of symmetry, and the directions of motion in the vortices are indicated by arrows. It is especially to be remarked that the velocity of the vortical motion is independent of μ', so that this effect is not to be obviated by taking the viscosity infinitely small. In that way the tendency to generate the vortices may indeed be diminished, but in the same proportion the maintenance of the vortices is facilitated, so that when the motion has reached a final state the vortices are as important with a small as with a large viscosity. The fact that when viscosity is neglected from the first no such vortices make their appearance in the solution shews what extreme care is required in dealing with problems relating to the behaviour of slightly viscous fluid in contact with solid bodies.

In estimating the mean motion to the second order there is another point to be considered which has not yet been touched upon. The values of u_1 and v_1 in (24) are, it is true, strictly periodic, but the same property does not attach to the motions thereby defined of the particles of the fluid. In our notation u is not the velocity of any individual particle of the fluid, but of the particle, whichever it may be, that at the moment under consideration occupies the position x, y, (§ 237). If $x + \xi$, $y + \eta$ define the actual position at time t of the particle whose mean position during several vibrations is (x, y), then the actual velocities of the particle at time t are, not u_1, v_1, but

$$u_1 + \frac{du_1}{dx}\xi + \frac{du_1}{dy}\eta, \qquad v_1 + \frac{dv_1}{dx}\xi + \frac{dv_1}{dy}\eta :$$

and thus the mean velocity parallel to x is not necessarily zero, but is equal to the mean value of

$$\xi\, du_1/dx + \eta\, du_1/dy \dots\dots\dots\dots\dots\dots(42),$$

in which again

$$\xi = \int u_1 dt, \qquad \eta = \int v_1 dt \dots\dots\dots\dots\dots (43).$$

In the present case the mean value of (42) is

$$- u_0{}^2/4a \,.\, \sin 2kx\, e^{-\beta y} (e^{-\beta y} - \cos \beta y) \dots\dots\dots (44),$$

which is to be regarded as an addition to (37). However, at a short distance from the wall (44) may be neglected, so that (39) remains adequate.

We have seen that the width of the direct current along the wall $y = 0$ is $\cdot 423\, y_1$, and that of the return current, measured up to the plane of symmetry, is $\cdot 577\, y_1$. The ratio of these widths is not altered by the inclusion of the second half of the channel lying beyond the plane of symmetry; so that the direct current is distinctly narrower than the return current. This disproportion will be increased in the case of a tube of circular section. The point under consideration depends in fact only upon a complementary function analogous to (32), and is so simple that it may be worth while briefly to indicate the steps of the calculation.

The equation for ψ_2 is[1]

$$\left(\frac{d^2}{dr^2} - \frac{1}{r}\frac{d}{dr} - 4k^2\right)^2 \psi_2 = 0 \dots\dots\dots\dots\dots (45);$$

but, if we suppose that the radius of the tube is small in comparison with λ, k^2 may be omitted. The general solution is

$$\psi_2 = \{A + Br^2 + B'r^2 \log r + Cr^4\} \sin 2kx \dots\dots (46),$$

so that

$$u_2 = d\psi_2/r\, dr = \{2B + B'(2 \log r + 1) + 4Cr^2\} \sin 2kx \dots (47),$$

---

[1] Stokes, *Trans. Camb. Phil. Soc.*, vol. IX. 1856; Basset's *Hydrodynamics*, § 485.

whence $B' = 0$, by the condition at $r = 0$. Again,

$$v_2 = - d\psi_2/r\,dx = - 2k\{Ar^{-1} + Br + Cr^3\}\cos 2kx \dots (48),$$

whence $A = 0$.

We may therefore take

$$\left.\begin{aligned}u_2 &= \{2B + 4Cr^2\} \sin 2kx\\ v_2 &= - 2k\{Br + Cr^3\}\cos 2kx\end{aligned}\right\} \quad \dots\dots\dots\dots (49).$$

If, as in (40), $v_2 = 0$, when $r = R$, $B + CR^2 = 0$, and

$$u_2 = 2C(2r^2 - R^2) \sin 2kx \dots\dots\dots\dots\dots (50).$$

Thus u_2 vanishes, when

$$r = R/\sqrt{2} = \cdot707\,R, \qquad R - r = \cdot293\,R.$$

The direct current is thus limited to an annulus of thickness $\cdot293\,R$, the return current occupying the whole interior and having therefore a diameter of $2 \times \cdot707\,R$, or $1\cdot414\,R$.

CHAPTER XX.

353. THE subject of the present chapter is the behaviour of inviscid incompressible fluid vibrating under the action of gravity and capillary force, more especially the latter. In virtue of the first condition we may assume the existence of a velocity-potential (ϕ), which by the second condition must satisfy (§ 241) the equation

$$\nabla^2 \phi = 0 \dots\dots\dots\dots\dots\dots(1),$$

throughout the interior of the fluid. In terms of ϕ the equation for the pressure is (§ 244)

$$\delta p / \rho = R - d\phi/dt \dots\dots\dots\dots(2),$$

if we assume that the motion is so small that its square may be neglected. The only impressed force, acting upon the interior of the fluid, which we have occasion to consider is that due to gravity; so that, if z be measured vertically downwards, $R = gz$, and (2) becomes

$$\delta p / \rho = gz - d\phi/dt \dots\dots\dots\dots(3).$$

Let us now consider the propagation of waves upon the horizontal surface ($z = 0$) of water, or other liquid, of uniform depth l, limiting our attention to the case of two dimensions, where the motion is confined to the plane zx. The general solution of (1) under this condition, and that the motion is proportional to e^{ikx}, is

$$\phi = e^{ikx} (A e^{kz} + B e^{-kz});$$

or, with regard to the condition that the vertical velocity must vanish at the bottom where $z = l$,

$$\phi = C \cosh k (z - l) . e^{ikx} \dots\dots\dots\dots(4).$$

If the motion be proportional also to e^{int}, and we throw away the imaginary part in (4), we get as the expression for waves propagated in the negative direction

$$\phi = C \cosh k (z - l) \, \cos (nt + kx) \dots\dots\dots\dots(5),$$

in which it remains to find the connection between n and k.

If h denote the elevation of the water surface at the point x, and T the constant tension, the pressure at the surface due to capillarity is $- T d^2 h/dx^2$, and (3) becomes

$$\frac{T}{\rho} \frac{d^2 h}{dx^2} = gh + \frac{d\phi}{dt} :$$

or, if we differentiate with respect to t and remember that $dh/dt = - d\phi/dz$,

$$\frac{T}{\rho} \frac{d^3 \phi}{dx^2 dz} = g \frac{d\phi}{dz} - \frac{d^2 \phi}{dt^2} \dots\dots\dots\dots\dots(6).$$

Applying this equation to (5) where $z = 0$, we get for the velocity of propagation

$$V^2 = n^2/k^2 = (g/k + Tk/\rho) \tanh kl \dots\dots\dots\dots(7)[1],$$

where, as usual,

$$k = 2\pi/\lambda \dots\dots\dots\dots\dots\dots (8).$$

In many cases the depth of liquid is sufficient to allow us to take $\tanh kl = 1$; and then

$$V^2 = \frac{g\lambda}{2\pi} + \frac{2\pi T}{\rho\lambda} \dots\dots\dots\dots\dots (9)[2],$$

gives the relation between V and λ. When λ is great, the waves move mainly under gravity and with velocity approximately equal to $\sqrt{(g\lambda/2\pi)}$. On the other hand, when λ is small, the influence of capillarity becomes predominant and the expression for the velocity assumes the form

$$V = \sqrt{(2\pi T/\rho\lambda)}\dots\dots\dots\dots\dots\dots(10).$$

Since $\lambda = V\tau$, the relation between wave-length and periodic time corresponding to (10) is

$$\lambda^3/\tau^2 = 2\pi T/\rho \dots\dots\dots\dots\dots\dots(11).$$

Except as regards the numerical factor, the relations (10), (11) can be deduced by considerations of dimensions from the fact that the dimensions of T are those of a force divided by a line.

---

[1] A more general formula for the velocity of propagation (n/k) at the interface between two liquids is given in (7) § 365.

[2] Kelvin, *Phil. Mag.* vol. XLII. p. 375, 1871.

If we inquire what values of λ correspond to a given value of V, we obtain from the quadratic (9)

$$\lambda = \pi V^2/g \pm \pi/g \cdot \sqrt{(V^4 - 4Tg/\rho)} \ldots\ldots\ldots\ldots (12),$$

which shews that for no wave-length can V be less than V_0, where

$$V_0 = (4Tg/\rho)^{\frac{1}{4}} \ldots\ldots\ldots\ldots\ldots\ldots (13).$$

The values of λ and of τ corresponding to the minimum velocity are given by

$$\lambda_0 = 2\pi (T/g\rho)^{\frac{1}{2}}, \qquad \tau_0 = 2\pi (T/4g^3\rho)^{\frac{1}{4}} \ldots\ldots\ldots (14).$$

If we take in C.G.S. measure $g = 981$, and for water $\rho = 1$, $T = 76$, we have $V_0 = 23\cdot1$, $\lambda_0 = 1\cdot71$, $1/\tau = 13\cdot6$.

The accompanying table gives a few corresponding values of wave-length, velocity, and frequency in the neighbourhood of the critical point :—

Wave-length	·5	1·0	1·7	2·5	3·0	5·0
Velocity	31·5	24·7	23·1	23·9	24·9	29·5
Frequency	63·0	24·7	13·6	9·6	8·3	5·9

A comparison of Kelvin's formula (9) with observation has been effected by Matthiessen[1], the ripples being generated by touching the surface of the various liquids with dippers attached to vibrating forks of known pitch. Among the liquids tried were water, mercury, alcohol, ether, bisulphide of carbon; and the agreement was found to be satisfactory. The observations include frequencies as high as 1832, and wave-lengths as small as ·04 cm.

Somewhat similar experiments have been carried out by the author[2] with the view of determining T by a method independent of any assumption respecting angles of contact between fluid and solid, and admitting of application to surfaces purified to the utmost from grease. In order to see the waves well, the light was made intermittent in a period equal to that of the waves (§ 42), and Foucault's optical method was employed for rendering visible small departures from truth in plane or spherical reflecting

[1] *Wied. Ann.* vol. XXXVIII. p. 118, 1889.

[2] On the Tension of Water Surfaces, clean and contaminated. *Phil. Mag.* vol. XXX. p. 386, 1890.

surfaces. From the measured values of τ and λ, T may be determined by (11), corrected, if necessary, for any small effect of gravity. The values thus found were for clean water 74·0 C.G.S., for a surface greasy to the point where camphor motions nearly cease 53·0, for a surface saturated with olive-oil 41·0, and for one saturated with oleate of soda 25·0. It should be remembered that the tension of contaminated surfaces is liable to variations dependent upon the extension which has taken place, or is taking place; but it is not necessary for the purposes of this work to enter further upon the question of "superficial viscosity."

354. Another way of generating capillary waves, or crispations as they were termed by Faraday, depends upon the principle discussed in § 68 b. If a glass plate, held horizontally and made to vibrate as for the production of Chladni's figures, be covered with a thin layer of water or other mobile liquid, the phenomena in question may be readily observed[1]. Over those parts of the plate which vibrate sensibly the surface is ruffled by minute waves, the degree of fineness increasing with the frequency of vibration. The same crispations are observed upon the surface of liquid in a large wine-glass or finger-glass which is caused to vibrate in the usual manner by carrying the moistened finger round the circumference (§ 234). All that is essential to the production of crispations is that a body of liquid with a free surface be constrained to execute a vertical vibration. It is indifferent whether the origin of the motion be at the bottom, as in the first case, or, as in the second, be due to the alternate advance and retreat of a lateral boundary, to accommodate itself to which the neighbouring surface must rise and fall.

More than sixty years ago the nature of these vibrations was examined by Faraday[2] with great ingenuity and success. The conditions are simplest when the motion of the vibrating horizontal plate on which the liquid is spread is a simple up and down motion without rotation. To secure this Faraday attached the plate to the centre of a strip of glass or lath of deal, supported at the nodes, and caused to vibrate by friction. Still more convenient is a large iron bar, maintained in vibration electrically, to which the plate may be attached by cement.

---

[1] On the Crispations of Fluid resting upon a Vibrating Support. *Phil. Mag.* vol. XVI. p. 50, 1883.

[2] *Phil. Trans.* 1831, p. 299.

The vibrating liquid standing upon the plate presents appearances which at first are rather difficult to interpret, and which vary a good deal with the nature of the liquid in respect of transparency and opacity, and with the incidence of the light. The vibrations are too quick to be followed by the eye; and thus the effect observed is an average, due to the superposition of an indefinite number of elementary impressions corresponding to the various phases.

If the plate be rectangular, the motion of the liquid consists of two sets of stationary vibrations superposed, the ridges and furrows of the two sets being perpendicular to one another and usually parallel to the edges of the plate. Confining our attention for the moment to one set of stationary waves, let us consider what appearance it might be expected to present. At one moment the ridges form a set of parallel and equidistant lines, the interval being λ. Midway between these are the lines which represent at that moment the position of the furrows. After the lapse of a $\frac{1}{4}$ period the surface is flat; after another $\frac{1}{4}$ period the ridges and furrows are again at their maximum developement, but the positions are exchanged. Now, since only an average effect can be perceived, it is clear that no distinction is recognizable between the ridges and the furrows, and that the observed effect must be periodic within a distance equal to $\frac{1}{2}\lambda$. If the liquid on the plate be rendered moderately opaque by addition of aniline blue, and be seen by diffused transmitted light, the lines of ridge and furrow will appear bright in comparison with the intermediate nodal lines where the normal depth is preserved throughout the vibration. The gain of light when the thickness is small will, in accordance with the law of absorption, outweigh the loss of light which occurs half a period later when the furrow is replaced by a ridge.

The actual phenomenon is more complicated in consequence of the coexistence of the two sets of ridges and furrows in perpendicular directions (x, y). In the adjoining figure (Fig. 66) the thick lines represent the ridges, and the thin lines the furrows, of the two systems at a moment of maximum excursion. One quarter period later the surface is flat, and one half period later the ridges and furrows are interchanged. The places of maximum elevation and depression are the intersections of the thick lines with one another and of the thin lines with one another, places not distin-

guishable by ordinary vision. They appear like holes in the sheet
of colour. The nodal lines where the normal depth of colour is
preserved throughout the vibration are shewn dotted; they are
inclined at 45°, and pass through the intersections of the thin
lines with the thick lines. The pattern is recurrent in the

Fig. 66.

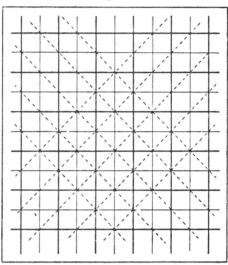

directions both of x and y, and in each case with an interval
equal to the real wave-length (λ). The distance between the
bright spots measured parallel to x or y is thus λ; but the
shortest distance between these spots is in directions inclined at
45°, and is equal to $\lambda/\sqrt{2}$.

As in all similar cases, these stationary waves may be resolved
into their progressive components by a suitable motion of the eye.
Consider, for example, the simple set of waves represented by

$$2 \cos kx \cos nt = \cos (nt + kx) + \cos (nt - kx).$$

This is with reference to an origin fixed in space. But let us
refer the phenomenon to an origin moving forward with the velocity
(n/k) of the waves, so as to obtain the impression that would be
produced upon the eye, or in a photographic camera, carried
forward in this manner. Writing $kx' + nt$ for kx, we get

$$\cos (kx' + 2nt) + \cos kx'.$$

Now the average effect of the first term is independent of x', so that what is seen is simply that set of progressive waves which moves with the eye.

In order to see the progressive waves it is not necessary to move the head as a whole, but only to turn the eye as when we follow the motion of a real object. To do this without assistance is not very easy at first, especially if the area of the plate be somewhat small. By moving a pointer at various speeds until the right one is found, the eye may be guided to do what is required of it; and after a few successes repetition becomes easy.

Faraday's assertion that the waves have a period double that of the support has been disputed, but it may be verified in various ways. Observation by stroboscopic methods is perhaps the most satisfactory. The violence of the vibrations and the small depth of the liquid interfere with an accurate calculation of frequency on the basis of the observed wave-length. The theory of vibrations in the sub-octave has already been considered (§ 68 *b*).

355. Typical stationary waves are formed by the superposition of equal positive and negative progressive waves of like frequency. If the one set be derived from the other by reflection, the equality of frequencies is secured automatically; but if the two sets of waves originate in different sources, the unison is a matter of adjustment, and a question arises as to the effect of a slight error. We may take as the expression for the two sets of progressive waves of equal amplitude and of approximately equal frequency

$$\cos(kx - nt) + \cos(k'x + n't),$$

or, which is the same,

$$2\cos\left\{\tfrac{1}{2}(k+k')x + \tfrac{1}{2}(n'-n)t\right\} \times \cos\left\{\tfrac{1}{2}(k'-k)x + \tfrac{1}{2}(n'+n)t\right\}$$
$$\dots\dots\dots(1).$$

If $n' = n$, $k' = k$, the waves are absolutely stationary; but we have now to interpret (1) when $(n'-n)$, $(k'-k)$ are merely small.

The position at any time t of the crests and hollows of the nearly stationary waves represented by (1) is given by

$$\tfrac{1}{2}(k+k')x + \tfrac{1}{2}(n'-n)t = m\pi \dots\dots\dots(2),$$

where m is an integer. The velocity of displacement U is accordingly

$$U = (n - n')/(k + k'),$$

or approximately

$$U = (n - n')/2k \quad \dots\dots\dots\dots\dots\dots(3)[1],$$

from which it appears that in every case the shifting takes place in the direction of waves of higher pitch, or towards the source of graver pitch. If V be the velocity (n/k) of propagation of the progressive waves, (3) may be written

$$U/V = (n - n')/2n \quad \dots\dots\dots\dots\dots\dots (4).$$

The slow travel under these circumstances of the places where the maximum displacements occur is a general phenomenon, not dependent upon the peculiarities of any particular kind of waves; but the most striking example is that afforded by capillary waves and described by Lissajous[2]. In his experiment two nearly unisonant forks touch the surface of water so as to form approximately stationary waves in the region between the points of contact. Since the crests and troughs cannot be distinguished, the pattern seen has an apparent wave-length half that of the real waves, and it travels slowly towards the graver fork. A frequency of about 50 will be found suitable for convenient observation.

If the waves be aerial, there is no difference of velocity; but (4) still holds good, and gives the rate at which the ear must travel in order to remain continually in a loop or in a node.

356. One of the best opportunities for the examination of capillary waves occurs when they are reduced to rest by a contrary movement of the water. Waves of this kind are sometimes described as standing waves, and they may usually be observed when the uniform motion of a stream is disturbed by obstacles. Thus when the surface is touched by a small rod, or by a fishing-line, or is displaced by the impact of a gentle stream of air from a small nozzle, a beautiful pattern is often displayed, stationary with respect to the obstacle. This was described and figured by Scott Russell[3], who remarked that the purity of the water had much to do with the extent and range of the phenomenon. On the up-stream side of the obstacle the wave-length is short, and, as was first clearly shewn by Kelvin, the force governing the vibra-

---

[1] *Phil. Mag.* vol. XVI. p. 57, 1883.

[2] *Compt. Rend.* vol. LXVII. p. 1187, 1868.

[3] *Brit. Ass. Rep.* 1844, p. 375, Plate 57. See also Poncelet, *Ann. d. Chim.* vol. XLVI. p. 5, 1831.

tions is principally cohesion. On the down-stream side the waves are longer and are governed principally by gravity. Both sets of waves move with the same velocity relatively to the water (§ 353); namely, that required in order that they may maintain a fixed position relatively to the obstacle. The same condition governs the velocity and therefore the wave-lengths of those parts of the pattern where the fronts are oblique to the direction of motion. If the angle between this direction and the normal to the wave-front be called θ, the velocity of propagation must be equal to $v_0 \cos \theta$, where v_0 represents the velocity of the water.

If v_0 be less than 23 cm. per sec., no wave-pattern is possible, for no waves can then move over the surface so slowly as to maintain a stationary position with respect to the obstacle. When v_0 exceeds 23 cm. per sec., a pattern is formed; but the angle θ has a limit defined by $v_0 \cos \theta = 23$, and the curved wave-front has a corresponding asymptote.

It would lead us too far to go further into the matter here, but it may be mentioned that the problem in two dimensions admits of analytical treatment[1], and that the solution explains satisfactorily one of the peculiar features of the case, namely, the limitation of the smaller capillary waves to the up-stream side, and of the larger (gravity) waves to the down-stream side of the obstacle.

357. A large class of phenomena, interesting not only in themselves but also as throwing light upon others yet more obscure, depend for their explanation upon the transformations undergone by a cylindrical body of liquid when slightly displaced from its equilibrium configuration and then left to itself. Such a cylinder is formed when liquid issues under pressure through a circular orifice, at least when gravity may be neglected; and the behaviour of the jet, as studied experimentally by Savart, Magnus, Plateau and others, is substantially independent of the forward motion common to all its parts. It will save repetition and be more in accordance with the general character of this work if we commence our investigation with the theory of an infinite cylinder of liquid, considered as a system in equilibrium under the action

---

[1] On the form of Standing Waves on the Surface of Running Water. *Proc. Lond. Math. Soc.* vol. xv. p. 69, 1883.

of the capillary force. With a solution of this mechanical problem most of the experimental results will easily be connected.

Taking cylindrical coordinates z, r, ϕ, the equation of the slightly disturbed surface may be written

$$r = a_0 + f(\phi, z) \dots\dots\dots\dots\dots\dots(1),$$

in which $f(\phi, z)$ is always a small quantity. By Fourier's theorem the arbitrary function f may be expanded in a series of terms of the type $\alpha_n \cos n\phi \cos kz$; and, as we shall see in the course of the investigation, each of these terms may be considered independently of the others. Either cosine may be replaced by a sine; and the summation extends to all positive values of k and to all positive integral values of n, zero included.

During the motion the quantity a_0 does not remain absolutely constant; its value must be determined by the condition that the enclosed *volume* is invariable. Now for the surface

$$r = a_0 + \alpha_n \cos n\phi \cos kz \dots\dots\dots\dots (2),$$

we find

$$\text{Volume} = \tfrac{1}{2} \iint r^2 \, d\phi \, dz = z \left(\pi a_0^2 + \tfrac{1}{4} \pi \alpha_n^2 \right);$$

so that, if a denote the radius of the section of the undisturbed cylinder,

$$a^2 = a_0^2 + \tfrac{1}{4} \alpha_n^2,$$

whence approximately

$$a_0 = a \left(1 - \tfrac{1}{8} \alpha_n^2 / a^2 \right) \dots\dots\dots\dots\dots (3).$$

This holds good when $n = 1, 2, 3 \dots$ If $n = 0$, (2) gives in place of (3)

$$a_0 = a \left(1 - \tfrac{1}{4} \alpha_0^2 / a^2 \right) \dots\dots\dots\dots\dots (4).$$

The potential energy of the system in any configuration, due to the capillary force, is proportional simply to the surface. Now in (2)

$$\text{Surface} = \iint \left\{ 1 + \left(\frac{dr}{dz} \right)^2 + \left(\frac{dr}{r d\phi} \right)^2 \right\}^{\frac{1}{2}} r \, d\phi \, dz$$

$$= z \left\{ 2\pi a_0 + \tfrac{1}{4} \pi k^2 \alpha_n^2 a + \tfrac{1}{4} \pi n^2 \alpha_n^2 / a \right\};$$

so that by (3), if σ denote the surface corresponding upon the average to unit of length,

$$\sigma = 2\pi a + \tfrac{1}{4} \pi \left(k^2 a^2 + n^2 - 1 \right) \alpha_n^2 / a \dots\dots\dots\dots (5).$$

The potential energy due to capillarity, estimated per unit length and from the configuration of equilibrium, is accordingly

$$P = \tfrac{1}{4}\pi T \left(k^2 a^2 + n^2 - 1\right) \alpha_n^2 / a \ \dots\dots\dots\dots (6),$$

T denoting, as usual, the superficial tension.

In (6) it is supposed that k and n are not zero. If k be zero, (6) requires to be doubled in order to give the potential energy corresponding to

$$r = a_0 + \alpha_n \cos n\phi \ \dots\dots\dots\dots\dots (7):$$

and again, if n be zero, we are to take

$$P = \tfrac{1}{2}\pi T \left(k^2 a^2 - 1\right) \alpha_0^2 / a \ \dots\dots\dots\dots (8),$$

corresponding to

$$r = a_0 + \alpha_0 \cos kz \dots\dots\dots\dots\dots(9).$$

From (6) it appears that when n is unity or any greater integer, the value of P is positive, shewing that for all displacements of these kinds the original equilibrium is stable. For the case of displacements symmetrical about the axis ($n = 0$), we see from (8) that the equilibrium is stable or unstable according as ka is greater or less than unity, i.e. according as the wave-length $(2\pi/k)$ of the symmetrical deformation is less or greater than the circumference of the cylinder, a proposition first established by Plateau.

If the expression for r in (2) involve a number of terms with various values of n and k, and with arbitrary substitution of sines for cosines, the corresponding expression for P is found by simple addition of the expressions relating to the component terms, and it contains the squares only (and not the products) of the quantities α.

We have now to consider the kinetic energy of the motion. Since the fluid is supposed to be inviscid, there is a velocity-potential ψ, and this in virtue of the incompressibility satisfies Laplace's equation. Thus, (4) § 241,

$$\frac{d^2\psi}{dr^2} + \frac{1}{r}\frac{d\psi}{dr} + \frac{1}{r^2}\frac{d^2\psi}{d\phi^2} + \frac{d^2\psi}{dz^2} = 0,$$

or, if in order to correspond with (2) we assume that the variable part is proportional to $\cos n\phi \cos kz$,

$$\frac{d^2\psi}{dr^2} + \frac{1}{r}\frac{d\psi}{dr} - \left(\frac{n^2}{r^2} + k^2\right)\psi = 0\dots\dots\dots\dots(10).$$

The solution of (10) under the condition that there is no introduction or abstraction of fluid along the axis of symmetry is § 200

$$\psi = \beta_n J_n (ikr) \cos n\phi \cos kz \dots\dots\dots\dots (11).$$

The constant β_n is to be found from the condition that the radial velocity when $r = a$ coincides with that implied in (2). Thus

$$ik\beta_n J_n{}' (ika) = d\alpha_n/dt \dots\dots\dots\dots (12).$$

If ρ be the density, the kinetic energy of the motion is by Green's theorem (2) § 242

$$\tfrac{1}{2}\rho \iint [\psi \, d\psi/dr]_{r=a} \, a\,d\phi\,dz = \tfrac{1}{4}\pi\rho z \,.\, ika \,.\, J_n (ika) \, J_n{}' (ika) \,.\, \beta_n{}^2 \,;$$

so that by (12), if K denote the kinetic energy per unit of length,

$$K = \tfrac{1}{4}\pi\rho a^2 \frac{J_n (ika)}{ika \,.\, J_n{}' (ika)} \left(\frac{d\alpha_n}{dt}\right)^2 \dots\dots\dots\dots (13).$$

When $n = 0$, we must take in place of (13)

$$K = \tfrac{1}{2}\pi\rho a^2 \frac{J_0 (ika)}{ika \,.\, J_0{}' (ika)} \left(\frac{d\alpha_0}{dt}\right)^2 \dots\dots\dots\dots (14).$$

The most general value of K is to be found by simple summation from the particular values expressed in (13), (14). Since the expressions for P and K involve the squares only, and not the products, of the quantities α, $d\alpha/dt$, and the corresponding quantities in which cosines are replaced by sines, it follows that the motions represented by (2) take place in perfect independence of one another, so long as the whole displacement is small.

For the free motion we get by Lagrange's method from (6), (13)

$$\frac{d^2\alpha_n}{dt^2} + \frac{T}{\rho a^3} \frac{ika \,.\, J_n{}' (ika)}{J_n (ika)} (n^2 + k^2 a^2 - 1) \, \alpha_n = 0 \dots\dots(15),$$

which applies without change to the case $n = 0$. Thus, if α_n varies as $\cos (pt - \epsilon)$,

$$p^2 = \frac{T}{\rho a^3} \frac{ika \,.\, J_n{}' (ika)}{J_n (ika)} (n^2 + k^2 a^2 - 1) \dots\dots\dots (16)[1],$$

giving the frequency of vibration in the cases of stability.

If $n = 0$, and $ka < 1$, the solution changes its form. If we suppose that α_0 varies as $e^{\pm qt}$,

$$q^2 = \frac{T}{\rho a^3} \frac{ika \,.\, J_0{}' (ika)}{J_0 (ika)} (1 - k^2 a^2) \dots\dots\dots\dots(17).$$

[1] *Proc. Roy. Soc.* vol. xxix. p. 94, 1879.

When n is greater than unity, the circumstances are usually such that the motion is approximately in two dimensions only. We may then advantageously introduce into (16) the supposition that ka is small. In this way we get, (5) § 200,

$$p^2 = n (n^2 - 1 + k^2 a^2) \frac{T}{\rho a^3} \left[1 + \frac{k^2 a^2}{n (2n + 2)} \right] \dots \dots (18),$$

or, if ka be neglected altogether,

$$p^2 = (n^3 - n) \frac{T}{\rho a^3} \quad \dots \dots \dots \dots \dots (19),$$

the two-dimensional formula. When $n = 1$, there is no force of restitution for a displacement purely in two dimensions. If λ denote the wave-length measured round the circumference, $\lambda = 2\pi a / n$. Thus in (19), if n and a are infinite,

$$p^2 = \frac{T}{\rho} \left(\frac{2\pi}{\lambda} \right)^3 \dots \dots \dots \dots \dots \dots (20),$$

in agreement with the theory of capillary waves upon a plane surface. Compare (7) § 353. A similar conclusion may be reached by the consideration of waves whose length is measured axially. Thus, if $\lambda = 2\pi/k$, and $a = \infty$, $n = 0$, (16) reduces to (20) in virtue of the relation, §§ 302, 350,

$$\text{Limit}_{z=\infty} \; i J_0{}'(iz)/J_0(iz) = 1.$$

358. Many years ago Bidone investigated by experiment the behaviour of jets of water issuing horizontally under considerable pressure from orifices in thin plates. If the orifice be circular, the section of the jet, though diminished in area, retains the circular form. But if the orifice be not circular, curious transformations ensue. The peculiarities of the orifice are exaggerated in the jet, but in an inverted manner. Thus in the case of an elliptical aperture, with major axis horizontal, the sections of the jet taken at increasing distances gradually lose their ellipticity until at a certain distance the section is circular. Further out the section again assumes ellipticity, but now with major axis vertical, and (in the circumstances of Bidone's experiments) the ellipticity increases until the jet is reduced to a flat sheet in the vertical plane, very broad and thin. This sheet preserves its continuity to a considerable distance (e.g. six feet) from the orifice, where finally it is penetrated by air. If the orifice be in the form of an equilateral triangle, the jet resolves itself into three sheets disposed symmetrically round the axis, the planes of the sheets being perpendicular to the sides of the orifice; and in like manner if the aperture be a regular polygon of any number of sides, there are developed a corresponding number of sheets perpendicular to the sides of the polygon.

Bidone explains the formation of these sheets by reference to simpler cases of meeting streams. Thus equal jets, moving in the same straight line with equal and opposite velocities, flatten themselves into a disc situated in the perpendicular plane. If the axes of the jets intersect obliquely, a sheet is formed symmetrically in the plane perpendicular to that of the impinging jets. Those portions of a jet which proceed from the outlying parts of a single unsymmetrical orifice are regarded as behaving in some degree like independent meeting streams.

In many cases, especially when the orifices are small and the pressures low, the extension of the sheets reaches a limit. Sections taken at still greater distances from the orifice shew a gradual gathering together of the sheets, until a compact form is regained similar to that at the first contraction. Beyond this point, if the jet retains its coherence, sheets are gradually thrown out again, but in directions bisecting the angles between the directions of the former sheets. These sheets may in their turn reach a limit of developement, again contract, and so on. The forms assumed in the case of orifices of various shapes including the rectangle, the equilateral triangle, and the square, have been carefully investigated and figured by Magnus. Phenomena of this kind are of every day occurrence, and may generally be observed whenever liquid falls from the lip of a moderately elevated vessel.

As was first suggested by Magnus[1] and Buff[2], the cause of the contraction of the sheets after their first developement is to be found in the capillary force, in virtue of which the fluid behaves as if enclosed in an envelope of constant tension; and the recurrent form of the jet is due to *vibrations* of the fluid column about the circular figure of equilibrium, superposed upon the general progressive motion. Since the phase of the vibration, initiated during passage through the aperture, depends upon the

[1] Hydraulische Untersuchungen, *Pogg. Ann.* vol. xcv, p. 1, 1855.

[2] *Pogg. Ann.* vol. c, p. 168, 1857.

time elapsed, it is always the same at the same point in space, and thus the motion is *steady* in the hydrodynamical sense, and the boundary of the jet is a fixed surface. Relatively to the water the waves here concerned are progressive, such as may be compounded of two stationary systems, and they move up stream with a velocity equal to that of the water so as to maintain a fixed position relatively to external objects, § 356.

If the departure from the circular form be small, the vibrations are those considered in § 357, of which the frequency is determined by equations (16), (18), (19). The distance between consecutive corresponding points of the recurrent figure, or, as it may be called, the wave-length of the figure, is the space travelled over by the stream during one vibration. Thence results a relation between wave-length and period. If the circumference of the jet be small in comparison with the wave-length, so that (19) § 357 is applicable, the periodic time is independent of the wave-length; and then the wave-length is directly proportional to the velocity of the jet, or to the square root of the pressure. The elongation of wave-length with increasing pressure was remarked by Bidone and by Magnus, but no definite law was arrived at.

In the experiments of the author[1] upon elliptical, triangular, and square apertures, the jets were caused to issue horizontally in order to avoid the complications due to gravity; and, if the pressure were not too high, the law above stated was found to be verified. At higher pressures the observed wave-lengths had a marked tendency to increase more rapidly than the velocity of the jet. This result points to a departure from the law of isochronous vibration. Strict isochronism is only to be expected when vibrations are infinitely small, that is when the section of the jet never deviates more than infinitesimally from the circular form. Under the high pressures in question the departures from circularity were very considerable, and there is no reason for expecting that such vibrations will be executed in precisely the same time as vibrations of infinitely small amplitude.

The increase of amplitude under high pressure is easily explained, inasmuch as the lateral velocities to which the vibrations are mainly due vary in direct proportion to the longitudinal velocity of the jet. Consequently the amplitude varies approxi-

---

[1] *Proc. Roy. Soc.* vol. XXIX, p. 71, 1879.

mately as the square root of the pressure, or as the wave-length. In general, the periodic time of a vibration is an even function of amplitude (§ 67); and thus, if h represent the head of liquid, the wave-length may be expected to be a function of h of the form $(M + Nh)\sqrt{h}$, where M and N are constants for a given aperture. It appears from experiment, and might perhaps have been expected, that N is here positive.

For a comparison with theory it is necessary to keep within the range of the law of isochronism; and it is convenient to employ in the calculations the area of the section of the jet in place of the mean radius. Thus, if $A = \pi a^2$, (19) § 357 may be written

$$p = \pi^{\frac{3}{4}} T^{\frac{1}{2}} \rho^{-\frac{1}{2}} A^{-\frac{3}{4}} \sqrt{(n^3 - n)} \dots\dots\dots\dots (1),$$

in which A is to be determined by experiments upon the rate of total discharge. For the case of water (§ 353) we may take in c.g.s. measure $T = 74$, $\rho = 1$; so that for the frequency of the gravest vibration ($n = 2$) we get from (1)

$$p/2\pi = 7{\cdot}91 A^{-\frac{3}{4}} \dots\dots\dots\dots\dots (2).$$

For a sectional area of one square centimetre there are thus about 8 vibrations per second. A pitch of 256 would correspond to a diameter of about one millimetre.

For the general value of n, we have

$$p/2\pi = 3{\cdot}23 A^{-\frac{3}{4}} \sqrt{(n^3 - n)} \dots\dots\dots\dots (3).$$

If h be the head of water to which the velocity of the jet is due and λ the wave-length,

$$\lambda = \frac{\sqrt{(2gh)} \cdot A^{\frac{3}{4}}}{3{\cdot}23 \sqrt{(n^3 - n)}} \dots\dots\dots\dots\dots (4).$$

In one experiment with an elliptical aperture ($n = 2$) the observed value of λ was $3{\cdot}95$ while the value calculated from (4) is $3{\cdot}93$. In the case of a triangular aperture ($n = 3$) the observed value of λ was $2{\cdot}3$ and the calculated was $2{\cdot}1$. Again the observed value for a square aperture ($n = 4$) was $1{\cdot}85$ and the calculated $1{\cdot}78$. The excess of the observed over the calculated values in the last two cases may perhaps have been due to excessive departure from the circular figure.

The general theory, unrestricted to small amplitudes, would doubtless involve great complications; but some information respecting it may be obtained with facility by the method of dimensions. If the *shape* of the orifice be given, λ may be regarded as a function of T, ρ, A, and H the pressure under which the jet escapes. Of these T is a force divided by a line, so that its dimensions are 1 in mass, 0 in length, and -2 in time; ρ is of dimensions 1 in mass, -3 in length, 0 in time; A is of dimensions 0 in mass, 2 in length, 0 in time; and finally H is of dimensions 1 in mass, -1 in length, and -2 in time. If we assume

$$\lambda \propto T^x \rho^y A^z H^u,$$

then $\quad x + y + u = 0, \quad -3y + 2z - u = 1, \quad -2x - 2u = 0,$

whence $\quad\quad u = -x, \quad y = 0, \quad z = \tfrac{1}{2}(1 - x);$

so that $\quad\quad\quad \lambda \propto A^{\frac{1}{2}} (TA^{-\frac{1}{2}} H^{-1})^x.$

The exponent x is here undetermined; and, since any number of terms with different values of x may occur simultaneously, all that we can infer is that λ is of the form

$$\lambda = A^{\frac{1}{2}} \cdot f(TA^{-\frac{1}{2}} H^{-1}),$$

or, if we prefer it,

$$\lambda = T^{-\frac{1}{2}} H^{\frac{1}{2}} A^{\frac{3}{4}} \cdot F(HA^{\frac{1}{2}} T^{-1}) \dots \dots \dots \dots (5),$$

where f and F are arbitrary functional symbols. Thus for a given liquid and shape of orifice there is complete dynamical similarity if the pressure be taken inversely proportional to the linear dimension. The simple case previously considered where the departures from circularity are small, and the vibrations take place approximately in two dimensions, corresponds to $F = \text{constant}$.

The method of determining T by observations upon λ is scarcely delicate enough to compete with others that may be employed for the same purpose when the tension is constant. But the possibility of thus experimenting upon surfaces which have been formed but a fraction of a second earlier is of considerable interest. In this way it may be proved with great ease that the tension of a soapy solution immediately after the formation of a free surface differs comparatively little from that of pure water, whereas when a few seconds have elapsed the difference becomes very great[1].

---

[1] On the Tension of Recently Formed Liquid Surfaces, *Proc. Roy. Soc.* vol. XLVII, p. 281, 1890.

Hitherto it has been supposed for the sake of simplicity that the jet after its issue from the nozzle is withdrawn from the action of gravity. If the direction of projection be vertically downwards, as is often convenient, the velocity of flow (v) continually increases, while at the same time the area of the section diminishes, the relation being $vA = \text{constant}$. But, so far as regards λ, the disturbance which thus ensues is less than might have been expected, for the changes in v and A compensate one another to a considerable extent. By (1)

$$\lambda \propto v/p \propto v^{\frac{1}{4}} \propto h^{\frac{1}{8}},$$

if h denote the whole difference of level between the surface of liquid in the reservoir and the place where λ is measured.

359. In § 358 the motion of the liquid is regarded as steady, every portion as in turn it passes the orifice being similarly affected. Under these circumstances no term corresponding to $n = 0$ can appear in the mathematical expressions; but it must not be forgotten that for certain disturbances of this type the cylindrical form is unstable and that therefore the jet cannot long preserve its integrity. The minute disturbances required to bring the instability into play are such as act differently at different moments of time, and have their origin in eddying motions of the fluid due to friction, and especially in vibration communicated to the nozzle and of such a character as to render the rate of discharge subject to a slight periodic variation. If v be the velocity of the jet and τ the period of the vibration, the cylindrical column issuing from a circular orifice is launched subject to a disturbance of wave-length (λ) equal to $v\tau$. If this wave-length exceed the circumference of the jet ($2\pi a$), the disturbance grows exponentially, until finally the column of liquid is divided into detached masses separated by the common interval λ, and passing a fixed point with velocity v and frequency $1/\tau$. Even though no regular vibration has access to the nozzle, the instability cannot fail to assert itself, and casual disturbances of a complex character will bring about disintegration. It will be convenient to discuss in the first place somewhat in detail the theory of the case of $n = 0$ in (16), (17) § 357, and then to consider its application to the beautiful phenomena described by Savart and to a large extent explained by Plateau.

If $ka = z$, and we introduce the notation of § 221 a, (17) § 357 becomes

$$q^2 = \frac{T}{\rho a^3} \frac{z I_1(z)}{I_0(z)} (1 - z^2)\dots\dots\dots\dots\dots(1).$$

In this equation $I_1(z)$ and $I_0(z)$ are both positive, so that as z decreases (or as λ increases) q first becomes real when $z = 1$. At this point instability commences, and at first the degree of instability is infinitely small. Also when z is very small, or λ is very great,

$$q^2 = \frac{T}{\rho a^3} \frac{z^2}{2}$$

ultimately, so that q is again small. For some value of z between 0 and 1, q is a maximum, and the investigation of this value is a matter of importance, because, as has already been shewn § 87, the unstable equilibrium will give way by preference in the mode so characterized.

The function to be made a maximum is

$$z (1 - z^2) I_1(z)/I_0(z) \dots\dots\dots\dots\dots (2),$$

or, expanded in powers of z,

$$\frac{1}{2}\left(z^2 - \frac{9}{8} z^4 + \frac{7}{2^4.3} z^6 - \frac{25}{2^{10}} z^8 + \frac{91}{2^{11}.3.5} z^{10} + \dots\right).$$

Hence, to find the maximum, we obtain on differentiation

$$1 - \frac{9}{4} z^2 + \frac{7}{2^4} z^4 - \frac{100}{2^{10}} z^6 + \frac{91}{2^{11}.3} z^8 + \dots = 0.$$

If the last terms be neglected, the quadratic gives $z^2 = \cdot 4914$. If this value be substituted in the small terms, the equation becomes

$$\cdot 98928 - \tfrac{9}{4}z^2 + \tfrac{7}{16}z^4 = 0,$$

whence
$$z^2 = \cdot 486, \qquad z = \cdot 679[1].$$

The values of expression (2), or of its square root, to which q is proportional, may be calculated from tables of I_0 and I_1, § 221 a. We have

z	$\{(2)\}^{\frac{1}{2}}$	z	$\{(2)\}^{\frac{1}{2}}$
0·0	·0000	0·6	·3321
0·1	·0703	0·7	·3433
0·2	·1382	0·8	·3269
0·3	·2012	0·9	·2647
0·4	·2567	1·0	·0000
0·5	·3015		

[1] On the Instability of Jets, *Proc. Lond. Math. Soc.* vol. x, p. 7, 1878.

From these values we find for the maximum by Lagrange's interpolation formula $z = ·696$, corresponding to

$$\lambda = 2\pi a/z = 4·51 \times 2a \dots\dots\dots\dots\dots(3).$$

Hence the maximum instability occurs when the wave-length of disturbance is about half as great again as that at which instability first commences.

Taking for water in C.G.S. units $T = 73$, $\rho = 1$, we get for the case of maximum instability

$$q^{-1} = \frac{(2a)^{\frac{3}{2}}}{73^{\frac{1}{2}}.2^{\frac{3}{2}} \times ·343} = ·120 \,(2a)^{\frac{3}{2}} \dots\dots\dots\dots (4).$$

This is the time in which the disturbance is multiplied in the ratio $e : 1$. Thus in the case of a diameter of one centimetre the disturbance is multiplied $2·7$ times in about $\frac{1}{8}$ second. If the disturbance be multiplied 1000 fold in time t, $qt = 3 \log_e 10 = 6·9$, so that $t = ·828 \,(2a)^{\frac{3}{2}}$. For example, if the diameter be one millimetre, the disturbance is multiplied 1000 fold in about $\frac{1}{40}$ second. In view of these estimates the rapid disintegration of a jet of water will not cause surprise.

The above theory of the instability of a cylindrical surface separating liquid from gas may be extended to meet the case where the liquid is outside and the gas, whose inertia is neglected, is inside the surface. This represents a jet of gas discharged under liquid; and it appears that the degree of maximum instability is even higher than before, and that it occurs when $\lambda = 6·48 \times 2a$[1]. But it is scarcely necessary for our purpose to pursue this part of the subject further.

360. The application of our mathematical results to actual jets presents no great difficulty. The disturbances, by which equilibrium is upset, are impressed upon the fluid as it leaves the aperture, and the continuous portion of the jet represents the distance travelled over during the time necessary to produce disintegration. Thus the length of the continuous portion necessarily depends upon the character of the disturbances in respect of amplitude and wave-length. It may be increased considerably, as Savart shewed[2], by a suitable isolation of the reservoir from

---

[1] On the Instability of Cylindrical Fluid Surfaces, *Phil. Mag.* vol. XXXIV, p. 177, 1892.

[2] *Ann. de Chimie*, LIII, p. 337, 1833.

tremors, whether due to external sources or to the impact of the jet itself in the vessel placed to receive it. Nevertheless it does not appear possible to carry the prolongation very far. Whether the residual disturbances are of external origin or are due to friction, or to some peculiarity of the fluid motion within the reservoir, has not been satisfactorily determined. On this point Plateau's explanations are not very clear, and he sometimes expresses himself as if the time of disintegration depended only upon the capillary tension without reference to initial disturbances at all.

Two laws were formulated by Savart with respect to the length of the continuous portion of a jet, and have been to a certain extent explained by Plateau[1]. For a given fluid and a given orifice the length is approximately proportional to the square root of the head. This follows at once from theory, if it can be assumed that the disturbances remain always of the same character, so that the *time* of disintegration is constant. When the head is given, Savart found the length to be proportional to the diameter of the orifice. From (4) § 359 it appears that the time in which a small disturbance is multiplied in a given ratio varies not as a, but as $a^{\frac{3}{2}}$. Again, when the fluid is changed, the time varies as $\rho^{\frac{1}{2}}T^{-\frac{1}{2}}$. But it may well be doubted whether the length of the continuous portion obeys any very simple laws, even when external disturbances are avoided as far as possible.

When a jet falls vertically downwards, the circumstances upon which its stability or instability depend are continually changing, more especially if the initial velocity be very small. The kind of disturbance to which the jet is most sensitive as it leaves the nozzle is one which impresses upon it undulations of length equal to about $4\frac{1}{2}$ times the initial diameter. But as the jet falls, its velocity increases, with consequent lengthening of the undulations, and its diameter diminishes, so that the degree of instability soon becomes much reduced. On the other hand, the kind of disturbance which will be effective in a later stage is altogether ineffective in the earlier stages. The change of conditions during fall has thus a protective influence, and the continuous part tends to become longer than would be the case were the velocity constant, the initial disturbances being unaltered.

[1] Statique experimentale et théorique des Liquides soumis aux seules forces moléculaires, Paris, 1873.

When the circumstances are such that the reservoir is influenced by the shocks due to the impact of the jet, the disintegration often assumes a complete regularity and is attended by a musical note (Savart). The impact of the regular series of drops, which at any moment strike the receiving vessel, determines the rupture into similar drops of the portion of the jet at the same moment passing the orifice. The pitch of the note, though not definite, cannot differ greatly from that which corresponds to the division of the column into wave-lengths of maximum instability; and in fact Savart found that the frequency was directly as the square root of the head, inversely as the diameter of the orifice, and independent of the nature of the fluid—laws which follow immediately from Plateau's theory.

From the observed pitch of the note due to a jet of given diameter, and issuing under a given head, the wave-length of the nascent divisions can be at once deduced. Reasoning from some observations of Savart, Plateau found in this way 4·38 as the ratio of the length of a division to the diameter of the jet. Now that the length of a division can be estimated *a priori*, it is preferable to reverse Plateau's calculation and to exhibit the frequency of vibration in terms of the other data of the problem. Thus

$$\text{frequency} = \frac{v}{4 \cdot 5 \, l \times 2a} \dots \dots \dots \dots \dots (1),$$

and in many cases, where the jet is not too fine, v may be replaced by $\sqrt{(2gh)}$ with sufficient accuracy.

But the most certain method of attaining complete regularity of resolution is to bring the reservoir under the influence of an external vibrator, whose pitch is approximately the same as that proper to the jet. Magnus[1] employed a Neef's hammer, attached to the frame which supported the reservoir. Perhaps an electrically maintained tuning-fork is still better. Magnus shewed that the most important part of the effect is due to the forced vibration of that side of the vessel which contains the orifice, and that but little of it is propagated through the air. With respect to the limits of pitch, Savart found that the note might be a fifth above, and more than an octave below, that proper to the jet. According to theory there is no well defined lower limit; while, on the other side the external vibration cannot be efficient if it tends to produce divisions

[1] *Pogg. Ann.* vol. cvi, p. 1, 1859.

whose length is less than the circumference of the jet. This gives for the interval defining the upper limit $\pi : 4\cdot51$, or about a fifth. In the case of Plateau's numbers ($\pi : 4\cdot38$) the discrepancy is a little greater.

361. The question of the influence of vibrations of low frequency is difficult to treat experimentally in consequence of the complications which arise from the almost universal presence of harmonic overtones. It is evident that the octave, for example, of the principal tone, though present in a very subordinate degree, may nevertheless be the more important agent of the two in determining the behaviour of the jet, if its pitch happen to lie in the neighbourhood of that of maximum instability. In my own experiments[1] tuning-forks were employed as sources of vibration, and in every case the behaviour of the jet on its horizontal course was examined not only by direct inspection, but also by the method of intermittent illumination (§ 42) so arranged that there was one view for each complete period of the phenomenon to be observed. Except when it was important to eliminate the octave as far as possible, the vibration was communicated to the reservoir through the table on which it stood. The forks were either screwed to the table and vibrated by a bow, or maintained electrically, the former method being adequate when only one fork was required at a time. The circumstances of the jet were such that the pitch of maximum sensitiveness, as determined by calculation, was 259, and that forming the transition between stability and instability 372.

With pitches varying downwards from 370 to about 180, the observed phenomena agreed perfectly with the unambiguous predictions of theory. From the point—decidedly below 370—at which a regular effect was first observed, there was always one drop for each complete vibration of the fork, and a single stream, each drop breaking away under precisely the same conditions as its predecessor. After passing 180 it becomes a question whether the octave of the fork's note may not produce an effect as well as the prime. If this effect be sufficient, the number of drops is doubled, and when the prime is very subordinate indeed, there is a double stream, alternate drops breaking away under different conditions and (under the action of gravity) taking sensibly

[1] *Proc. Roy. Soc.* vol. xxxiv, p. 133, 1882.

different courses. In these experiments the influence of the prime was usually sufficient to determine the number of drops, even in the neighbourhood of pitch 128. Sometimes, however, the octave became predominant and doubled the number of drops. When the octave is not strong enough actually to double the drops, it often produces an effect which is very apparent to an observer examining the transformation through the revolving holes. On one occasion a vigorous bowing of the fork, which favours the octave, gave at first a double stream, but this after a few seconds passed into a single one. Near the point of resolution those consecutive drops which ultimately coalesce as the fork dies down are connected by a ligament. If the octave is strong enough, this ligament subsequently breaks, and the drops are separated; otherwise the ligament draws the half-formed drops together, and the stream becomes single. The transition from the one state of things to the other could be watched with facility.

In order to get rid entirely of the influence of the octave a different arrangement was necessary. It was found that the desired result could be arrived at by holding a 128 fork in the hand over a resonator of the same pitch resting upon the table. The transformation was now quite similar in appearance to that effected by a fork of frequency 256, the only differences being that the drops were bigger and twice as widely spaced, and that the *spherule*, which results from the gathering together of the ligament, was much larger. We may conclude that the cause of the doubling of a jet by the sub-octave of the note natural to it is to be found in the presence of the second component from which hardly any musical notes are free.

When two forks of pitches 128 and 256 were sounded together, the single or double stream could be obtained at pleasure by varying the relative intensities. Any imperfection in the tuning is rendered very evident by the behaviour of the jet, which performs evolutions synchronous with the audible beats. This observation, which does not require the aid of the stroboscopic disc, suggests that the effect depends in some degree upon the relative phases of the two tones, as might be expected *a priori*. In some cases the influence of the sub-octave is shewn more in making the alternate drops unequal in magnitude than in projecting them into very different paths.

Returning now to the case of a single fork screwed to the table, it was found that as the pitch was lowered below 128, the double stream was regularly established. The action of the twelfth ($85\frac{1}{3}$) below the principal note demands special attention. At this pitch we might expect the first three components of a compound note to influence the result. If the third component were pretty strong, it would determine the number of drops, and the result would be a three-fold stream. In the case of a fork screwed to the table the third component of the note must be extremely weak if not altogether missing; but the second (octave) component is fairly strong, and in fact determined the number of drops ($190\frac{2}{3}$). At the same time the influence of the prime ($85\frac{1}{3}$) is sufficient to cause the alternate drops to pursue different paths, so that a double stream is observed.

By the addition of a 256 fork there was no difficulty in obtaining a triple stream; but it was of more interest to examine whether it were possible to reduce the double stream to a single one with only $85\frac{1}{3}$ drops per second. In order to secure as strong and as pure a fundamental tone as possible and to cause it to act upon the jet in the most favourable manner, the air space in the reservoir (an aspirator bottle) above the water was tuned to the note of the fork by sliding a plate of glass over the neck so as partially to cover it (§ 305). When the fork was held over the resonator thus formed, the pressure which expels the jet was rendered variable with a frequency of $85\frac{1}{3}$, and overtones were excluded as far as possible. To the unaided eye, however, the jet still appeared double, though on more attentive examination one set of drops was seen to be decidedly smaller than the other. With the revolving disc, giving about 85 views per second, the real state of the case was made clear. The smaller drops were the *spherules*, and the stream was single in the same sense as the streams given by pure tones of frequencies 128 and 256. The increased size of the spherule is of course to be attributed to the greater length of the ligament, the principal drops being now three times as widely spaced as when the jet is under the influence of the 256 fork.

With still graver forks screwed to the table the number of drops continued to correspond to the second component of the note. The double octave of the principal note (64) gave 128 drops per second, and the influence of the prime was so feeble that the duplicity of the stream was only just recognisable. Below 64 the observations were not carried, and even at this pitch attempts to attain a single stream of drops were unsuccessful.

362.   Savart's experiments upon this subject have been further developed by Mr C. A. Bell, who shewed that a jet may be made to play the part of a telephonic receiver[1]. The external vibrations may be conveyed to the nozzle through a string telephone (§ 156 a). An india rubber membrane, stretched over the upper end of a metal tube, receives the jet and communicates the vibration due to the varying impact to the cavity behind, with which the ear may be connected. The diameter and velocity of the jet require to be accommodated to the general character as to pitch of the sounds to be dealt with. " When the membrane is held close under the jet orifice, no sound will be audible in the ear-piece ; but as the receiving tube is gradually withdrawn along the jet path, a sound will be heard corresponding in pitch and quality to the disturbing sound—provided, of course, that the jet is at such pressure as to be capable of responding to all the higher tones to which the disturbing sound may owe its timbre. The intensity of this sound grows as the distance between jet orifice and membrane is increased. Finally, while the jet is still continuous above the membrane, a point of maximum intensity and purity of tone will be reached ; and if the membrane be carried beyond this point the sound heard will at first increase in loudness, becoming harsh in character at the same time, and at a still lower point will degenerate into an unmusical roar. In the latter case the jet will be seen to break above the membrane."

From the fact that small jets travelling at high speeds respond equally to sounds whose pitch varies over a wide range Mr Bell argues that Plateau's theory is inadequate, and he looks rather to vortex motion, dependent upon unequal velocity at the centre and at the exterior of the column, as the real cause of the phenomena presented by these jets.

As an example of a jet self-excited, the interrupter of § 235 r may be referred to. In this case the machinery by which the effect is carried back to the nozzle is electric. But ordinary mechanical devices answer the purpose equally well. The introduction of a resonator, such as the fork of § 235 r, or the telephone

---

[1] *Phil. Trans.* vol. 177, p. 383, 1886.

plate which may be made to take its place, if the telephone be brought in contact with the nozzle, gives greater regularity to the process, and usually allows also of a greater latitude in respect of pitch. It should not be forgotten that in all these cases of self-excitation a certain condition as to phase needs to be satisfied. If for instance in the interrupter of § 235 r, supposed to be working well, the platinum points be displaced through half the interval between consecutive drops, it is evident that the action will cease until some fresh accommodation is brought about.

363. When a small jet is projected upwards in a nearly vertical direction, there are complications dependent upon the collisions of the drops with one another. Such collisions are inevitable in consequence of the different velocities acquired by the drops as they break away irregularly from the continuous portion of the column. Even when the resolution is regularized by the action of external vibrations of suitable frequency, the drops must still come into contact before they reach the summit of their parabolic path. In the case of a continuous jet the "equation of continuity" shews that as the jet loses velocity in ascending, it must increase in section. When the stream consists of drops following the same path in single file, no such increase of section is possible; and then the constancy of the total stream demands a gradual approximation of the drops, which in the case of a nearly vertical direction of motion cannot stop short of actual contact. Regular vibration has, however, the effect of postponing the collisions and consequent scattering of the drops, and in the case of a direction of motion less nearly vertical may prevent them altogether.

The behaviour of a nearly vertical fountain is influenced in an extraordinary manner by the neighbourhood of an electrified body. The experiment may be tried with a jet from a nozzle of 1 mm. diameter rising about 50 centims. In its normal state the jet resolves itself into drops, which even before passing the summit, and still more after passing it, are scattered through a considerable width. When a feebly electrified body is presented to it, the jet undergoes a remarkable transformation, and appears to become coherent; but under more powerful electrical action the scattering becomes even greater than at first. The second effect is readily attributed to the mutual repulsion of the electrified drops, but the action of feeble electricity in producing apparent coherence depends upon a different principle.

It has been shewn by Beetz[1] that the coherence is apparent only, and that the place where the jet breaks into drops is not perceptibly shifted by the electricity. By screening various parts with metallic plates connected to earth, Beetz further proved that, contrary to the opinion of earlier observers, the seat of sensitiveness is not at the root of the jet where it leaves the orifice, but at the place of resolution into drops. As in Lord Kelvin's water-dropping apparatus for atmospheric electricity, the drops carry away with them an electric charge, which may be collected by receiving them in an insulated vessel.

It may be proved by instantaneous illumination that the normal scattering is due to the rebound of the drops when they come into collision. Under moderate electrical influence there is no material change in the resolution into drops nor in the subsequent motion of the drops up to the moment of collision. The difference begins here. Instead of rebounding after collision, as the unelectrified drops of clean water generally do, the electrified drops *coalesce*, and thus the jet is no longer scattered about[2]. An elaborate discussion of this subject would be out of place here. It must suffice to say that the effect depends upon a *difference* of potential between the drops at the moment of collision, and that when this difference is too small to cause coalescence there is complete electrical insulation between the contiguous masses.

When the jet is projected upwards at a moderate obliquity, the scattering is confined to the vertical plane. Under these circumstances there are few or no collisions, as the drops have room to clear one another, and moderate electrical influence is without effect. At a higher obliquity the drops begin to be scattered out of the vertical plane, which is a sign that collisions are taking place. Moderate electrical influence will reduce the scattering to the vertical plane by causing coalescence of drops which come into contact.

If, as in Savart's beautiful experiments, the resolution into drops is regularized by external vibrations of suitable frequency,

---

[1] *Pogg. Ann.* vol. CXLIV. p. 443, 1872.

[2] The influence of Electricity on Colliding Water Drops, *Proc. Roy. Soc.* vol. XXVIII. p. 406, 1879.

the principal drops follow the same course, and unless the projection is nearly vertical there are no collisions between them. But it sometimes happens that the spherules are thrown out laterally in a distinct stream, making a considerable angle with the main stream. This is the result of collisions between the spherules and the principal drops. It may even happen that the former are reflected backwards and forwards several times until at last they escape laterally. In all cases the behaviour under feeble electrical influence is a criterion of the occurrence of collisions.

In an experiment, due to Magnus[1], the spherules are diverted from the main stream without collisions by electrical attraction. Advantage may be taken of this to obtain a regular procession of drops finer than would otherwise be possible.

364. The detached masses of liquid into which a jet is resolved do not at once assume and retain a spherical figure, but execute a series of vibrations, being alternately compressed and elongated in the direction of the axis of symmetry. When the resolution is effected in a perfectly periodic manner, each drop is in the same phase of its vibration as it passes through a given point of space; and thence arises the remarkable appearance of alternate swellings and contractions described by Savart. The interval from one swelling to the next is the space described by the drop during one complete vibration about its figure of equilibrium, and is therefore, as Plateau shewed, proportional *cæteris paribus* to the square root of the head.

The time of vibration is of course itself a function of the nature of the fluid (T, ρ) and of the size of the drop, to the calculation of which we now proceed. It may be remarked that the argument from dimensions is sufficient to shew that the time (τ) of an infinitely small vibration of any type is proportional to $\sqrt{(\rho V/T)}$, where V is the volume of the drop.

In the mathematical investigation of the small vibrations of a liquid mass about its spherical figure of equilibrium, we will confine ourselves to modes of vibration symmetrical about an axis, which suffice for the problem in hand. These modes require for their expression only Legendre's functions P_n; the more general

[1] *Pogg. Ann.* vol. cvi. p. 27, 1859.

problem, involving Laplace's functions, may be treated in the same way and leads to the same results.

The radius r of the surface bounding the liquid may be expanded at any time t in the series (§ 336)

$$r = a_0 + a_1 P_1(\mu) + \dots + a_n P_n(\mu) + \dots\dots\dots (1),$$

where $a_1, a_2\dots$ are small quantities relatively to a_0, and μ represents, as usual, the cosine of the colatitude (θ).

For the volume (V) included within the surface (1) we have

$$V = \tfrac{2}{3}\pi \int_{-1}^{+1} r^3 d\mu = \tfrac{4}{3}\pi a_0^3 \left[1 + 3\Sigma (2n+1)^{-1} a_n^2/a_0^2\right] \dots (2),$$

the summation commencing at $n = 1$. Thus, if a be the radius of the sphere of equilibrium,

$$a = a_0 \left[1 + \Sigma (2n+1)^{-1} a_n^2/a_0^2\right] \dots\dots\dots\dots (3).$$

The potential energy of capillarity is the product of the tension T and of the surface S. To calculate S we have

$$S = 2\pi \int r \sin\theta \left\{r^2 + \left(\frac{dr}{d\theta}\right)^2\right\}^{\frac{1}{2}} d\theta = 2\pi \int \left\{r^2 + \tfrac{1}{2}\left(\frac{dr}{d\theta}\right)^2\right\} \sin\theta\, d\theta.$$

For the first part

$$\int_{-1}^{+1} r^2 d\mu = 2a_0^2 + 2\Sigma (2n+1)^{-1} a_n^2.$$

For the second part

$$\tfrac{1}{2}\int\left(\frac{dr}{d\theta}\right)^2 \sin\theta\, d\theta = \tfrac{1}{2}\int_{-1}^{+1} (1-\mu^2)\left[\Sigma a_n \frac{dP_n}{d\mu}\right]^2 d\mu.$$

The value of the quantity on the right may be found with the aid of the formula

$$\int_{-1}^{+1} (1-\mu^2) \frac{dP_m}{d\mu}\frac{dP_n}{d\mu}\, d\mu = n(n+1)\int_{-1}^{+1} P_m P_n\, d\mu,$$

in which m is an integer equal to or different from n. Thus

$$\tfrac{1}{2}\int_{-1}^{+1}\left(\frac{dr}{d\theta}\right)^2 \sin\theta\, d\theta = \tfrac{1}{2}\int_{-1}^{+1} (1-\mu^2)\Sigma a_n^2 \left(\frac{dP_n}{d\mu}\right)^2 d\mu$$

$$= \tfrac{1}{2}\Sigma n(n+1) a_n^2 \int_{-1}^{+1} P_n^2 d\mu = \Sigma n(n+1)(2n+1)^{-1} a_n^2.$$

Accordingly

$$S = 4\pi a_0^2 + 2\pi \Sigma (2n+1)^{-1}(n^2+n+2) a_n^2$$

$$= 4\pi a^2 + 2\pi \Sigma (n-1)(n+2)(2n+1)^{-1} a_n^2 \dots\dots (4)$$

by (3).

Thus, if T be the cohesive tension, the potential energy (P) corresponding thereto may be taken to be

$$P = 2\pi T . \Sigma (n-1)(n+2)(2n+1)^{-1} a_n{}^2 \ldots\ldots\ldots (5).$$

We have now to calculate the kinetic energy of the motion. The velocity-potential ψ may be expanded in the series

$$\psi = \beta_0 + \beta_1 r P_1(\mu) + \ldots + \beta_n r^n P_n(\mu) + \ldots\ldots\ldots (6);$$

and thus for the kinetic energy we get

$$K = \tfrac{1}{2}\rho \iint \psi \, d\psi/dr . a^2 \, d\phi \, d\mu$$

$$= 2\pi\rho a^2 . \Sigma (2n+1)^{-1} n a^{2n-1}\beta_n{}^2.$$

But by comparison of the value of $d\psi/dr$ from (6) with (1) we find

$$n a^{n-1}\beta_n = da_n/dt ;$$

and thus

$$K = 2\pi\rho a^3 . \Sigma (2n^2 + n)^{-1} (da_n/dt)^2 \ldots\ldots\ldots\ldots (7).$$

Since the products of the quantities a_n and da_n/dt do not occur in the expressions for P and K, the motions represented by the various terms take place independently of one another. The equation for a_n is by Lagrange's method (§ 87)

$$\frac{d^2 a_n}{dt^2} + n(n-1)(n+2)\frac{T}{\rho a^3} a_n = 0 \ldots\ldots\ldots\ldots (8);$$

so that, if $a_n \propto \cos(pt+\epsilon)$,

$$p^2 = n(n-1)(n+2)\frac{T}{\rho a^3}\ldots\ldots\ldots\ldots\ldots(9).[1]$$

The periodic time is equal to $2\pi/p$, so that in terms of V (equal to $\tfrac{4}{3}\pi a^3$)

$$\tau = \frac{\surd\{3\pi\rho V/T\}}{\surd\{n(n-1)(n+2)\}} \ldots\ldots\ldots\ldots (10),$$

or in the particular case of n equal to 2

$$\tau = \surd\{3\pi\rho V/8T\} \ldots\ldots\ldots\ldots\ldots (11).$$

To find the radius of the sphere of water which vibrates seconds, we put in (9) $p = 2\pi$, $T = 74$, $\rho = 1$, $n = 2$. Thus $a = 2\cdot47$ centims., or a little less than one inch.

An attempt to compare (11) with the phenomena observed in a jet did not bring out a good agreement. A stream of $19\cdot7$ cub.

---

[1] *Proc. Roy. Soc.* vol. **xxix.** p. 97, 1879; Webb, *Mess. of Math.* vol. **ix.** p. 177, 1880.

cent. per second was broken up under the action of a fork making 128 vibrations per second. Neglecting the mass of the small spherules, we may take for the volume of each principal drop 19·7/128, or ·154 cub. cent. Thence by (11), putting $\rho = 1$, $T = 74$, we have $\tau = ·0494$ second. This is the calculated value. By observation of the vibrating jet the distance between the first and second swellings, corresponding to the maximum oblateness of the drops, was 16·5 centims. The level of the contraction midway between the two swellings was 36·8 centims. below the surface of the liquid in the reservoir, corresponding to a velocity of 26 centims. per second. These data give for the time of vibration

$$\tau = 16·5/269 = ·0612 \text{ second.}$$

The discrepancy between the two values of τ is probably attributable to excessive amplitude, entailing a departure from the law of isochronism. Observations upon the vibrations of drops delivered singly from pipettes have been made by Lenard[1].

The tendency of the capillary force is always towards the restoration of the spherical figure of equilibrium. By electrifying the drop we may introduce a force operative in the opposite direction. It may be proved[2] that if Q be the charge of electricity in electrostatic measure, the formula corresponding to (9) is

$$p^2 = \frac{n(n-1)}{\rho a^3}\left\{(n+2)T - \frac{Q^2}{4\pi a^3}\right\} \quad \ldots\ldots\ldots\ldots (12).$$

If $T > Q^2/16\pi a^3$, the spherical form is stable for all displacements. When Q is great, the spherical form becomes unstable for all values of n below a certain limit, the maximum instability corresponding to a great, but still finite, value of n. Under these circumstances the liquid is thrown out in fine jets, whose fineness, however, has a limit.

Observations upon the swellings and contractions of a regularly resolved jet may be made stroboscopically, one view corresponding to each complete period of the vibrator; or photographs may be taken by the instantaneous illumination furnished by a powerful electric spark[3].

---

[1] *Wied. Ann.* vol. xxx. p. 209, 1887.

[2] *Phil. Mag.* vol. xiv. p. 184, 1882.

[3] Some Applications of Photography, *Proc. Roy. Soc. Inst.* vol. xiii. p. 261, 1891; *Nature*, vol. xliv. p. 249, 1891.

In the mathematical investigations of this chapter no account has been taken of viscosity. Plateau held the opinion that the difference between the wave-length of spontaneous division of a jet $(4\cdot5 \times 2a)$ and the critical wave-length $(\pi \times 2a)$ was an effect of viscosity; but we have seen that it is sufficiently accounted for by *inertia*. The inclusion of viscosity considerably complicates the mathematical problem[1], and it will not here be attempted. The result is to shew that, when viscosity is paramount, long threads do not tend to divide themselves into drops at mutual distances comparable with the diameter of the thread, but rather to give way by attenuation at few and distant places. This appears to be in agreement with the observed behaviour of highly viscous threads of glass, or treacle, when supported only at the terminals. A separation into numerous drops, or a varicosity pointing to such a resolution, may thus be taken as evidence that the fluidity has been sufficient to bring inertia into play.

A still more general investigation, in which the influence of electrification is considered, has been given by Basset[2].

[1] *Phil. Mag.* vol. xxxiv. p. 145, 1892.

[2] *Amer. Journ. of Math.* vol. xvi. No. 1.

CHAPTER XXI.

VORTEX MOTION AND SENSITIVE JETS.

365. A LARGE and important group of acoustical phenomena have their origin in the instability of certain fluid motions of the kind classified in hydrodynamics as steady. A motion, the same at all times, satisfies the dynamical conditions, and is thus in a sense possible; but the smallest departure from the ideal so defined tends spontaneously to increase, and usually with great rapidity according to the law of compound interest. Examples of such instability are afforded by sensitive jets and flames, æolian tones, and by the flute pipes of the organ. These phenomena are still very imperfectly understood; but their importance is such as to demand all the consideration that we can give them.

So long as we regard the fluid as absolutely inviscid there is nothing to forbid a finite slip at the surface where two masses come into contact. At such a surface the vorticity (§ 239) is infinite, and the surface may be called a vortex sheet. The existence of a vortex sheet is compatible with the dynamical conditions for steady motion; but, as was remarked at an early date by v. Helmholtz[1], the steady motion is unstable. The simplest case occurs when a plane vortex sheet separates two masses of fluid which move with different velocities, but without internal relative motion—a problem considered by Lord Kelvin in his investigation of the influence of wind upon waves[2]. In the following discussion the method of Lord Kelvin is applied to determine the law of falling away from steady motion in some of the simpler cases of a plane surface of separation.

---

[1] *Phil. Mag.* vol. xxxvi. p. 337, 1868.

[2] *Phil. Mag.* vol. xLII. p. 368, 1871. See also *Proc. Math. Soc.* vol. x. p. 4, 1878; Basset's *Hydrodynamics*, § 391, 1888; Lamb's *Hydrodynamics*, § 224, 1895.

Let us suppose that below the plane $z = 0$ the fluid is of constant density ρ and moves parallel to x with velocity V, and that above that plane the density is ρ' and the velocity V'. As in § 353, let z be measured downwards, and let there be rigid walls bounding the lower fluid at $z = l$ and the upper fluid at $z = -l'$. The disturbance is supposed to involve x and t only through the factors e^{ikx}, e^{int}. The velocity potential $(Vx + \phi)$ in the lower fluid satisfies Laplace's equation, and thus ϕ by the condition at $z = l$ takes the form

$$\phi = C \cosh k (z - l) \cdot e^{i(nt+kx)} \dots\dots\dots\dots\dots(1);$$

and a similar expression,

$$\phi' = C' \cosh k (z + l') \cdot e^{i(nt+kx)} \dots\dots\dots\dots (2),$$

applies to the lower fluid, if the whole velocity-potential be there $(V'x + \phi')$. The connection between ϕ and the elevation (h) at the common surface is

$$-\frac{d\phi}{dz}(z = 0) = \frac{dh}{dt} + V\frac{dh}{dx};$$

so that, if

$$h = He^{i(nt+kx)} \dots\dots\dots\dots\dots\dots (3),$$

$$kC \sinh kl = i(n + kV) H \dots\dots\dots\dots (4).$$

In like manner, $-kC' \sinh kl' = i(n + kV') H \dots\dots\dots\dots (5).$

We have now to express the condition relating to pressures at $z = 0$. The general equation (2), § 244, gives for the lower fluid

$$\frac{\delta p}{\rho} = -gh - \frac{d\phi}{dt} - \tfrac{1}{2}\left(V + \frac{d\phi}{dx}\right)^2 - \tfrac{1}{2}\left(\frac{d\phi}{dz}\right)^2$$

$$= -gh - in\phi - ikV\phi,$$

squares of small quantities being neglected. In like manner for the upper fluid at $z = 0$

$$\frac{\delta p'}{\rho'} = -gh - in\phi' - ikV'\phi'.$$

If there be no capillary tension, δp and $\delta p'$ are equal. If the capillary tension be T, the difference is

$$\delta p - \delta p' = -T \, d^2h/dx^2 = k^2 Th,$$

so that

$$g(\rho - \rho') h + k^2 Th = i\rho'(n + kV') \phi' - i\rho(n + kV)\phi\dots\dots\dots(6).$$

When the values of ϕ, ϕ' at $z = 0$ are introduced from (1), (2), (4), (5), the condition becomes

$$g (\rho - \rho') + k^2 T = k\rho (V + n/k)^2 \coth kl + k\rho' (V' + n/k)^2 \coth kl' \qquad \dots\dots\dots\dots(7).$$

This is the equation which determines the values of n/k. If the roots of the quadratic are real, waves are propagated with the corresponding real velocities; if on the other hand the roots are imaginary, exponential functions of the time enter into the solution, indicating that the steady motion is unstable. The criterion of stability is accordingly

$$(\rho \coth kl + \rho' \coth kl') \{g (\rho - \rho') + Tk^2\}$$
$$- k\rho\rho' \coth kl \coth kl' (V - V')^2 > 0 \dots\dots(8).$$

If g and T both vanish, the motion is unstable for all disturbances, that is, whatever may be the value of k. If T vanish, the operation of gravity may be to secure stability for certain values of k, but it cannot render the steady motion stable on the whole. For when k is infinitely great, that is, when the corrugations are infinitely fine, $\coth kl = \coth kl' = 1$, and the term in g disappears from the criterion. In spite of the impressed forces tending to stability the motion is necessarily unstable for waves of infinitesimal length; and this conclusion may be extended to vortex sheets of any form and to impressed forces of any kind.

If T be finite, then on the contrary there is of necessity stability for waves of infinitesimal length, although there may be instability for waves of finite length.

For further examination we may take the simpler conditions which arise when l and l' are infinite. The criterion of stability then becomes

$$(\rho + \rho') \{g (\rho - \rho') + Tk^2\} - k\rho\rho' (V - V')^2 > 0 \dots\dots (9),$$

and the critical case is determined by equating the left-hand member to zero. This gives a quadratic in k. If the roots of the quadratic are imaginary, the criterion (9) is satisfied for all intermediate values of k, as well as for the infinitely small and infinitely large values by which it is satisfied in all cases, provided that $\rho > \rho'$. The condition of complete stability is thus

$$4g (\rho - \rho') T > \frac{\rho^2 \rho'^2 (V - V')^4}{(\rho + \rho')^2} \dots\dots\dots (10).$$

Let W denote the minimum velocity (§ 353) of waves when $V = 0$, $V' = 0$. Then by (7)

$$(\rho + \rho')^2 W^4 = 4g (\rho - \rho') T \dots\dots\dots\dots (11),$$

and (10) may be written

$$W^4 > \frac{\rho\rho' (V - V')^2}{(\rho + \rho')^2} \dots\dots\dots\dots\dots (12).$$

If $(V - V')$ do not exceed the value thus determined, the steady motion is stable for all disturbances; otherwise there will be some finite wave-lengths for which disturbances increase exponentially.

If we now omit the terms in (7) dependent upon gravity and upon capillarity, the equation becomes

$$\rho (n + kV)^2 \coth kl + \rho' (n + kV')^2 \coth kl' = 0\dots\dots(13).$$

When $l = l'$, or when both these quantities are infinite, we have simply

$$\rho (n + kV)^2 + \rho' (n + kV')^2 = 0\dots\dots\dots\dots(14),$$

or

$$\frac{n}{k} = -\frac{\rho V + \rho' V' \pm i \sqrt{(\rho\rho')} \cdot (V - V')}{\rho + \rho'} \dots\dots (15).$$

We see from (15) that, as was to be expected, a motion common to both parts of the liquid has no dynamical significance. An equal addition to V and V' is equivalent to a deduction of like amount from n/k. If $\rho = \rho'$, (15) becomes

$$n/k = -\tfrac{1}{2} (V + V') \pm \tfrac{1}{2} i (V - V') \dots\dots\dots\dots (16).$$

The essential features of the case are brought out by the simple case where $V' = -V$, so that the steady motions of the two masses of fluid are equal and opposite. We have then

$$n/k = \pm iV \dots\dots\dots\dots\dots\dots (17);$$

and for the elevation,

$$h = H e^{\pm kVt} \cos (kx + \epsilon) \dots\dots\dots\dots (18),$$

corresponding to

$$h = H \cos (nx + \epsilon) \dots\dots\dots\dots\dots (19),$$

initially.

If when $t = 0$, $dh/dt = 0$,

$$h = H \cosh kVt \cos (nx + \epsilon) \dots\dots\dots\dots (20),$$

indicating that the waves upon the surface of separation are stationary, and increase in amplitude with the time according to the law of the hyperbolic cosine. The rate of increase of the term with the positive exponent is extremely rapid. Since $k = 2\pi/\lambda$, the amplitude is multiplied by e^{π}, or about 23, in the time occupied by either stream in passing over a distance λ.

If $V' = V$, the roots (16) are equal, but the general solution may be obtained by the usual method. Thus, if we put

$$V' = V(1 + \alpha),$$

where α is ultimately to vanish,

$$n/k = -V \pm \tfrac{1}{2} i\alpha V;$$

and $\qquad h = e^{ik(x - Vt)} \left\{ A e^{\frac{1}{2}ikVt.\alpha} + B e^{-\frac{1}{2}ikVt.\alpha} \right\},$

where A, B are arbitrary constants. Passing now to the limit where $\alpha = 0$, and taking new arbitrary constants, we get

$$h = e^{ik(x - Vt)} \{ C + Dt \},$$

or in real quantities,

$$h = \{ C + Dt \} \cos k (x - Vt + \epsilon).$$

If initially $\qquad h = \cos kx, \quad dh/dt = 0,$

$$h = \cos k(Vt - x) + kVt \sin k(Vt - x) \dots\dots (21).$$

The peculiarity of this case is that previous to the displacement there is no real surface of separation at all.

The general solution involving l and l' may be adapted to represent certain cases of disturbance of a two-dimensional jet of width $2l$ playing into stationary fluid. For if the disturbance be *symmetrical*, so that the median plane is a plane of symmetry, the conditions are the same as if a fixed wall were there introduced. If the surrounding fluid be unlimited, $l' = \infty$, $\coth kl' = 1$; and the equation determining n becomes, if $V' = 0$, $\rho' = \rho$,

$$(n + kV)^2 \coth kl + n^2 = 0 \dots\dots\dots\dots (22),$$

of which the solution is

$$\frac{n}{kV} = \frac{-1 \pm i \sqrt{(\tanh kl)}}{1 + \tanh kl} \dots\dots\dots\dots (23).$$

Thus $\qquad h = H e^{\pm \mu kVt} \cos k \left\{ x - \frac{Vt}{1 + \tanh kl} \right\} \dots\dots\dots (24),$

where $\qquad \mu = \dfrac{\sqrt{(\tanh kl)}}{1 + \tanh kl} \dots\dots\dots\dots (25).$

This represents the progression of symmetrical disturbances in a jet of width $2l$ playing into a stationary environment of the same density.

If kl be very small, so that the wave-length is large in comparison with the thickness of the jet,

$$h = He^{\pm \sqrt{(kl)} \cdot kVt} \cos k \{x - Vt\} \dots\dots\dots\dots (26).$$

The investigation of the asymmetrical disturbance of a jet requires the solution of the problem of a single vortex sheet when the condition to be satisfied at $z = l$ is $\phi = 0$, instead of as hitherto $d\phi/dz = 0$. The value of ϕ is

$$\phi = -i(n + kV) H \frac{\sinh k(z - l)}{k \cosh kl} e^{int} e^{ikx} \dots\dots\dots\dots (27);$$

from which, if as before $d\phi'/dz = 0$ when $z = -l'$,

$$\rho(n + kV)^2 \tanh kl + \rho'(n + kV')^2 \coth kl' = 0 \dots (28).$$

If $l' = \infty$, $\rho' = \rho$, $V' = 0$,

$$(n + kV)^2 \tanh kl + n^2 = 0 \dots\dots\dots\dots (29).$$

This is applicable to a jet of width $2l$, moving with velocity V in still fluid and displaced in such a manner that the sinuosities of its two surfaces are parallel.

When kl is small, we have approximately

$$h = He^{\pm \sqrt{(kl)} \cdot kVt} \cos k(x - kl \cdot Vt) \dots\dots\dots (30).$$

By a combination of the solutions represented by (26), (30), we may determine the consequences of any displacements in two dimensions of the two surfaces of a thin jet moving with velocity V in still fluid of its own density.

366. The investigations of § 365 may be considered to afford an adequate general explanation of the sensitiveness of jets. In the ideal case of abrupt transitions of velocity, constituting vortex sheets, in frictionless fluid, the motion is always unstable, and the degree of instability increases as the wave-length of the disturbance diminishes.

The direct application of this result to actual jets would lead us to the conclusion that their sensitiveness increases indefinitely with pitch. It is true that, in the case of certain flames, the pitch of the most efficient sounds is very high, not far from the upper limit of human hearing; but there are other kinds of sensitive jets on which these high sounds are without effect, and which require for their excitation a moderate or even a grave pitch.

A probable explanation of the discrepancy readily suggests itself. The calculations are founded upon the supposition that the changes of velocity are discontinuous—a supposition that cannot possibly agree with reality. In consequence of fluid friction a surface of discontinuity, even if it could ever be formed, would instantaneously disappear, the transition from the one velocity to the other becoming more and more gradual, until the layer of transition attained a sensible width. When this width is comparable with the wave-length of a sinuous disturbance, the solution for an abrupt transition ceases to be applicable, and we have no reason for supposing that the instability would increase for much shorter wave-lengths.

A general idea of the influence of viscosity in broadening a jet may be obtained from Fourier's solution of the problem where the initial width is supposed to be infinitesimal. Thus, if in the general equations v and w vanish, while u is a function of y only, the equation satisfied by u is (as in § 347)

$$\frac{du}{dt} = \frac{\mu}{\rho} \frac{d^2u}{dy^2} \dots\dots\dots\dots\dots(1).$$

The solution of this equation for the case where u is initially sensible only at $y = 0$ is

$$u = U_1 \frac{e^{-y^2/4\nu t}}{2\sqrt{(\pi\nu t)}} \dots\dots\dots\dots(2),$$

where $\nu = \mu/\rho$, and U_1 denotes the initial value of $\int u\, dy$. When $y^2 = 4\nu t$, the value of u is less than that to be found at the same time at $y = 0$ in the ratio $e : 1$. For air $\nu = 16$ C.G.S., and thus after a time t the thickness $(2y)$ of the jet is comparable in magnitude with $1\cdot6\sqrt{t}$; for example, after one second it may be considered to be about $1\frac{1}{2}$ cm.

There is therefore ample foundation for the suspicion that the phenomena of sensitive jets may be greatly influenced by fluid friction, and deviate materially from the results of calculations based upon the supposition of discontinuous changes of velocity. Under these circumstances it becomes important to investigate the character of the equilibrium of stratified motion in cases more nearly approaching what is met with in practice. A complete investigation which should take account of all the effects of viscosity would encounter many formidable difficulties. For the present purpose we shall treat the fluid as frictionless and be content to obtain solutions for laws of stratification which are free from discontinuity. For the undisturbed motion the component velocities v, w are zero, and u is a function of y only, which we will denote by U. A curve in which U is ordinate and y is abscissa represents the law of stratification, and may be called for brevity the velocity curve. The vorticity Z (§ 239) of the steady motion is equal to $\frac{1}{2} dU/dy$.

If in the disturbed motion, assumed to be in two dimensions, the velocities be denoted by $U + u$, v, and the vorticity by $Z + \zeta$, the general equation (4), § 239, takes the form

$$\frac{d(Z+\zeta)}{dt} + (U+u)\frac{d(Z+\zeta)}{dx} + v\frac{d(Z+\zeta)}{dy} = 0,$$

in which $\qquad dZ/dt = 0, \quad dZ/dx = 0.$

Thus, if the square of the disturbances be neglected, the equation may be written

$$\frac{d\zeta}{dt} + U\frac{d\zeta}{dx} + v\frac{dZ}{dy} = 0 \dots\dots\dots\dots\dots(3);$$

and the equation of continuity for an incompressible fluid gives

$$\frac{du}{dx} + \frac{dv}{dy} = 0 \dots\dots\dots\dots\dots\dots(4).$$

If the values of Z and ζ in terms of the velocities be substituted in (3),

$$\left(\frac{d}{dt} - U\frac{d}{dx}\right)\left(\frac{du}{dy} - \frac{dv}{dx}\right) + v\frac{d^2 U}{dy^2} = 0 \dots\dots\dots(5).$$

We now introduce the supposition that as functions of x and t, u and v are proportional to $e^{int} \cdot e^{ikx}$. From (4)

$$iku + dv/dy = 0 \dots\dots\dots\dots\dots\dots(6);$$

and if this value of u be substituted in (5), we obtain

$$\left(\frac{n}{k} + U\right)\left(\frac{d^2 v}{dy^2} - k^2 v\right) - \frac{d^2 U}{dy^2} v = 0 \dots\dots\dots(7)[1].$$

---

[1] *Proc. Math. Soc.* vol. XI. p. 68, 1880.

In (7) k may be regarded as real, and in any particular problem that may be proposed the principal object is to determine the corresponding value of n, and especially whether it is real or imaginary. One general proposition of importance relates to the case where d^2U/dy^2 is of one sign, so that the velocity curve is wholly convex, or wholly concave, throughout the entire space between two fixed walls at which the condition $v = 0$ is satisfied. Let $n/k = p + iq$, $v = \alpha + i\beta$, where p, q, α, β are real. Substituting in (7) we get

$$\frac{d^2\alpha}{dy^2} + i\frac{d^2\beta}{dy^2} = \left[k^2 + \frac{d^2U}{dy^2}\frac{p + U - iq}{(p + U)^2 + q^2} \right](\alpha + i\beta) = 0;$$

or, on equating separately to zero the real and imaginary parts,

$$\frac{d^2\alpha}{dy^2} = k^2\alpha + \frac{d^2U}{dy^2}\frac{(p + U)\alpha + q\beta}{(p + U)^2 + q^2} \quad\dots\dots\dots\dots (8),$$

$$\frac{d^2\beta}{dy^2} = k^2\beta + \frac{d^2U}{dy^2}\frac{-q\alpha + (p + U)\beta}{(p + U)^2 + q^2} \quad\dots\dots\dots (9).$$

Multiplying (8) by β, (9) by α, and subtracting, we get

$$\beta\frac{d^2\alpha}{dy^2} - \alpha\frac{d^2\beta}{dy^2} = \frac{d}{dy}\left(\beta\frac{d\alpha}{dy} - \alpha\frac{d\beta}{dy}\right) = \frac{d^2U}{dy^2}\frac{q(\alpha^2 + \beta^2)}{(p + U)^2 + q^2}\dots(10).$$

At the limits v, and therefore both α and β, are by hypothesis zero. Hence integrating (10) between the limits, we see that q must be zero, if d^2U/dy^2 is of one sign throughout the range of integration. Accordingly n is real, and the motion, if not absolutely stable, is at any rate not exponentially unstable.

Another general conclusion worthy of notice can be deduced from (7). Writing it in the form

$$\frac{d^2v}{dy^2} = \left\{ k^2 + \frac{d^2U/dy^2}{U + n/k} \right\}v,$$

we see that, if n be real, v cannot pass from one zero value to another zero value, unless d^2U/dy^2 and $(n + kU)$ be somewhere of contrary signs. Thus if we suppose that U is positive and d^2U/dy^2 negative throughout, and that V is the greatest value of U, we find that $n + kV$ must be positive.

367. A class of problems admitting of fairly simple solution is obtained by supposing the vorticity Z to be constant throughout layers of finite thickness and to change its value only in passing a limited number of planes, for each of which y is constant. In such cases the velocity curve is composed of portions of straight lines which meet one another at finite angles. This state of things is supposed to be disturbed by bending the surfaces of transition.

Throughout any layer of constant vorticity $d^2U/dy^2 = 0$, and thus by (7), § 366, wherever $n + kU$ is not equal to zero,

$$\frac{d^2v}{dy^2} - k^2v = 0 \dots\dots\dots\dots\dots\dots(1),$$

of which the solution is

$$v = A e^{ky} + B e^{-ky} \dots\dots\dots\dots\dots(2).$$

If there are several layers in each of which Z is constant, the various solutions of the form (2) are to be fitted together, the arbitrary constants being so chosen as to satisfy certain boundary conditions. The first of these conditions is evidently the continuity of v, or as it may be expressed,

$$\Delta v = 0 \dots\dots\dots\dots\dots\dots\dots\dots(3).$$

The other necessary condition may be obtained by integrating (7), § 366, across the surface of transition. Thus

$$\left(\frac{n}{k} + U\right) . \Delta \left(\frac{dv}{dy}\right) - \Delta \left(\frac{dU}{dy}\right) . v = 0 \dots\dots\dots (4).$$

These are the conditions that the velocity shall be continuous at the places where dU/dy changes its value.

In the problems which we shall consider the fluid is either bounded by a fixed plane at which y is constant, or else extends to infinity. For the former the condition is simply $v = 0$. If there be a layer extending to infinity in the positive direction, A must vanish in the expression (2) applicable to this layer; if a layer extend to infinity in the negative direction, the corresponding B must vanish.

Under the first head we will consider a problem of some generality, where the stratified steady motion takes place between fixed walls at $y = 0$ and at $y = b_1 + b' + b_2$.

The vorticity is constant throughout each of the three layers bounded by $y = 0$, $y = b_1$; $y = b_1$, $y = b_1 + b'$; $y = b_1 + b'$, $y = b_1 + b' + b_2$ (Fig. 67). There are thus two internal surfaces where the vorticity changes. The values of U at these surfaces may be denoted by U_1, U_2.

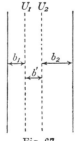

Fig. 67.

In conformity with (3) and with the condition that $v = 0$ when $y = 0$, we may take in the first layer

$$v = v_1 = \sinh ky \dots\dots\dots\dots\dots (5);$$

in the second layer

$$v = v_2 = v_1 + M_1 \sinh k (y - b_1) \dots\dots\dots\dots (6);$$

in the third layer

$$v = v_3 = v_2 + M_2 \sinh k (y - b_1 - b') \dots\dots\dots (7).$$

The condition that $v = 0$, when $y = b_1 + b' + b_2$, now gives

$$0 = M_2 \sinh kb_2 + M_1 \sinh k (b_2 + b') + \sinh k (b_2 + b' + b_1)\dots(8).$$

We have still to express the other two conditions (4) at the surfaces of transition. At the first surface

$$v = \sinh kb_1, \qquad\qquad \Delta (dv/dy) = kM_1;$$

at the second surface

$$v = M_1 \sinh kb' + \sinh k (b_1 + b'), \quad \Delta (dv/dy) = kM_2.$$

If we denote the values of $\Delta (dU/dy)$ at the two surfaces respectively by Δ_1, Δ_2, our conditions become

$$(n + kU_1) M_1 - \Delta_1 \sinh kb_1 = 0 \dots\dots\dots\dots (9),$$

$$(n + kU_2) M_2 - \Delta_2 \{M_1 \sinh kb' + \sinh k (b_1 + b')\} = 0\dots(10).$$

By (8), (9), (10) the values of M_1, M_2, n are determined.

The equation for n is found by equating to zero the determinant of the three equations. It may be written

$$An^2 + Bn + C = 0 \dots\dots\dots\dots\dots (11),$$

where

$$A = \sinh k (b_2 + b' + b_1) \dots\dots\dots\dots\dots\dots (12),$$

$$\begin{aligned} B = {}& k (U_1 + U_2) \sinh k (b_2 + b' + b_1) \\ & + \Delta_2 \sinh kb_2 \, \sinh k (b_1 + b') + \Delta_1 \sinh kb_1 \, \sinh k (b_2 + b')\dots(13), \end{aligned}$$

$$\begin{aligned} C = {}& k^2 U_1 U_2 \sinh k (b_2 + b' + b_1) \\ & + kU_1 \Delta_2 \sinh kb_2 \, \sinh k (b_1 + b') + kU_2 \Delta_1 \sinh kb_1 \, \sinh k (b_2 + b') \\ & + \Delta_1 \Delta_2 \sinh kb_1 \, \sinh kb_2 \, \sinh kb' \dots\dots\dots\dots\dots (14). \end{aligned}$$

To find the character of the roots we have to form the expression for $B^2 - 4AC$. On reduction we get

$$\begin{aligned} B^2 - 4AC = {}& \{k (U_1 - U_2) \sinh k (b_2 + b' + b_1) \\ & + \Delta_1 \sinh kb_1 \, \sinh k (b_2 + b') - \Delta_2 \sinh kb_2 \, \sinh k (b_1 + b')\}^2 \\ & + 4\Delta_1 \Delta_2 \sinh^2 kb_1 \, \sinh^2 kb_2 \dots\dots\dots\dots\dots(15). \end{aligned}$$

Hence if Δ_1, Δ_2 have the same sign, that is, if the velocity curve (§ 366) be of one curvature throughout, $B^2 - 4AC$ is positive, and the two values of n are real. Under these circumstances the disturbed motion is stable.

We will now suppose that the surfaces at which the vorticity changes are symmetrically situated, so that $b_1 = b_2 = b$.

In this case we find

$$A = \sinh k\,(2b + b')\dots\dots\dots\dots\dots\dots\dots\dots\dots\dots(16),$$

$$B = k(U_1 + U_2)\sinh k(2b+b') + (\Delta_1 + \Delta_2)\sinh kb\ \sinh k(b+b')\dots(17),$$

$$C = k^2 U_1 U_2 \sinh k\,(2b + b') + k\,(U_1\Delta_2 + U_2\Delta_1)\sinh kb \sinh k\,(b + b')$$
$$+ \Delta_1\Delta_2 \sinh^2 kb\ \sinh kb' \dots\dots\dots\dots\dots(18),$$

$$B^2 - 4AC = 4\Delta_1\Delta_2 \sinh^4 kb$$
$$+ \{k(U_1 - U_2)\sinh k(2b+b') + (\Delta_1 - \Delta_2)\sinh kb\ \sinh k(b+b')\}^2\dots(19).$$

Under this head there are two sub-cases which may be especially noted. The first is that in which the values of U are the same on both sides of the median plane, so that the middle layer is a region of constant velocity without vorticity, and the velocity curve is that shewn in Fig. 68. We may suppose that $U = V$ in the middle layer, and that $U = 0$ at the walls, without loss of generality, since any constant velocity (U_0) superposed upon this system merely alters n by the corresponding quantity $- kU_0$, as is evident from (7), § 366.

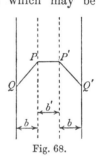

Fig. 68.

Thus $U_1 = U_2 = V, \quad \Delta_2 = \Delta_1 = \Delta = - V/b\,;$

and $B^2 - 4AC = 4\Delta^2 \sinh^4 kb.$

Hence $n + kV = \dfrac{V}{b} \dfrac{\sinh kb \sinh k\,(b + b') \pm \sinh^2 kb}{\sinh k\,(2b + b')} \dots\dots(20).$

As was to be expected, since the curvature of the velocity curve is of one sign, the values of n in (20) are real. It is easy from the symmetry to see that the two normal disturbances are such that the values of v at the surfaces of separation are either equal or opposite for a given value of x. In the first case the surfaces are bent towards the same side, and (as may be found from the equations or inferred from the particular case presently to be mentioned) the corresponding value of n in (20) has the upper sign. In the second case the motion is symmetrical with respect to the median plane which behaves as a fixed wall.

If the middle layer be absent $(b' = 0)$, one value of n, that corresponding to the symmetrical motion, vanishes. The remaining value is given by

$$n + kV = \frac{2 \sinh^2 kb}{\sinh 2kb} = \frac{V \tanh kb}{b} \quad \dots\dots\dots\dots (21).$$

The other case which we shall consider is that in which the velocities U on the two sides of the median plane are opposite to one another; so that

$$U_1 = - U_2 = V, \quad \Delta_2 = - \Delta_1 = - \mu V \dots\dots\dots (22).$$

Here $B = 0$, and

$$C = - k^2 V^2 \sinh k (2b + b') - 2k\mu V^2 \sinh kb \ \sinh k (b + b')$$
$$- \mu^2 V^2 \sinh^2 kb \ \sinh kb'.$$

For the sake of brevity we will write $kb = \beta$, $kb' = \beta'$; so that the equation for n becomes

$$\frac{n^2}{k^2 V^2} = \frac{k^2 \sinh(2\beta + \beta') + 2k\mu \sinh \beta \ \sinh(\beta + \beta') + \mu^2 \sinh^2 \beta \ \sinh^2 \beta'}{k^2 \sinh (2\beta + \beta')}.$$

$$= \frac{\{\mu \sinh \beta \ \sinh \beta' + k \sinh (\beta + \beta')\}^2 - k^2 \sinh^2 \beta}{k^2 \sinh \beta' \sinh (2\beta + \beta')} \quad \dots\dots(23).$$

Here the two values of n are equal and opposite; and, since Δ_1, Δ_2 are of opposite signs, the question is open as to whether n is real or imaginary.

It is at once evident that n is real if μ be positive, that is, if Δ_1 and V are of the same sign as in Fig. 69.

Even when μ is negative, n^2 is necessarily positive for great values of k, that is, for small wave-lengths. For we have ultimately from (23) $\quad n = \pm kV.$

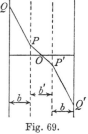

Fig. 69.

We may now inquire for what values of μ n^2 may be negative when k is very small, that is, when the wave-length is very great. Equating the numerator of (23) to zero, and expanding the hyperbolic sines, we get as a quadratic in μ,

$$\mu^2 b^2 b' + 2\mu b (b + b') + 2b + b' = 0,$$

whence $\quad \mu = 1/b, \quad$ or $\quad \mu = - 1/b - 2/b' \dots\dots\dots\dots (24).$

When μ lies between these limits (and then only), n^2 is negative, and the disturbance (of great wave-length) increases exponentially with the time.

We may express these results by means of the velocity V_0 at the wall where $y = 0$. We have

$$V_0 = V \frac{b + \frac{1}{2}b'}{\frac{1}{2}b'} + \Delta_1 b = V \left(\frac{b + \frac{1}{2}b'}{\frac{1}{2}b'} + \mu b \right).$$

The limiting values of V_0 are therefore $bV/\frac{1}{2}b'$ and 0. The velocity curve corresponding to the first limit is shewn in Fig. 70 by the line $QPOP'Q'$, the point Q being found by drawing a line AQ parallel to OP to meet the wall in Q. If $b' = 2b$, QP is parallel to OA, or the velocity is constant in each of the extreme layers.

At the second limit $V_0 = 0$, and the velocity curve is that shewn in Fig. 71.

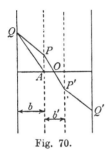
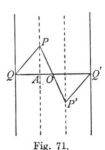

Fig. 70.   Fig. 71.

It is important to notice that motions represented by velocity curves intermediate between these limits are unstable in a manner not possible to motions in which the velocity curve, as in Fig. 68, is of one curvature throughout.

According to the first approximation, the motion of Fig. 71 is on the border-line between stability and instability for disturbances of great wave-length; but, if we pursue the calculation, we find that it is really unstable. Taking in (23)

$$\mu = -1/b - 2/b',$$

we get, after reduction,

$$\frac{n^2}{k^2 V^2} = -\frac{k^2 b^2}{3} \dots\dots\dots\dots\dots\dots(25),$$

indicating instability.

From the second form of (23) we see that, whatever may be the value of k, it is possible so to determine μ that the disturbance shall be unstable. The condition is simply that μ must be between the limits

$$- k\frac{\sinh k(b + b') \pm \sinh kb}{\sinh kb \ \sinh kb'},$$

or $- k\{\coth kb + \coth \tfrac{1}{2}kb'\}, \ - k\{\coth kb + \tanh \tfrac{1}{2}kb'\}\ldots(26),$
of which the first corresponds to the superior limit to the *numerical value* of μ.

When k is very large, the limits are very great and very close. When k is small, they become

$$- 1/b - 2/b' \quad \text{and} \quad - 1/b,$$

as has already been proved. As k increases from 0 to ∞, the numerical value of the upper limit increases continuously from $1/b + 2/b'$ to ∞, and in like manner that of the inferior limit from $1/b$ to ∞. The motion therefore cannot be stable for *all values of k*, if μ (being negative) exceed numerically $1/b$. The final condition of complete stability is therefore that algebraically

$$\mu > - 1/b \ldots\ldots\ldots\ldots\ldots\ldots(27).$$

In the transition case

$$V_0 = \left(\mu - \frac{1}{b} + \frac{2}{b'}\right) Vb = \frac{2Vb}{b'} \ldots\ldots\ldots\ldots(28);$$

it is that represented in Fig. 70. If PQ be bent more downwards than is there shewn, as for example in Fig. 71, the steady motion is certainly unstable.

Reverting to the general equations (11), (12), (13), (14), (15), let us suppose that $\Delta_2 = 0$, amounting to the abolition of the corresponding surface of discontinuity. We get

$$B = k(U_1 + U_2)\sinh k(b_2 + b' + b_1) + \Delta_1 \sinh kb_1 \ \sinh k(b_2 + b'),$$
$$B^2 - 4AC = \{k(U_1 - U_2)\sinh k(b_2 + b' + b_1)$$
$$+ \Delta_1 \sinh kb_1 \ \sinh k(b_2 + b')\}^2;$$

so that $$n = - kU_2\ldots\ldots\ldots\ldots\ldots\ldots\ldots(29),$$

or $$n = - kU_1 - \frac{\Delta_1 \sinh kb_1 \ \sinh k(b_2 + b')}{\sinh k(b_1 + b' + b_2)}\ldots\ldots\ldots(30).$$

The latter is the general solution for two layers of constant vorticity of breadths b_1 and $b' + b_2$. An equivalent result may be obtained by supposing in (11) &c. that $b' = 0$, or that $b_1 = 0$.

The occurrence of (29) suggests that any value of $-kU$ is admissible as a value of n, and the meaning of this is apparent from the fundamental equation (7), § 366. For, at the place where $n + kU = 0$, (1) need not be satisfied, that is, the arbitrary constants in (2) may change their values. It is evident that, with the prescribed values of n and k, a solution may be found satisfying the required conditions at the walls and at the surfaces where dU/dy changes value, as well as equation (3) at the plane where $n + kU = 0$. In this motion an additional vorticity is supposed to be communicated to the fluid at the plane in question, and it moves with the fluid at velocity U.

We may inquire what occurs at a second place in the fluid where the velocity happens to be the same as at the first place of added vorticity. The second place may be either within a layer of originally uniform vorticity, or upon a surface of transition. In the first case nothing very special presents itself. If there be no new vorticity at the second place, the value of v is definite as usual, save as to one arbitrary multiplyer. But, consistently with the given value of n, there may be new vorticity at the second as well as at the first place, and then the complete value of v for the given n may be regarded as composed of two parts, each proportional to one of the new vorticities and each affected by an arbitrary multiplyer.

If the second place lie upon a surface of transition, it follows from (4) that $v = 0$, since $\Delta (dU/dy)$ is finite. From this fact we might be tempted to infer that the surface in question behaves like a fixed wall, but a closer examination shews that the inference would be unwarranted. In order to understand this, it may be well to investigate the relation between v and the displacement of the surface, supposed also to be proportional to $e^{int} . e^{ikx}$. Thus, if the equation of the surface be

$$F = y - h e^{int+ikx} = 0 \dots\dots\dots\dots\dots(31),$$

the condition to be satisfied is[1]

$$\frac{dF}{dt} + U \frac{dF}{dx} + v \frac{dF}{dy} = 0 \dots\dots\dots\dots\dots (32),$$

so that $- ih (n + kU) + v = 0 \dots\dots\dots\dots\dots (33)$

[1] Lamb's *Hydrodynamics*, § 10.

is the required relation.   A finite h is thus consistent with an evanescent v.

368.   In the problems of § 367 the fluid is bounded by fixed walls; in those to which we now proceed, it will be considered to be unlimited.   As a first example, let us suppose that on the upper side of a layer of thickness b the undisturbed velocity U is equal to $+V$, and on the lower side to $-V$, while inside the layer

Fig. 72.   Fig. 73.   Fig. 74.

it changes uniformly, Fig. 72.   The vorticity within the layer is V/b, and outside the layer it is zero.

The most straightforward method of attacking this problem is perhaps on the lines of § 367.   From $y = -\infty$ to $y = 0$, we should assume an expression of the form $v_1 = e^{ky}$, satisfying the necessary condition when $y = -\infty$.   Then from $y = 0$ to $y = b$,

$$v_2 = v_1 + M_1 \sinh ky;$$

and from $y = b$ to $y = +\infty$,

$$v_3 = v_2 + M_2 \sinh k(y - b).$$

But by the conditions at $+\infty$, v_3 must be of the form e^{-ky}, so that

$$1 + M_1 + M_2 e^{-kb} = 0.$$

The two other conditions may then be formed as in § 367, and the two constants M_1, M_2 eliminated, giving finally an equation for n. But it will be more appropriate and instructive to follow a different course, suggested by vortex theory.

If we write the fundamental equation

$$\left(\frac{n}{k} + U\right)\left(\frac{d^2v}{dy^2} - k^2v\right) - \frac{d^2U}{dy^2} v = 0 \dots\dots\dots\dots (1),$$

in the form

$$d^2v/dy^2 - k^2v = Y \dots\dots\dots\dots\dots\dots(2),$$

we see that, if $Y = 0$ from $y = -\infty$ to $y = +\infty$, then $v = 0$.   Any value that v may have may thus be regarded as dependent upon Y, and further, in virtue of the linearity, as compounded by simple addition of the values corresponding to the partial values of Y.

In the applications which we have in view Y vanishes, except at certain definite places—the surfaces of discontinuity—where alone d^2U/dy^2 differs from zero. The complete value of v may thus be found by summation of partial values, each corresponding to a single surface of discontinuity.

To find the partial value corresponding to a surface of discontinuity situate at $y = y_1$, we have to suppose in (2) that Y vanishes at all other places, while v vanishes at $\pm \infty$. Thus, when $y > y_1$, v must be proportional to $e^{-k(y-y_1)}$, and when $y < y_1$, v must be proportional to $e^{+k(y-y_1)}$. Moreover, since v itself must be continuous at $y = y_1$, the coefficients of the exponentials must be equal, so that the value may be written

$$v = C e^{\pm k(y-y_1)} \dots\dots\dots\dots\dots\dots(3),$$

when C is some constant.

In the particular problem above proposed there are two surfaces of discontinuity, at $y = 0$ and at $y = b$; and accordingly the complete value of v may be written in the form

$$v = A e^{\pm ky} + B e^{\pm k(y-b)} \dots\dots\dots\dots\dots(4).$$

We have now to satisfy at each surface the equation of condition (4), § 367. When $y = 0$, we have from (4)

$$v_0 = A + B e^{-kb}, \qquad\qquad \Delta (dv/dy)_0 = -2kA,$$

while $U = -V$, $\qquad \Delta (dU/dy) = +2V/b$;

and when $y = b$,

$$v_b = A e^{-kb} + B, \qquad\qquad \Delta (dv/dy)_b = -2kB,$$

while $U = +V$, $\qquad \Delta (dU/dy) = -2V/b$.

The conditions to be satisfied by $B : A$ and n are thus

$$A \{n - kV + V/b\} + B \{V e^{-kb}/b\} = 0 \dots\dots\dots\dots(5),$$

$$A \{V e^{-kb}/b\} - B \{n + kV - V/b\} = 0 \dots\dots\dots\dots(6);$$

from which by elimination of $B : A$,

$$n^2 = \frac{V^2}{b^2} \{(kb - 1)^2 - e^{-2kb}\} \dots\dots\dots\dots(7).$$

When kb is small, that is, when the wave-length is great in comparison with b, the case approximates to that of a sudden transition from the velocity $-V$ to the velocity $+V$. Then from (7)

$$n^2 = -k^2 V^2 \dots\dots\dots\dots\dots\dots(8),$$

in agreement with the value already found (17), § 365. In this case the steady motion is unstable. On the other hand, when kb is great, we find from (7)

$$n^2 = k^2 V^2 \dots\dots\dots\dots\dots\dots\dots(9);$$

and, since the two values of n are real, the motion is stable. It appears, therefore, that so far from the instability increasing indefinitely with vanishing wave-length, as happens when the transition from $-V$ to $+V$ is sudden, a diminution of wave-length below a certain value is accompanied by an instability which gradually decreases, and is finally exchanged for actual stability. The following table exhibits more in detail the progress of $b^2 n^2 / V^2$ as a function of kb :—

kb	$b^2 n^2 / V^2$	kb	$b^2 n^2 / V^2$
·2	− ·03032	1·0	− ·13534
·4	− ·08933	1·2	− ·05072
·6	− ·14120	1·3	+ ·01573
·8	− ·16190	2·0	+ ·98168

We see that the instability is greatest when $kb = \cdot8$ nearly, that is, when $\lambda = 8b$; and that the passage from instability to stability takes place when $kb = 1\cdot3$ nearly, or $\lambda = 5b$.

Corresponding with the two values of n, there are two ratios of $B : A$ determined by (5) or (6), each of which gives a normal mode of disturbance, and by means of these normal modes arbitrary initial circumstances may be represented. It will be seen that for the stable disturbances the ratio $B : A$ is real, indicating that the sinuosities of the two surfaces are at every moment in the same phase.

We may next take an example from a jet of thickness $2b$ moving in still fluid, supposing that the velocity in the middle of the jet is V, and that it falls uniformly to zero on either side, (Fig. 73). Taking the origin of y in the middle line, we may write

$$U = V (1 \mp y/b) \dots\dots\dots\dots\dots (10),$$

in which the − sign applies to the upper, and the + sign to the lower half of the jet (Fig. 73). There are now three surfaces $y = -b$, $y = 0$, $y = +b$, at which the form of v suffers discontinuity. As in (4) we may take

$$v = A e^{\pm k(y+b)} + B e^{\pm ky} + C e^{\pm k(y-b)} \dots\dots\dots (11);$$

so that, when

$$y = -b, \quad U = 0, \quad \Delta (dU/dy) = V/b,$$
$$v = A + Be^{-kb} + Ce^{-2kb}, \quad \Delta (dv/dy) = -2kA \; ;$$

when　　$y = 0, \quad U = V, \quad \Delta (dU/dy) = -2V/b,$
$$v = Ae^{-kb} + B + Ce^{-kb}, \quad \Delta (dv/dy) = -2kB \; ;$$

when　　$y = b, \quad U = 0, \quad \Delta (dU/dy) = V/b,$
$$v = Ae^{-2kb} + Be^{-kb} + C, \quad \Delta (dv/dy) = -2kC.$$

The introduction of these values into the equations of condition (4), § 367 gives

$$mA + \gamma B + \gamma^2 C = 0 \quad\ldots\ldots\ldots\ldots\ldots\ldots\ldots (12),$$

$$\gamma A + (\tfrac{3}{2} - \tfrac{1}{2} m - kb) B + \gamma C = 0 \quad\ldots\ldots\ldots\ldots (13),$$

$$\gamma^2 A + \gamma B + mC = 0 \quad\ldots\ldots\ldots\ldots\ldots\ldots\ldots (14),$$

which are the equations determining $A : B : C$ and n.

By the symmetries of the case, or by inspection of (12), (13), (14), we see that one of the normal disturbances is defined by

$$B = 0, \quad A + C = 0 \ldots\ldots\ldots\ldots\ldots\ldots (15),$$

and that the corresponding value of m is γ^2. Thus for the symmetrical disturbance

$$n = -\frac{V}{2b} (1 - e^{-2kb}) \ldots\ldots\ldots\ldots\ldots\ldots (16),$$

indicating stability, so far as this mode is concerned.

The general determinant of the system of three equations may be put into the form

$$(m - \gamma^2) \{ m^2 + (\gamma^2 + 2kb - 3) m + \gamma^2 (1 + 2kb) \} = 0 \ldots (17),$$

in which the first factor corresponds to the symmetrical disturbance already considered. The two remaining values of n are real, if

$$(\gamma^2 + 2kb - 3)^2 - 4\gamma^2 (1 + 2kb) > 0 \ldots\ldots\ldots\ldots (18),$$

but not otherwise. When kb is infinite, $\gamma = 0$, and (18) is satisfied; so that the motion is stable when the wave-length of disturbance is small in comparison with the thickness ($2b$) of the jet. On the other hand, as may be proved without difficulty by expanding γ, or e^{-kb}, in (18), the motion is unstable, when the wave-length is great in comparison with the thickness of the jet.

The values of the left-hand member of (18) can be more easily computed when it is thrown into the form

$$(5 + 2kb - e^{-2kb})^2 - 16 (1 + 2kb) \ldots\ldots\ldots\ldots (19).$$

Some corresponding values of (19) and $2kb$ are tabulated below :—

$2kb$	(19)	$2kb$	(19)
·5	− ·054	2·5	− ·975
1·0	− ·279	3·0	− ·794
1·5	− ·599	3·5	− ·263
2·0	− ·876	4·0	+ ·671

The imaginary part of n, when such exists, is proportional to the square root of (19). The wave-length of maximum instability is thus determined approximately by $2kb = 2·5$, or $\lambda = 2·5 \times 2b$. The critical wave-length is given by $2kb = 3·5$ nearly, or $\lambda = 1·8 \times 2b$, smaller wave-lengths than this leading to stability, and greater wave-lengths to instability. In these respects there is a fairly close analogy with cylindrical columns of liquid under capillary force (§ 357), although the nature of the equilibrium itself and the manner in which it is departed from are so entirely different.

One more step in the direction of generality may be taken by supposing the maximum velocity V to extend through a layer of finite thickness b' in the middle of the jet (Fig. 74). In this layer accordingly there is no vorticity, while in the adjacent layers of thickness b the vorticity and velocity remain as before.

Taking, as in (11), four constants A, B, C, D to represent the discontinuities at the four surfaces considered in order, and writing $\gamma = e^{-kb}$, $\gamma' = e^{-kb'}$, we have at the first surface

$$U = 0, \qquad \Delta (dU/dy) = + V/b,$$
$$v = A + \gamma B + \gamma\gamma'C + \gamma^2\gamma'D, \qquad \Delta (dv/dy) = - 2kA \;;$$

at the second surface

$$U = V, \qquad \Delta (dU/dy) = - V/b,$$
$$v = \gamma A + B + \gamma'C + \gamma\gamma'D, \qquad \Delta (dv/dy) = - 2kB \;;$$

at the third surface

$$U = V, \qquad \Delta (dU/dy) = - V/b,$$
$$v = \gamma\gamma'A + \gamma'B + C + \gamma D, \qquad \Delta (dv/dy) = - 2kC \;;$$

at the fourth surface

$$U = 0, \qquad \Delta (dU/dy) = + V/b,$$
$$v = \gamma^2\gamma'A + \gamma\gamma'B + \gamma C + D, \qquad \Delta (dv/dy) = - 2kD.$$

Using these values in (4) § 367, we get

$$A \{1 + 2bn/V\} + \gamma B + \gamma\gamma'C + \gamma^2\gamma'D = 0\ldots\ldots(20),$$

$$\gamma A + B \{1 - 2b (k + n/V)\} + \gamma'C + \gamma\gamma'D = 0\ldots\ldots(21),$$

$$\gamma\gamma'A + \gamma'B + C \{1 - 2b (k + n/V)\} + \gamma D = 0\ldots\ldots(22),$$

$$\gamma^2\gamma'A + \gamma\gamma'B + \gamma C + D \{1 + 2bn/V\} = 0\ldots\ldots(23).$$

The elimination of the ratios $A : B : C : D$ would give a bi-quadratic in n, which, however, may be split into two quadratics, one relating to symmetrical disturbances for which $A + D = 0$, $B + C = 0$; and the other to disturbances for which $A - D = 0$, $B - C = 0$. The resulting equation in n may be written

$$\left(\frac{2bn}{V}\right)^2 + (\pm \gamma' \mp \gamma'\gamma^2 + 2kb) \frac{2bn}{V}$$
$$\pm \gamma' - 1 + 2kb + \gamma^2 (1 \mp \gamma' \mp 2kb\gamma') = 0\ldots\ldots(24).$$

In (24) the upper signs of the ambiguities correspond to the symmetrical disturbances. The roots are real, and the corresponding disturbances are stable, if

$$(\pm \gamma' \mp \gamma'\gamma^2 + 2kb)^2 - 4 [\pm \gamma' - 1 + 2kb + \gamma^2 (1 \mp \gamma' \mp 2kb\gamma')]\ldots(25),$$

be positive.

In what follows we will limit our attention to the symmetrical disturbances, that is, to the upper signs in (25), and to terms of orders not higher than the first in b'. The expression (25) may then be reduced to

$$(1 - \gamma^2 - 2kb)^2 + 2kb' (1 + \gamma^2) (1 - \gamma^2 - 2kb)\ldots\ldots(26).$$

If kb be very small, this becomes

$$4k^4b^4 - 8kb' . k^2b^2 \ldots\ldots\ldots\ldots\ldots (27).$$

If b' is zero (27) is positive, and the disturbance is stable, as we found before; but, if b and b' be of the same order of magnitude and both small compared with λ, it follows from (27) that the disturbance is unstable, although it be symmetrical.

If in (24) we suppose that $b' = 0$, we fall back upon the suppositions of the previous problem. For the symmetrical disturbances, putting $\gamma' = 1$ in (24), we get

$$\left(\frac{2bn}{V}\right)^2 + (1 - \gamma^2 + 2kb)\frac{2bn}{V} + 2kb\,(1 - \gamma^2) = 0,$$

shewing that the values of $2bn/V$ are $\gamma^2 - 1$ and $-2kb$. The former agrees with (16), and the latter gives $n + kV = 0$. We have already seen that any value of $-kU$ is a possible solution for n.

If on the other hand we suppose that $b = 0$, we fall back upon the case of a jet of uniform velocity V and thickness b' moving in still fluid. The equation for n becomes, after division by b^2,

$$n^2 + (1 \pm \gamma')\,kV.\,n + \tfrac{1}{2}(1 \pm \gamma')\,k^2V^2 = 0,$$

or $$\qquad (n + kV)^2\frac{1 \pm \gamma'}{1 \mp \gamma'} + n^2 = 0 \dots\dots\dots\dots\dots(28).$$

In (28) $\quad \dfrac{1 + \gamma'}{1 - \gamma'} = \coth \tfrac{1}{2}kb', \quad \dfrac{1 - \gamma'}{1 + \gamma'} = \tanh \tfrac{1}{2}kb' ;$

so that the result is in harmony with (22), (29), § 365, where l corresponds with $\tfrac{1}{2}b'$.

Another particular case of (24), comparable with previous results, is obtained by supposing b' to be infinite.

369. When d^2U/dy^2 is finite, we must fall back upon the general equation § 366

$$\left(\frac{n}{k} + U\right)\left(\frac{d^2v}{dy^2} - k^2v\right) - \frac{d^2U}{dy^2}\,v = 0\dots\dots\dots\dots(1),$$

from which the curve representing v as a function of y can theoretically be constructed when n (being real) is known. In fact we may regard (1) as determining the curvature with which we are to proceed in tracing the curve through any point. At a place when $n + kU$ vanishes, that is, where the stream-velocity is equal to the wave-velocity, the curvature becomes infinite, unless v vanishes. The character of the infinity at such a place (suppose $y = 0$) would be most satisfactorily investigated by means of the complete solution of some particular case. It is, however, sufficient to examine the form of solution in the neighbourhood of $y = 0$, and for this purpose the differential equation may be simplified. Thus, when y is small, $n + kU$ may be treated as proportional to y, and

d^2U/dy^2 as approximately constant. In comparison with the large term, $k^2 v^2$ may be neglected, and it suffices to consider

$$d^2v/dy^2 + y^{-1}v = 0 \dots\dots\dots\dots\dots(2),$$

a known constant multiplying y being omitted for the sake of brevity. This falls under the head of Ricati's equation

$$d^2v/dy^2 + y^\mu v = 0 \dots\dots\dots\dots\dots(3),$$

of which the solution is in general (m fractional)[1]

$$v = \sqrt{y} \cdot \{AJ_m(\xi) + BJ_{-m}(\xi)\} \dots\dots\dots\dots(4),$$

where $m = 1/(\mu + 2)$, $\xi = 2my^{1/2m} \dots\dots\dots\dots(5).$

When, as in the present case, m is integral, $J_{-m}(\xi)$ is to be replaced (§ 341) by the function of the second kind $Y_m(\xi)$. The general solution of (2) is accordingly

$$v = \sqrt{y} \cdot \{AJ_1(2\sqrt{y}) + BY_1(2\sqrt{y})\} \dots\dots\dots\dots(6).$$

In passing through zero y changes sign and with it the character of the functions. If we regard (6) as applicable on the positive side, then on the negative side we may write

$$v = \sqrt{y} \cdot \{CJ_1(2\sqrt{y}) + DY_1(2\sqrt{y})\} \dots\dots\dots\dots(7),$$

the argument of the functions in (7) being pure imaginaries.

From the known forms of the functions (§ 341) we may deduce, as applicable when y is small,

$$v = A\{y - \tfrac{1}{2}y^2\}$$
$$+ B\{\tfrac{1}{2}(1 - y + \tfrac{1}{4}y^2) - \log(2\sqrt{y}) \cdot (y - \tfrac{1}{2}y^2) + y - \tfrac{3}{4}y^2\} \dots\dots(8);$$

so that ultimately

$$v = \tfrac{1}{2}B, \quad \frac{dv}{dy} = A - \tfrac{1}{2}B\log y, \quad \frac{d^2v}{dy^2} = -A - \tfrac{1}{2}By^{-1} \dots\dots(9),$$

v remaining finite in any case.

We will now shew that any value of $-kU$ is an admissible value of n in (1). The place where $n + kU = 0$ is taken as origin of y; and in the first instance we will suppose that $n + kU$ vanishes nowhere else. In the immediate neighbourhood of $y = 0$ the solutions applicable upon the two sides are (6), (7), and they are subject to the condition that v shall be continuous. Hence by

---

[1] Lommel, *Studien über die Bessel'schen Functionen* § 31, Leipzig, 1868; Gray and Matthews' *Bessel Functions*, p. 233, 1895.

(9), $B = D$, leaving three constants arbitrary. The manner in which the functions start from $y = 0$ being thus ascertained, their further progress is subject to the original equation (1), which completely defines them when the three arbitraries are known. In the present case two relations are given by the conditions to be satisfied at the fixed walls or other boundaries of the fluid, and thus is determined the entire form of v, save as to a constant multiplyer. If B and D are finite, there is infinite vorticity at the origin.

Any other places at which $n + kU = 0$ may be treated in a similar manner, and the most general solution will contain as many arbitrary constants as there are places of infinite vorticity. But the vorticity need not be infinite merely because $n + kU = 0$; and in fact a particular solution may be obtained with only one infinite vorticity. At any other of the critical places, such for example as we may now suppose the origin to be, B and D may vanish, so that $v = 0$, $d^2v/dy^2 = A$, or C.

From this discussion it would seem that the infinities which present themselves when $n + kU = 0$ do not seriously interfere with the application of the general theory, so long as the square of the disturbance from steady motion is neglected.

A large part of the preceding paragraphs is taken from certain papers by the author[1]. The reader should also consult Lord Kelvin's writings[2] in which the effects of viscosity are dealt with.

370. It remains to describe the phenomena of sensitive flames and to indicate, so far as can be done, the application of theoretical principles. In a sense the combination of flame and resonator described in § 322 h may be called sensitive, but in this case it is rather the resonator to which the name attaches, the office of the flame being to maintain by a periodic supply of heat the vibration of the resonator when once started. Following Tyndall, we may conveniently limit the term to naked flames and jets, where the origin of the sensitiveness is undoubtedly to be found in the instability which accompanies vortex motion.

The earliest observation upon this subject was that of Prof.

---

[1] *Proc. Math. Soc.*, vol. xi. p. 57, 1880; vol. xix. p. 67, 1887. It is hoped shortly to communicate a supplement.

[2] *Phil. Mag.* vol. xxiv. pp. 188, 272, 1887.

Leconte [1], who noticed the jumping of the flame from an ordinary fishtail burner in response to certain notes of a violoncello. The sensitive condition demanded that in the absence of sound the flame should be on the point of flaring. When the pressure of gas was reduced, the sensitiveness was lost.

An independent observation of the same nature drew the attention of Prof. Barrett to sensitive flames; and he investigated the kind of burner best suited to work with the ordinary pressure of the gas mains [2]. "It is formed of glass tubing about $\frac{3}{8}$ of an inch (1 cm.) in diameter, contracted to an orifice $\frac{1}{16}$ of an inch (·16 cm.) in diameter. It is very essential that this orifice should be slightly V-shaped....Nothing is easier than to form such a burner; it is only necessary to draw out a piece of glass tubing in a gas flame, and with a pair of scissors snip the contraction into the shape indicated."

But the most striking by far is the high-pressure flame employed by Tyndall. The gas is supplied from a special holder under a pressure of say 25 cm. of water to a pinhole steatite burner, and the flame rises to a height of about 40 cm. Under the influence of a sound of suitable (very high) pitch the flame roars, and drops down to perhaps half its original height [3]. Tyndall shewed that the seat of sensitiveness is at the root of the flame. Sound coming along a tube is ineffective when presented to the flame a little higher up, and also when caused to impinge upon the burner below the place of issue.

It is to Tyndall that we owe also the demonstration that it is not to the flame as such that these extraordinary effects are to be ascribed. Phenomena substantially the same are obtained when a jet of unignited gas, of carbonic acid, hydrogen, or even air itself, issues from an orifice under proper pressure. They may be rendered visible in two ways. By association with smoke the whole course of the jet may be made apparent; and it is found that suitable smoke jets can surpass even flames in delicacy. "The notes here effective are of much lower pitch than those which are most efficient in the case of flames." Another way of making the sensitiveness of an air-jet visible to the eye is to cause

---

[1] On the Influence of Musical Sounds on the Flame of a Jet of Coal-gas. *Phil. Mag.* vol. xv. p. 235, 1858.

[2] *Phil. Mag.* vol. xxxiii. p. 216, 1867.

[3] *Phil. Mag.* vol. xxxiii. pp. 92, 375, 1867; *Sound*, 3rd ed. ch. vi.

it to impinge upon a flame, such as a candle flame, which plays merely the part of an indicator.

In the sensitive flame of Prof. Govi [1] and of Mr Barry [2] the gas is unignited at the burner, but catches fire on the further side of wire-gauze held at a suitable distance. On the same principle is an arrangement employed by the author [3]. A jet of coal gas from a pinhole burner rises vertically in the interior of a cavity from which air is excluded. It then passes into a brass tube a few inches long, and on reaching the top burns in the open. The front wall of the cavity is formed of a flexible membrane of tissue-paper, through which external sounds can reach the burner. In these cases the sensitive agent is the unignited part of the jet. Used in this way a given burner requires a much less pressure of gas than is necessary when the flame is allowed to reach it, and the sounds which have the most influence are graver.

Struck by the analogy between these phenomena and those of water-jets investigated by Savart and Plateau, the earlier observers seem to have leaped to the conclusion that the manner of disintegration was also similar—symmetrical, that is, about the axis; and Prof. Leconte went so far as to deduce the existence of a cohesive force in gases. A surface tension, however, requires a very abrupt transition between the properties of the matter on the two sides, such as could have only a momentary existence when there is a tendency to mix, so that it appears extremely unlikely that capillarity plays here any sensible part.

The question of the manner of disintegration, whether it be by gradually increasing *varicosity* or by gradually increasing *sinuosity*, is of the greatest importance, and the answer is still, perhaps, in some cases open to doubt. But that the latter is predominant in general follows from a variety of arguments. The necessity, as remarked by Barrett, for an unsymmetrical orifice points strongly in this direction. The same conclusion is drawn by Ridout [4] from the results of some ingenious experiments. The latter observer found further that fishtail flames, formed by the union at a small angle of jets from two perfectly similar glass nozzles, shewed a

[1] Torino, *Atti Acad. Sci.* vol. v. p. 396, 1869.

[2] Tyndall's *Sound*, 3rd edition, p. 240.

[3] *Camb. Phil. Soc. Proc.* vol. iv. p. 17, 1883.

[4] *Nature*, vol. xviii. p. 604, 1878.

sensitiveness dependent upon the direction of the sound. If this direction lie in the plane of symmetry containing the flame (that perpendicular to the plane of the nozzles), there is no response.

Even in the case of the tall high-pressure flame from a pin-hole burner, where to all appearance both the nozzle and the flame (when undisturbed) are perfectly symmetrical, there is reason to believe that the manner of disintegration is sinuous, or unsymmetrical. Perhaps the easiest road to this conclusion is by examining the behaviour of the flame when exposed to stationary sonorous waves, such as may be derived by superposing upon direct waves from a source giving a pure tone the waves reflected perpendicularly from a flat obstacle, e.g. a sheet of glass. According to the analogy with capillary jets, an analogy pushed further than it will bear by most writers upon this subject, the flame should be excited when the nozzle is situated at a node, where the pressure varies most, and remain unaffected at a loop where the pressure does not vary at all. There was no difficulty in proving experimentally [1] that the facts are precisely the opposite. The source of sound was a bird-call (§ 371), and the observations were made by moving the burner to and fro in front of the reflector until the positions were found in which the flame was least disturbed. These positions were very well defined, and the measurements shewed distances from the reflector proportional to the series of numbers 1, 2, 3, &c., and therefore corresponding to nodes. If the positions had coincided with loops, the distances would have formed a series proportional to the *odd* numbers 1, 3, 5, &c. The wave-length of the sound, determined by the doubled interval between consecutive minima, was $31\cdot2$ mm., corresponding to pitch $f\sharp^{r\prime}$.

A few observations were made at the same time on the positions of the silences as estimated by the ear listening through a tube. As was to be expected, they coincided with the loops, bisecting the intervals given by the flame. When the flame was in a position of minimum effect, and the free end of the tube was held close to the burner at an equal distance from the reflector, the sound heard was a maximum, and diminished when the end of the tube was displaced a little in either direction. It was thus established that the flame is affected where the ear would not be affected, and *vice versâ*.

[1] *Phil. Mag.* vol. VII. p. 153, 1879.

Flames from pinhole burners, which perform well in other respects, seem always to shew a marked difference according to the direction in which the sound arrives. If, while a bird-call is in operation, the burner be turned steadily round its axis, two positions differing by 180° are found, in which there is little or no response. This peculiarity may sometimes be turned to account in experiment[1]. Thus after such an adjustment has been made that the direct sound has no effect, vigorous flaring may yet result from the impact of sound from the same source after reflection from a small pane of glass, the pane being held so that the direction of arrival is at 90° to that of the direct sound, and this although the distance travelled by the reflected sound is the greater.

Tyndall[2] lays it down as an essential condition of complete success in the more delicate experiments with these flames, "that a free way should be open for the transmission of the vibrations from the flame, *backwards,* through the gaspipe which feeds it. The orifices of the stopcocks near the flame ought to be as wide as possible." The recommendation is probably better justified than the reason given for it. Prof. Barrett[3] attributes the evil effect of a partially opened stopcock to the irregular flow and consequent ricochetting of the current of gas from side to side of the pipe. In some experiments of my own[4] the introduction of a glass nozzle into the supply pipe, making the flow of gas in the highest degree irregular, did not interfere, nor did other obstructions unless attended by hissing sounds. The prejudicial action of a partially opened stopcock was thus naturally attributed to the production of internal sounds of the kind to which the flame is sensitive, and this view of the matter was confirmed by some observations of the pressure of the gas in the neighbourhood of the burner. "In the path of the gas there were inserted two stopcocks, one only a little way behind the manometer junction, the other separated from it by a long length of india-rubber tubing. When the first cock was fully open, and the flame was brought near the flaring-point by adjustment of the distant cock, the sensitiveness to external sounds was great,

---

[1] *Proc. Roy. Inst.* vol. XII. p. 192, 1888; *Nature,* vol. XXXVIII. p. 208, 1888.

[2] *Phil. Mag.* vol. XXXIII. p. 99, 1867.

[3] *Phil. Mag.* vol. XXXIII. p. 288, 1867.

[4] *Phil. Mag.* vol. XIII. p. 345, 1882.

and the manometer indicated a pressure of 10 inches (25·4 cm.) of water. But when the distant cock stood fully open and the adjustment was effected at the other, high sensitiveness could not be obtained; and the reason was obvious, because the flame flared without external excitation while the pressure was still an inch (2·54 cm.) short of that which had been borne without flinching in the former arrangement. On opening again the neighbouring cock to its full extent, and adjusting the distant one until the pressure at the manometer measured 9 inches (22·9 cm.), the flame was found comparatively insensitive."

The most direct and satisfactory evidence as to the manner of disintegration is of course that of actual observation. Using a jet of phosphorus smoke from a glass nozzle and a stroboscopic disc, I was able (in 1879) to see the sinuosities when the jet was disturbed by a fork of pitch 256 vibrating in its neighbourhood[1]. Moreover by placing the nozzle exactly in the plane of symmetry between the prongs of the tuning-fork, it could be verified that the disturbance required is motion transverse to the jet. In this position there was but little effect; but the slightest displacement led to an early rupture of the jet.

"In order to exalt the sensitiveness of jets to notes of moderate pitch, I found the use of resonators advantageous. These may be of Helmholtz's pattern; but suitably selected wide-mouth bottles answer the purpose. What is essential is that the jet should issue from the nozzle in the region of rapid reciprocating motion at the mouth of the resonator, and in a transverse direction.

"Good results were obtained at a pitch of 256. When two forks of about this pitch, and slightly out of tune with one another, were allowed to sound simultaneously, the evolutions of the smoke-jet in correspondence with the audible beats were very remarkable. By gradually raising the pressure at which the smoke is supplied, in the manner usual in these experiments, a high degree of sensitiveness may be attained, either with a drawn-out glass nozzle or with the steatite pinhole burner used by Tyndall. In some cases (even at pitch 256) the combination of jet and resonator proved almost as sensitive to sound as the ear itself.

"The behaviour of the sensitive jet does not depend upon the smoke-particles, whose office is merely to render the effects more

[1] Phil. Mag. vol. XVII. p. 188, 1884.

easily visible. I have repeated these observations without smoke by simply causing air-jets from the same nozzles to impinge upon the flame of a candle placed at a suitable distance. In such cases, as has been pointed out by Tyndall, the flame acts merely as an indicator of the condition of the otherwise invisible jet. Even without a resonator the sensitiveness of such jets to hissing sounds may be taken advantage of to form a pretty experiment.

"The combination of jet, resonator, and flame shows sometimes a tendency to *speak* on its own account ; but I did not succeed in getting a well-sustained sound. Such as it is, the effect probably corresponds to one observed by Savart and Plateau with water-jets breaking up under the operation of the capillary tension and, when resolved into drops, impinging upon a solid obstacle, such as the bottom of a sink, in mechanical connection with the nozzle from which the jet originally issues. In virtue of the connexion, any regular cycle in the mode of disintegration is able, as it were, to propagate itself."

"In the hope of being able to make better observations upon the transformations of unstable jets, I next had recourse to coloured water issuing under water. In this form the experiment is more manageable than in the case of smoke-jets, which are difficult to light, and liable to be disturbed by the slightest draught. Permanganate of potash was preferred as a colouring agent, and the colour may be discharged by mixing with the general mass of liquid a little acid ferrous sulphate. The jets were usually projected downwards into a large beaker or tank of glass, and were lighted from behind through a piece of ground glass.

"The notes of maximum sensitiveness of these liquid jets were found to be far graver than for smoke-jets or for flames. Forks vibrating from 20 to 50 times per second appeared to produce the maximum effect, to observe which it is only necessary to bring the stalk of the fork into contact with the table supporting the apparatus. The general behaviour of the jet could be observed without stroboscopic appliances by causing the liquid in the beaker to vibrate from side to side under the action of gravity. The line of colour proceeding from the nozzle is seen to become gradually more and more sinuous, and a little further down presents the appearance of a rope bent backwards and forwards upon itself. I have followed the process of disintegration with gradually increasing frequencies of vibrational disturbance from 1 or 2 per second up to about 24 per second, using electro-magnetic interruptors to send intermittent currents through an electro-magnet which acted upon a soft-iron armature attached to the nozzle. At each stage the pressure at which the jet is supplied should be adjusted so as to give the right degree of sensitiveness. If the pressure be too great, the jet flares independently of the imposed vibration, and the transformations become irregular: in the contrary case the phenomena, though usually observable, are not so well marked as when a suitable adjustment is made. After a little practice it is possible to interpret pretty well what is seen directly; but in order to have before the eye an image of what is really going on, we must have recourse to intermittent vision. The best results are obtained with two forks slightly out of tune, one of which is used to effect the disintegration of the jet, and the other (by means of perforated plates attached to its prongs) to give an intermittent view. The difference of frequencies should be about one per second. When the means of obtaining uniform rotation are at hand, a stroboscopic disk may be substituted for the second fork[1].

" The carrying out of these observations, especially when it is desired to make a drawing, is difficult unless we can control the plane of the bendings. In order to see the phases properly it is necessary that the plane of bendings should be perpendicular to the line of vision; but with a symmetrical nozzle this would occur only by accident. The difficulty may be got over by slightly nicking the end of the drawn-out glass nozzle at two opposite points (Barrett). In this way the plane of bending is usually rendered determinate, being that which includes the nicks, so that by turning the nozzle round its axis the sinuosities of the jet may be properly presented to the eye.

" Occasionally the jet appears to divide itself into two parts imperfectly connected by a sort of sheet. This seems to correspond to the duplication of flames and smoke-jets under powerful sonorous action, and to be due to what we may regard as the broken waves taking alternately different courses."

" It has already been noticed that the notes appropriate to water-jets are far graver than for air-jets from the same nozzles.

[1] In the original paper (*Phil. Mag.* vol. XVII. p. 188, 1884) drawings by Mrs Sidgwick are given. See also *Proc. Roy. Inst.* vol. XIII. p. 261, 1891, for reproductions of instantaneous photographs.

Moreover, the velocities suitable in the former case are much less than in the latter. This difference relates not, as might perhaps be at first supposed, to the greater density, but to the smaller viscosity of the water, measured of course kinematically. It is not difficult to see that the density, presumed to be the same for the jet and surrounding fluid, is immaterial, except of course in so far as a denser fluid requires a greater pressure to give it an assigned velocity. The influence of fluid viscosity upon these phenomena is explained in a former paper on the Stability or Instability of certain Fluid Motions[1]; and the laws of dynamical similarity with regard to fluid friction, laid down by Prof. Stokes[2], allow us to compare the behaviour of one fluid with another. The dimensions of the kinematic coefficient of viscosity are those of an area divided by a time. If we use the same nozzle in both cases, we must keep the same standard of length; and thus the times must be taken inversely, and the velocities directly, as the co-efficients of viscosity. In passing from air to water the pitch and velocity are to be reduced some ten times. But, in spite of the smaller velocity, the water-jet will require the greater pressure behind it, inasmuch as the densities differ in a ratio exceeding' 100 : 1."

Guided by these considerations, I made experiments to try whether the jets would behave differently in warm (less viscous) water, and as to the effect of substituting for water a mixture of alcohol and water in equal parts, a fluid known to be more viscous than either of its constituents. The effect of varying the viscosity was found to be very distinct. A jet which would not bear a pressure of more than $\frac{1}{4}$ inch ($\cdot$63 cm.) of water without flaring when the liquid was water at a temperature under the boiling-point required about 25 inches (63 cm.) pressure to make it flare when the alcoholic mixture was substituted. The importance of viscosity in these phenomena was thus abundantly established.

The manner in which viscosity operates is probably as follows. At the root of the jet, just after it issues from the nozzle, there is a near approach to discontinuous motion, and a high degree of instability. If a disturbance of sufficient intensity and of

---

[1] *Math. Soc. Proc.* Feb. 12, 1880. See § 366.

[2] *Camb. Phil. Trans.* 1850, " On the Effect of Internal Friction of Fluids on the Motion of Pendulums," § 5. See also Helmholtz, *Wied. Ann.* Bd. vii. p. 337 (1879) or Reprint, vol. i. p. 891.

suitable period have access, the regular motion is lost and cannot afterwards be recovered. But the instability has a very short time in which to produce its effect. Under the influence of viscosity the changes of velocity become more gradual, and the instability decreases rapidly if it does not disappear altogether. Thus if the disturbance be insufficient to cause disintegration during the brief period of instability, the jet may behave very much as though it had not been disturbed at all, and may reach the full developement observed in long flames and smoke-jets. This temporary character of the instability is a second feature differentiating strongly these jets from those of Savart, in which capillarity has an unlimited time of action.

When a flame is lighted at the burner, there are further complications of which it is difficult to give an adequate explanation. The high temperature leads indeed to increased viscosity, and this tends to explain the higher pressure then admissible, and the graver notes which then become operative. But it is probable that the change due to ignition is of a still more fundamental nature.

An ingenious method of observation, due to Mr C. Bell [1], may be applied so as to give valuable information with regard to the disintegration of jets; but the results obtained by the author are· not in harmony with the views of Mr Bell, who favours the symmetrical theory. In this method a second similar nozzle faces directly the nozzle from which the air issues, and is connected with the ear of the observer by means of rubber tubing. Suitable means are provided whereby the position of the hearing nozzle may be adjusted with accuracy, both longitudinally and laterally. When the distance is properly chosen, small disturbances acting upon the jet are perceived upon a magnified scale. Thus a fork vibrating feebly and presented to the jet is loudly heard; and that the effect is due to the peculiar properties of the jet is proved at once by cutting off the supply of air, when the sound becomes feeble, if not inaudible. Mr Bell proved that the efficacy of the arrangement requires a *small* area in the hearing nozzle; if the latter be large enough to receive the whole stream of air accompanying the jet, comparatively little is heard.

In the following experiments an air-jet from a well-regulated bellows issued from a glass nozzle and impinged upon a similar

[1] *Phil. Trans.* vol. CLXXVII. p. 383, 1886.

hearing nozzle. It was excited by forks (c' or c'') held in the neighbourhood.

If the position of the fork was such that the plane of its prongs was perpendicular to the jet, and that the prolongation of the axis of the stalk intersected the delivery end of the nozzle, the sound perceived was much less than when the fork was displaced laterally in its own plane so as to bring the nozzle nearer to one prong. This appears to prove that here again the effect is due, not to variation of pressure, but to transverse motion, causing the jet to become sinuous.

Confirmatory evidence may be drawn from observations upon the effect of slight movements of the hearing nozzle. When this is adjusted axially, but little is perceived of the fundamental tone of a fork presented laterally to the jet nozzle, but the octave tone is heard and often very strongly. When, however, the hearing nozzle is displaced laterally, the fundamental tone of the fork comes in loudly.

371. In that very convenient source of sounds of high pitch, the "bird-call," a stream of air issuing from a circular hole in a thin plate impinges centrically upon a similar hole in a parallel plate held at a little distance. The circumstances upon which the pitch depends have been investigated by Sondhauss[1], but much remains obscure as regards the manner in which the vibrations are excited.

According to Sondhauss the pitch is comparatively independent of the size and shape of the plates, varying directly as the velocity (v) of the jet and inversely as the distance (d) between the plates. If we assume independence of other elements, and that the frequency (n) is a function only of v, d, and b the diameter of the jet, it follows from dynamical similarity that

$$n = v/d \cdot f(b/d) \quad \dots\dots\dots\dots\dots\dots(1),$$

where f is an arbitrary function. Thus, if b/d be constant, Sondhauss' law must hold. From the very small dimensions employed it might fairly be argued that the action must be nearly independent of the velocity of sound, and therefore (v being given) of the density of the gas; but the question arises whether viscosity may not be an element of importance. If we suppose

[1] *Pogg. Ann.* vol. XCI. p. 126, 1854.

that geometrical similarity is maintained (b proportional to d), the theoretical form, when viscosity is retained, is

$$n = v/d \, . \, F(\nu/vd) \dots\dots\dots\dots\dots\dots (2),$$

ν being the kinematic coefficient of viscosity, of dimensions 2 in space and -1 in time. But when we take a numerical example, it appears improbable that the degree of viscosity can play much part in determining the pitch. In C.G.S. measure $\nu = ·16$ for air; and if the pressure propelling the jet be 1 cm. of mercury, $v = 4000$ (cm./sec.). Thus, if we take $d = ·1$ cm., we have $\nu/vd = ·0004$, so that $F(\nu/vd)$ could hardly differ much from $F(0)$.

Bird-calls are very easily made. The first plate, of 1 or 2 cm. in diameter, is cemented, or soldered, to the end of a short supply tube. The second plate may conveniently be made triangular, the turned down corners being soldered to the first plate. For calls of medium pitch the holes may be made in tin plate, but when it is desired to attain a very high pitch thin brass, or sheet silver, is more suitable. The holes may then be as small as $\frac{1}{2}$ mm. in diameter, and the distance between them as little as 1 mm. In any case the edges of the holes should be sharp and clean[1].

In order to test a bird-call it should be connected with a well-regulated supply of wind and with a manometer by which the operative pressure can be measured with precision. When it is found to speak well, the pressure and corresponding wave-length should be recorded. If the tones are high or inaudible, a high-pressure sensitive flame is required, the wave-length being deduced from the interval between the positions in which a reflector must be held in order that the flame may shew the least disturbance (§ 370). There is no difficulty in obtaining wave-lengths (complete) as low as 1 cm., and with care wave-lengths of ·6 cm. may be reached, corresponding to about 50,000 vibrations per second. In experimenting upon minimum wave-lengths, the distance between the call and the flame should not exceed 50 cm., and the flame should be adjusted to the verge of flaring.

In many cases a bird-call, which otherwise will not speak, may be made to do so by a reflecting plate held at a short distance in front. In practice the reflector is with advantage reduced to a

---

[1] Prof. A. M. Mayer has constructed beautifully finished bird-calls in which the distance between the plates is adjustable by a screw motion.

strip of metal, e.g. 1 cm. wide ; and, when this assistance is required, the right distance is an (even or odd) multiple of the half wave-length. In some cases the necessary position of the strip is very sharply defined.

On the question whether the disturbance of the jet accompanying the production of the sound is varicose or sinuous, some evidence may be derived from observations upon the manner in which the sound radiates. Upon the latter view we might expect that the sound would fall off, or even disappear altogether, in the axial direction, as happens, for example, in the case of the sound radiated from a bell (§ 282). But, so far as I have been able to observe, the sound emitted from a bird-call, speaking without the aid of a reflecting strip, is uniform through a wide angle ; and this fact may be regarded as telling strongly in favour of the view that the disturbance is here symmetrical, or varicose, in character. Other evidence tending in the same direction is afforded by the behaviour of resonating pipes made to speak with the aid of bird-calls. The pair of perforated plates is mounted symmetrically at one end of a pipe 40 or 50 cm. long. The other end of the pipe is acoustically open, and a gentle stream of air is made to pass the bird-call, most easily with the aid of a very narrow tube inserted into the open end and supplied from the mouth. By careful regulation of the force of the blast, the pipe may be made to speak in various harmonics, and the fact that it speaks at all seems to shew that the issue of air through the bird-call is variable.

The manner of action is perhaps somewhat as follows. When a symmetrical excrescence reaches the second plate, it is unable to pass the hole with freedom, and the disturbance is thrown back, probably with the velocity of sound, to the first plate, where it gives rise to a further disturbance, to grow in its turn during the progress of the jet. But the elucidation of this and many kindred phenomena remains still to be effected.

372. Æolian tones, as in the æolian harp, are generated when wind plays upon a stretched wire capable of vibration at various speeds, and their production also is doubtless connected with the instability of vortex sheets. It is not essential, however, that the wire should partake in the vibration, and the general phenomenon has been investigated by Strouhal[1], under the name of *reibungstöne*.

---

[1] *Wied. Ann.* vol. v. p. 216, 1878. See also W. Kohlrausch, *Wied. Ann.* vol. XIII. p. 545, 1881.

In Strouhal's experiments a vertical wire attached to a suitable frame was caused to revolve with uniform velocity about a parallel axis. The pitch of the æolian tone generated by the relative motion of the wire and of the air was found to be independent of the length and of the tension of the wire, but to vary with the diameter (d) and with the speed (v) of the relative motion. Within certain limits the relation between the frequency (n) and these data was expressible by

$$n = \cdot185 \; v/d \dots\dots\dots\dots\dots\dots\dots(1),$$

the centimetre and second being units.

When the speed is such that the æolian tone coincides with one of the proper tones of the wire, supported so as to be capable of free independent vibration, the sound is greatly reinforced, and with this advantage Strouhal found it possible to extend the range of the observations. Under the more extreme conditions then practicable the observed pitch deviated sensibly from the value given by (1). He shewed further that with a given diameter and a given speed a rise of temperature was attended by a fall in pitch.

Observations[1] upon a string, vibrating after the manner of the æolian harp under the stimulus of a chimney draught, have shewn that, contrary to the opinion generally expressed, the vibrations are effected in a plane perpendicular to the direction of the wind. According to (1) the distance travelled over by the wind during one complete vibration is about 6 times the diameter of the wire.

If, as appears probable, the compressibility of the fluid may be left out of account, we may regard n as a function of v, d, and ν the kinematic coefficient of viscosity. In this case n is necessarily of the form

$$n = v/d \cdot f(\nu/vd) \dots\dots\dots\dots\dots\dots(2),$$

where f represents an arbitrary function; and there is dynamical similarity, if $\nu \propto v\,d$. In observations upon air at one temperature ν is constant; and, if d vary inversely as v, nd/v should be constant, a result fairly in harmony with the observations of Strouhal. Again, if the temperature rises, ν increases, and in order to accord with observation, we must suppose that the function f diminishes with increasing argument.

[1] *Phil. Mag.* vol. vii. p. 161, 1879.

An examination of the actual values in Strouhal's experiments shew that ν/vd was always small; and we are thus led to represent f by a few terms of Mac Laurin's series. If we take

$$f(x) = a + bx + cx^2,$$

we get

$$n = a\,\frac{v}{d} + b\,\frac{\nu}{d^2} + c\,\frac{\nu^2}{vd^3} \dotfill (3).$$

If the third term in (3) may be neglected, the relation between n and v is linear. This law was formulated by Strouhal, and his diagrams shew that the coefficient b is negative, as is also required to express the observed effect of a rise of temperature. Further

$$d\,.\,\frac{dn}{dv} = a - \frac{c\nu^2}{v^2 d^2} \dotfill (4),$$

so that $d\,.\,dn/dv$ is very nearly constant, a result also given by Strouhal on the basis of his measurements.

On the whole it would appear that the phenomena are satisfactorily represented by (2) or (3), but a dynamical theory has yet to be given. It would also be of interest to extend the experiments to liquids.

CHAPTER XXII.

VIBRATIONS OF SOLID BODIES.

373. It is impossible in the present work to attempt anything approaching to a full consideration of the problems suggested by vibrating solid bodies: and yet the simpler parts of the theory seem to demand our notice. We shall limit ourselves entirely to the case of *isotropic* matter.

The general equations of equilibrium have already been stated in § 345. If ρ be the density, and

$$a^2 = (\kappa + \tfrac{4}{3} n)/\rho, \qquad b^2 = n/\rho \dots\dots\dots\dots(1),$$

we have
$$(a^2 - b^2)\frac{d\delta}{dx} + b^2 \nabla^2 \alpha + X' = 0, \text{ etc.} \dots\dots\dots\dots(2),$$

where X', Y', Z' are the impressed forces reckoned per unit of *mass*.

If from these we separate the reactions against acceleration, we obtain by D'Alembert's principle

$$\frac{d^2\alpha}{dt^2} = (a^2 - b^2)\frac{d\delta}{dx} + b^2 \nabla^2 \alpha + X' \dots\dots\dots\dots(3),$$

and two similar equations. In (3) δ is the *dilatation*, related to α, β, γ according to

$$\delta = d\alpha/dx + d\beta/dy + d\gamma/dz \dots\dots\dots\dots(4).$$

If α, β, γ, etc. be proportional to e^{ipt}, $d^2\alpha/dt^2 = -p^2\alpha$, and (3) becomes

$$(a^2 - b^2)\frac{d\delta}{dx} + b^2 \nabla^2 \alpha + p^2 \alpha + X' = 0 \dots\dots\dots\dots(5).$$

Differentiating equation (3) and its companions with respect to x, y, z, and adding, we obtain by (4)

$$\frac{d^2\delta}{dt^2} = a^2\nabla^2\delta + \frac{dX'}{dx} + \frac{dY'}{dy} + \frac{dZ'}{dz} \ldots\ldots\ldots\ldots(6).$$

Similar equations may be obtained for the *rotations* (compare § 239), defined by

$$\frac{d\gamma}{dy} - \frac{d\beta}{dz} = 2\varpi', \quad \frac{d\alpha}{dz} - \frac{d\gamma}{dx} = 2\varpi'', \quad \frac{d\beta}{dx} - \frac{d\alpha}{dy} = 2\varpi'''\ldots(7).$$

Thus, if we differentiate the third of equations (3) with respect to y, the second with respect to z, and subtract,

$$\frac{d^2\varpi'}{dt^2} = b^2\nabla^2\varpi' + \tfrac{1}{2}\frac{dZ'}{dy} - \tfrac{1}{2}\frac{dY'}{dz} \ldots\ldots\ldots\ldots(8);$$

and there are two similar equations relative to ϖ'', ϖ'''. It is to be observed that ϖ', ϖ'', ϖ''' are subject to the relation

$$d\varpi'/dx + d\varpi''/dy + d\varpi'''/dz = 0 \ldots\ldots\ldots\ldots(9).$$

We will now consider briefly certain cases of the propagation of plane waves in the absence of impressed forces. In (6), if X', Y', Z' vanish, and δ be a function of x only,

$$d^2\delta/dt^2 = a^2 d^2\delta/dx^2 \ldots\ldots\ldots\ldots\ldots(10),$$

of which the solution is, as in § 245,

$$\delta = f(x - at) + F(x + at) \ldots\ldots\ldots\ldots\ldots(11).$$

In this wave $\delta = d\alpha/dx$, while β and γ vanish; so that the case is similar to that of the propagation of waves in a compressible fluid. It should be observed, however, that by (1) the velocity depends upon the constant of rigidity (n) as well as upon that of compressibility (κ).

In the dilatational wave (11) the rotations ϖ', ϖ'', ϖ''' vanish, as appears at once from their expressions in (7). We have now to consider a wave of transverse vibration for which δ vanishes. If, for example, we suppose that α and β vanish and that γ is a function of x (and t) only, we have

$$\delta = 0, \qquad \varpi' = \varpi''' = 0, \qquad 2\varpi'' = -d\gamma/dx.$$

The equation for ϖ'' is

$$d^2\varpi''/dt^2 = b^2 d^2\varpi''/dx^2 \ldots\ldots\ldots\ldots\ldots(12),$$

of the same form as (10); and the same equation obtains for γ. The transverse vibration is thus propagated in plane waves with velocity b, a velocity *less* than that (a) of the dilatational waves.

The formation of stationary waves by superposition of positive and negative progressive waves of like wave-length need not be dwelt upon. If $k = 2\pi/\lambda$, where λ is the wave-length, the superposition of the positive wave $\gamma = \Gamma \cos k\,(bt - x)$ upon the negative wave $\gamma = \Gamma \cos k\,(bt + x)$ gives

$$\gamma = 2\Gamma \cos kbt \,.\, \cos kx \dots\dots\dots\dots\dots (13).$$

The second progressive wave may be the *reflection* of the first at a bounding surface impenetrable to energy. This may be either a free surface, or one at which γ is prevented from varying.

374. The problem of the propagation in three dimensions of a disturbance initially limited to a finite region of the solid was first successfully considered by Poisson, and the whole subject has been exhaustively treated by Stokes[1]. By (6), (8) § 373 the dilatation and the rotations satisfy the equations

$$d^2\delta/dt^2 = a^2\nabla^2\delta, \qquad d^2\varpi/dt^2 = b^2\nabla^2\varpi\dots\dots\dots\dots (1),$$

the solutions of which, applicable to the present purpose, have already been fully discussed in §§ 273, 274. It appears that distinct waves of dilatation and distortion are propagated outwards with different velocities, so that at a sufficient distance from the source they become separated. If we consider what occurs at a distant point, we see that at first there is neither dilatation nor distortion. When the wave of dilatation arrives this effect commences, but there is no distortion. After a while the wave of dilatation passes, and there is an interval of no dilatation and no distortion. Then the wave of distortion arrives and for a time produces its effect, after which there is never again either dilatation or distortion.

The complete discussion requires the expressions for the displacements in terms of δ, ϖ_1, ϖ_2, ϖ_3, for the derivation of which we have not space. From these it may be proved that before the arrival of the wave of dilatation and subsequently to the passage of the wave of distortion, the medium remains at rest. Between the two waves the medium is not absolutely undisturbed, although there is neither dilatation nor distortion.

If the initial disturbances be of such a character that there is no wave of distortion, the whole disturbance is confined to the wave of dilatation.

[1] "Dynamical Theory of Diffraction," *Camb. Phil. Trans.* Vol. IX. p. 1, 1849.

375. The subject of § 374 was the free propagation of waves resulting from a disturbance initially given. A problem at least equally important is that of divergent waves maintained by harmonic forces operative in the neighbourhood of a given centre.

We may take first the case of a harmonic force of such a character as to generate waves of dilatation. By equation (6) § 373 we may suppose that at all points except the origin of coordinates

$$d^2\delta/dt^2 = a^2\nabla^2\delta \dots\dots\dots\dots\dots(1);$$

or, if δ as a function of x, y, z depend upon r, or $\sqrt{(x^2 + y^2 + z^2)}$, only, and as a function of the time be proportional to e^{ipt}, § 241,

$$\frac{d^2\delta}{dr^2} + \frac{2}{r}\frac{d\delta}{dr} + h^2\delta = 0 \dots\dots\dots\dots\dots(2),$$

where $h = p/a$. The solution of (2) is, as in § 277,

$$\delta = \frac{e^{-ihr}}{r} \dots\dots\dots\dots\dots\dots(3).$$

In terms of real quantities

$$\delta = \frac{A \cos(pt - hr + \epsilon)}{r}\dots\dots\dots\dots\dots(4),$$

in which A and ϵ are arbitrary.

By transformation of (4) § 373, the relation between δ and the rad. l displacement w may be shewn to be

$$\delta = r^{-2}d(r^2w)/dr\dots\dots\dots\dots\dots(5),$$

or at a great distance from the origin simply

$$\delta = dw/dr \dots\dots\dots\dots\dots(6).$$

Thus, when r is great, corresponding to (4)

$$w = -\frac{A}{hr} \sin(pt - hr + \epsilon)\dots\dots\dots\dots(7).$$

In these purely dilatational waves the motion is radial, that is, parallel to the direction of propagation, and the distribution is symmetrical with respect to the origin.

The theory of forced waves of distortion proceeding outwards from a centre is of still greater interest. The simplest case is when the waves are due to a periodic force, say Z', acting through a space T at the origin. If we suppose in (8) § 373 that X', Y' vanish, and that all the quantities are proportional to e^{ipt}, we find

$$\nabla^2\varpi' + k^2\varpi' + \tfrac{1}{2}b^{-2}dZ'/dy = 0\ldots\ldots\ldots\ldots(8),$$

$$\nabla^2\varpi'' + k^2\varpi'' - \tfrac{1}{2}b^{-2}dZ'/dx = 0\ldots\ldots\ldots\ldots(9),$$

$$\nabla^2\varpi''' + k^2\varpi''' \qquad\qquad = 0\ldots\ldots\ldots\ldots(10),$$

k being written for p/b.

These equations are solved as in § 277. We get $\varpi''' = 0$, and

$$\varpi' = \frac{1}{8\pi b^2}\iiint \frac{dZ'}{dy}\frac{e^{-ikr}}{r}\,dx\,dy\,dz,$$

r denoting the distance between the element at x, y, z near the origin (O) and the point (P) under consideration. If we integrate partially with respect to y, we find

$$\varpi' = -\frac{1}{8\pi b^2}\iiint Z'\frac{d}{dy}\left(\frac{e^{-ikr}}{r}\right)dx\,dy\,dz\ldots\ldots\ldots(11),$$

the integrated term vanishing in virtue of the condition that Z' is finite only within the space T. Moreover, since the dimensions of T are supposed to be very small in comparison with the wavelength, $d(r^{-1}e^{-ikr})/dy$ may be removed from under the integral sign. It will be convenient also to change the meaning of x, y, z, so that they shall now represent the coordinates of P relatively to O. Thus, if Z' now stand for the mean value of Z' throughout the space T,

$$\varpi' = +\frac{TZ'}{8\pi b^2}\frac{d}{dr}\left(\frac{e^{-ikr}}{r}\right)\cdot\frac{y}{r}\ldots\ldots\ldots\ldots(12).$$

In like manner

$$\varpi'' = -\frac{TZ'}{8\pi b^2}\frac{d}{dr}\left(\frac{e^{-ikr}}{r}\right)\cdot\frac{x}{r}\ldots\ldots\ldots\ldots(13);$$

and

$$\varpi''' = 0\ldots\ldots\ldots\ldots\ldots\ldots\ldots\ldots(14).$$

In virtue of the symmetry round the axis of z it suffices to consider points which lie in the plane ZX. Then ϖ' vanishes, so that the rotation takes place about an axis perpendicular both to the direction of propagation (r) and to that of the force (z). If θ denote the angle between these directions, the resultant rotation, coincident with ϖ'', is

$$\varpi = -\frac{TZ'\sin\theta}{8\pi b^2}\frac{d}{dr}\left(\frac{e^{-ikr}}{r}\right)\ldots\ldots\ldots\ldots(15).$$

If we confine our attention to points at a great distance, this becomes simply

$$\varpi = \frac{ik\,TZ'\sin\theta}{8\pi b^2}\,\frac{e^{-ikr}}{r} \qu\dots\dots\dots\dots(16).$$

The displacement, corresponding to (16), is perpendicular to r and in the plane zr. Its value is given by

$$-2\!\int\!\varpi\,dr = \frac{TZ'\sin\theta}{4\pi b^2}\,\frac{e^{-ikr}}{r}\,;$$

or, if we restore the factor e^{ikbt}, and reject the imaginary part of the solution,

$$-2\!\int\!\varpi\,dr = \frac{TZ'\sin\theta}{4\pi b^2}\,\frac{\cos k\,(bt-r)}{r} \qu\dots\dots\dots(17).$$

If $Z_1\cos kbt$ denote the whole force applied at the origin,

$$Z_1 = TZ'\,.\,\rho \qu\dots\dots\dots\dots\dots(18),$$

so that (17) may be written

$$-2\!\int\!\varpi\,dr = \frac{Z_1\sin\theta}{4\pi\rho b^2}\,\frac{\cos k\,(bt-r)}{r} \qu\dots\dots\dots\dots(19).$$

The amplitude of the vibration radiated outwards is thus inversely as the distance, and directly as the sine of the angle between the ray and the direction in which the force acts. In the latter direction itself there is no transverse vibration propagated.

These expressions may be applied to find the secondary vibration dispersed in various directions when plane waves impinge upon a small obstacle of density different from that of the rest of the solid. We may suppose that the plane waves are expressed by

$$\gamma = \Gamma\cos k\,(bt-x) \qu\dots\dots\dots\dots(20),$$

and that they impinge at the origin upon an obstacle of volume T and density ρ'. The additional inertia of the solid at this place would be compensated by a force $(\rho'-\rho)\ddot\gamma$, or $-(\rho'-\rho)k^2b^2\Gamma\cos kbt$, acting throughout T; and, if this force be actually applied, the primary waves would proceed without interruption. The secondary waves may thus be regarded as due to a force equal to the opposite of this, acting at O parallel to Z. The whole amount of the force is given by

$$Z_1\cos kbt = (\rho'-\rho)k^2b^2\,T\Gamma\cos kbt\qu\dots\dots\dots(21);$$

so that by (19) the secondary displacement at a distant point (r, θ) is

$$\frac{(\rho'-\rho)k^2T\Gamma\sin\theta}{4\pi\rho}\,.\,\frac{\cos k\,(bt-r)}{r} \qu\dots\dots\dots(22).$$

The intensity of the scattered vibration is thus inversely as the fourth power of the wave-length (Γ being given), and as the square of the sine of the angle between the scattered ray and the direction of vibration in the primary waves. Thus, if the primary ray be along x and the secondary ray along z, there are no secondary vibrations if (as above supposed) the primary vibrations are parallel to z; but if the primary vibrations are parallel to y, there are secondary vibrations of full amplitude ($\sin\theta = 1$), and these vibrations are themselves executed in a direction parallel to y.[1]

376. In § 375 we have examined the effect of a periodic force $Z_1\cos kbt$, localized at the origin. We now proceed to consider the case of a force uniformly distributed along an infinite line.

Of this there are two principal sub-cases: the first where the force, itself always parallel to z, is distributed along the axis of z, the second where the distribution is along the axis of y. In the first, with which we commence, the entire motion is in two dimensions, symmetrical with respect to OZ, and further is such that α and β vanish, while γ is a function of $(x^2 + y^2)$ only. If, as suffices, we limit ourselves to points situated along OX, ϖ', ϖ''' vanish, and we have only to find ϖ''.

The simplest course to this end is by integration of the result given in (16) § 375. $\rho TZ'$ will be replaced by $Z_{11}dz$, the amount of the force distributed on dz; r denotes the distance between P on OX and dz on OZ; θ the angle between r and z. The rotation ϖ'' about an axis parallel to y and due to this element of the force is thus

$$\frac{ikZ_{11}dz}{8\pi b^2\rho}\frac{xe^{-ikr}}{r^2}\quad\dots\dots\dots\dots\dots\dots(1).$$

In the integration x is constant, and $r^2 = x^2 + z^2$, so that we have to consider

$$\int\frac{ie^{-ikr}dr}{r\sqrt{(r^2-x^2)}}\quad\text{or}\quad\int\frac{ie^{-ikx}e^{-ikh}dh}{(x+h)\,.\,\sqrt{(2x+h)}\,.\,\sqrt{h}}\dots\dots(2),$$

if we write $r - x = h$.

---

[1] "On the Light from the Sky, its Polarization and Colour." *Phil. Mag.* Vol. XLI. pp. 107, 274, 1871; see also *Phil. Mag.* Vol. XLI. p. 447, 1871, for an investigation of the case where the obstacle differs in elastic quality, as well as in density, from the remainder of the medium.

From this integral a rigorous solution may be developed, but, as in § 342, we may content ourselves with the limiting form assumed when kx is very great. Thus, as the equivalent of (2), we get

$$\frac{2e^{-ikx}}{x \cdot \sqrt{(2x)}} \int_0^\infty \frac{ie^{-ikh}dh}{\sqrt{h}} = \frac{2\sqrt{\pi} \cdot e^{-i(kx-\frac{1}{4}\pi)}}{x \cdot \sqrt{(2kx)}} \dots\dots\dots (3);$$

so that as the integral of (1)

$$\varpi'' = \frac{kZ_{11}}{4\pi b^2 \rho} \frac{\sqrt{\pi}}{\sqrt{(2kx)}} e^{-i(kx-\frac{1}{4}\pi)} \dots\dots\dots\dots (4).$$

From this γ may be at once deduced. We have

$$\gamma = -2\int\varpi'' dx = \frac{Z_{11}}{2\pi b^2 \rho} \frac{\sqrt{\pi}}{\sqrt{(2kx)}} e^{-i(kx+\frac{1}{4}\pi)} \dots\dots\dots (5);$$

or, if we restore the time-factor, and omit the imaginary part of the solution,

$$\gamma = \frac{Z_{11}}{2\pi b^2 \rho} \frac{\sqrt{\pi}}{\sqrt{(2kx)}} \cos k(bt - x - \tfrac{1}{8}\lambda) \dots\dots\dots (6).$$

This corresponds to the force $Z_{11} \cos kbt$ per unit of length of the axis of z. In virtue of the symmetry we may apply (6) to points not situated upon the axis of x, if we replace x by $\sqrt{(x^2 + y^2)}$. That the value of γ would be inversely as the square root of the distance from the axis of z might have been anticipated from the principle of energy.

The solution might also be investigated directly in terms of γ without the aid of the rotations ϖ.

It now remains to consider the case in which the applied force, still parallel to z, is distributed along OY, instead of along OZ. The point P, at which the effect is required, may be supposed to be situated in the plane ZX at a great distance R from O and in such a direction that the angle ZOP is θ.

In virtue of the two-dimensional character of the force, $\beta = 0$, while α, γ are independent of y. Hence ϖ', ϖ''' vanish. But, although these component rotations vanish as regards the resultant effect, the action of a single element of the force $Z_{11}dy$, situated at y, would be more complicated. Into this, however, we need not enter, because, as before, the effect in reality depends only upon the elements in the neighbourhood of O. Thus, in place of (1), we may take

$$\frac{ikZ_{11}dy \cdot \sin \theta}{8\pi b^2 \rho} \frac{e^{-ikr}}{r} \dots\dots\dots\dots\dots (7),$$

r being the distance between dy and P, so that

$$dy/r = dr/y = dr/\sqrt{(r^2 - R^2)}.$$

Writing $r - R = h$, we get, as in (2), (3), (4),

$$\varpi'' = \frac{kZ_{11} \sin \theta}{4\pi b^2 \rho} \frac{\sqrt{\pi}}{\sqrt{(2kR)}} e^{-i(kR - \frac{1}{4}\pi)} \quad\ldots\ldots\ldots\ldots(8);$$

and for the displacement, perpendicular to R,

$$- 2\int \varpi'' dR = \frac{Z_{11} \sin \theta}{2\pi b^2 \rho} \frac{\sqrt{\pi}}{\sqrt{(2kR)}} e^{-i(kR + \frac{1}{4}\pi)} \quad\ldots\ldots\ldots\ldots(9).$$

Hence, corresponding to the force $Z_{11} \cos kbt$ per unit of length of the axis of y, we have the displacement perpendicular to R at the point (R, θ)

$$\frac{Z_{11} \sin \theta}{2\pi b^2 \rho} \frac{\sqrt{\pi}}{\sqrt{(2kR)}} \cos k(bt - R - \tfrac{1}{8}\lambda)\ldots\ldots\ldots(10).$$

377.　As in § 375, we may employ the results of § 376 to form expressions for the secondary waves dispersed from a small cylindrical obstacle, coincident with OZ and of density ρ', upon which primary parallel waves impinge. If the expression for the primary waves be (20) § 375, we have

$$Z_{11} = (\rho' - \rho) k^2 b^2 . \pi c^2 . \Gamma \ldots\ldots\ldots\ldots\ldots(1),$$

πc^2 being the area of the cross section of the obstacle. Thus, if we denote $\sqrt{(x^2 + y^2)}$ by r, we have from (6) § 376 as the expression of the secondary waves,

$$\gamma = \frac{(\rho' - \rho) k^2 . \pi c^2 . \Gamma}{2\pi \rho} \frac{\sqrt{\pi}}{\sqrt{(2kr)}} \cos k(bt - r - \tfrac{1}{8}\lambda)$$

$$= \frac{\pi (\rho' - \rho) . \pi c^2 . \Gamma}{\rho \lambda^{\frac{3}{2}} r^{\frac{1}{2}}} \cos \frac{2\pi}{\lambda} (bt - r - \tfrac{1}{8}\lambda) \quad\ldots\ldots(2),$$

k being replaced by its equivalent $(2\pi/\lambda)$. In this case the secondary waves are symmetrical, and their intensity varies inversely as the distance and as the *cube* of the wave-length.

The solution expressed by (10) § 376 shews that if primary waves

$$\beta = B \cos k(bt - x)\ldots\ldots\ldots\ldots\ldots(3)$$

impinge upon the same small cylindrical obstacle, the displacement perpendicular to the secondary ray, viz. r, will be

$$\frac{\pi (\rho' - \rho) . \pi c^2 . B . \cos \theta}{\rho \lambda^{\frac{3}{2}} r^{\frac{1}{2}}} \cos \frac{2\pi}{\lambda} (bt - r - \tfrac{1}{8}\lambda)\ldots\ldots(4),$$

θ denoting the angle between the direction of the primary ray (x) and the secondary ray (r). In this case the secondary disturbance vanishes in one direction, that is along a ray parallel to the primary vibration.

Returning to the first case, in which α and β vanish throughout, while γ is a function of x and y only, let us suppose that the material composing the cylindrical obstacle differs from its surroundings in rigidity (n') as well as in density (ρ'). The conditions to be satisfied at the cylindrical surface are

$$\gamma \text{ (inside)} = \gamma \text{ (outside)},$$

$$n' d\gamma/dr \text{ (inside)} = n\, d\gamma/dr \text{ (outside)}.$$

In the exterior space γ satisfies the equation (§ 373)

$$d^2\gamma/dx^2 + d^2\gamma/dy^2 + k^2\gamma = 0,$$

where $k = p/b$; and in the space interior to the cylinder γ satisfies

$$d^2\gamma/dx^2 + d^2\gamma/dy^2 + k'^2\gamma = 0,$$

where $k' = p/b'$ and b' denotes the velocity of transverse vibrations in the material composing the cylinder. The investigation of the secondary waves thrown off by the obstacle when primary plane waves impinge upon it is then analogous to that of § 343, and the conclusion is that, corresponding to primary waves

$$\gamma = \Gamma \cos \frac{2\pi}{\lambda} (bt - x) \dots\dots\dots\dots\dots (5),$$

the secondary waves thrown off by a small cylinder in a direction making an angle θ with x are given by

$$\gamma = \frac{2\pi \cdot \pi c^2 \cdot \Gamma}{\lambda^{\frac{3}{2}} r^{\frac{1}{2}}} \left\{ \frac{\rho' - \rho}{2\rho} - \frac{n' - n}{n' + n} \cos \theta \right\} \cos \frac{2\pi}{\lambda} (bt - r - \tfrac{1}{8}\lambda)\dots(6),$$

which includes (2) as a particular case.

378. We now return to the fundamental problem, already partially treated in § 375, of the vibrations in an unlimited solid due to the application of a periodic force at the origin of coordinates. Equations (12), (13), (14) § 375 give the solution so far as to specify the values of the component rotations. If, as we shall ultimately suppose, the solid be incompressible, we have in addition $\delta = 0$. On this basis the solution might be completed, but it may be more instructive to give an independent investigation.

Since in the notation of § 373 $X' = Y' = 0$, we have by (5)

$$(a^2 - b^2)\, d\delta/dx + b^2\nabla^2\alpha + p^2\alpha = 0 \ldots\ldots\ldots\ldots(1),$$
$$(a^2 - b^2)\, d\delta/dy + b^2\nabla^2\beta + p^2\beta = 0 \ldots\ldots\ldots\ldots(2),$$
$$(a^2 - b^2)\, d\delta/dz + b^2\nabla^2\gamma + p^2\gamma = -Z' \ldots\ldots\ldots (3).$$

Let us assume

$$\alpha = d^2\chi/dx\,dz, \quad \beta = d^2\chi/dy\,dz, \quad \gamma = d^2\chi/dz^2 + w\ldots\ldots(4),$$

and accordingly

$$\delta = d\,(\nabla^2\chi)/dz + dw/dz\ldots\ldots\ldots\ldots\ldots\ldots\ldots(5).$$

The substitution of these values in (1) gives

$$\frac{d^2}{dx\,dz}\{a^2\nabla^2\chi + p^2\chi + (a^2 - b^2)\,w\} = 0\,;$$

so that (1) and (2) are satisfied if

$$a^2\nabla^2\chi + p^2\chi + (a^2 - b^2)\,w = 0 \ldots\ldots\ldots\ldots\ldots(6).$$

The same substitutions in (3) give

$$\frac{d^2}{dz^2}\{a^2\nabla^2\chi + p^2\chi + (a^2 - b^2)\,w\} + b^2\nabla^2 w + p^2 w + Z' = 0,$$

or in virtue of (6)

$$b^2\nabla^2 w + p^2 w + Z' = 0\ldots\ldots\ldots\ldots\ldots\ldots\ldots(7).$$

By this equation w is determined, and thence χ by (6).

In the notation of § 375, $k = p/b$, $h = p/a$. Since $Z' = 0$ at all points other than the origin, (7) becomes

$$(\nabla^2 + k^2)\,w = 0 \ldots\ldots\ldots\ldots\ldots\ldots\ldots(8),$$

whence by (6) $\quad (\nabla^2 + h^2)\,(\nabla^2 + k^2)\,\chi = 0\ldots\ldots\ldots\ldots\ldots\ldots(9)$

is to be satisfied everywhere except at the origin. The solution of (9) is

$$\chi = A\,\frac{e^{-ikr}}{r} + B\,\frac{e^{-ihr}}{r} \ldots\ldots\ldots\ldots\ldots (10),$$

where A and B are constants. The corresponding values of w and δ are by (6) and (5)

$$w = k^2 A\,\frac{e^{-ikr}}{r}\,, \qquad \delta = -h^2 B\,\frac{d}{dz}\left(\frac{e^{-ihr}}{r}\right)\ldots\ldots\ldots(11).$$

To connect A and B with Z', we have from (7), as in § 375,

$$w = \frac{1}{4\pi b^2}\iiint Z'\,\frac{e^{-ikr}}{r}\,dx\,dy\,dz = \frac{Z_1}{4\pi b^2 \rho}\,\frac{e^{-ikr}}{r}\,,$$

so that

$$A = \frac{Z_1}{4\pi k^2 b^2 \rho} \ldots\ldots\ldots\ldots\ldots\ldots\ldots(12).$$

Again, by (6) § 373,

$$\nabla^2 \delta + h^2 \delta + a^{-2} \, dZ'/dz = 0 \; ;$$

so that, as in § 375,

$$\delta = \frac{1}{4\pi a^2} \iiint \frac{dZ'}{dz} \frac{e^{-ihr}}{r} \, dx \, dy \, dz = \frac{Z_1}{4\pi a^2 \rho} \frac{d}{dz}\left(\frac{e^{-ihr}}{r}\right).$$

Thus, by comparison with (11),

$$B = -\frac{Z_1}{4\pi h^2 a^2 \rho} = -A \; \dots\dots\dots\dots\dots (13) ;$$

and

$$\chi = \frac{Z_1}{4\pi k^2 b^2 \rho} \frac{e^{-ikr} - e^{-ihr}}{r} \; \dots\dots\dots\dots (14).$$

From the values of χ and w thus fully determined α, β, γ are found by simple differentiations, as indicated in (4). We have

$$\frac{d^2}{dx\,dz}\left(\frac{e^{-ikr}}{r}\right) = \frac{xz\,e^{-ikr}}{r^2}\left(-\frac{k^2}{r} + \frac{3ik}{r^2} + \frac{3}{r^3}\right) \dots\dots\dots(15),$$

$$\frac{d^2}{dz^2}\left(\frac{e^{-ikr}}{r}\right) = e^{-ikr}\left[\frac{z^2}{r^2}\left(-\frac{k^2}{r} + \frac{3ik}{r^2} + \frac{3}{r^3}\right) - \frac{ik}{r^2} - \frac{1}{r^3}\right] \dots(16).$$

As the complete expressions are rather long, we will limit ourselves to the case of incompressibility ($h = 0$). Thus, if we restore the time-factor (e^{ipt}) and throw away the imaginary part of the solution, we get

$$\alpha = \frac{A k^2 xz}{r^3}\left[\left(-1 + \frac{3}{k^2 r^2}\right)\cos(pt - kr) - \frac{3}{kr}\sin(pt - kr) - \frac{3}{k^2 r^2}\cos pt\right]$$
$$\dots\dots\dots(17),$$

$$\gamma = \frac{A k^2}{r}\left[\left(1 - \frac{z^2}{r^2} + \frac{3z^2}{k^2 r^4} - \frac{1}{k^2 r^2}\right)\cos(pt - kr)\right.$$
$$\left. + \left(\frac{1}{kr} - \frac{3z^2}{kr^3}\right)\sin(pt - kr) - \left(\frac{3z^2}{k^2 r^4} - \frac{1}{k^2 r^2}\right)\cos pt\right] \dots(18),$$

the value of β differing from that of α merely by the substitution of y for x. The value of A is given in (12), and $Z_1 \cos pt$ is the whole force operative at the origin at time t.

At a great distance from the origin (17), (18) reduce to

$$\alpha = -\frac{Z_1}{4\pi b^2 \rho} \frac{xz}{r^2} \frac{\cos(pt - kr)}{r} \; \dots\dots\dots\dots (19),$$

$$\gamma = \frac{Z_1}{4\pi b^2 \rho}\left(1 - \frac{z^2}{r^2}\right)\frac{\cos(pt - kr)}{r} \; \dots\dots\dots (20),$$

in agreement with (19) § 375.

W. König[1] has remarked upon the non-agreement of the complete solution (17), (18), first given in a different form by Stokes[2], with the results of a somewhat similar investigation by Hertz[3], in which the terms involving $\cos pt$, $\sin pt$ do not occur, and he seems disposed to regard Stokes' results as affected by error. But the fact is that the problems treated are essentially different, that of Hertz having no relation to elastic solids. The source of the discrepancy is in the first terms of (1) &c., which are omitted by Hertz in his theory of the ether. But assuredly in a theory of elastic solids these terms must be retained. Even when the material is supposed to be incompressible, so that δ vanishes, the retention is still necessary, because, as was fully explained by Stokes in the memoir referred to, the factor $(a^2 - b^2)$ is infinite at the same time.

If we suppose in (17), (18) that p and k are very small, and trace the limiting form, we obtain the solution of the statical problem of the deformation of an incompressible solid by a force localized at a point in its interior.

379. In § 373 we saw that in a uniform medium plane waves of transverse vibration

$$\alpha = 0, \quad \beta = 0, \quad \gamma = \Gamma \cos (pt - kx) \ldots\ldots\ldots\ldots (1)$$

may be propagated without limit. We will now suppose that on the positive side of the plane $x = 0$ the medium changes, so that the density becomes ρ_1 instead of ρ, while the rigidity becomes n_1 instead of n. In the transmitted wave p remains the same, but k is changed to k_1, where

$$k_1^2/k^2 = n\rho_1/n_1\rho \ldots\ldots\ldots\ldots\ldots\ldots\ldots(2).$$

Assuming, as will be verified presently, that no change of phase need be allowed for, we may take as the expressions for the transmitted and reflected waves

$$\gamma_1 = \Gamma_1 \cos (pt - k_1x), \qquad \gamma = \Gamma' \cos (pt + kx) \ldots\ldots(3),$$

so that altogether the value of γ in the first medium is

$$\gamma = \Gamma \cos (pt - kx) + \Gamma' \cos (pt + kx) \ldots\ldots\ldots(4),$$

and in the second

$$\gamma_1 = \Gamma_1 \cos (pt - k_1x) \ldots\ldots\ldots\ldots\ldots\ldots(5).$$

[1] *Wied. Ann.* vol. xxxvii. p. 651, 1889.

[2] *Camb. Phil. Trans.* vol. ix. p. 1, 1849; *Collected Works*, vol. ii. p. 243.

[3] *Wied. Ann.* vol. xxxvi. p. 1, 1889.

The conditions to be satisfied at the interface ($x = 0$), upon which no external force acts, are

$$\gamma_1 = \gamma, \qquad n_1 d\gamma_1/dx = n\,d\gamma/dx \quad\ldots\ldots\ldots\ldots(6)\,;$$

so that $$\Gamma + \Gamma' = \Gamma_1, \qquad nk\,(\Gamma - \Gamma') = n_1 k_1 \Gamma_1\ldots\ldots\ldots\ldots(7).$$

If, as can plainly be done, Γ', Γ_1 be determined in accordance with (7), the conditions are all satisfied. We have

$$\frac{\Gamma'}{\Gamma} = \frac{nk - n_1 k_1}{nk + n_1 k_1} = \frac{\sqrt{(n\rho)} - \sqrt{(n_1\rho_1)}}{\sqrt{(n\rho)} + \sqrt{(n_1\rho_1)}} \quad\ldots\ldots\ldots\ldots(8),$$

$$\frac{\Gamma_1}{\Gamma} = \frac{\Gamma + \Gamma'}{\Gamma} = \frac{2\sqrt{(n\rho)}}{\sqrt{(n\rho)} + \sqrt{(n_1\rho_1)}} \quad\ldots\ldots\ldots\ldots(9),$$

by which the reflected and transmitted waves are determined. The particular cases in which $\rho_1 = \rho$, or $n_1 = n$, may be specially noted.

When the incidence upon the plane separating the two bodies is oblique, the problem becomes more complicated, and divides itself into two parts according as the vibrations (always perpendicular to the incident ray) are executed in the plane of incidence, or in the perpendicular plane. Into these matters, which have been much discussed from an optical point of view, we shall not enter. The method of investigation, due mainly to Green, is similar to that of § 270. A full account with the necessary references is given in Basset's *Treatise on Physical Optics*, Ch. XII.

380. The vibrations of solid bodies bounded by free surfaces which are plane, cylindrical, or spherical, can be investigated without great difficulty, but the subject belongs rather to the Theory of Elasticity. For an infinite plate of constant thickness the functions of the coordinates required are merely circular and exponential[1]. The solution of the problem for an infinite cylinder[2] depends upon Bessel's functions, and is of interest as giving a more complete view of the longitudinal and flexural vibrations of a thin rod.

The case of the sphere is important as of a body limited in all directions. The symmetrical radial vibrations, purely dilatational in their character, were first investigated by Poisson and

---

[1] *Proc. Lond. Math. Soc.* vol. XVII. p. 4, 1885 ; vol. XX. p. 225, 1889.

[2] Pochhammer, *Crelle*, vol. LXXXI. 1876 ; Chree, *Quart. Journ.* 1886. See also Love's *Theory of Elasticity*, ch. XVII.

Clebsch[1]. The complete theory is due to Jaerisch[2] and especially to Lamb[3]. An exposition of it will be found in Love's treatise already cited.

The calculations of frequency are complicated by the existence of *two* elastic constants κ and n § 373, or q and μ § 214. From the principle of § 88 we may infer, as Lamb has remarked, that the frequency increases with any rise either of κ or of n, for as appears from (1) § 345 either change increases the potential energy of a given deformation.

381[4]. In the course of this work we have had frequent occasion to notice the importance of the conclusions that may be arrived at by the method of dimensions. Now that we are in a position to draw illustrations from a greater variety of acoustical phenomena relating to the vibrations of both solids and fluids, it will be convenient to resume the subject, and to develope somewhat in detail the principles upon which the method rests.

In the case of systems, such as bells or tuning-forks, formed of uniform isotropic material, and vibrating in virtue of elasticity, the acoustical elements are the shape, the linear dimension c, the constants of elasticity q and μ (§ 149), and the density ρ. Hence, by the method of dimensions, the periodic time varies *cæteris paribus* as the linear dimension, at least if the amplitude of vibration be in the same proportion; and, if the law of isochronism be assumed, the last-named restriction may be dispensed with. In fact, since the dimensions of q and ρ are respectively $[ML^{-1}T^{-2}]$ and $[ML^{-3}]$, while μ is a mere number, the only combination capable of representing a time is $q^{-\frac{1}{2}} \cdot \rho^{\frac{1}{2}} \cdot c$.

The argument which underlies this mathematical shorthand is of the following nature. Conceive two geometrically similar bodies, whose mechanical constitution at corresponding points is the same, to execute similar movements in such a manner that the corresponding changes occupy times[5] which are proportional to the

---

[1] *Theorie der Elasticität Fester Körper*, Leipzig, 1862.

[2] *Crelle*, vol. LXXXVIII. 1879.

[3] *Proc. Lond. Math. Soc.* vol. XIII. p. 189, 1882.

[4] This section appeared in the First Edition as § 348.

[5] The conception of an alteration of scale in space has been made familiar by the universal use of maps and models, but the corresponding conception for time is often less distinct. Reference to the case of a musical composition performed at different speeds may assist the imagination of the student.

linear dimensions—in the ratio, say, of $1 : n$. Then, if the one movement be possible as a consequence of the elastic forces, the other will be also. For the masses to be moved are as $1 : n^3$, the accelerations as $1 : n^{-1}$, and therefore the necessary forces are as $1 : n^2$; and, since the strains are the same, this is in fact the ratio of the elastic forces due to them when referred to corresponding areas. If the elastic forces are competent to produce the supposed motion in the first case, they are also competent to produce the supposed motion in the second case.

The dynamical similarity is disturbed by the operation of a force like gravity, proportional to the cubes, and not to the squares, of corresponding lines; but in cases where gravity is the sole motive power, dynamical similarity may be secured by a different relation between corresponding spaces and corresponding times. Thus if the ratio of corresponding spaces be $1 : n$, and that of corresponding times be $1 : n^{\frac{1}{2}}$, the accelerations are in both cases the same, and may be the effects of forces in the ratio $1 : n^3$ acting on masses which are in the same ratio. As examples coming under this head may be mentioned the common pendulum, sea-waves, whose velocity varies as the square root of the wave-length, and the whole theory of the comparison of ships and their models by which Froude predicted the behaviour of ships from experiments made on models of moderate dimensions.

The same comparison that we have employed above for elastic solids applies also to aerial vibrations. The pressures in the cases to be compared are the same, and therefore when acting over areas in the ratio $1 : n^2$, give forces in the same ratio. These forces operate on masses in the ratio $1 : n^3$, and therefore produce accelerations in the ratio $1 : n^{-1}$, which is the ratio of the actual accelerations when both spaces and times are as $1 : n$. Accordingly the periodic times of similar resonant cavities, filled with the same gas, are directly as the linear dimension—a very important law first formulated by Savart.

Since the same method of comparison applies both to elastic solids and to elastic fluids, an extension may be made to systems into which both kinds of vibration enter. For example, the scale of a system compounded of a tuning-fork and of an air resonator may be supposed to be altered without change in the motion other than that involved in taking the times in the same ratio as the linear dimensions.

Hitherto the alteration of scale has been supposed to be uniform in all dimensions, but there are cases, not coming under this head, to which the principle of dynamical similarity may be most usefully applied. Let us consider, for example, the flexural vibrations of a system composed of a thin elastic lamina, plane or curved. By §§ 214, 215 we see that the thickness of the lamina b, and the mechanical constants q and ρ, will occur only in the combinations qb^3 and $b\rho$, and thus a comparison may be made even although the alteration of thickness be not in the same proportion as for the other dimensions. If c be the linear dimension when the thickness is disregarded, the times must vary *cæteris paribus* as $q^{-\frac{1}{2}} \cdot \rho^{\frac{1}{2}} \cdot c^2 \cdot b^{-1}$. For a given material, thickness, and shape, the times are therefore as the *squares* of the linear dimension. It must not be forgotten, however, that results such as these, which involve a law whose truth is only approximate, stand on a different level from the more immediate consequences of the principle of similarity.

CHAPTER XXIII.

FACTS AND THEORIES OF AUDITION.

382. THE subject of the present chapter has especial relation to the ear as the organ of hearing, but it can be considered only from the physical side. The discussion of anatomical or physiological questions would accord neither with the scope of this book nor with the qualifications of the author. Constant reference to the great work of Helmholtz is indispensable[1]. Although, as we shall see, some of the positions taken by the author have been relinquished, perhaps too hastily, by subsequent writers, the importance of the observations and reasonings contained in it, as well as the charm with which they are expounded, ensure its long remaining the starting point of all discussions relating to sound sensations.

383. The range of pitch over which the ear is capable of perceiving sounds is very wide. Naturally neither limit is well defined. From his experiments Helmholtz concluded that the sensation of musical tone begins at about 30 vibrations per second, but that a determinate musical pitch is not perceived till about 40 vibrations are performed in a second. Preyer[2] believes that he heard pure tones as low as 15 per second, but it seems doubtful whether the octave was absolutely excluded. On a recent review of the evidence and in the light of some fresh experiments, Van Schaik[3] sees no reason for departing greatly from Helmholtz's estimate, and fixes the limit at about 24 vibrations per second.

---

[1] *Tonempfindungen*, 4th edition, 1877; *Sensations of Tone*, 2nd English edition translated from the 4th German edition by A. J. Ellis. Citations will be made from this English edition, which is further furnished by the translator with many valuable notes.

[2] *Physiologische Abhandlungen*, Jena, 1876.

[3] *Arch. Néerl.* vol. xxix. p. 87, 1895.

On the upper side the discrepancies are still greater. Much no doubt depends upon the intensity of the vibrations. In experiments with bird-calls (§ 371) nothing is heard above 10,000, although sensitive flames respond up to 50,000. But forks carefully bowed, or metal bars struck with a hammer, appear to give rise to audible sounds of still higher frequencies. Preyer gives 20,000 as near the limit for normal ears.

In the case of very high sounds there is little or no appreciation of pitch, so that for musical purposes nothing over 4000 need be considered.

The next question is how accurately can we estimate pitch by the ear only ? The sounds are here supposed to be heard in succession, for (§ 59) when two uniformly sustained notes are sounded together there is no limit to the accuracy of comparison attainable by the method of beats. From a series of elaborate experiments Preyer[1] concludes that at no part of the scale can ·20 vibration per second be distinguished with certainty. The sensitiveness varies with pitch. In the neighbourhood of 120, ·4 vibration per second can be just distinguished; at 500 about ·3 vibration; and at 1000 about ·5 vibration per second. In some cases where a difference of pitch was recognised, the observer could not decide which of the two sounds was the graver.

384. In determinations of the limits of pitch, or of the perceptible differences of pitch, the sounds are to be chosen of convenient intensity. But a further question remains behind as to the degree of intensity at given pitch necessary for audibility. The earliest estimate of the amplitude of but just audible sounds appears to be that of Toepler and Boltzmann[2]. It depends upon an ingenious application of v. Helmholtz's theory of the open organ-pipe (§ 313) to data relating to the maximum condensation within the pipe, as obtained by the authors experimentally (§ 322 d). They conclude that plane waves, of pitch 181, in which the maximum condensation (s) is $6 \cdot 5 \times 10^{-8}$, are just audible.

It is evident that a superior limit to the amplitude of waves giving an audible sound may be derived from a knowledge of the energy which must be expended in a given time in order to

---

[1] An account of Preyer's work was given by A. J. Ellis in the *Proceedings of the Musical Association*, 3rd session, p. 1, 1877.

[2] *Pogg. Ann.* vol. cxli. p. 321, 1870.

generate them and of the extent of surface over which the waves so generated are spread at the time of hearing. An estimate founded on these data will necessarily be too high, both because sound-waves must suffer some dissipation in their progress and also because a part, and in some cases a large part, of the energy expended never takes the form of sound-waves at all.

In the first application of the method[1], the source of sound was a whistle, mounted upon a Wolfe's bottle, in connection with which was a siphon manometer for the purpose of measuring the pressure of the wind. The apparatus was inflated from the lungs, and with a little practice there was no difficulty in maintaining a sufficiently constant blast of the requisite duration. The most suitable pressure was determined by preliminary trials, and was measured by a column of water $9\frac{1}{2}$ cm. high.

The first point to be determined was the distance from the source to which the sound remained clearly audible. The experiment was tried upon a still winter's day and it was ascertained that the whistle could be heard without effort (in both directions) to a distance of 820 metres.

The only remaining datum necessary for the calculation is the quantity of air which passes through the whistle in a given time. This was determined by a laboratory experiment from which it appeared that the consumption was 196 cub. cents. per second.

In working out the result it is most convenient to use consistently the C. G. S. system. On this system of measurement the pressure employed was $9\frac{1}{2} \times 981$ dynes per sq. cent., and therefore the work expended per second in generating the waves was $196 \times 9\frac{1}{2} \times 981$ ergs[2].

Now (§ 245) the mechanical value of a series of progressive waves is the same as the kinetic energy of the whole mass of air concerned, supposed to be moving with the maximum velocity (v) of vibration; so that, if S denote the area of the wave-front considered, a the velocity of sound, ρ the density of air, the mechanical value of the waves passing in a unit of time is expressed by $S \cdot a \cdot \rho \cdot \frac{1}{2}v^2$, in which the numerical value of a is about 34100, and that of ρ about ·0013. In the present application S is the area of the surface of a hemisphere whose radius is

---

[1] *Proc. Roy. Soc.* vol. XXVI. p. 248, 1877.

[2] Nearly 2×10^6 ergs.

82000 centimetres; and thus, if the whole energy of the escaping air were converted into sound and there were no dissipation on the way, the value of v at a distance of 82000 centimetres would be given by the equation

$$v^2 = \frac{2 \times 196 \times 9\frac{1}{2} \times 981}{2\pi \, (82000)^2 \times 34100 \times \cdot 0013},$$

whence $\quad v = \cdot 0014 \, \dfrac{\text{cm.}}{\text{sec.}}, \quad s = \dfrac{v}{a} = 4\cdot 1 \times 10^{-8}.$

This result does not require a knowledge of the pitch of the sound. If the period be τ, the relation between the maximum excursion x and the maximum velocity v is $x = v\tau/2\pi$. In the experiment under discussion the note was f^{IV}, with a frequency of about 2730. Hence

$$x = \frac{\cdot 0014}{2\pi \times 2730} = 8\cdot 1 \times 10^{-8} \, \text{cm.},$$

or the amplitude of the aerial particles was less than a ten-millionth of a centimetre. It was estimated that under favourable conditions an amplitude of 10^{-8} cm. would still have been audible.

It is an objection to the above method that when such large distances are concerned it is difficult to feel sure that the disturbing influence of atmospheric refraction is sufficiently excluded Subsequently experiments were attempted with pipes of lower pitch which should be audible to a less distance, but these were not successful, and ultimately recourse was had to tuning-forks.

"A fork of known dimensions, vibrating with a known amplitude, may be regarded as a store of energy of which the amount may readily be calculated. This energy is gradually consumed by internal friction and by generation of sound. When a resonator is employed the latter element is the more important, and in some cases we may regard the dying down of the amplitude as sufficiently accounted for by the emission of sound. Adopting this view for the present, we may deduce the rate of emission of sonorous energy from the observed amplitude of the fork at the moment in question and from the rate at which the amplitude decreases. Thus if the law of decrease be $e^{-\frac{1}{2}kt}$ for the amplitude of the fork, or e^{-kt} for the energy, and if E be the total energy at time t, the rate at which energy is emitted at that time is $-dE/dt$, or kE. The value of k is deducible from observations of the rate of decay, e. g. of the time during which the amplitude is halved. With these arrangements there is no difficulty in converting energy into sound upon a small scale, and thus in reducing the distance of audibility to such a figure as 30 metres. Under these circumstances the observations are much more manageable than when the operators are separated by half a mile, and there is no reason to fear disturbance from atmospheric refraction.

The fork is mounted upon a stand to which is also firmly attached the observing-microscope. Suitable points of light are obtained from starch grains, and the line of light into which each point is extended by the vibration is determined with the aid of an eyepiece-micrometer. Each division of the micrometer-scale represents ·001 centim. The resonator, when in use, is situated in the position of maximum effect, with its mouth under the free ends of the vibrating prongs.

The course of an experiment was as follows:—In the first place the rates of dying down were observed, with and without the resonator, the stand being situated upon the ground in the middle of a lawn. The fork was set in vibration with a bow, and the time required for the double amplitude to fall to half its original value was determined. Thus in the case of a fork of frequency 256, the time during which the vibration fell from 20 micrometer-divisions to 10 micrometer-divisions was 16^s without the resonator, and 9^s when the resonator was in position. These times of halving were, as far as could be observed, independent of the initial amplitude. To determine the minimum audible, one observer (myself) took up a position 30 yards (27·4 metres) from the fork, and a second (Mr. Gordon) communicated a large vibration to the fork. At the moment when the double amplitude measured 20 micrometer-divisions the second observer gave a signal, and immediately afterwards withdrew to a distance. The business of the first observer was to estimate for how many seconds after the signal the sound still remained audible. In the case referred to the time was 12^s. When the distance was reduced to 15 yards (13·7 metres), an initial double amplitude of 10 micrometer-divisions was audible for almost exactly the same time.

These estimates of audibility are not made without some difficulty. There are usually 2 or 3 seconds during which the observer is in doubt whether he hears or only imagines, and different individuals decide the question in opposite ways. There is also of course room for a real difference of hearing, but this has not obtruded itself much. A given observer on a given day will often agree with himself surprisingly well, but the accuracy thus suggested is, I think, illusory. Much depends upon freedom from disturbing noises. The wind in the trees or the twittering of birds embarrasses the observer, and interferes more or less with the accuracy of results.

The equality of emission of sound in various horizontal directions was tested, but no difference could be found. The sound issues almost entirely from the resonator, and this may be expected to act as a simple source.

When the time of audibility is regarded as known, it is easy to deduce the amplitude of the vibration of the fork at the moment when the sound ceases to impress the observer. From this the rate of emission of sonorous energy and the amplitude of the aerial vibration as it reaches the observer are to be calculated.

The first step in the calculation is the expression of the total energy of the fork as a function of the amplitude of vibration measured at the extremity of one of the prongs. This problem is considered in § 164. If l be the length, ρ the density, and ω the sectional area of a rod damped at one end and free at the other, the kinetic energy T is connected with the displacement η at the free end by the equation (10)

$$T = \tfrac{1}{8}\rho l\omega (d\eta/dt)^2.$$

At the moment of passage through the position of equilibrium $\eta = 0$ and $d\eta/dt$ has its maximum value, the whole energy being then kinetic. The maximum value of $d\eta/dt$ is connected with the maximum value of η by the equation

$$(d\eta/dt)_{\max.} = 2\pi/\tau \,.\, (\eta)_{\max.};$$

so that if we now denote the double amplitude by 2η, the whole energy of the vibrating bar is $\tfrac{1}{8}\rho\omega l\pi^2/\tau^2 \,.\, (2\eta)^2$,

or for the two bars composing the fork

$$E = \tfrac{1}{4}\rho\omega l\pi^2/\tau^2 \,.\, (2\eta)^2, \dots\dots\dots\dots\dots\dots\dots (A)$$

where $\rho\omega l$ is the mass of each prong.

The application of (A) to the 256-fork, vibrating with a double amplitude of 20 micrometer-divisions, is as follows. We have

$$l = 14\cdot0 \text{ cm.}, \qquad \omega = \cdot6 \times 1\cdot1 = \cdot66 \text{ sq. cm.},$$
$$1/\tau = 256, \qquad \rho = 7\cdot8, \qquad 2\eta = \cdot050 \text{ cm.};$$

and thus $\qquad\qquad E = 4\cdot06 \times 10^3 \text{ ergs.}$

This is the whole energy of the fork when the actual double amplitude at the ends of the prongs is ·050 centim.

As has already been shewn, the energy lost per second is kE, if the amplitude vary as $e^{-\frac{1}{2}kt}$. For the present purpose k must be regarded as made up of two parts, one k_1 representing the dissipation which occurs in the absence of the resonator, the other k_2 due to the resonator. It is the latter part only which is effective towards the production of sound. For when the resonator is out of use the fork is practically silent; and, indeed, even if it were worth while to make a correction on account of the residual sound, its phase would only accidentally agree with that of the sound issuing from the resonator.

The values of k_1 and k are conveniently derived from the times, t_1 and t, during which the amplitude falls to *one half*. Thus

$$k = 2\log_e 2 . /t, \quad k_1 = 2\log_e 2 . /t_1 ;$$

so that

$$k_2 = 2\log_e 2 . (1/t - 1/t_1) = 1\cdot386\,(1/t - 1/t_1).$$

And the energy converted into sound per second is k_2E.

We may now apply these formulæ to the case, already quoted, of the 256-fork, for which $t = 9$, $t_1 = 16$. Thus t_2, the time which would be occupied in halving the amplitude were the dissipation due entirely to the resonator, is 20·6; and $k_2 = \cdot0674$. Accordingly,

$$k_2E = 267 \text{ ergs per second,}$$

corresponding to a double amplitude represented by 20 micrometer-divisions. In the experiment quoted the duration of audibility was 12 seconds, during which the amplitude would fall in the ratio $2^{12/9} : 1$, and the energy in the ratio $4^{12/9} : 1$. Hence at the moment when the sound was just becoming inaudible the energy emitted as sound was 42·1 ergs per second[1].

---

[1] It is of interest to compare with the energy-emission of a source of light. An incandescent electric lamp of 200 candles absorbs about a horse-power, or say 10^{10} ergs per second. Of the total radiation only about $\frac{1}{100}$ part acts effectively upon the eye; so that radiation of suitable quality consuming 5×10^5 ergs per second corresponds to a candle-power. This is about 10^4 times that emitted as sound by the fork in the experiment described above. At a distance of $10^2 \times 30$, or 3000 metres, the stream of energy from the ideal candle would be about equal to the stream of energy just audible to the ear. It appears that the streams of energy required to influence the eye and the ear are of the same order of magnitude, a conclusion already drawn by Toepler and Boltzmann.

The question now remains, What is the corresponding amplitude or condensation in the progressive aerial waves at $27\cdot4$ metres from the source? If we suppose, as in my former calculations, that the ground reflects well, we are to treat the waves as hemispherical. On the whole this seems to be the best supposition to make, although the reflexion is doubtless imperfect. The area S covered at the distance of the observer is thus $2\pi \times 2740^2$ sq. centim., and since (§ 245)

$$S \cdot \tfrac{1}{2}a\rho v^2 = S \cdot \tfrac{1}{2}\rho a^3 s^2 = 42\cdot1,$$

we find

$$s^2 = \frac{42\cdot1}{\pi \times 2740^2 \times \cdot00125 \times 34100^3},$$

and

$$s = 6\cdot0 \times 10^{-9}.$$

The condensation s is here reckoned in atmospheres; and the result shews that the ear is able to recognize the addition and subtraction of densities far less than those to be found in our highest vacua.

The amplitude of aerial vibration is given by $as\tau/2\pi$, where $1/\tau = 256$, and is thus equal to $1\cdot27 \times 10^{-7}$ cm.

It is to be observed that the numbers thus obtained are still somewhat of the nature of superior limits, for they depend upon the assumption that all the dissipation *due to the resonator* represents production of sound. This may not be strictly the case even with the moderate amplitudes here in question, but the uncertainty under this head is far less than in the case of resonators or organpipes caused to speak by wind. From the nature of the calculation by which the amplitude or condensation in the aerial waves is deduced, a considerable loss of energy does not largely influence the final numbers.

Similar experiments have been tried at various times with forks of pitch 384 and 512. The results were not quite so accordant as was at first hoped might be the case, but they suffice to fix with some approximation the condensation necessary for audibility. The mean results are as follows:—

$$c', \quad \text{frequency} = 256, \quad s = 6\cdot0 \times 10^{-9},$$
$$g', \quad \quad\quad\;\; = 384, \quad s = 4\cdot6 \times 10^{-9},$$
$$c'', \quad \quad\quad\;\; = 512, \quad s = 4\cdot6 \times 10^{-9},$$

no reliable distinction appearing between the two last numbers. Even the distinction between $6\cdot0$ and $4\cdot6$ should be accepted with reserve; so that the comparison must not be taken to prove much more than that the condensation necessary for audibility varies but slowly in the singly dashed octave[1]."

Results of the same order of magnitude have been obtained also by Wien[2], who used an entirely different method.

385. For most purposes of experiment and for many of ordinary life it makes but little difference whether we employ one ear only, or both; and yet there can be no doubt that we can derive most important information from the simultaneous use of the two ears. How this is effected still remains very obscure.

Although the utmost precautions be taken to ensure separate action, it is certain that a sound led into one ear is capable of giving beats with a second sound of slightly different pitch led into the other ear. There is, of course, no approximation to such silence as would occur at the moment of antagonism were the two sounds conveyed to the same ear; but the beats are perfectly distinct, and remain so as the sounds die away so as to become singly all but inaudible[3]. It is found, however, that combination tones (§ 391) are not produced under these conditions[4]. Some curious observations with the telephone are thus described by Prof. S. P. Thompson[5]. "Almost all persons who have experimented with the Bell telephone, when using a pair of instruments to receive the sound, one applied to each ear, have at some time or other noticed the apparent localization of the sounds of the telephone at the back of the head. Few, however, seemed to be aware that this was the result of either reversed order in the connection of the terminals of the instruments with the circuit, or reversed order in the polarity of the magnet of one of the receiving instruments. When the two vibrating discs execute similar vibrations, both advancing or both receding at once, the sound is heard as usual in the ears; but if the action of one instrument be reversed, so that when one disc advances the other recedes, and the vibrations have opposite phases, the sound apparently changes its place from the interior of the ear, and is heard as if proceeding from the back of the head, or, as I would rather say, from the top

---

[1] *Phil. Mag.* vol. xxxviii. p. 366, 1894.

[2] *Wied. Ann.* vol. xxxvi. p. 834, 1889.

[3] S. P. Thompson, *Phil. Mag.* vol. iv. p. 274, 1877.

[4] See also Dove, *Pogg. Ann.* vol. cvii. p. 652, 1859.

[5] *Phil. Mag.* vol. vi. p. 385, 1878.

of the cerebellum."....."I arranged a Hughes's microphone with two cells of a Fuller's battery and two Bell telephones, one of them having a commutator under my control. Placing the telephones to my ears, I requested my assistant to tap on the wooden support of the microphone. The result was deafening. I felt as if simultaneous blows had been given to the tympana of my ears. But on reversing the current through one telephone, I experienced a sensation only to be described as of some one tapping with a hammer *on the back of the skull from the inside.*"

In our estimation of the direction in which a sound comes to us we are largely dependent upon the evidence afforded by binaural audition. This is one of those familiar and instinctive operations which often present peculiar difficulties to scientific analysis. A blindfold observer in the open air is usually able to indicate within a few degrees the direction of a sound, even though it be of short duration, such as a single vowel or a clap of the hands. The decision is made with confidence and does not require a movement of the head.

To obtain further evidence experiments were made with the approximately pure tones emitted from forks in association with resonators; but in order to meet the objection that the first sound of the fork, especially when struck, might give a clue, and so vitiate the experiment, *two* similar forks and resonators, of pitch 256, were provided. These were held by two assistants, between whom the observer stood midway. In each trial both forks were struck, and afterwards *one* only was held to its resonator. The results were perfectly clear. When the forks were to the right and to the left, the observer could distinguish them instinctively and without fail. But when he turned through a right angle, so as to bring the forks to positions in front and behind him, no discrimination was possible, and an attempt to pronounce was felt to be only guessing.

That it should be impossible to distinguish whether a pure tone comes from in front or from behind is intelligible enough. On account of the symmetry the two ears would be affected alike in both cases, and any difference of intensity due to the position could not avail in the absence of information as to the original intensity. The difficulty is rather to understand how the discrimination between front and rear is effected in other cases, *e.g.* of the voice, where it is found to be easy. It can only be conjectured that the quality of a compound sound is liable to modification by the external ear, which is differently presented in the two cases.

The ready discrimination between right and left, even when pure tones are concerned, is naturally attributed to the different intensities with which the sound would be perceived by the two ears. But this explanation is not so complete as might be supposed. It is true that very high sounds, such as a hiss, are ill heard with the averted ear: but when the pitch is moderate, *e.g.* 256 per second, the difference of intensity on the two sides does not seem very great. The experiment may easily be tried roughly by stopping one ear with the finger and turning round backwards and forwards while listening to a sound held steadily. Calculation (§ 328) shews, moreover, that the human head, considered as an obstacle to the waves of sound, is scarcely big enough in relation to the wave length to throw a distinct shadow. As an illustration I have calculated the intensity of sound due to a distant source at various points on the surface of a fixed spherical obstacle. The result depends upon the ratio (kc) between the circumference of the sphere and the length of the wave. If we call the point upon the spherical surface nearest to the source the anterior pole, and the opposite point (where the shadow might be expected to be most intense) the posterior pole, the results on three suppositions as to the relative magnitudes of the sphere and wave length are as follows :—

	$kc = 2$	$kc = 1$	$kc = \frac{1}{2}$
Anterior pole	·69	·50	·29
Posterior pole	·32	·28	·26
Equator	·36	·24	·23

When for example the circumference of the sphere is but half the wave length, the intensity at the posterior pole is only about a tenth part less than at the anterior pole, while the intensity is least of all in a lateral direction. When kc is less than $\frac{1}{2}$, the difference of the intensities at the two poles is still less important, amounting to about 1 per cent. when $kc = \frac{1}{4}$.

The case of the head and a pitch c' would correspond to $kc = ·4$ about, so that the differences of intensity indicated by theory are decidedly small. The explanation of the power of discrimination actually observed would be easier, if it were possible to suppose account taken of the different phases of the vibrations by which the two ears are attacked[1].

386. Passing on to another branch of our subject, we have now to consider more closely the impression produced upon the ear by an arbitrary sequence of aerial pressures fluctuating about a certain mean value. According to the literal statement of Ohm's law (§ 27) the ear is capable of hearing as separate tones all the simple vibrations into which the sequence of pressures may be analysed by Fourier's theorem, provided that the pitch of these components lies between certain limits. Components whose pitch lies outside the limits would be ignored. Moreover, within the limits of audibility the relative phases of the various components would be a matter of indifference.

To the law stated in this extreme form there must obviously be exceptions. It is impossible to suppose that the ear would hear as separate tones simple components of extremely nearly the same frequency. Such components, it is well known, give rise to beats, and their relative phase is a material element in the question. Again, it will be evident that the corresponding tone will not be heard unless a vibration reaches a certain intensity. A finite intensity would be demanded, even if the vibration stood by itself; and we should expect that the intensity necessary for audibility would be greater in the presence of other vibrations, especially perhaps when these correspond to harmonic undertones. It will be advisable to consider these necessary exceptions to the universality of Ohm's law a little more in detail.

The course of events, when the interval between two simple vibrations is gradually increased, has been specially studied by Bosanquet[2]. As in §§ 30, 65a, if the components be $\cos 2\pi n_1 t$, $\cos 2\pi n_2 t$, we have for the resultant,

$$u = \cos 2\pi n_1 t + \cos 2\pi n_2 t$$
$$= 2 \cos \pi(n_2 - n_1)t \cdot \cos \pi(n_2 + n_1)t \ldots \ldots (1);$$

shewing that the resultant of two simple vibrations of equal amplitudes and of frequencies n_1, n_2 can be represented mathematically by a single vibration whose frequency is the mean, viz.

[1] *Nature*, vol. xiv. p. 32, 1876. *Phil. Mag.* vol. iii. p. 456, 1877; vol. vii. p. 149, 1879.

[2] *Phil. Mag.* vol. xi. p. 420, 1881.

$\frac{1}{2}(n_1 + n_2)$, and whose amplitude varies according to the cosine law, involving a change of sign (§ 65a), with a frequency $\frac{1}{2}(n_2 - n_1)$. This single vibration is not *simple*. The question now arises under which of the two forms in (1) will the ear perceive the sound. According to the strict reading of Ohm's law the two tones n_1 and n_2 would be perceived separately. We know that when n_1 and n_2 are nearly enough equal this does not and could not happen. The second form then represents the phenomenon; and it indicates *beats*, the tone $\frac{1}{2}(n_1 + n_2)$ having an *intensity* which varies between 0 and 4 with a frequency $(n_2 - n_1)$ equal to the difference of frequencies of the original tones. Mr Bosanquet found that "(α) the critical interval at which two notes begin to be heard beside their beats, or resultant displacements, is about two commas, throughout that medium portion of the scale which is used in practical music; (β) this critical interval appears to be not exactly the same for all ears." But in both the cases examined the beats alone were heard with an interval of one comma, and the two notes were quite clear beside the beats with an interval of three commas. "As the interval increases, the separate notes become more and more prominent, and the beats diminish in loudness and distinctness, till, by the time that a certain interval is reached, which is about a minor third in the middle of the scale, the beats practically disappear and the two notes alone survive."

On the second question as to the strength in which a component simple vibration, of sufficiently distinct pitch, must be present in order to assert itself as a separate tone there is but little evidence, and that not very accordant. According to the experiments of Brandt and Helmholtz (§ 130) Young's law as to the absence in certain cases of particular components from the sound of a plucked string is verified. Observations of this kind are easily made with resonators; but for the present purpose the use of resonators is inadmissible, the question relating to the behaviour of the unassisted ear.

On the other hand A. M. Mayer[1] found that sounds of considerable intensity when heard by themselves were liable to be completely obliterated by *graver* sounds of sufficient force. In some experiments the graver note was from an open organ-pipe which sounded steadily, while the higher was that of a fork, excited vigorously and then allowed to die down. The action of the fork could be

[1] *Phil. Mag.* vol. II. p. 500, 1876.

made intermittent by moving the hand to and fro over the mouth of its resonance box. The results are thus described. "At first every time that the mouth of the box is open the sound of the fork is distinctly heard and changes the quality of the note of the open pipe. But as the vibrations of the fork run down in amplitude the sensations of its effect become less and less till they soon entirely vanish, and not the slightest change can be observed in the quality or intensity of the note of the organ-pipe, whether the resonance box of the fork be open or closed. Indeed at this stage of the experiment the vibrations of the fork may be suddenly and totally stopped without the ear being able to detect the fact. But if instead of stopping the fork when it becomes inaudible we stop the sound of the organ-pipe, it is impossible not to feel surprised at the strong sound of the fork which the open pipe had smothered and had rendered powerless to affect the ear."

But "no sound, even when very intense, can diminish or obliterate the sensation of a concurrent sound which is lower in pitch. This was proved by experiments similar to the last, but differing in having the more intense sound higher (instead of lower) in pitch. In this case, when the ear decides that the sound of the (lower and feebler) tuning-fork is just extinguished, it is generally discovered on stopping the higher sound that the *fork*, which should produce the lower sound, *has ceased to vibrate.* This surprising experiment must be made in order to be appreciated. I will only remark that very many similar experiments, ranging through four octaves, have been made, with consonant and dissonant intervals, and that scores of different hearers have confirmed this discovery."

These results, which are not difficult to verify[1], involve a serious deduction from the universality of Ohm's law, and must have an important bearing upon other unsettled questions relating to audition. It is to be observed that in Mayer's experiments the question is not merely whether a particular tone can be heard as such. The higher sound of feebler intensity is not heard *at all.*

The audibility of a sound, even when isolated, is influenced by the state of the ear as regards fatigue. The effect is especially

---

[1] Instead of a box screwed to the fork, I found it better to use an independent resonator, to the mouth of which the fork is made to approach and recede in a definite manner.

apparent with the very high notes of bird-calls (§ 371). "A bird-call was mounted in connection with a loaded gas-bag and a water-manometer, by which means the pressure could be kept constant for a considerable time. When the ear is placed at a moderate distance from the instrument, a disagreeable sound is heard at first, but after a short interval, usually not more than three or four seconds, fades away and disappears altogether. A very short intermission suffices for at any rate a partial recovery of the power of hearing. A pretty rapid passage of the hand, screening the ear for a fraction of a second, allows the sound to be heard again[1]."

But although Ohm's law is subject to important limitations, it can hardly be disputed that the ear is capable of making a rough analysis of a compound vibration into its simple parts. The nature of the difficulty commonly met with has already been referred to (§§ 25, 26), but a few further remarks may here be made.

In resolving compound notes a certain control over the attention is the principal requisite, and Helmholtz found that the advantage does not always lie with musically trained ears. Before a particular tone is listened for, it ought to be sounded so as to become fixed in the memory, but not too loudly, lest the sensitiveness of the ear be unduly impaired. As a rule the uneven component tones, twelfth, higher third, &c., are more easily recognised than the octaves.

On the pianoforte, for example, let g' be first gently given, and as soon as the key is released, let c be sounded strongly. The tone g' on which the attention should be kept rivetted throughout, may now be heard as part of the compound note c. A similar experiment may be made with the higher third e'', and an acute ear may detect a slight fall in pitch. This is a consequence of the equal temperament tuning (§ 19), and shews clearly that the apparent prolongation of the tone is not due to imagination. In modern pianos the seventh and ninth component tones are often weak or altogether absent, but on the harmonium these tones may usually be heard.

It is still better when the tone to be listened for is first obtained as a harmonic from the string c itself. In the case of

---

[1] *Phil. Mag.* vol. XIII. p. 344, 1882.

the twelfth, for example, strike the key gently while the string is lightly touched at one-third of its length, and then after removal of the finger more strongly. The proper point may be conveniently found by sliding the finger slowly along the string, while the key is continually struck. When a point of aliquot division is reached, the corresponding harmonic rings out clearly; otherwise the sound is feeble and muffled. In this way Helmholtz succeeded in hearing the overtones of thin strings as far as the sixteenth. From this point they lie too close together to be easily distinguished.

A further slight modification of this method is especially recommended by Helmholtz. Instead of using the finger, the nodal point is touched with a small camel's hair brush. This allows the degree of damping to be varied at pleasure, and a gradual transition to be made from the pure harmonic, free from all admixture of components which have not a node at the point touched, to the natural note of the string.

But it is with the assistance of resonators that overtones are most easily heard in the first instance. For this purpose a resonator is chosen, tuned, say, to g', and the ear is placed in communication with its cavity. When c is sounded, either on the piano or harmonium, or with the human voice, the tone g' may usually be heard very loud and distinct. Indeed on many pianofortes a tone g' may be heard as loudly from its harmonic undertones g or c as from the string g' itself. When an overtone has once been heard, the assistance of the resonator should be gradually withdrawn, which may be done either by removing it from the ear, or putting it out of tune by an obstacle (such as the finger) held near its mouth.

387. If it be admitted that the ear is capable of analysing a musical note into components, or partials, it follows almost of necessity that these more elementary sensations correspond to simple vibrations. So long as we keep within the range of the principle of superposition, this is the kind of analysis effected by mechanical appliances, such as resonators, and all the more patent facts go to prove that the ear resolves according to the same laws. Moreover, the *à priori* probabilities of the case seem to tend in the same direction. It is difficult to suppose that physiological effects—electrical, chemical, or of some unknown character,—are produced directly by the impact of sonorous waves involving merely a variable fluid pressure. Helmholtz's theory of audition is based upon the more natural supposition that the immediate effect of the waves is to set into ordinary mechanical vibration certain internal vibrators[1], and that nervous excitation follows as a secondary consequence.

The *modus operandi* is conceived to be as follows. When a simple tone finds access to the ear, all the parts capable of motion vibrate in synchronism with the source. If there be any part, approximately isolated, whose natural period nearly agrees with that of the sound, then the vibration of that part is far more intense than it would otherwise be. Practically this part of the system may be said to respond only to tones whose pitch lies within somewhat narrow limits. Now it is supposed that the auditory nerves are in communication with vibrating parts of the kind described, whose natural pitch ranges at small intervals between the limits of hearing in such a manner that when any part vibrates the corresponding nerve is excited and conveys the impression to the brain. In the case of a simple tone, one (or at most a relatively small number) of the whole series of nerves is excited, the excitation of the nerve being the proximate cause of the hearing of the tone.

At this point the question presents itself whether more than one simple vibration may not have the power of exciting the same nerve? *A priori*, this might well be the case; for the vibrating parts might be susceptible of more than one mode of vibration, and therefore of more than one natural period. If we were to suppose that the natural periods of any vibrating part formed a harmonic scale, so that the same auditory nerve was excited by a tone and its octave, the supposition would certainly give a very ready explanation of the remarkable resemblance of octaves, and would tend to mitigate some of the difficulties which at present stand in the way of accepting Helmholtz's theory as a complete account of the facts of audition[2]. As we shall see presently,

---

[1] The drum-skin and its attachments are here regarded as external to the true auditory mechanism. However important may be the part they play, it is analogous rather to that of a hearing tube or of the disc of a mechanical telephone.

[2] A curious question suggests itself as to what would happen in case the vibrations capable of exciting the same nerve deviated sensibly or considerably from the harmonic scale. In this way ears naturally confused in their appreciation of musical relations may easily be imagined.

Helmholtz would admit, or rather assert, that *when the sounds are strong* two originally simple vibrations, such as c and c', would excite to some extent the same nerve, but he regards this as depending upon a failure in the law of superposition, due to excessive vibration.

388. It is evident that Helmholtz's theory gives a very natural account of Ohm's law, as well as of the limitation to which it is subject when two simple vibrations are in operation of nearly the same pitch. Some of the internal vibrators are then within the influence of both disturbing causes, and are accordingly excited in an intermittent manner, giving rise to beats, when the period is long, and to a sensation of roughness as the beats become too quick to be easily perceived separately. But when the interval between the two vibrations is increased, a point is soon reached after which no internal vibrator is sensibly affected by both disturbing causes, so that from this point onwards the resulting sensation is free from beats or roughnesses, or at least should be so according to the strict interpretation of the law. To this point we shall return later.

The magnitude of the interval, over which a single internal vibrator will respond sensibly, is an element of considerable importance in the theory. It has already been shewn (§ 49) that there is a relation between this interval and the number of free vibrations which can be executed by the vibrating body. Thus, if the interval between the natural and the forced vibration required to reduce the resonance to $\frac{1}{10}$ of the maximum be a semitone, this implies that after 9·5 free vibrations the intensity would be reduced to $\frac{1}{10}$ of its original value, and so on for other intervals. From a consideration of the effect of trills in music, Helmholtz concludes that the case of the ear corresponds somewhat to that above specified, and he gives the accompanying table shewing the

Difference of pitch	Intensity of resonance	Difference of pitch	Intensity of resonance
0·0	100	0·6	7·2
0·1	74	0·7	5·4
0·2	41	0·8	4·2
0·3	24	0·9	3·3
0·4	15	Whole tone	2·7
Semitone	10		

relation obtaining in this case between the difference of free and forced pitch and the intensity of resonance, measured by the square of the amplitude of vibration.

Although according to Helmholtz's theory the sensation of dissonance is caused by intermittent excitation of those vibrating parts which are within the range of two or more elements of the sound, it is not to be inferred that the number of beats is a sufficient measure of the dissonance. On the contrary it is found that if the number of beats be retained constant (e.g. 33 per second), the effect is more and more free from roughness as the sounds are made deeper, the *intervals* being correspondingly increased.

The experiments of A. M. Mayer[1] extend over a considerable range of pitch, and have been made by two methods. In the first method a sound, which would otherwise be a pure tone, is rendered intermittent, and the rate of intermittence is gradually raised to the point at which the effect upon the ear again becomes smooth. The results are shewn in the accompanying table, in which the first column gives the pitch of the sound and the second the minimum number of intermittences per second required to eliminate the roughness.

Pitch (n)	Frequency of Intermittence (m)
64	23·1
128	36
256	62
320	73
384	88
512	108
640	126
768	143
1023	170

The theory of intermittent vibrations has already been given § 65 *a*. It is to be remembered that by the nature of the case an intermittent vibration cannot be simple. To a first approximation it may be supposed to be equivalent to *three* simple vibrations of frequencies, $n - m$, n, $n + m$, and the roughness experienced by

[1] *Phil. Mag.* vol. XLIX. p. 352, 1875; vol. XXXVII. p. 259, 1894.

the ear may be looked upon as due to the beats of these three tones.

Mayer has experimented also upon the "smallest consonant intervals among simple tones," i.e. upon the intervals at which the roughness due to beats just disappears, the plural being used since it is found that the necessary interval varies at different parts of the scale.

Pitch (n_1)	Additional vibrations required $(n_2 - n_1)$	Smallest consonant intervals in semitones
48	......	none
64	......	none
96	41	6·15
128	38	4·50
192	48	3·86
256	58	3·53
316	68	3·34
432	85	3·12
575	107	2·95
766	130	2·70
1707	210	2·00
2304	245	1·76
2560	256	1·64
2806	266	1·54

Different observers agreed very closely as to the point at which roughness disappeared.

According to the theory of intermittent sounds it is to be expected that for a given pitch m in the first set of experiments should be nearly the same as $(n_2 - n_1)$ in the second, and this is pretty well verified by Mayer's numbers, at least over the middle region of the scale.

389. From the degree of damping above determined it follows that the natural pitch of the internal vibrators, which respond sensibly to a given simple sound, ranges over about a whole tone, and it may excite surprise that we are able to compare with such accuracy the pitch of musical sounds heard in succession. The explanation probably depends a good deal upon the symmetry of the effects on the two sides of the maximum. A comparison with the capabilities of the eye in a similar case may be instructive. In setting the cross wires of a telescope upon the centre of a symmetrical luminous band, e.g. an interference band, it is found that the error need not exceed $\frac{1}{100}$ of the width. A similarly accurate judgment as to the centre of the region excited by a given musical note would lead to an estimation of pitch accurate to about $\frac{1}{1000}$, agreeing well enough with the facts to be explained.

In the light of the same principle we may consider how far the perception of pitch would be prejudiced by a limitation of the number of vibrations executed during the continuance of a sound. According to the estimate of Helmholtz already employed (§ 388) the internal vibrations, excited and then left to themselves, would remain sensible over about 10 periods. The number of impulses required to produce nearly the full effect is of this order of magnitude. If the number were increased beyond 20 or 30, there would be little further concentration of effect in the neighbourhood of the maximum, and therefore little foundation for greater accuracy in the estimation of pitch.

Experiments upon this subject have been made by Seebeck[1], Pfaundler[2], S. Exner[3], Auerbach[4], and W. Kohlrausch[5], those of the last being the most extensive. An arc of a circle carrying a limited number of teeth was attached to a pendulum, which could be let go under known conditions. In their passage the teeth struck against a card suitably held ; and the sound thus generated was compared with that of a monochord. By varying the length in the usual manner the chord was tuned until the pitch was just perceptibly higher, or just perceptibly lower, than that proceeding from the card, and the interval between the two, called the characteristic interval, determined the precision with which the pitch could be estimated in the case of a given total number of vibrations. The best results were obtained only after considerable practice and in the entire absence of extraneous sounds.

Sixteen teeth appeared to define the pitch with all the precision attainable, the characteristic interval (on the mean of a number of experiments) being- in this case ·9922. Even with 9

[1] *Pogg. Ann.* vol. LIII. p. 417, 1841.
[2] *Wien. Ber.* vol. LXXVI. p. 561, 1877.
[3] *Pflüger's Archiv*, vol. XIII. p. 228, 1876. .
[4] *Wied. Ann.* vol. VI. p. 591, 1879.
[5] *Wien. Ann.* vol. X. p. 1, 1880.

teeth the characteristic interval was 9903, shewing that this small number of vibrations was capable of defining the pitch to within one part in 200. But the most surprising results were those obtained with a very low number of teeth. For 3 teeth the characteristic interval was ·9790, and for 2 teeth ·9714.

The fact that pitch can be defined with considerable accuracy by so small a sequence of vibrations has sometimes been regarded as an objection to Helmholtz's theory of audition. I do not think that there is any ground for this opinion. So far as there is a difficulty, it is one that would tell equally against any other theory that could be proposed.

It would seem that the delimitation of pitch in Kohlrausch's experiments may have been greatly favoured by the approximate discontinuity of the impulses. For it is to be remembered that the internal vibrators concerned are not those only whose period ranges round the interval between the taps, but also those whose periods are approximately submultiples of this quantity. As regards the vibrators in the octave, the number of impulses is practically doubled, for the twelfth trebled, and so on, just as in optics the resolving power of a grating with a limited number of lines is increased in the spectra of the second and higher orders.

Vibrations limited to a moderate number of periods are sometimes generated by reflection of short sounds from railings or steps. At Terling there is a flight of about 20 steps which returns an echo of a clap of the hands as a note resembling the chirp of a sparrow. In all such cases the action is exactly analogous to that of a grating in optics.

390. When two sounds nearly in unison are compound, we have to consider not only the beats of their first partials, or primes, but also the beats of the overtones. The beats of the octave components are twice, and those of the twelfth three times, as quick as the simultaneous beats of the primes. In some cases, especially where the pitch is very low, mistakes may easily be made by overlooking the prime beats, which affect the ear but feebly. If the octave beats be reckoned as though they were the beats of the primes, the difference of pitch will be taken to be the double of its true value.

But it is in the case of disturbed consonances other than the unison that the importance of upper partials, or overtones, makes itself specially felt. For example, take the Fifth *c—g*. The third partial of *c* and the second partial of *g* coincide at *g'*. If the interval be true, there are no beats; but if it be slightly disturbed from the true ratio 3 : 2, the two previously coincident tones separate from one another and give rise to beats. The frequency of the beats follows at once from the manner of their genesis. Thus if the lower note be disturbed from its original frequency by one vibration per second, its third partial is changed by 3 vibrations per second, and 3 per second is accordingly the frequency of the beats. But if the upper note undergoes a disturbance of one vibration per second, while the lower remains unaltered, the frequency of the beats is 2. This rule is evidently general. If the consonance be such that the *h*th partial of the lower note coincides with the *k*th partial of the upper note, then when the lower note is altered by one vibration per second, the frequency of the beats is *h*, and when the upper note is altered by the same quantity, the frequency of the beats is *k*.

" We have stated that the beats heard are the beats of those partial tones of both compounds which nearly coincide. Now it is not always very easy on hearing a Fifth or an Octave which is slightly out of tune, to recognise clearly with the unassisted ear which part of the whole sound is beating. On listening we are apt to feel that the whole sound is alternately reinforced and weakened. Yet an ear accustomed to distinguish upper partial tones, after directing its attention upon the common upper partials concerned, will easily hear the strong beats of these particular tones, and recognise the continued and undisturbed sound of the primes. Strike the note (*c*), attend to its upper partial (*g'*), and then strike a tempered Fifth (*g*); the beats of (*g'*) will be clearly heard. To an unpractised ear the resonators already described will be of great assistance. Apply the resonator for (*g'*), and the above beats will be heard with great distinctness. If, on the other hand, a resonator, tuned to one of the primes (*c*) or (*g*), be employed, the beats are heard much less distinctly, because the continuous part of the tone is then reinforced[1]."

Experiments of this kind are conveniently made on the harmonium. Small changes of pitch may be obtained by only partially (instead of fully) depressing the key, the effect of which is to flatten the note. The beats of the common overtone are

[1] *Sensations of Tone*, 2nd ed. p. 181.

easily heard when a (tempered) Fifth is sounded; those of the equal temperament Third are somewhat rapid.

The harmonium is also a suitable instrument for experiments illustrative of just intonation. A reed may be flattened by loading the free end of the tongue with a fragment of wax, and sharpened by a slight filing at the same place. It is easy, especially with the aid of resonators, to tune truly the chords $c'—e'—g'$, $f'—a'—c''$, whose consonance will then contrast favourably with the unaltered tempered chord $g'—b'—d''$. It is not consistent with the plan of this work to enter at length into questions of temperament and just intonation. Full particulars will be found in the English edition of Helmholtz (with Ellis's notes) and in Mr Bosanquet's treatise.

According to Helmholtz's theory it is mainly the beats of the upper partials which determine the ordinary consonant intervals, any departure from which is made evident by the beats of the previously coincident overtones. But even when the notes are truly tuned, the various consonances differ in degree, on account of the disturbances which may arise from overtones which approach one another too nearly.

The unison, octave, twelfth, double octave, etc., may be regarded as absolute consonances, the second component introducing no new element but merely reinforcing a part of the other.

The remaining consonant intervals, such as the Fifth and the Major Third, are in a manner disturbed by their neighbourhood to other consonant intervals. In the case of the truly tuned Fifth, for example, with frequencies represented by 3 and 2, there is indeed coincidence between the second partial of the higher note and the third partial of the graver note, but the partials which define the Fourth, of pitch $3 \times 3 = 9$ and $4 \times 2 = 8$, are within a whole Tone of one another and accordingly near enough to produce disturbance. In like manner the Major Third may be regarded as disturbed by its neighbourhood to the Fourth, and so on in the case of other intervals.

The importance of these disturbances, and consequently the order in which the various intervals stand in respect to their degree of consonance, varies with the quality of the sounds. As an example where overtones are present in considerable strength, Helmholtz has estimated the degree of consonance of various intervals on the violin, and has exhibited the results in the form of a curve[1].

391. The principle of superposition (§ 83), assumed in ordinary acoustical discussions, depends for its validity upon the assumption that the vibrations concerned are infinitely small, or at any rate similar in their character to infinitely small vibrations, and it is only upon this supposition that Ohm's law finds immediate application. One apparent exception to the law has long been known. This is the combination-tone discovered by Sorge and Tartini in the last century. If two notes, at the interval for example of a Major Third, be sounded together strongly, there is heard a grave sound in addition to the two others. In the case specified, where the primary sounds, or generators, as they may conveniently be called, are represented by the numbers 4 and 5, the combination-tone is represented by 1, being thus two octaves below the graver generator.

In the above example the new tone has the period of the cycle of the generating tones; but Helmholtz found that this rule fails in many cases. The following table[2] exhibits his results as obtained by means of tuning-forks:

Generators		Combination tone	Relative Frequency	
			Generators	Combination tone
b	f'	B	2 : 3	1
f'	b'	B	3 : 4	1
b	d'	B_{-1}	4 : 5	1
d'	f'	B_{-1}	5 : 6	1
f'	as'	B_{-1}	6 : 7	1
b	g'	es	3 : 5	2
d'	as'	B	5 : 7	2
d'	b'	f	5 : 8	3

In the last three cases the tones heard were not those in the period of the complete cycle, but their frequencies are the differences of the frequencies of the generators. In virtue of this rule, which was found to apply in all cases[3], the combination-tones in question are called difference-tones.

---

[1] *Sensations of Tone*, p. 193.

[2] *Berlin Monatsber.*, 1856.

[3] It is, however, disputed by other writers.

According to Helmholtz it is necessary to the distinct audibility of combination-tones that the generators be *strong*. We shall see presently that this statement has been contested. "They are most easily heard when the two generating tones are less than an octave apart, because in that case the differential is deeper than either of the two generating tones. To hear it at first, choose two tones which can be held with great force for some time, and form a justly intoned harmonic interval. First sound the low tone and then the high one. On properly directing attention, a weaker low tone will be heard at the moment that the higher note is struck; this is the required combinational tone. For particular instruments, as the harmonium, the combinational tones can be made more audible by properly tuned resonators. In this case the tones are generated in the air contained within the instrument. But in other cases where they are generated solely within the ear, the resonators are of little or no use[1]."

On the strength of some observations by Bosanquet and Preyer, doubts have been expressed as to the correctness of Helmholtz's statement that combination-tones may exist outside the ear, and strangely enough they have been adopted by Ellis. The question has an important bearing upon the theory of combination-tones; and it has recently been examined by Rücker and Edser[2], who used apparatus entirely independent of the ear. They conclude that "Helmholtz was correct in stating that the siren produces two objective notes the frequencies of which are respectively equal to the sum and difference of the frequencies of the fundamentals." My own observations have been made upon the harmonium, and leave me at a loss to understand how two opinions are possible. The resonator is held with its mouth as near as may be to the reeds which sound the generating notes, and is put in and out of tune to the difference-tone by slight movements of the finger. When the tuning is good, the difference-tone swells out with considerable strength, but a slight mistuning (probably of the order of a semitone) reduces it almost to silence. In some cases, e.g. when the interval between the generators is a (tempered) Fifth, the difference-tone is heard to *beat*.

The last observation proves that in some cases there exist two difference-tones of nearly the same pitch. Helmholtz finds the

---

[1] *Sensations of Tone*, p. 153.
[2] *Phil. Mag.* vol. xxxix. p. 357, 1895.

explanation of this in the compound nature of the sounds. Thus in the case of the Fifth, represented by the numbers 2 and 3, we have not only the primes to consider, but the overtones 2×2, 3×2, etc., 2×3, 3×3, etc. Accordingly the difference-tone 1 may be derived from $2 \times 2 = 4$ and 3, as well as from 3 and 2, and since the octave partial is usually strong, the one source may be as important as the other. But if we substitute the Major Third $(5:4)$ for the Fifth, we do not get a second difference-tone 1 until we come to the fourth partial (16) of the graver note and the third (15) of the higher, and these would usually be too feeble to produce much effect.

As regards the frequency of the beats, let us return to the case of the Fifth, supposing it to be so disturbed that the frequencies are 200 and 301. The difference tone due to the primes is $301 - 200 = 101$, and that due to the octave partial is

$$2 \times 200 - 301 = 99;$$

and these difference-tones sounding together will give beats with frequency 2. This, it will be observed, is the same number of beats as is due to the common overtone, viz. $2 \times 301 - 3 \times 200$; but while the latter beats are those of the tone 600, the beats of the combination-tone are at pitch 100.

392. According to the views of the older theorists Chladni, Lagrange, Young, etc., the explanation of the difference-tone presented no particular difficulty. As the generators separate in pitch, the beats quicken and at last become too rapid for appreciation as such, passing into a difference-tone, whose frequency is continuous with the frequency of the beats. This view of the matter, which has commended itself to many writers, was rejected by Helmholtz, as inconsistent with Ohm's law; and that physicist has elaborated an alternative theory, according to which the failure is not in Ohm's law, but in the principle of superposition.

Helmholtz's calculation of the effect of a want of symmetry in the forces of restitution, when the vibrations of a system cannot be regarded as infinitely small, has already been given (§ 68). It appears that in addition to the terms in pt, qt, corresponding to the generating forces, there must be added other terms of the second order in $2pt$, $2qt$, $(p+q)t$, $(p-q)t$, the last of which represents the difference-tone. This explanation depends, as Hermann

[1] *Pflüger's Archiv*, vol. XLIX. p. 507, 1891.

has remarked, upon the assumed failure of symmetry. If, as in § 67, we suppose a force of restitution proportional partly to the first power and partly to the cube of the displacement, we do *not* obtain a term in $(p-q)t$, but in place of it terms of the *third* order involving $(2p-q)t$, $(2q-p)t$, etc. This objection, however, is of little practical importance, because the failure of symmetry almost always occurs. It may suffice to instance the all important case of aerial vibrations. Whether we are considering progressive waves advancing from a source, or the stationary vibrations of a resonator, there is an essential want of symmetry between condensation and rarefaction, and the formation in some degree of octaves and combination-tones is a mathematical necessity.

The production of external, or objective, combination-tones demands the coexistence of the generators at a place where they are strong[1]. This will usually occur only when the generating sounds are closely associated, as in the polyphonic siren and in the harmonium. In these cases the conditions are especially favourable, because the limited mass of air included within the instrument is necessarily strongly affected by both tones. When the generating sources are two organ-pipes, even though they stand pretty near together, the difference-tone is not appreciably strengthened by a resonator, from which we may infer that but little of it exists externally to the ear.

We have as yet said nothing about the summation-tone, corresponding to the term in $(p+q)t$. The existence of this tone was deduced by Helmholtz theoretically; and he afterwards succeeded in hearing it, not only from the siren and harmonium, where it exists objectively and is reinforced by resonators, but also from tuning-forks and organ-pipes. Helmholtz narrates also an experiment in which he caused a membrane to vibrate in response to the summation-tone, and similar experiments have recently been carried out with success by Rücker and Edser (l. c.).

Nevertheless, it must be admitted that summation-tones are extremely difficult to hear. Hermann (l. c.) asserts that he can neither hear them himself nor find any one able to do so; and he regards this difficulty as a serious objection to Helmholtz's theory, according to which the summation and the difference tone should be about equally strong.

---

[1] The estimates of condensation (§ 384) for sounds just audible make it highly improbable that the principle of superposition could fail to apply to sounds of that order of magnitude.

An objection of another kind has been raised by König[1]. He remarks that even if a tone exist of the pitch of the summation-tone, it may in reality be a difference-tone, derived from the upper partials of the generators. As a matter of arithmetic this argument cannot be disputed; for if p and q be commensurable, it will always be possible to find integers h and k, such that

$$p + q = hp - kq.$$

But this explanation is plausible only when h and k are *small* integers.

It seems to me that the comparative difficulty with which summation-tones are heard is in great measure, if not altogether, explained by the observations of Mayer (§ 386). These tones are of necessity higher in pitch than their generators, and are accordingly liable to be overwhelmed and rendered inaudible. On the other hand the difference-tone, being usually graver, and often much graver, than either of its generators, is able to make itself felt in spite of them. And even as regards difference-tones, it had already been remarked by Helmholtz that they become more difficult to hear when they cease to constitute the gravest element of the sound by reason of the interval between the generators exceeding an octave.

393. In the numerous cases where differential tones are audible which are not reinforced by resonators, it is necessary in order to carry out Helmholtz's theory to suppose that they have their origin in the vibrating parts of the outer ear, such as the drum-skin and its attachments. Helmholtz considers that the structure of these parts is so unsymmetrical that there is nothing forced in such a supposition. But it is evident that this explanation is admissible only when the generating sounds are loud, *i.e.* powerful as they reach the ear. Now, the opponents of Helmholtz's views, represented by Hermann, maintain that this condition is not at all necessary to the perception of difference-tones. Here we have an issue as to facts, the satisfactory resolution of which demands better experiments, preferably of a quantitative nature, than any yet executed. My own experience tends rather to support the view of Helmholtz that loud generators are necessary. On several occasions stopped organ-pipes d''', e''', were blown with

[1] *Pogg. Ann.* vol. 157, p. 177, 1876.

a steady wind, and were so tuned that the difference-tone gave slow beats with an electrically maintained fork, of pitch 128, mounted in association with a resonator of the same pitch. When the ear was brought up close to the mouths of the pipes, the difference-tone was so loud as to require all the force of the fork in order to get the most distinct beats. These beats could be made so slow as to allow the momentary disappearance of the grave sound, when the intensities were rightly adjusted, to be observed with some precision. In this state of things the two tones of pitch 128, one the difference-tone and the other derived from the fork, were of equal strength as they reached the observer; but as the ear was withdrawn so as to enfeeble both sounds by distance, it seemed that the combination-tone fell off more quickly than the ordinary tone from the fork. It might be possible to execute an experiment of this kind which should prove decisively whether the combination-tone is really an effect of the second order, or not.

In default of decisive experiments we must endeavour to balance the *a priori* probabilities of the case. According to the views of the older theorists, adopted by König, Hermann, and other critics of Helmholtz, the beats of the generators, with their alternations of swellings and pauses, pass into the differential tone of like frequency, without any such failure of superposition as is invoked by Helmholtz. The critics go further, and maintain that the ear is capable of recognising as a tone any periodicity within certain limits of frequency[1].

Plausible as this doctrine is from certain points of view, a closer examination will, I think, shew that it is encumbered with difficulties. Among these is the ambiguity, referred to in § 12, as to what exactly is meant by period. A periodicity with frequency 128 is also periodicity with frequency 64. Is the latter tone to be heard as well as the former? So far as theory is concerned, such questions are satisfactorily answered by Ohm's law. Experiment may compel us to abandon this law, but it is well to remember that there is nothing to take its place. Again, by consideration of particular cases it is not difficult to prove that the general doctrine above formulated cannot be true. Take the example above mentioned in which two organ-pipes gave a difference-tone of pitch 128. There is periodicity with frequency 128, and the corresponding tone is heard[1]. So far, so good. But experiment proves also that it is only necessary to superpose upon this another tone of frequency 128, obtained from a fork, in order to neutralize the combination-tone and reduce it to silence. The periodicity of 128 remains, if anything in a more marked manner than before, but the corresponding tone is *not* heard.

I think it is often overlooked in discussions upon this subject that a difference-tone is not a mere sensation, but involves a *vibration* of definite amplitude and phase. The question at once arises, how is the phase determined? It would seem natural to suppose that the maximum swell of the beats corresponds to one or other extreme elongation in the difference-tone, but upon the principles under discussion there seems to be no ground for a selection between the alternatives. Again, how is the amplitude determined? The tone certainly vanishes with either of the generators. From this it would seem to follow that its amplitude must be proportional to the product of the amplitudes of the generators, exactly as in Helmholtz's theory. If so, we come back to difference-tones of the second order, and their asserted easy audibility from feeble generators is no more an objection to one theory than to another.

An observation, of great interest in itself, and with a possible bearing upon our present subject, has been made by König and Mayer[2]. Experimenting both with forks and bird-calls, they have found that audible difference-tones may arise from generators whose pitch is so high that they are separately inaudible. Perhaps an interpretation might be given in more than one way, but the passage of an inaudible beat into an audible difference-tone seems to be more easily explicable upon the basis of Helmholtz's theory.

Upon the whole this theory seems to afford the best explanation of the facts thus far considered, but it presupposes a more ready departure from superposition of vibrations within the ear than would have been expected.

394. In § 390 we saw that in the case of ordinary compound sounds, containing upper partials fairly developed, the recognised consonant intervals are distinguished from neighbouring intervals

---

[1] In strictness, the periodicity is incomplete, unless p and q are multiples of $(p-q)$.

[2] Mayer, *Rep. Brit. Ass.* p. 573, 1894.

by well marked phenomena, of which there was no difficulty in rendering a satisfactory account. We have now to consider the more difficult subject of consonance among pure tones; and we shall have to encounter considerable differences of opinion, not only as to theoretical explanations, but as to matters of observation. Here, as elsewhere, it will be convenient to begin with a statement of Helmholtz's views[1] according to which, in a word, the beats of such mistuned consonances are due to combination-tones.

" If combinational tones were not taken into account, two simple tones, as those of tuning-forks, or stopped organ-pipes, could not produce beats unless they were very nearly of the same pitch, and such beats are strong when their interval is a minor or major second, but weak for a Third, and then only recognisable in the lower parts of the scale, and they gradually diminish in distinctness as the interval increases, without shewing any special differences for the harmonic intervals themselves. For any larger interval between two simple tones there would be absolutely no beats at all, if there were no upper partial or combinational tones, and hence the consonant intervals...would be in no way distinguished from adjacent intervals; there would in fact be no distinction at all between wide consonant intervals and absolutely dissonant intervals.

Now such wider intervals between simple tones are known to produce beats, although very much weaker than those hitherto considered, so that even for such tones there is a difference between consonances and dissonances, although it is very much more imperfect than for compound tones[2]."

Experiments upon this subject are difficult to execute satisfactorily. In the first place it is not easy to secure simple tones. As sources recourse is usually had to stopped organ-pipes or to tuning-forks, but much precaution is required. From the free ends of the vibrating prongs of a fork many overtones may usually be heard[3]. Again, if a fork be employed after the manner of musicians with its stalk pressed against a resonating board, the octave is loud and often predominant[4]. The best way is to hold

---

[1] Ascribed by him to Hällström and Scheibler.

[2] *Sensations of Tone*, p. 199.

[3] König's experiments shew that this is especially the case when the prongs are thin. *Wied. Ann.* vol. XIV. p. 373, 1881.

[4] The prime tone may even disappear altogether. If in their natural position the prongs of a fork are closest below, an outward movement during the vibration

the free ends of the prongs over a suitably tuned resonator. But even then we cannot be sure that a loud sound thus obtained is absolutely free from the octave partial.

In the case of the octave the differential tone already considered suffices. "If the lower note makes 100 vibrations per second, while the imperfect octave makes 201, the first differential tone makes $201 - 100 = 101$, and hence nearly coincides with the lower note of 100 vibrations, producing one beat for each 100 vibrations. There is no difficulty in hearing these beats, and hence it is easily possible to distinguish imperfect octaves from perfect ones, even for simple tones, by the beats produced by the former."

The frequency of the beats is the same as if it were due to overtones; but there is one important difference between the two cases noted by Ellis though scarcely, if at all, referred to by Helmholtz. In the latter the beats would affect the octave tone, whereas according to the above theory the beats will belong to the lower tone. Bosanquet, König and others are agreed that in this respect the theory is verified.

Again, if the beats were due to combination-tones, they must tend to disappear as the sounds die away. The experiment is very easily tried with forks, and according to my experience the facts are in harmony. When the sounds are much reduced, the mistuning fails to make itself apparent.

"For the Fifth, the first order of differential tones no longer suffices. Take an imperfect Fifth with the ratio $200 : 301$; then the differential tone of the first order is 101, which is too far from either primary to generate beats. But it forms an imperfect Octave with the tone 200, and, as just seen, in such a case beats ensue. Here they are produced by the differential tone 99 arising from the tone 101 and the tone 200, and this tone 99 makes two beats in a second with the tone 101. These beats then serve to distinguish the imperfect from the justly intoned Fifth, even in the case of two simple tones. The number of these beats is also exactly the same as if they were the beats due to

will depress the centre of inertia, the stalk being immovable, but if the prongs are closest above, the contrary result may ensue. There must be some intermediate construction for which the centre of inertia will remain at rest during the vibration. In this case the sound from a resonance board is of the second order, and is destitute of the prime tone.

the upper partial tones. But to observe these beats the two primary tones must be loud, and the ear must not be distracted by any extraneous noise. Under favourable circumstances, however, they are not difficult to hear."

It is important to be clear as to the order of magnitude of the various differential tones concerned. If the primary tones, with frequencies represented by p and q, have amplitudes e and f respectively, quantities of the first order, then (§ 68) the first difference and summation tones have frequencies corresponding to

$$2p, \quad 2q, \quad p+q, \quad p-q,$$

and are of the second order in e and f. A complete treatment of the second differential tones requires the retention of another term βu^3 (§ 67) in the expression of the force of restitution. From this will arise terms of the third order in e and f with frequencies corresponding to

$$3p, \quad 2p \pm q, \quad p \pm 2q, \quad 3q; [1]$$

and there are in addition other terms of the same frequencies and order of magnitude, independent of β, arising from the full development to the third order of au^2. In the case of the disturbed Fifth above taken, the beats are between the tone $2q - p = 99$, which is of the third order of magnitude, and $p - q = 101$ of the second order. The exposition, quoted from Helmholtz, refers to the terms last mentioned, which are independent of β.

The beats of a disturbed Fourth or major Third depend upon difference-tones of a still higher order of magnitude, and according to Helmholtz's observations they are scarcely, if at all, audible, even when the primary tones are strong. This is no more than would have been expected; the difficulty is rather to understand how the beats of the disturbed Fifth are perceptible and those of the disturbed Octave so easy to hear.

When more than two simple tones are sounded together, fresh conditions arise. "We have seen that Octaves are precisely limited even for simple tones by the beats of the first differential tone with the lower primary. Now suppose that an Octave has been tuned perfectly, and that then a third tone is interposed to act as a Fifth. Then if the Fifth is not perfect, beats will ensue from the first differential tone.

[1] Bosanquet, *Phil. Mag.* vol. xi. p. 497, 1881.

Let the tones forming the perfect Octave have the pitch numbers 200 and 400, and let that of the imperfect Fifth be 301. The differential tones are

$$400 - 301 = 99$$
$$301 - 200 = \underline{101}$$
$$\text{Number of beats}\quad 2.$$

These beats of the Fifth which lies between two Octaves are much more audible than those of the Fifth alone without its Octave. The latter depend on the weak differential tones of the second order, the former on those of the first order. Hence Scheibler some time ago laid down the rule for tuning tuning-forks, first to tune two of them as a perfect Octave, and then to sound them both at once with the Fifth, in order to tune the latter. If Fifth and Octave are both perfect, they also give together the perfect Fourth.

The case is similar, when two simple tones have been tuned to a perfect Fifth, and we interpose a new tone between them to act as a major Third. Let the perfect Fifth have the pitch numbers 400 and 600. On intercalating the impure major Third with the pitch number 501 in lieu of 500, the differential tones are

$$600 - 501 = 99$$
$$500 - 400 = \underline{101}$$
$$\text{Number of beats}\quad 2.\text{''}$$

395. In Helmholtz's theory of imperfect consonances the cycles heard are regarded as risings and fallings of intensity of one or more of the constituents of the sound, whether these be present from the first, or be generated by transformation, to use Bosanquet's phrase, in the transmitting mechanism of the ear. According to Ohm's law, such changes of intensity are the only thing that could be heard, for the relative phases of the constituents (supposed to be sufficiently removed from one another in pitch) are asserted to be matters of indifference.

This question of independence of phase-relation was examined by Helmholtz in connection with his researches upon vowel sounds (§ 397). Various forks, electrically driven from one interrupter (§ 64), could be made to sound the prime tone, octave, twelfth etc., of a compound note, and the intensities and phases of the constituents could be controlled by slight modifications in the (natural) pitch of the forks and associated resonators. According to Helmholtz's observations changes of phase were without distinct effect upon the quality of the compound sound.

It is evident, however, that the question of the effect, if any, upon the ear of a change in the phase relationship of the various components of a sound can be more advantageously examined by the method of slightly mistuned consonances. If, for example, an Octave interval between two pure tones be a very little imperfect, the effect upon the ear at any particular moment will be that of a true interval with a certain relation of phases, but after a short time, the phase relationship will change, and will pass in turn through every possible value. The audibility of the cycle is accordingly a criterion for the question whether or not the ear appreciates phase relationship; and the results recorded by Helmholtz himself, and easily to be repeated, shew that in a certain sense the answer must be in the affirmative. Otherwise slow beats of an imperfect Octave would not be heard. The explanation by means of combination-tones does not alter the fact that the ear appreciates the phase relationship of two originally simple tones, at any rate when they are moderately loud[1].

According to the observations of Lord Kelvin[2] the "beats of imperfect harmonies," other than the Octave and Fifth, are not so difficult to hear as Helmholtz supposed. The tuning-forks employed were mounted upon box resonators, and it might indeed be argued that the sounds conveyed down the stalks were not thoroughly purged from Octave partials. But this consideration would hardly affect the result in some of the cases mentioned. It appeared that the beats on approximations to each of the harmonies 2 : 3, 3 : 4, 4 : 5, 5 : 6, 6 : 7, 7 : 8, 1 : 3, 3 : 5 could be distinctly heard, and that they all "fulfil the condition of having the whole period of the imperfection, and not any sub-multiple of it, for their period," the same rule as would apply were the beats due to nearly coincident overtones. As regards the necessity for loud notes, Kelvin found that the beats of an imperfectly tuned chord 3 : 4 : 5 were sometimes the very last sound heard, as the vibrations of the forks died down, when the intensities of the three notes chanced at the end to be suitably proportioned.

[1] König, *Wied. Ann.* vol. XIV. p. 375, 1881.
[2] *Proc. Roy. Soc. Edin.* vol. IX. p. 602, 1878.

The last observation is certainly difficult to reconcile with a theory which ascribes the beats to combination-tones. But on the other side it may be remarked that the relatively easy audibility of the beats from a disturbed Octave and from a disturbed chord of three notes (3 : 4 : 5), which would depend upon the first differential tone, is in good accord with that theory, and (so far as appears) is not explained by any other.

396. But the observations most difficult of reconciliation with the theory of Helmholtz are those recorded by König[1], who finds tones, described as beat-tones, not included among the combination-tones; and these observations, coming from so skilful and so well equipped an investigator, must carry great weight. The principal conclusions are thus summarised by Ellis[2]. "If two simple tones of either very slightly or greatly different pitches, called generators, be sounded together, then the upper pitch number necessarily lies between two multiples of the lower pitch number, one smaller and the other greater, and the differences between these multiples of the pitch number of the lower generator and the pitch number of the upper generator give two numbers which either determine the frequency of the two sets of beats which may be heard or the pitch of the beat-notes which may be heard in their place.

The frequency arising from the lower multiple of the lower generator is called the frequency of the *lower* beat or lower beat-note, that arising from the higher multiple is called the frequency of the *higher* beat or beat-note, without at all implying that one set of beats should be greater or less than the other, or that one beat-note should be sharper or flatter than the other. They are in reality sometimes one way and sometimes the other.

Both sets of beats, or both beat-notes, are not usually heard at the same time. If we divide the intervals examined into groups (1) from 1 : 1 to 1 : 2, (2) from 1 : 2 to 1 : 3, (3) from 1 : 3 to 1 : 4, (4) from 1 : 4 to 1 : 5, and so on, the lower beats and beat-tones extend over little more than the lower half of each group, and the upper beats and beat-tones over little more than the upper half. For a short distance in the middle of each period both sets of beats, or both beat-notes are audible, and these beat-notes beat with each

[1] *Pogg. Ann.* vol. CLVII. p. 177, 1876.
[2] *Sensations of Tone*, p. 529.

other, forming secondary beats, or are replaced by new or secondary beat-notes."

In certain cases the beat-notes coincide with the differential tone, but König considers that the existence of combinational tones has not been proved with certainty. It is to be observed that in these experiments the generating tones were as simple as König could make them; but the possibility remains that overtones, not audible except through their beats, may have arisen within the ear by transformation. This is the view favoured by Bosanquet, who has also made independent observations with results less difficult of accommodation to Helmholtz's views.

It will be seen that König adopts in its entirety the opinion that beats, when quick enough, pass into tones. Some objections to this idea have already been pointed out; and the question must be regarded as still an open one. Experiments upon these subjects have hitherto been of a merely qualitative character. The difficulties of going further are doubtless considerable; but I am disposed to think that what is most wanted at the present time is a better reckoning of the intensities of the various tones dealt with and observed. If, for example, it could be shewn that the intensity of a beat-tone is proportional to that of the generators, it would become clear that something more than combination-tones is necessary to explain the effects.

König has also examined the question of the dependence of quality upon phase relation, using a special siren of his own construction[1]. His conclusion is that while quality is mainly determined by the number and relative intensity of the harmonic tones, still the influence of phase is not to be neglected. A variation of phase produces such differences as are met with in different instruments of the same class, or in various voices singing the same vowel. A ready appreciation of such minor differences requires a series of notes, upon which a melody can be executed, and they may escape observation when only a single note is available. To me it appears that these results are in harmony with the view that would ascribe the departure from Ohm's law, involved in any recognition of phase relations, to secondary causes.

397. The dependence of the quality of musical sounds of given pitch upon the proportions in which the various partial tones are

[1] *Wied. Ann.* vol. XIV. p. 392, 1881.

present has been investigated by Helmholtz in the case of several musical instruments. Further observations upon wind instruments will be found in a paper by Blaikley [1]. But the most interesting, and the most disputed, application of the theory is to the vowel sounds of human speech.

The acoustical treatment of this subject may be considered to date from a remarkable memoir by Willis [2]. His experiments were conducted by means of the free reed, invented by Kratzenstein (1780) and subsequently by Greniè, which imitates with fair accuracy the operation of the larynx. Having first repeated successfully Kempelen's experiment of the production of vowel sounds by shading in various degrees the mouth of a funnel-shaped cavity in association with the reed, he passed on to examine the effect of various lengths of cylindrical tube, the mounting being similar to that adopted in organ-pipes. The results shewed that the vowel quality depended upon the length of the tube. From these and other experiments he concluded that cavities yielding (when sounded independently) an identical note " will impart the same vowel quality to a given reed, or indeed to any reed, provided the note of the reed be flatter than that of the cavity." Willis proceeds (p. 243) : " A few theoretical considerations will shew that some such effects as we have seen, might perhaps have been expected. According to Euler, if a single pulsation be excited at the bottom of a tube closed at one end, it will travel to the mouth of this tube with the velocity of sound. Here an echo of the pulsation will be formed which will run back again, be reflected from the bottom of the tube, and again present itself at the mouth where a new echo will be produced, and so on in succession till the motion is destroyed by friction and imperfect reflection....The effect therefore will be a propagation from the mouth of the tube of a succession of equidistant pulsations alternately condensed and rarefied, at intervals corresponding to the time required for the pulse to travel down the tube and back again ; that is to say, a short burst of the musical note corresponding to a stopped pipe of the length in question, will be produced.

Let us now endeavour to apply this result of Euler's to the case before us, of a vibrating reed, applied to a pipe of any length,

[1] *Phil. Mag.* vol. VI. p. 119, 1878.
[2] On the Vowel Sounds, and on Reed Organ-pipes. *Camb. Phil. Trans.* vol. III. p. 231, 1829.

and examine the nature of the series of pulsations that ought to be produced by such a system upon this theory.

The vibrating tongue of the reed will generate a series of pulsations of equal force, at equal intervals of time, but alternately condensed and rarefied, which we may call the primary pulsations; on the other hand each of these will be followed by a series of secondary pulsations of decreasing strength, but also at equal intervals from their respective primaries, the interval between them being, as we have seen, regulated by the length of the attached pipe."

And further on (p. 247): " Experiment shews us that the series of effects produced are characterized and distinguished from each other by that quality we call the vowel, and it shews us more, it shews us not only that the pitch of the sound produced is always that of the reed or primary pulse, but that the vowel produced is always identical for the same value of s [the period of the secondary pulses]. Thus, in the example just adduced, g'' is peculiar to the vowel A° [as in *Paw, Nought*]; when this is repeated 512 times in a second, the pitch of the sound is c', and the vowel is A°: if by means of another reed applied to the same pipe it were repeated 340 times in a second, the pitch would be f, but the vowel still A°. Hence it would appear that the ear in losing the consciousness of the pitch of s, is yet able to identify it by this *vowel* quality."

From the importance of his results and from the fact that the early volumes of the Cambridge Transactions are not everywhere accessible, I have thought it desirable to let Willis speak for himself. It will be seen that so far as general principles are concerned, he left little to be effected by his successors. Somewhat later in the same memoir (p. 249) he gives an account of a special experiment undertaken as a test of his theory. " Having shewn the probability that a given vowel is merely the rapid repetition of its peculiar note, it should follow that if we can produce this rapid repetition in any other way, we may expect to hear vowels. Robison and others had shewn that a quill held against a revolving toothed wheel, would produce a musical note by the rapid equidistant repetition of the snaps of the quill upon the teeth. For the quill I substituted a piece of watch-spring pressed lightly against the teeth of the wheel, so that each snap became the musical note of the spring. The spring being at the same time grasped in a pair of pincers, so as to admit of any alteration in length of the vibrating portion. This system evidently produces a compound sound similar to that of the pipe and reed, and an alteration in the length of the spring ought therefore to produce the same effect as that of the pipe. In effect the sound produced retains the same pitch as long as the wheel revolves uniformly, but puts on in succession all the vowel qualities, as the effective length of the spring is altered, and that with considerable distinctness, when due allowance is made for the harsh and disagreeable quality of the sound itself."

In his presentation of vowel theory Helmholtz, following Wheatstone[1], puts the matter a little differently. The aerial vibrations constituting natural or artificial vowels are, when a uniform regime has been attained (§§ 48, 66, 322 k), truly periodic, and the period is that of the reed. According to Fourier's theorem they are susceptible of analysis into simple vibrations, whose periods are accurately submultiples of the reed period. The effect of an associated resonator can only be to modify the intensity and phase of the several components, whose periods are already prescribed. If the note of the resonating cavity—the mouth-tone—coincide with one of the partial tones of the voice- or larynx-note, the effect must be to exalt in a special degree the intensity of that tone; and whether there be coincidence or not, those partial tones whose pitch approximates to that of the mouth-tone will be favoured.

This view of the action of a resonator is of course perfectly correct; but at first sight it may appear essentially different from, or even inconsistent with, the account of the matter given by Willis. For example, according to the latter the mouth-tone may be, and generally will be, inharmonic as regards the larynx-tone. In order to understand this matter we must bear in mind two things which are often imperfectly appreciated. The first is the distinction between forced and free vibrations. Although the *natural* vibrations of the oral cavity may be inharmonic, the *forced* vibrations can include only harmonic partials of the larynx note. And again, it is important to remember the definition of simple vibrations, according to which no vibrations can be simple that are not permanently maintained without variation of amplitude or phase. The secondary vibrations of Willis, which

---

[1] *London and Westminster Review*, Oct. 1837; *Wheatstone's Scientific Papers*, London, 1879, p. 348.

·die down after a few periods, are not simple. When the complete succession of them is resolved by Fourier's theorem, it is represented, not by one simple vibration, but by a large or infinite number of such.

From these considerations it will be seen that both ways of regarding the subject are legitimate and not inconsistent with one another. When the relative pitch of the mouth-tone is low, so that, for example, the partial of the larynx note most reinforced is the second or the third, the analysis by Fourier's series is the proper treatment. But when the pitch of the mouth-tone is high, and each succession of vibrations occupies only a small fraction of the complete period, we may agree with Hermann that the resolution by Fourier's series is unnatural, and that we may do better to concentrate our attention upon the actual form of the curve by which the complete vibration is expressed. More especially shall we be inclined to take this course if we entertain doubts as to the applicability of Ohm's law to partials of high order.

Since the publication of Helmholtz's treatise the question has been much discussed whether a given vowel is characterized by the prominence of partials of given *order* (the relative pitch theory), or by the prominence of partials of given *pitch* (the fixed pitch theory), and every possible conclusion has been advocated. We have seen that Willis decided the question, without even expressly formulating it, in favour of the fixed pitch theory. Helmholtz himself, if not very explicitly, appeared to hold the same opinion, perhaps more on *a priori* grounds than as the result of experiment. If indeed, as has usually been assumed by writers on phonetics, a particular vowel quality is associated with a given oral configuration, the question is scarcely an open one. Subsequently under Helmholtz's superintendence the matter was further examined by Auerbach[1], who along with other methods employed a direct analysis of the various vowels by means of resonators associated with the ear. His conclusion on the question under discussion was the intermediate one that *both* characteristics were concerned. The analysis shewed also that in all cases the first, or fundamental tone, was the strongest element in the sound.

A few years later Edison's beautiful invention of the phono-

[1] *Pogg. Ann.* Ergänzung-band VIII. p. 177, 1876.

graph stimulated anew inquiry upon this subject by apparently affording easy means of making an *experimentum crucis*. If vowels were characterized by fixed pitch, they should undergo alteration with the speed of the machine; but if on the other hand the relative pitch theory were the true one, the vowel quality should be preserved and only the pitch of the note be altered. But, owing probably to the imperfection of the earlier instruments, the results arrived at by various observers were still discrepant. The balance of evidence inclined perhaps in favour of the fixed pitch theory[1]. Jenkin and Ewing[2] analysed the impressions actually made upon the recording cylinder, and their results led them to take an intermediate view, similar to that of Auerbach. It is clear, they say, " that the quality of a vowel sound does not depend either on the absolute pitch of reinforcement of the constituent tones alone, or on the simple grouping of relative partials independently of pitch. Before the constituents of a vowel can be assigned, the pitch of the prime must be given; and, on the other hand, the pitch of the most strongly reinforced partial is not alone sufficient to allow us to name the vowel."

With the improved phonographs of recent years the question can be attacked with greater advantage, and observations have been made by McKendrick and others, but still with variable results. Especially to be noted are the extensive researches of Hermann published in *Pflüger's Archiv*. Hermann pronounces unequivocally in favour of the fixed pitch characteristic as at any rate by far the more important, and his experiments apparently justify this conclusion. He finds that the vowels sounded by the phonograph are markedly altered when the speed is varied.

Hermann's general view, to which he was led independently, is identical with that of Willis. " The vowel character consists in a mouth-tone of amplitude variable in the period of the larynx tone[3]." The propriety of this point of view may perhaps be considered to be established, but Hermann somewhat exaggerates the difference between it and that of Helmholtz.

His examination of the automatically recorded curves was effected in more than one way. In the case of the vowel A[4] the

---

[1] Graham Bell, *Ann. Journ. of Otology*, vol. I. July, 1879.

[2] *Edin. Trans.* vol. xxviii. p. 745, 1878.

[3] *Pflüg. Arch.* vol. xlvii. p. 351, 1890.

[4] The vowel signs refer of course to the continental pronunciation.

amplitudes of the various partials, as given by the Fourier analysis, are set forth in the annexed table, from which it appears that the favoured partial lies throughout between e^2 and g^2.

VOWEL A.

Note	\multicolumn{10}{c}{Ordinal number of partial.}									
	1	2	3	4	5	6	7	8	9	10
G						·12 d^2	·37 $< f^2$	·42 g^2	·11 a^2	·12 h^2
A				·13 cis^2	·30 e^2	·33 $< g^2$	·10 a^2	·09 h^2	·08 cis^3	
H H	·05		·09 fis'	·22 h'	·37 dis^2	·45 fis^2	·10 $< a^2$	·15 h^2		
c c	·11			·19 c^2	·54 e^2	·38 g^2	·16 $<ais^2$	·09 c^3	·10 d^3	
d				·29 d^2	·52 fis^2	·08 a^2	·18 $< c^3$		·06 e^3	
e			·13 h'	·55 e^2	·28 gis^2	·24 h^2	·07 $< d^2$			
fis			·30 cis^2	·61 fis^2	·07 ais^2	·11 cis^2	·11 $< e^2$			
g g	·11		·39 d^2	·55 g^2	·21 h^2	·11 d^3	·08 $< f^3$			
a			·71 e^2	·18 a^2	·18 cis^3	·09 e^3				
h			·74 fis^2	·17 h^2	·13 dis^3					
c'		·41 c^3	·54 g^2	·40 c^3	·11 e^3					
d'		·71 d^2	·31 a^2	·26 d^3						

The analysis of the curves into their Fourier components involves a great deal of computation, and Hermann is of opinion that the principal result, the pitch of the vowel characteristic, can be obtained as accurately and far more simply by direct measurement on the diagram of the wave-lengths of the intermittent vibrations. The application of this method to the curves for *A* before used gave

VOWEL A.

Note	L mm.	l mm.	Characteristic tone	
			Frequency	Note
G 98	18·5	2·4	756	$> \mathrm{fis}^2$ (740)
A 110	16·3	2·5	717	$> \mathrm{f}^2$ (698·5)
H 123·5	14·9	2·6	708	$> \mathrm{f}^2$ (698·5)
c 130·8	13·6	2·55	698	f^2
d 146·8	11·6	2·4	710	$> \mathrm{f}^2$ (698·5)
e 164·8	10·9	2·3	781	$< \mathrm{g}^2$ (784)
fis 185	9·8	2·5	725	$< \mathrm{fis}^2$ (744)
g 196	9·1	2·5	714	$> \mathrm{f}^2$ (698·5)
a 220	8·2	2·5	714	$> \mathrm{f}^2$ (698·5)
h 246·9	7·3	2·6	693	$< \mathrm{f}^2$ (698·5)
c′ 261·7	6·8	?	?	
d′ 293·7	6·2	?	?	

Here L is the double period of the complete vibration and l the double period of the vowel characteristic. It appears plainly that l preserves a nearly constant value when L varies over a considerable range.

A general comparison of his results with those obtained by other methods has been given by Hermann, from which it will be seen that much remains to be done before the perplexities involving this subject can be removed. Some of the discrepancies

Vowel	Characteristic tone from graphical records Hermann	Mouth-tones according to			
		Donders	Helmholtz	König	Auerbach
A	$\mathrm{e}^2 - \mathrm{gis}^2$	b'	b^2	b^2	$\mathbf{f}^2$
E	$\mathrm{h}^3 - \mathrm{c}^4$	cis^3	$\mathrm{f}_{,}, \mathrm{b}^3$	b^3	$\mathrm{a}' (-\mathrm{g}')$
I	$\mathrm{d}^4 - \mathrm{g}^4$	f^3	f, d^4	b^4	$\mathbf{f}'$
O	$\mathrm{d}^2 - \mathrm{e}^2$	d'	b'	b'	a'
U	$\mathrm{c}^2 - \mathrm{d}^2$	f'	f	b	f'

that have been encountered may probably have their origin in real differences of pronunciation to which only experts in phonetics are sufficiently alive [1]. Again, the question of double resonance has to be considered, for the known shape of the cavities concerned

---

[1] Lloyd, *Phonetische Studien*, vol. III. part J.

renders it not unlikely that the complete characterization of a vowel is of a multiple nature (§ 310). It should be mentioned that in Lloyd's view the double characteristic is essential, and that the identity of a vowel depends not upon the absolute pitch of one or more resonances, but upon the relative pitch of two or more. In this way he explains the difficulty arising from the fact that the articulation for a given vowel appears to be the same for an infant and for a grown man, although on account of the great difference in the size of the resonating cavities the absolute pitch must vary widely.

It would not be consistent with the plan of this work to go further into details with regard to particular vowels; but one remarkable discrepancy between the results of Hermann and Auerbach must be alluded to. The measurements by the former of graphical records shew in all cases a nearly complete absence of the first, or fundamental, tone from the general sound, which Auerbach on the contrary, using resonators, found this tone the most prominent of all. Hermann, while admitting that the tone is heard, regards it as developed within the ear after the manner of combination-tones (§ 393). I have endeavoured to repeat some of Auerbach's observations, and I find that for all the principal vowels (except perhaps A) the fundamental tone is loudly reinforced, the contrast being very marked as the resonator is put in and out of tune by a movement of the finger over its mouth. This must be taken to prove that the tone in question does exist externally to the ear, as indeed from the manner in which the sound is produced could hardly fail to be the case; and the contrary evidence from the records must be explained in some other way.

An important branch of the subject is the artificial imitation of vowel sounds. The actual synthesis by putting together in suitable strengths the various partials was effected by Helmholtz[1]. For this purpose he used tuning-forks and resonators, the forks being all driven from a single interrupter (§§ 63, 64). These experiments are difficult, and do not appear to have been repeated. Helmholtz was satisfied with the reproduction in some cases, although in others the imitation was incomplete. Less satisfactory results were attained when organ-pipes were substituted for the forks.

[1] *Sensations of Sound*, ch. VI.

Vowel sounds have been successfully imitated by Preece and Stroh[1], who employed an apparatus upon the principle of the phonograph, in which the motion of the membrane was controlled by specially shaped teeth, cut upon the circumference of a revolving wheel. They found that the vowel quality underwent important changes as the speed of rotation was altered.

For artificial vowels, illustrative of his special views, Hermann recommends the polyphonic siren (§ 11). If when the series of 12 holes is in operation and a suitable velocity has been attained, the series of 18 holes be put alternately into and out of action, the difference-tone (6) is heard with great loudness and it assumes distinctly the character of an O. At a greater speed the vowel is Ao, and at a still higher speed an unmistakable A.

With the use of double resonators, suitably proportioned, Lloyd has successfully imitated some of the *whispered* vowels.

In the account here given of the vowel question it has only been possible to touch upon a few of the more general aspects of it. The reader who wishes to form a judgment upon controverted points and to pursue the subject into detail must consult the original writings of recent workers, among whom may be specially mentioned Hermann, Pipping, and Lloyd. The field is an attractive one; but those who would work in it need to be well equipped, both on the physical and on the phonetic side.

---

[1] *Proc. Roy. Soc.* vol. xxviii. p. 358, 1879.

NOTE TO § 86 [1].

It may be observed that the motion of any point belonging to a system of n degrees of freedom, which executes a harmonic motion, is in general linear. For, if x, y, z be the space coordinates of the point, we have

$$x = X \cos nt, \qquad y = Y \cos nt, \qquad z = Z \cos nt,$$

where X, Y, Z are certain constants; so that at all times

$$x : y : z = X : Y : Z.$$

If there be more than one mode of the frequency in question, the coordinates are not necessarily in the same phase. The most general values of x, y, z, subject to the given periodicity, are then

$$x = X_1 \cos nt + X_2 \sin nt,$$
$$y = Y_1 \cos nt + Y_2 \sin nt,$$
$$z = Z_1 \cos nt + Z_2 \sin nt,$$

equations which indicate elliptic motion in the plane

$$x\,(Y_1 Z_2 - Z_1 Y_2) + y\,(Z_1 X_2 - X_1 Z_2) + z\,(X_1 Y_2 - Y_1 X_2) = 0.$$

---

[1] This note appears now for the first time.

APPENDIX TO CHAPTER V[1].

ON THE VIBRATIONS OF COMPOUND SYSTEMS WHEN THE AMPLITUDES
ARE NOT INFINITELY SMALL.

In §§ 67, 68 we have found second approximations for the vibrations of systems of one degree of freedom, both in the case where the vibrations are free and where they are due to the imposition of given forces acting from without. It is now proposed to extend the investigation to cases where there is more than one degree of freedom.

In the absence of dissipative and of impressed forces, everything may be expressed (§ 80) by means of the functions T and V. In the case of infinitely small motion in the neighbourhood of the configuration of equilibrium, T and V reduce themselves to quadratic functions of the velocities and displacements with constant coefficients, and by a suitable choice of coordinates the terms involving *products* of the several coordinates may be made to disappear (§ 87). Even though we intend to include terms of higher order, we may still avail ourselves of this simplification, choosing as coordinates those which have the property of reducing the terms of the second order to sums of squares. Thus we may write

$$T = \tfrac{1}{2} A_{11} \dot{\phi}_1{}^2 + \tfrac{1}{2} A_{22} \dot{\phi}_2{}^2 + \ldots + A_{12} \dot{\phi}_1 \dot{\phi}_2 + A_{13} \dot{\phi}_1 \dot{\phi}_3 + \ldots\ldots\ldots(1),$$

in which A_{11}, A_{22},... are functions of ϕ_1, ϕ_2,...including constant terms a_1, a_2, ..., while A_{12}, A_{13}, ... are functions of the same variables *without constant terms* :

$$V = \tfrac{1}{2} c_1 \phi_1{}^2 + \tfrac{1}{2} c_2 \phi_2{}^2 + \ldots + V_3 + V_4 + \ldots\ldots\ldots\ldots (2),$$

where V_3, V_4, ... denote the parts of V which are of degree 3, 4, ... in ϕ_1, ϕ_2, ...

For the first approximation, applicable to infinitely small vibrations, we have

$$A_{11} = a_1, \quad A_{22} = a_2, \ldots A_{12} = 0, \quad A_{13} = 0 \ldots, \quad V_3 = 0, \quad V_4 = 0, \ldots ;$$

[1] This appendix appears now for the first time.

so that (§ 87) Lagrange's equations are

$$a_1 \ddot{\phi}_1 + c_1 \phi_1 = 0, \quad a_2 \ddot{\phi}_2 + c_2 \phi_2 = 0, \text{ \&c. } \quad \ldots\ldots\ldots\ldots(3),$$

in which the coordinates are separated. The solution relative to ϕ_1 may be taken to be

$$\phi_1 = H_1 \cos nt, \qquad \phi_2 = 0, \qquad \phi_3 = 0, \ldots \text{ \&c. } \ldots\ldots\ldots(4),$$

where

$$c_1 - n^2 a_1 = 0 \ldots\ldots\ldots\ldots\ldots\ldots\ldots\ldots\ldots(5).$$

Similar solutions exist relative to the other coordinates.

The second approximation, to which we now proceed, is to be founded on (4), (5); and thus ϕ_2, ϕ_3, ... are to be regarded as small quantities relatively to ϕ_1.

For the coefficients in (1) we write

$$A_{11} = a_1 + a_{11}\phi_1 + a_{12}\phi_2 + \ldots, \quad A_{12} = a_2\phi_1 + \ldots, \quad A_{13} = a_3\phi_1 + \ldots$$
$$\ldots\ldots\ldots(6),$$

and in (2)
$$V_3 = \gamma_1 \phi_1^3 + \gamma_2 \phi_1^2 \phi_2 + \ldots \ldots\ldots\ldots\ldots\ldots\ldots\ldots\ldots(7) ;$$

so that for a further approximation

$$dT/d\dot{\phi}_1 = (a_1 + a_{11}\phi_1)\dot{\phi}_1 + a_2\phi_1\dot{\phi}_2 + a_3\phi_1\dot{\phi}_3 + \ldots,$$

$$\frac{d}{dt}\left(\frac{dT}{d\dot{\phi}_1}\right) = (a_1 + a_{11}\phi_1)\ddot{\phi}_1 + a_{11}\dot{\phi}_1^2$$
$$+ a_2\phi_1\ddot{\phi}_2 + a_2\dot{\phi}_1\dot{\phi}_2 + a_3\phi_1\ddot{\phi}_3 + a_3\dot{\phi}_1\dot{\phi}_3 + \ldots,$$

$$dT/d\phi_1 = \tfrac{1}{2}a_{11}\dot{\phi}_1^2 + a_2\dot{\phi}_1\dot{\phi}_2 + a_3\dot{\phi}_1\dot{\phi}_3 + \ldots$$

Thus as the equation (§ 80) for ϕ_1, terms of the order ϕ_1^2 being retained, we get

$$(a_1 + a_{11}\phi_1)\ddot{\phi}_1 + \tfrac{1}{2}a_{11}\dot{\phi}_1^2 + c_1\phi_1 + 3\gamma_1\phi_1^2 = 0 \ldots\ldots\ldots\ldots(8).$$

To this order of approximation the coordinate ϕ_1 is separated from the others, and the solution proceeds as in the case of but one degree of freedom (§ 67). We have from (4)

$$\phi_1\ddot{\phi}_1 = -n^2 H_1^2 \cos^2 nt = -\tfrac{1}{2}n^2 H_1^2 (1 + \cos 2nt),$$
$$\dot{\phi}_1^2 = n^2 H_1^2 \sin^2 nt = \tfrac{1}{2}n^2 H_1^2 (1 - \cos 2nt),$$
$$\phi_1^2 = H_1^2 \cos^2 nt = \tfrac{1}{2}H_1^2 (1 + \cos 2nt) ;$$

so that (8) becomes

$$a_1 \ddot{\phi}_1 + c_1 \phi_1 + (-\tfrac{1}{4}n^2 a_{11} + \tfrac{3}{2}\gamma_1) H_1^2 + (-\tfrac{3}{4}n^2 a_{11} + \tfrac{3}{2}\gamma_1) H_1^2 \cos 2nt = 0$$
$$\ldots\ldots\ldots(9).$$

The solution of (9) may be expressed in the form

$$\phi_1 = H_0 + H_1 \cos nt + H_2 \cos 2nt + \ldots\ldots\ldots\ldots\ldots(10),$$

and a comparison gives

$$c_1 H_0 = (\tfrac{1}{4}n^2 a_{11} - \tfrac{3}{2}\gamma_1) H_1^2,$$
$$(c_1 - n^2 a_1) H_1 = 0,$$
$$(c_1 - 4n^2 a_1) H_2 = (\tfrac{3}{4}n^2 a_{11} - \tfrac{3}{2}\gamma_1) H_1^2.$$

Thus to a second approximation

$$\phi_1 = \frac{(\frac{1}{4}n^2 a_{11} - \frac{3}{2}\gamma_1) H_1^2}{c_1} + H_1 \cos nt + \frac{(\frac{3}{4}n^2 a_{11} - \frac{3}{2}\gamma_1) H_1^2}{c_1 - 4n^2 a_1} \cos 2nt \dots (11);$$

and the value of n is the same, i.e. $\sqrt{(c_1/a_1)}$, as in the first approximation.

We have now to express the corresponding values of ϕ_2, ϕ_3 ....
From (6)

$$dT/d\dot{\phi}_2 = a_2 \phi_1 \dot{\phi}_1 + a_2 \dot{\phi}_2 + \dots,$$
$$dT/d\phi_2 = \frac{1}{2} a_{12} \dot{\phi}_1^2 + \dots,$$

and Lagrange's equation becomes, terms of order ϕ_1^2 being retained,

$$a_2 \ddot{\phi}_2 + c_2 \phi_2 + a_2 \phi_1 \ddot{\phi}_1 + (a_2 - \frac{1}{2} a_{12}) \dot{\phi}_1^2 + \gamma_2 \phi_1^2 = 0,$$

or on substitution from (4) in the small terms

$$a_2 \ddot{\phi}_2 + c_2 \phi_2 + (-\frac{1}{4} n^2 a_{12} + \frac{1}{2} \gamma_2) H_1^2 + (-n^2 a_2 + \frac{1}{4} n^2 a_{12} + \frac{1}{2} \gamma_2) H_1^2 \cos 2nt = 0$$
$$\dots\dots\dots(12).$$

Accordingly, if

$$\phi_2 = K_0 + K_1 \cos nt + K_2 \cos 2nt + \dots \qquad \dots\dots\dots(13),$$

we find on comparison with (12)

$$c_2 K_0 = (\frac{1}{4} n^2 a_{12} - \frac{1}{2} \gamma_2) H_1^2 \dots\dots\dots\dots\dots(14),$$

$$(c_2 - n^2 a_2) K_1 = 0 \dots\dots\dots\dots\dots\dots\dots\dots\dots(15),$$

$$(c_2 - 4n^2 a_2) K_2 = (n^2 a_2 - \frac{1}{4} n^2 a_{12} - \frac{1}{2} \gamma_2) H_1^2 \dots\dots\dots(16).$$

Thus $K_1 = 0$, and the introduction of the values of K_0 and K_2 from (14), (16) into (13) gives the complete value of ϕ_2 to the second approximation.

The values of ϕ_3, ϕ_4, &c. are obtained in a similar manner, and thus we find to a second approximation the complete expression for those vibrations of a system of any number of degrees of freedom which to a first approximation are expressed by (4).

The principal results of the second approximation are (i) that the motion remains periodic with frequency unaltered, (ii) that terms, constant and proportional to $\cos 2nt$, are added to the value of that coordinate which is finite in the first approximation, as well as to those which in the first approximation are zero.

We now proceed to a third approximation; but for brevity we will confine ourselves to the case (a) where there are but two degrees of freedom, and (β) where the kinetic energy is completely expressed as a sum of squares of the velocities with constant coefficients. This will include the vibrations of a particle moving in two dimensions in the neighbourhood of a place of equilibrium.

We have

$$T = \tfrac{1}{2}a_1\dot{\phi}_1{}^2 + \tfrac{1}{2}a_2\dot{\phi}_2{}^2, \qquad V = \tfrac{1}{2}c_1\phi_1{}^2 + \tfrac{1}{2}c_2\phi_2{}^2 + V_3 + V_4,$$

where

$$V_3 = \gamma_1\phi_1{}^3 + \gamma_2\phi_1{}^2\phi_2 + \gamma'\phi_1\phi_2{}^2 + \dots \dots \dots \dots \dots (17),$$

$$V_4 = \delta_1\phi_1{}^4 + \delta_2\phi_1{}^3\phi_2 + \dots \quad \dots \dots \dots \dots \dots (18);$$

so that Lagrange's equations are

$$a_1\ddot{\phi}_1 + c_1\phi_1 + 3\gamma_1\phi_1{}^2 + 2\gamma_2\phi_1\phi_2 + 4\delta_1\phi_1{}^3 = 0 \quad \dots \dots . \ (19),$$

$$a_2\ddot{\phi}_2 + c_2\phi_2 + \gamma_2\phi_1{}^2 + 2\gamma'\phi_1\phi_2 + \delta_2\phi_1{}^3 = 0 \quad \dots \dots \dots (20).$$

As before, we are to take for the first approximation

$$\phi_1 = H_1 \cos nt, \qquad \phi_2 = 0 \dots \dots \dots \dots \dots (21).$$

For the solution of (19), (20) we may write

$$\phi_1 = H_0 + H_1 \cos nt + H_2 \cos 2nt + H_3 \cos 3nt + \dots \dots \dots(22),$$

$$\phi_2 = K_0 + K_1 \cos nt + K_2 \cos 2nt + K_3 \cos 3nt + \dots \dots \dots(23).$$

In (22), (23) H_0, H_2, K_0, K_2 are quantities of the second order in H_1, whose values have already been given, while K_1, H_3, K_3 are of the third order. Retaining terms of the third order, we have

$$\phi_1{}^2 = \tfrac{1}{2}H_1{}^2 + (2H_0H_1 + H_1H_2)\cos nt + \tfrac{1}{2}H_1{}^2 \cos 2nt + H_1H_2 \cos 3nt,$$

$$\phi_1\phi_2 = (H_1K_0 + \tfrac{1}{2}H_1K_2)\cos nt + \tfrac{1}{2}H_1K_2 \cos 3nt,$$

$$\phi_1{}^3 = \tfrac{3}{4}H_1{}^3 \cos nt + \tfrac{1}{4}H_1{}^3 \cos 3nt.$$

Substituting these values in the small terms of (19), (20), and from (22), (23) in the two first terms, we get the following 8 equations, correct to the third order,

$$c_1H_0 + \tfrac{3}{2}\gamma_1H_1{}^2 = 0 \dots \dots \dots \dots \dots(24),$$

$$c_1 - n^2a_1 + 3\gamma_1(2H_0 + H_2) + 2\gamma_2(K_0 + \tfrac{1}{2}K_2) + 3\delta_1H_1{}^2 = 0 \dots(25),$$

$$(c_1 - 4n^2a_1)H_2 + \tfrac{3}{2}\gamma_1H_1{}^2 = 0 \dots \dots \dots \dots \dots(26),$$

$$(c_1 - 9n^2a_1)H_3 + 3\gamma_1H_1H_2 + \gamma_2H_1K_2 + \delta_1H_1{}^3 = 0 \dots(27);$$

$$c_2K_0 + \tfrac{1}{2}\gamma_2H_1{}^2 = 0 \dots \dots \dots \dots \dots(28),$$

$$(c_2 - n^2a_2)K_1 + \gamma_2H_1(2H_0 + H_2) + \gamma'H_1(2K_0 + K_2) + \tfrac{3}{4}\delta_2H_1{}^3 = 0 \dots(29),$$

$$(c_2 - 4n^2a_2)K_2 + \tfrac{1}{2}\gamma_2H_1{}^2 = 0 \dots \dots \dots \dots \dots(30),$$

$$(c_2 - 9n^2a_2)K_3 + \gamma_2H_1H_2 + \gamma'H_1K_2 + \tfrac{1}{4}\delta_2H_1{}^3 = 0 \dots(31).$$

Of these (24), (26), (28), (30) give immediately the values of H_0, H_2, K_0, K_2, which are the same as to the second order of approximation, and the substitution of these values in (27), (29), (31) determines H_3, K_1, K_3 as quantities of the third order. The remaining equation (25) serves to determine n. We find as correct to this order

$$\frac{n^2a_1 - c_1}{H_1{}^2} = -\tfrac{3}{2}\gamma_1{}^2\left(\frac{6}{c_1} + \frac{3}{c_1 - 4n^2a_1}\right) - \tfrac{1}{2}\gamma_2{}^2\left(\frac{2}{c_2} + \frac{1}{c_2 - 4n^2a_2}\right) + 3\delta_1\dots(32).$$

If $\gamma_2 = 0$, this result will be found to harmonize with (9) § 67, when the differences of notation are allowed for, and the first approximation to n is substituted in the small terms.

The vibration above determined is that founded upon (21) as first approximation. The other mode, in which approximately $\phi_1 = 0$, can be investigated in like manner.

If V be an even function both of ϕ_1 and ϕ_2, γ_1, γ_2, γ', δ_2 vanish, and the third approximation is expressed by

$$H_0 = 0, \qquad H_2 = 0, \qquad H_3 = -\delta_1 H_1^3/(c_1 - 9n^2 a_1) ;$$
$$K_0 = 0, \qquad K_1 = 0, \qquad K_2 = 0, \qquad K_3 = 0 ;$$
$$n^2 a_1 - c_1 = 3\delta_1 H_1^2.$$

Indeed under this condition ϕ_2 vanishes to any order of approximation.

These examples may suffice to elucidate the process of approximation. An examination of its nature in the general case shews that the following conclusions hold good however far the approximation may be carried.

(a) The solution obtained by this process is periodic, and the frequency is an *even* function of the amplitude of the principal term (H_1).

(b) The Fourier series expressive of each coordinate contains cosines only, without sines, of the multiples of nt. Thus the whole system comes to rest at the same moment of time, e.g. $t = 0$, and then retraces its course.

(c) The coefficient of $\cos rnt$ in the series for any coordinate is of the rth order (at least) in the amplitude (H_1) of the principal term. For example, the series of the third approximation, in which higher powers of H_1 than H_1^3 are neglected, stop at $\cos 3nt$.

(d) There are as many types of solution as degrees of freedom but, it need hardly be said, the various solutions are not superposable.

One important reservation has yet to be made. It has been assumed that all the factors, such as $(c_2 - 4n^2 a_2)$ in (30), are finite, that is, that no coincidence occurs between a harmonic of the actual frequency and the natural frequency of some other mode of infinitesimal vibration. Otherwise, some of the coefficients, originally assumed to be subordinate, e.g. K_2 in (30), become infinite, and the approximation breaks down. We are thus precluded from obtaining a solution in some of the cases where we should most desire to do so.

As an example of this failure we may briefly notice the gravest vibrations in one dimension of a gas, obeying Boyle's law, and contained in a cylindrical tube with stopped ends. The equation to be satisfied throughout, (4) § 249, is of the form

$$\left(\frac{dy}{dx}\right)^2 \frac{d^2y}{dt^2} = \frac{d^2y}{dx^2},$$

and the procedure suggested by the general theory is to assume

$$y = x + y_0 + y_1 \cos nt + y_2 \cos 2nt + \ldots,$$

where

$$y_0 = H_{01} \sin x + H_{02} \sin 2x + H_{03} \sin 3x + \ldots,$$

$$y_1 = H_{11} \sin x + H_{12} \sin 2x + H_{13} \sin 3x + \ldots,$$

$$y_2 = H_{21} \sin x + H_{22} \sin 2x + H_{23} \sin 3x + \ldots,$$

and so on. In the first approximation

$$y = x + H_{11} \sin x \cos nt,$$

with $n = 1$. But when we proceed to a second approximation, we find

$$(4n^2 - 4) H_{22} = -\tfrac{1}{2} n^2 H_{11}{}^2,$$

still with n equal to 1, so that the method breaks down. The term $H_{22} \sin 2x \cos 2nt$ in the value of y, originally supposed to be subordinate, enters with an infinite coefficient.

It is possible that we have here an explanation of the difficulty of causing long narrow pipes to speak in their gravest mode.

The behaviour of a system vibrating under the action of an impressed force may be treated in a very similar manner. Taking, for example, the case of two degrees of freedom already considered in respect of its free vibrations, let us suppose that the impressed forces are

$$\Phi_1 = E_1 \cos pt, \qquad \Phi_2 = 0 \quad \ldots\ldots\ldots\ldots\ldots(33),$$

so that the solution to a first approximation is

$$\phi_1 = \frac{E_1 \cos pt}{c_1 - p^2 a_1}, \qquad \phi_2 = 0 \quad \ldots\ldots\ldots\ldots\ldots(34).$$

With substitution of p for n equations (22), (23) are still applicable, and also the resulting equations (24) to (31), except that in (25) the left-hand member is to be multiplied by H_1 and that on the right E_1 is to be substituted for zero. This equation now serves to determine H_1, instead of, as before, to determine n.

It is evident that in this way a truly periodic solution can always be built up. The period is that of the force, and the phases are such that the entire system comes to rest at the moment when the force is at a maximum (positive or negative). After this the previous course is retraced, as in the case of free vibrations, each series of cosines remaining unchanged when the sign of t is reversed.

NOTE TO § 273[1].

A method of obtaining Poisson's solution (8) given by Liouville[2] is worthy of notice.

If r be the polar radius vector measured from any point O, and the general differential equation be integrated over the volume included between spherical surfaces of radii r and $r + dr$, we find on transformation of the second integral by Green's theorem

$$\frac{d^2(r\lambda)}{dt^2} = a^2 \frac{d^2(r\lambda)}{dr^2} \quad\dotfill (a),$$

in which $\lambda = \iint \phi \, d\sigma$, that is to say is proportional to the mean value of ϕ reckoned over the spherical surface of radius r. Equation (a) may be regarded as an extension of (1) § 279; it may also be proved from the expression (5) § 241 for $\nabla^2 \phi$ in terms of the ordinary polar co-ordinates r, θ, ω.

The general solution of (a) is

$$r\lambda = \chi(at + r) + \theta(at - r) \quad\dotfill (\beta),$$

where χ and θ are arbitrary functions; but, as in § 279, if the pole be not a source, $\chi(at) + \theta(at) = 0$, so that

$$r\lambda = \chi(at + r) - \chi(at - r) \dotfill (\gamma).$$

It appears from (γ) that at O, when $r = 0$, $\lambda = 2\chi'(at)$, which is therefore also the value of $4\pi\phi$ at O at time t. Again from (γ)

$$2\chi'(at + r) = \frac{d(r\lambda)}{d(at)} + \frac{d(r\lambda)}{dr},$$

so that

$$2\chi'(r) = \left[\frac{d(r\lambda)}{d(at)} + \frac{d(r\lambda)}{dr}\right]_{(t=0)},$$

or in the notation of § 273

$$2\chi'(r) = \frac{r}{a} \iint F(r) \, d\sigma + \frac{d}{dr}\left[r \iint f(r) \, d\sigma\right] \dotfill (\delta).$$

By writing at in place of r in (δ) we obtain the value of $2\chi'(at)$, or $4\pi\phi$, which agrees with (8) § 273.

---

[1] This note appeared in the first edition.

[2] Liouville, tom. I. p. 1, 1856.

APPENDIX A. (§ 307[1].)

CORRECTION FOR OPEN END.

THE problem of determining the correction for the open end of a tube is one of considerable difficulty, even when there is an infinite flange. It is proved in the text (§ 307) that the correction a is greater than $\frac{1}{4}\pi R$, and less than $(8/3\pi)\,R$. The latter value is obtained by calculating the energy of the motion on the supposition that the velocity parallel to the axis is constant over the plane of the mouth, and comparing this energy with the square of the total current. The actual velocity, no doubt, increases from the centre outwards, becoming infinite at the sharp edge; and the assumption of a constant value is a some-what violent one. Nevertheless the value of a so calculated turns out to be not greatly in excess of the truth. It is evident that we should be justified in expecting a very good result, if we assume an axial velocity of the form

$$1 + \mu r^2/R^2 + \mu' r^4/R^4,$$

r denoting the distance of the point considered from the centre of the mouth, and then determine μ and μ' so as to make the whole energy a minimum. The energy so calculated, though necessarily in excess, must be a very good approximation to the truth.

In carrying out this plan we have two distinct problems to deal with, the determination of the motion (1) outside, and (2) inside the cylinder. The former, being the easier, we will take first.

The conditions are that ϕ vanish at infinity, and that when $x = 0$, $d\phi/dx$ vanish, except over the area of the circle $r = R$, where

$$d\phi/dx = 1 + \mu r^2/R^2 + \mu' r^4/R^4 \quad \ldots\ldots\ldots\ldots\ldots(1).$$

Under these circumstances we know (§ 278) that

$$\phi = -\frac{1}{2\pi} \iint \frac{d\phi}{dx} \frac{d\sigma}{\rho} \quad \ldots\ldots\ldots\ldots\ldots\ldots\ldots(2),$$

where ρ denotes the distance of the point where ϕ is to be estimated from the element of area $d\sigma$. Now

$$2\,(\text{kinetic energy})^2 = -\tfrac{1}{2} \iint \phi \, \frac{d\phi}{dx} \, d\sigma = \frac{1}{2\pi} \iint \frac{d\phi}{dx} \cdot \iint \frac{d\phi}{dx} \frac{d\sigma}{\rho} \cdot d\sigma = \frac{P}{\pi},$$

[1] This appendix appeared in the first edition.
[2] The density of the fluid is supposed to be unity.

if P represent the potential on itself of a disc of radius R, whose

$$\text{density} = 1 + \mu r^2/R^2 + \mu' r^4/R^4.$$

The value of P is to be calculated by the method employed in the text (§ 307) for a uniform density. At the edge of the disc, when cut down to radius a, we have the potential

$$V = 4a + \frac{20}{9}\frac{\mu a^3}{R^2} + \frac{356}{225}\frac{\mu' a^5}{R^4} \dots\dots\dots\dots\dots(3),$$

and thus

$$P = \int_0^R 2\pi a\, da\, V \left\{ 1 + \mu\frac{a^2}{R^2} + \mu'\frac{a^4}{R^4} \right\}$$

$$= \frac{8\pi R^3}{3}\left\{ 1 + \frac{14}{15}\mu + \frac{5}{21}\mu^2 + \frac{314}{525}\mu' + \frac{214}{675}\mu\mu' + \frac{89}{825}\mu'^2 \right\} \quad\dots\dots(4),$$

on effecting the integration. This quantity divided by π gives twice the kinetic energy of the motion defined by (1).

The total current

$$= \int_0^R 2\pi r\, dr \left(1 + \mu\frac{r^2}{R^2} + \mu'\frac{r^4}{R^4} \right) = \pi R^2 \left(1 + \tfrac{1}{2}\mu + \tfrac{1}{3}\mu' \right) \dots\dots (5).$$

We have next to consider the problem of determining the motion of an incompressible fluid within a rigid cylinder under the conditions that the axial velocity shall be uniform when $x = -\infty$, and when $x = 0$ shall be of the form

$$d\phi/dx = 1 + \mu r^2/R^2 + \mu' r^4/R^4.$$

It will conduce to clearness if we separate from ϕ, that part of it which corresponds to a uniform flow. Thus, if we take

$$d\phi/dx = 1 + \tfrac{1}{2}\mu + \tfrac{1}{3}\mu' + d\psi/dx,$$

ψ will correspond to a motion which vanishes when x is numerically great. When $x = 0$,

$$d\psi/dx = \mu\left(r^2 - \tfrac{1}{2}\right) + \mu'\left(r^4 - \tfrac{1}{3}\right) \dots\dots\dots\dots (6),$$

if for the sake of brevity we put $R = 1$.

Now ψ may be expanded in the series

$$\psi = \Sigma a_p e^{px} J_0(pr) \dots\dots\dots\dots\dots\dots(7),$$

where p denotes a root of the equation

$$J_0'(p) = 0 \dots\dots\dots\dots\dots\dots\dots(8)^1.$$

Each term of this series satisfies the condition of giving no radial

[1] The numerical values of the roots are approximately

$$p_1 = 3\cdot831705, \qquad p_2 = 7\cdot015, \qquad p_3 = 10\cdot174,$$
$$p_4 = 13\cdot324, \qquad p_5 = 16\cdot471, \qquad p_6 = 19\cdot616.$$

velocity, when $r = 1$; and no motion of any kind, when $x = -\infty$. It remains to determine the coefficients a_p so as to satisfy (6), when $x = 0$. From $r = 0$ to $r = 1$, we must have

$$\Sigma p a_p J_0 (pr) = \mu \left(r^2 - \tfrac{1}{2}\right) + \mu' \left(r^4 - \tfrac{1}{3}\right),$$

whence multiplying by $J_0(pr) \, r dr$ and integrating from 0 to 1,

$$p \, a_p \left[J_0 (p)\right]^2 = 2 \int_0^1 r dr J_0 (pr) \left\{\mu \left(r^2 - \tfrac{1}{2}\right) + \mu' \left(r^4 - \tfrac{1}{3}\right)\right\},$$

every term on the left, except one, vanishing by the property of the functions. For the right-hand side we have

$$\int_0^1 r dr \, J_0 (pr) = 0,$$

$$\int_0^1 r^3 dr \, J_0 (pr) = \frac{2}{p^2} J_0 (p),$$

$$\int_0^1 r^5 dr \, J_0 (pr) = \left(\frac{4}{p^2} - \frac{32}{p^4}\right) J_0 (p) ;$$

so that

$$a_p = \frac{4}{p^3 J_0 (p)} \left\{\mu + 2\mu' \left(1 - \frac{8}{p^2}\right)\right\} \quad \ldots\ldots\ldots\ldots\ldots(9).$$

The velocity-potential ϕ of the whole motion is thus

$$\phi = \left(1 + \tfrac{1}{2} \mu + \tfrac{1}{3} \mu'\right) x + 4\Sigma \frac{\mu + 2\mu' (1 - 8p^{-2})}{p^3 J_0 (p)} \, e^{px} J_0 (pr) \ldots (10),$$

the summation extending to all the admissible values of p. We have now to find the energy of motion of so much of the fluid as is included between $x = 0$, and $x = -l$, where l is so great that the velocity is there sensibly constant.

By Green's theorem

$$2 \,(\text{kinetic energy}) = \int_0^1 \phi \frac{d\phi}{dx} 2\pi r \, dr \quad (x = 0) - \int_0^1 \phi \frac{d\phi}{dx} 2\pi r \, dr \quad (x = -l).$$

Now, when $x = -l$,

$$\phi = -\left(1 + \tfrac{1}{2} \mu + \tfrac{1}{3} \mu'\right) l,$$

$$d\phi/dx = 1 + \tfrac{1}{2} \mu + \tfrac{1}{3} \mu' ;$$

so that the second term is $\pi l (1 + \tfrac{1}{2} \mu + \tfrac{1}{3} \mu')^2$.

In calculating the first term, we must remember that if p_1 and p_2 be two different values of p,

$$\int_0^1 2\pi r \, dr \, J_0 (p_1 r) \, J_0 (p_2 r) = 0.$$

Thus

$$\int_0^1 2\pi r \, dr \, \phi \frac{d\phi}{dx} (x=0) = 16 \, \Sigma \, \frac{\{\mu + 2\mu'(1 - 8p^{-2})\}^2}{p^5 \, [J_0(p)]^2} \int_0^1 2\pi r \, dr \, [J_0(pr)]^2$$

$$= 16 \pi \, \Sigma \left\{\mu + 2\mu'\left(1 - \frac{8}{p^2}\right)\right\}^2 p^{-5}.$$

Accordingly, on restoring R,

$$2 \text{ (kinetic energy)} = \pi R^2 l \left(1 + \tfrac{1}{2}\mu + \tfrac{1}{3}\mu'\right)^2$$

$$+ 16\pi R^3 \Sigma \left\{\mu + 2\mu'\left(1 - \frac{8}{p^2}\right)\right\}^2 p^{-5}.$$

To this must be added the energy of the motion on the positive side of $x=0$. On the whole

$$\frac{2 \text{ kinetic energy}}{\text{(current)}^2} = \frac{l}{\pi R^2} + \frac{16}{\pi R \left(1 + \tfrac{1}{2}\mu + \tfrac{1}{3}\mu'\right)^2} \Sigma \left\{\mu + 2\mu'\left(1 - \frac{8}{p^2}\right)\right\}^2 p^{-5}$$

$$+ \frac{8}{3\pi^2 R} \frac{1 + \tfrac{14}{15}\mu + \tfrac{5}{21}\mu^2 + \tfrac{314}{525}\mu' + \tfrac{214}{875}\mu\mu' + \tfrac{89}{825}\mu'^2}{\left(1 + \tfrac{1}{2}\mu + \tfrac{1}{3}\mu'\right)^2}.$$

Hence, if a be the correction to the length,

$$3\pi a/8R = \left[1 + \tfrac{14}{15}\mu + \tfrac{314}{525}\mu' + \left(6\pi \, \Sigma p^{-5} + \tfrac{5}{21}\right)\mu^2\right.$$

$$+ \left\{24\pi \left(\Sigma p^{-5} - 8\Sigma p^{-7}\right) + \tfrac{214}{875}\right\}\mu\mu'$$

$$+ \left\{24\pi \left(\Sigma p^{-5} - 16\Sigma p^{-7} + 64\Sigma p^{-9}\right) + \tfrac{89}{825}\right\}\mu'^2\right] \div \left(1 + \tfrac{1}{2}\mu + \tfrac{1}{3}\mu'\right)^2.$$

By numerical calculation from the values of p

$$\Sigma p^{-5} = \cdot 00128266 \, ; \quad \Sigma p^{-5} - 8\Sigma p^{-7} = \cdot 00061255,$$

$$\Sigma p^{-5} - 16\Sigma p^{-7} + 64\Sigma p^{-9} = \cdot 00030351,$$

and thus
$$3\pi a/8R = \left[1 + \cdot 9333333\,\mu + \cdot 5980951\,\mu'\right.$$

$$+ \cdot 2622728\,\mu^2 + \cdot 363223\,\mu\mu' + \cdot 1307634\,\mu'^2\right] \div \left(1 + \tfrac{1}{2}\mu + \tfrac{1}{3}\mu'\right)^2$$

$$= 1 - \frac{\cdot 0666667\,\mu + \cdot 0685716\,\mu' - \cdot 0122728\,\mu^2 - \cdot 029890\,\mu\mu' - \cdot 0196523\,\mu'^2}{\left(1 + \tfrac{1}{2}\mu + \tfrac{1}{3}\mu'\right)^2}$$

$$\dots\dots\dots\dots\dots (11).$$

The fraction on the right is the ratio of two quadratic functions of μ, μ', and our object is to determine its maximum value. In general if S and S' be two quadratic functions, the maximum and minimum values of $z = S \div S'$ are given by the cubic equation

$$-\Delta z^{-3} + \Theta z^{-2} - \Theta' z^{-1} + \Delta' = 0,$$

where
$$S = a\mu^2 + b\mu'^2 + c + 2f\mu' + 2g\mu + 2h\mu\mu',$$

$$S' = a'\mu^2 + b'\mu'^2 + c' + 2f'\mu' + 2g'\mu + 2h'\mu\mu',$$

$$\Delta = abc + 2fgh - af^2 - bg^2 - ch^2 = \frac{(h^2 - ab)(g^2 - ac) - (hg - af)^2}{a},$$

$$\Theta = (bc - f^2)\,a' + (ca - g^2)\,b' + (ab - h^2)\,c'$$

$$+ 2\,(gh - af)\,f' + 2\,(hf - bg)\,g' + 2\,(fg - ch)\,h',$$

and Θ', Δ', are derived from Θ and Δ by interchanging the accented and unaccented letters.

In the present case, since S' is a product of linear factors, $\Delta' = 0$; and since the two factors are the same, $\Theta' = 0$, so that $z = \Delta \div \Theta$ simply. Substituting the numerical values, and effecting the calculations, we find $z = \cdot0289864$, which is the maximum value of the fraction consistent with real values of μ and μ'.

The corresponding value of a is $\cdot82422\,R$, than which the true correction cannot be greater.

If we assume $\mu' = 0$, the greatest value of z then possible is $\cdot024363$, which gives

$$a = \cdot828146\,R^1.$$

On the other hand if we put $\mu = 0$, the maximum value of z comes out $\cdot027653$, whence

$$a = \cdot825353\,R.$$

It would appear from this result that the variable part of the normal velocity at the mouth is better represented by a term varying as r^4, than by one varying as r^2.

The value $a = \cdot8242\,R$ is probably pretty close to the truth. If the normal velocity be assumed constant, $a = \cdot848826\,R$; if of the form $1 + \mu r^2$, $a = \cdot82815\,R$, when μ is suitably determined; and when the form $1 + \mu r^2 + \mu' r^4$, containing another arbitrary constant, is made the foundation of the calculation, we get $a = \cdot8242\,R$.

The true value of a is probably about $\cdot82\,R$.

In the case of $\mu = 0$, the minimum energy corresponds to $\mu' = 1\cdot103$, so that

$$d\phi/dx = 1 + 1\cdot103\,r^4/R^4.$$

On this supposition the normal velocity of the edge $(r = R)$ would be about double of that near the centre.

---

[1] Notes on Bessel's functions. *Phil. Mag.* Nov. 1872.

INDEX OF AUTHORS.

INDEX OF SUBJECTS.

CATALOG OF DOVER BOOKS

BOOKS EXPLAINING SCIENCE AND MATHEMATICS

THE COMMON SENSE OF THE EXACT SCIENCES, W. K. Clifford. Introduction by James Newman, edited by Karl Pearson. For 70 years this has been a guide to classical scientific and mathematical thought. Explains with unusual clarity basic concepts, such as extension of meaning of symbols, characteristics of surface boundaries, properties of plane figures, vectors, Cartesian method of determining position, etc. Long preface by Bertrand Russell. Bibliography of Clifford. Corrected, 130 diagrams redrawn. 249pp. 5⅜ x 8.
T61 Paperbound **$1.60**

SCIENCE THEORY AND MAN, Erwin Schrödinger. This is a complete and unabridged reissue of SCIENCE AND THE HUMAN TEMPERAMENT plus an additional essay: "What is an Elementary Particle?" Nobel Laureate Schrödinger discusses such topics as nature of scientific method, the nature of science, chance and determinism, science and society, conceptual models for physical entities, elementary particles and wave mechanics. Presentation is popular and may be followed by most people with little or no scientific training. "Fine practical preparation for a time when laws of nature, human institutions . . . are undergoing a critical examination without parallel," Waldemar Kaempffert, N. Y. TIMES. 192pp. 5⅜ x 8.
T428 Paperbound **$1.35**

PIONEERS OF SCIENCE, O. Lodge. Eminent scientist-expositor's authoritative, yet elementary survey of great scientific theories. Concentrating on individuals—Copernicus, Brahe, Kepler, Galileo, Descartes, Newton, Laplace, Herschel, Lord Kelvin, and other scientists—the author presents their discoveries in historical order adding biographical material on each man and full, specific explanations of their achievements. The clear and complete treatment of the post-Newtonian astronomers is a feature seldom found in other books on the subject. Index. 120 illustrations. xv + 404pp. 5⅜ x 8.
T716 Paperbound **$1.50**

THE EVOLUTION OF SCIENTIFIC THOUGHT FROM NEWTON TO EINSTEIN, A. d'Abro. Einstein's special and general theories of relativity, with their historical implications, are analyzed in non-technical terms. Excellent accounts of the contributions of Newton, Riemann, Weyl, Planck, Eddington, Maxwell, Lorentz and others are treated in terms of space and time, equations of electromagnetics, finiteness of the universe, methodology of science. 21 diagrams. 482pp. 5⅜ x 8.
T2 Paperound **$2.00**

THE RISE OF THE NEW PHYSICS, A. d'Abro. A half-million word exposition, formerly titled THE DECLINE OF MECHANISM, for readers not versed in higher mathematics. The only thorough explanation, in everyday language, of the central core of modern mathematical physical theory, treating both classical and modern theoretical physics, and presenting in terms almost anyone can understand the equivalent of 5 years of study of mathematical physics. Scientifically impeccable coverage of mathematical-physical thought from the Newtonian system up through the electronic theories of Dirac and Heisenberg and Fermi's statistics. Combines both history and exposition; provides a broad yet unified and detailed view, with constant reference to experiment and observation. "A must for anyone doing serious study in the physical sciences," JOURNAL OF THE FRANKLIN INSTITUTE. "Extraordinary faculty . . . to explain ideas and theories of theoretical physics in the language of daily life," ISIS. First part of set covers philosophy of science, drawing upon the practice of Newton, Maxwell, Poincaré, Einstein, others, discussing modes of thought, experiment, interpretations of causality, etc. In the second part, 100 pages explain grammar and vocabulary of mathematics, with discussions of functions, groups, series, Fourier series, etc. The remainder is devoted to concrete, detailed coverage of both classical and quantum physics, explaining such topics as analytic mechanics, Hamilton's principle, wave theory of light, electromagnetic waves, groups of transformations, thermodynamics, phase rule, Brownian movement, kinetics, special relativity, Planck's original quantum theory, Bohr's atom, Zeeman effect, Broglie's wave mechanics, Heisenberg's uncertainty, Eigen-values, matrices, scores of other important topics. Discoveries and theories are covered for such men as Alembert, Born, Cantor, Debye, Euler, Foucault, Galois, Gauss, Hadamard, Kelvin, Kepler, Laplace, Maxwell, Pauli, Rayleigh, Volterra, Weyl, Young, more than 180 others. Indexed. 97 illustrations. ix + 982pp. 5⅜ x 8.
T3 Volume 1, Paperbound **$2.00**
T4 Volume 2, Paperbound **$2.00**

CONCERNING THE NATURE OF THINGS, Sir William Bragg. Christmas lectures delivered at the Royal Society by Nobel laureate. Why a spinning ball travels in a curved track; how uranium is transmuted to lead, etc. Partial contents: atoms, gases, liquids, crystals, metals, etc. No scientific background needed; wonderful for intelligent child. 32pp. of photos, 57 figures. xii + 232pp. 5⅜ x 8.
T31 Paperbound **$1.35**

THE UNIVERSE OF LIGHT, Sir William Bragg. No scientific training needed to read Nobel Prize winner's expansion of his Royal Institute Christmas Lectures. Insight into nature of light, methods and philosophy of science. Explains lenses, reflection, color, resonance, polarization, x-rays, the spectrum, Newton's work with prisms, Huygens' with polarization, Crookes' with cathode ray, etc. Leads into clear statement of 2 major historical theories of light, corpuscle and wave. Dozens of experiments you can do. 199 illus., including 2 full-page color plates. 293pp. 5⅜ x 8.
S538 Paperbound **$1.85**

THE PRINCIPLES OF SCIENCE, A TREATISE ON LOGIC AND THE SCIENTIFIC METHOD, W. S. Jevons. Treating such topics as Inductive and Deductive Logic, the Theory of Number, Probability, and the Limits of Scientific Method, this milestone in the development of symbolic logic remains a stimulating contribution to the investigation of inferential validity in the natural and social sciences. It significantly advances Boole's logic, and contains a detailed introduction to the nature and methods of probability in physics, astronomy, everyday affairs, etc. In his introduction, Ernest Nagel of Columbia University says, "[Jevons] continues to be of interest as an attempt to articulate the logic of scientific inquiry." Index. liii + 786pp. 5⅜ x 8.   S446 Paperbound **$2.98**

Group theory, algebra, sets

LECTURES ON THE ICOSAHEDRON AND THE SOLUTION OF EQUATIONS OF THE FIFTH DEGREE, Felix Klein. The solution of quintics in terms of rotation of a regular icosahedron around its axes of symmetry. A classic & indispensable source for those interested in higher algebra, geometry, crystallography. Considerable explanatory material included. 230 footnotes, mostly bibliographic. 2nd edition, xvi + 289pp. 5⅜ x 8.   S314 Paperbound **$1.85**

LINEAR GROUPS, WITH AN EXPOSITION OF THE GALOIS FIELD THEORY, L. E. Dickson. The classic exposition of the theory of groups, well within the range of the graduate student. Part I contains the most extensive and thorough presentation of the theory of Galois Fields available, with a wealth of examples and theorems. Part II is a full discussion of linear groups of finite order. Much material in this work is based on Dickson's own contributions. Also includes expositions of Jordan, Lie, Abel, Betti-Mathieu, Hermite, etc. "A milestone in the development of modern algebra," W. Magnus, in his historical introduction to this edition. Index. xv + 312pp. 5⅜ x 8.   S482 Paperbound **$1.95**

INTRODUCTION TO THE THEORY OF GROUPS OF FINITE ORDER, R. Carmichael. Examines fundamental theorems and their application. Beginning with sets, systems, permutations, etc., it progresses in easy stages through important types of groups: Abelian, prime power, permutation, etc. Except 1 chapter where matrices are desirable, no higher math needed. 783 exercises, problems. Index. xvi + 447pp. 5⅜ x 8.   S299 Clothbound **$3.95**
                                                                                                    S300 Paperbound **$2.00**

THEORY OF GROUPS OF FINITE ORDER, W. Burnside. First published some 40 years ago, this is still one of the clearest introductory texts. Partial contents: permutations, groups independent of representation, composition series of a group, isomorphism of a group with itself, Abelian groups, prime power groups, permutation groups, invariants of groups of linear substitution graphical representation, etc. 45pp. of notes. Indexes. xxiv + 512pp. 5⅜ x 8.
                                                                                                    S38 Paperbound **$2.45**

THEORY AND APPLICATIONS OF FINITE GROUPS, G. A. Miller, H. F. Blichfeldt, L. E. Dickson. Unusually accurate and authoritative work, each section prepared by a leading specialist: Miller on substitution and abstract groups, Blichfeldt on finite groups of linear homogeneous transformations, Dickson on applications of finite groups. Unlike more modern works, this gives the concrete basis from which abstract group theory arose. Includes Abelian groups, prime-power groups, isomorphisms, matrix forms of linear transformations, Sylow groups, Galois' theory of algebraic equations, duplication of a cube, trisection of an angle, etc. 2 Indexes. 267 problems. xvii + 390pp. 5⅜ x 8.   S216 Paperbound **$2.00**

CONTINUOUS GROUPS OF TRANSFORMATIONS, L. P. Eisenhart. Intensive study of the theory and geometrical applications of continuous groups of transformations; a standard work on the subject, called forth by the revolution in physics in the 1920's. Covers tensor analysis, Riemannian geometry, canonical parameters, transitivity, imprimitivity, differential invariants, the algebra of constants of structure, differential geometry, contact transformations, etc. "Likely to remain one of the standard works on the subject for many years . . . principal theorems are proved clearly and concisely, and the arrangement of the whole is coherent," MATHEMATICAL GAZETTE. Index. 72-item bibliography. 185 exercises. ix + 301pp. 5⅜ x 8.
                                                                                                    S781 Paperbound **$1.85**

THE THEORY OF GROUPS AND QUANTUM MECHANICS, H. Weyl. Discussions of Schroedinger's wave equation, de Broglie's waves of a particle, Jordan-Hoelder theorem, Lie's continuous groups of transformations, Pauli exclusion principle, quantization of Maxwell-Dirac field equations, etc. Unitary geometry, quantum theory, groups, application of groups to quantum mechanics, symmetry permutation group, algebra of symmetric transformation, etc. 2nd revised edition. Bibliography. Index. xxii + 422pp. 5⅜ x 8.   S268 Clothbound **$4.50**
                                                                                                    S269 Paperbound **$1.95**

ALGEBRAIC THEORIES, L. E. Dickson. Best thorough introduction to classical topics in higher algebra develops theories centering around matrices, invariants, groups. Higher algebra, Galois theory, finite linear groups, Klein's icosahedron, algebraic invariants, linear transformations, elementary divisors, invariant factors; quadratic, bi-linear, Hermitian forms, singly and in pairs. Proofs rigorous, detailed; topics developed lucidly, in close connection with their most frequent mathematical applications. Formerly "Modern Algebraic Theories." 155 problems. Bibliography. 2 indexes. 285pp. 5⅜ x 8.   S547 Paperbound **$1.50**

ALGEBRAS AND THEIR ARITHMETICS, L. E. Dickson. Provides the foundation and background necessary to any advanced undergraduate or graduate student studying abstract algebra. Begins with elementary introduction to linear transformations, matrices, field of complex numbers; proceeds to order, basal units, modulus, quaternions, etc.; develops calculus of linear sets, describes various examples of algebras including invariant, difference, nilpotent, semi-simple. "Makes the reader marvel at his genius for clear and profound analysis," Amer. Mathematical Monthly. Index. xii + 241pp. 5⅜ x 8.   S616 Paperbound **$1.35**

THE THEORY OF EQUATIONS WITH AN INTRODUCTION TO THE THEORY OF BINARY ALGEBRAIC FORMS, W. S. Burnside and A. W. Panton. Extremely thorough and concrete discussion of the theory of equations, with extensive detailed treatment of many topics curtailed in later texts. Covers theory of algebraic equations, properties of polynomials, symmetric functions, derived functions, Horner's process, complex numbers and the complex variable, determinants and methods of elimination, invariant theory (nearly 100 pages), transformations, introduction to Galois theory, Abelian equations, and much more. Invaluable supplementary work for modern students and teachers. 759 examples and exercises. Index in each volume. Two volume set. Total of xxiv + 604pp. 5⅜ x 8.   S714 Vol I Paperbound **$1.85**
S715 Vol II Paperbound **$1.85**
The set **$3.70**

COMPUTATIONAL METHODS OF LINEAR ALGEBRA, V. N. Faddeeva, translated by **C. D. Benster.** First English translation of a unique and valuable work, the only work in English presenting a systematic exposition of the most important methods of linear algebra—classical and contemporary. Shows in detail how to derive numerical solutions of problems in mathematical physics which are frequently connected with those of linear algebra. Theory as well as individual practice. Part I surveys the mathematical background that is indispensable to what follows. Parts II and III, the conclusion, set forth the most important methods of solution, for both exact and iterative groups. One of the most outstanding and valuable features of this work is the 23 tables, double and triple checked for accuracy. These tables will not be found elsewhere. Author's preface. Translator's note. New bibliography and index. x + 252pp. 5⅜ x 8.   S424 Paperbound **$1.95**

ALGEBRAIC EQUATIONS, E. Dehn. Careful and complete presentation of Galois' theory of algebraic equations; theories of Lagrange and Galois developed in logical rather than historical form, with a more thorough exposition than in most modern books. Many concrete applications and fully-worked-out examples. Discusses basic theory (very clear exposition of the symmetric group); isomorphic, transitive, and Abelian groups; applications of Lagrange's and Galois' theories; and much more. Newly revised by the author. Index. List of Theorems. xi + 208pp. 5⅜ x 8.   S697 Paperbound **$1.45**

THEORY OF SETS, E. Kamke. Clearest, amplest introduction in English, well suited for independent study. Subdivision of main theory, such as theory of sets of points, are discussed, but emphasis is on general theory. Partial contents: rudiments of set theory, arbitrary sets and their cardinal numbers, ordered sets and their order types, well-ordered sets and their cardinal numbers. Bibliography. Key to symbols. Index. vii + 144pp. 5⅜ x 8.
S141 Paperbound **$1.35**

Number theory

INTRODUCTION TO THE THEORY OF NUMBERS, L. E. Dickson. Thorough, comprehensive approach with adequate coverage of classical literature, an introductory volume beginners can follow. Chapters on divisibility, congruences, quadratic residues & reciprocity, Diophantine equations, etc. Full treatment of binary quadratic forms without usual restriction to integral coefficients. Covers infinitude of primes, least residues, Fermat's theorem, Euler's phi function, Legendre's symbol, Gauss's lemma, automorphs, reduced forms, recent theorems of Thue & Siegel, many more. Much material not readily available elsewhere. 239 problems. Index. J figure. viii + 183pp. 5⅜ x 8.   S342 Paperbound **$1.65**

ELEMENTS OF NUMBER THEORY, I. M. Vinogradov. Detailed 1st course for persons without advanced mathematics; 95% of this book can be understood by readers who have gone no farther than high school algebra. Partial contents: divisibility theory, important number theoretical functions, congruences, primitive roots and indices, etc. Solutions to both problems and exercises. Tables of primes, indices, etc. Covers almost every essential formula in elementary number theory! Translated from Russian. 233 problems, 104 exercises. viii + 227pp. 5⅜ x 8.   S259 Paperbound **$1.60**

THEORY OF NUMBERS and DIOPHANTINE ANALYSIS, R. D. Carmichael. These two complete works in one volume form one of the most lucid introductions to number theory, requiring only a firm foundation in high school mathematics. "Theory of Numbers," partial contents: Eratosthenes' sieve, Euclid's fundamental theorem, G.C.F. and L.C.M. of two or more integers, linear congruences, etc. "Diophantine Analysis": rational triangles, Pythagorean triangles, equations of third, fourth, higher degrees, method of functional equations, much more. "Theory of Numbers": 76 problems. Index. 94pp. "Diophantine Analysis": 222 problems. Index. 118pp. 5⅜ x 8.   S529 Paperbound **$1.35**

CONTRIBUTIONS TO THE FOUNDING OF THE THEORY OF TRANSFINITE NUMBERS, Georg Cantor. These papers founded a new branch of mathematics. The famous articles of 1895-7 are translated, with an 82-page introduction by P. E. B. Jourdain dealing with Cantor, the background of his discoveries, their results, future possibilities. Bibliography. Index. Notes. ix + 211 pp. 5⅜ x 8.   S45 Paperbound **$1.25**

See also: **TRANSCENDENTAL AND ALGEBRAIC NUMBERS, A. O. Gelfond.**

Probability theory and information theory

A PHILOSOPHICAL ESSAY ON PROBABILITIES, Marquis de Laplace. This famous essay explains without recourse to mathematics the principle of probability, and the application of probability to games of chance, natural philosophy, astronomy, many other fields. Translated from the 6th French edition by F. W. Truscott, F. L. Emory, with new introduction for this edition by E. T. Bell. 204pp. 5⅜ x 8.   S166 Paperbound **$1.35**

MATHEMATICAL FOUNDATIONS OF INFORMATION THEORY, A. I. Khinchin. For the first time mathematicians, statisticians, physicists, cyberneticists, and communications engineers are offered a complete and exact introduction to this relatively new field. Entropy as a measure of a finite scheme, applications to coding theory, study of sources, channels and codes, detailed proofs of both Shannon theorems for any ergodic source and any stationary channel with finite memory, and much more are covered. Bibliography. vii + 120pp. 5⅜ x 8.
S434 Paperbound **$1.35**

SELECTED PAPERS ON NOISE AND STOCHASTIC PROCESS, edited by **Prof. Nelson Wax,** U. of Illinois. 6 basic papers for newcomers in the field, for those whose work involves noise characteristics. Chandrasekhar, Uhlenbeck & Ornstein, Uhlenbeck & Ming, Rice, Doob. Included is Kac's Chauvenet-Prize winning Random Walk. Extensive bibliography lists 200 articles, up through 1953. 21 figures. 337pp. 6⅛ x 9¼.   S262 Paperbound **$2.35**

THEORY OF PROBABILITY, William Burnside. Synthesis, expansion of individual papers presents numerous problems in classical probability, offering many original views succinctly, effectively. Game theory, cards, selections from groups; geometrical probability in such areas as suppositions as to probability of position of point on a line, points on surface of sphere, etc. Includes methods of approximation, theory of errors, direct calculation of probabilities, etc. Index. 136pp. 5⅜ x 8.   S567 Paperbound **$1.00**

Vector and tensor analysis, matrix theory

VECTOR AND TENSOR ANALYSIS, A. P. Wills. Covers the entire field of vector and tensor analysis from elementary notions to dyads and non-Euclidean manifolds (especially detailed), absolute differentiation, the Lamé operator, the Riemann-Christoffel and Ricci-Einstein tensors, and the calculation of the Gaussian curvature of a surface. Many illustrations from electrical engineering, relativity theory, astro-physics, quantum mechanics. Presupposes only a good working knowledge of calculus. Exercises at end of each chapter. Intended for physicists and engineers as well as pure mathematicians. 44 diagrams. 114 problems. Bibliography. Index. xxxii + 285pp. 5⅜ x 8.   S454 Paperbound **$1.75**

APPLICATIONS OF TENSOR ANALYSIS, A. J. McConnell. (Formerly APPLICATIONS OF THE ABSOLUTE DIFFERENTIAL CALCULUS.) An excellent text for understanding the application of tensor methods to familiar subjects such as dynamics, electricity, elasticity, and hydrodynamics. Explains the fundamental ideas and notation of tensor theory, the geometrical treatment of tensor algebra, the theory of differentiation of tensors, and includes a wealth of practical material. Bibliography. Index. 43 illustrations. 685 problems. xii + 381pp. 5⅜ x 8.   S373 Paperbound **$1.85**

VECTOR AND TENSOR ANALYSIS, G. E. Hay. One of the clearest introductions to this increasingly important subject. Start with simple definitions, finish the book with a sure mastery of oriented Cartesian vectors, Christoffel symbols, solenoidal tensors, and their applications. Complete breakdown of plane, solid, analytical, differential geometry. Separate chapters on application. All fundamental formulae listed & demonstrated. 195 problems, 66 figures. viii + 193pp. 5⅜ x 8.   S109 Paperbound **$1.75**

VECTOR ANALYSIS, FOUNDED UPON THE LECTURES OF J. WILLARD GIBBS, by E. B. Wilson. Still a first-rate introduction and supplementary text for students of mathematics and physics. Based on the pioneering lectures of Yale's great J. Willard Gibbs, can be followed by anyone who has had some calculus. Practical approach, stressing efficient use of combinations and functions of vectors. Worked examples from geometry, mechanics, hydrodynamics, gas theory, etc., as well as practice examples. Covers basic vector processes, differential and integral calculus in relation to vector functions, and theory of linear vector functions, forming an introduction to the study of multiple algebra and matrix theory. While the notation is not always modern, it is easily followed. xviii + 436pp. 5⅜ x 8.
S656 Paperbound **$2.00**

PROBLEMS AND WORKED SOLUTIONS IN VECTOR ANALYSIS, L. R. Shorter. More pages of fully-worked-out examples than any other text on vector analysis. A self-contained course for home study or a fine classroom supplement. 138 problems and examples begin with fundamentals, then cover systems of coordinates, relative velocity and acceleration, the commutative and distributive laws, axial and polar vectors, finite displacements, the calculus of vectors, curl and divergence, etc. Final chapter treats applications in dynamics and physics: kinematics of a rigid body, equipotential surfaces, etc. "Very helpful . . . very comprehensive. A handy book like this . . . will fill a great want," MATHEMATICAL GAZETTE. Index. List of 174 important equations. 158 figures. xiv + 356pp. 5⅜ x 8. S135 Paperbound **$2.00**

THE THEORY OF DETERMINANTS, MATRICES, AND INVARIANTS, H. W. Turnbull. 3rd revised, corrected edition of this important study of virtually all the salient features and major theories of the subject. Covers Laplace identities, linear equations, differentiation, symbolic and direct methods for the reduction of invariants, seminvariants, Hilbert's Basis Theorem, Clebsch's Theorem, canonical forms, etc. New appendix contains a proof of Jacobi's lemma, further properties of symmetric determinants, etc. More than 350 problems. New references to recent developments. xviii + 374pp. 5⅜ x 8. S699 Paperbound **$2.00**

Differential equations, ordinary and partial, and integral equations

INTRODUCTION TO THE DIFFERENTIAL EQUATIONS OF PHYSICS, L. Hopf. Especially valuable to the engineer with no math beyond elementary calculus. Emphasizing intuitive rather than formal aspects of concepts, the author covers an extensive territory. Partial contents: Law of causality, energy theorem, damped oscillations, coupling by friction, cylindrical and spherical coordinates, heat source, etc. Index. 48 figures. 160pp. 5⅜ x 8. S120 Paperbound **$1.25**

INTRODUCTION TO THE THEORY OF LINEAR DIFFERENTIAL EQUATIONS, E. G. Poole. Authoritative discussions of important topics, with methods of solution more detailed than usual, for students with background of elementary course in differential equations. Studies existence theorems, linearly independent solutions; equations with constant coefficients; with uniform analytic coefficients; regular singularities; the hypergeometric equation; conformal representation; etc. Exercises. Index. 210pp. 5⅜ x 8. S629 Paperbound **$1.65**

DIFFERENTIAL EQUATIONS FOR ENGINEERS, P. Franklin. Outgrowth of a course given 10 years at M. I. T. Makes most useful branch of pure math accessible for practical work. Theoretical basis of D.E.'s; solution of ordinary D.E.'s and partial derivatives arising from heat flow, steady-state temperature of a plate, wave equations; analytic functions; convergence of Fourier Series. 400 problems on electricity, vibratory systems, other topics. Formerly "Differential Equations for Electrical Engineers." Index. 41 illus. 307pp. 5⅜ x 8. S601 Paperbound **$1.65**

DIFFERENTIAL EQUATIONS, F. R. Moulton. A detailed, rigorous exposition of all the non-elementary processes of solving ordinary differential equations. Several chapters devoted to the treatment of practical problems, especially those of a physical nature, which are far more advanced than problems usually given as illustrations. Includes analytic differential equations; variations of a parameter; integrals of differential equations; analytic implicit functions; problems of elliptic motion; sine-amplitude functions; deviation of formal bodies; Cauchy-Lipschitz process; linear differential equations with periodic coefficients; differential equations in infinitely many variations; much more. Historical notes. 10 figures. 222 problems. Index. xv + 395pp. 5⅜ x 8. S451 Paperbound **$2.00**

LECTURES ON CAUCHY'S PROBLEM, J. Hadamard. Based on lectures given at Columbia, Rome, this discusses work of Riemann, Kirchhoff, Volterra, and the author's own research on the hyperbolic case in linear partial differential equations. It extends spherical and cylindrical waves to apply to all (normal) hyperbolic equations. Partial contents: Cauchy's problem, fundamental formula, equations with odd number, with even number of independent variables; method of descent. 32 figures. Index. iii + 316pp. 5⅜ x 8. S105 Paperbound **$1.75**

PARTIAL DIFFERENTIAL EQUATIONS OF MATHEMATICAL PHYSICS, A. G. Webster. A keystone work in the library of every mature physicist, engineer, researcher. Valuable sections on elasticity, compression theory, potential theory, theory of sound, heat conduction, wave propagation, vibration theory. Contents include: deduction of differential equations, vibrations, normal functions, Fourier's series, Cauchy's method, boundary problems, method of Riemann-Volterra. Spherical, cylindrical, ellipsoidal harmonics, applications, etc. 97 figures. vii + 440pp. 5⅜ x 8. S263 Paperbound **$2.00**

ORDINARY DIFFERENTIAL EQUATIONS, E. L. Ince. A most compendious analysis in real and complex domains. Existence and nature of solutions, continuous transformation groups, solutions in an infinite form, definite integrals, algebraic theory, Sturmian theory, boundary problems, existence theorems, 1st order, higher order, etc. "Deserves the highest praise, a notable addition to mathematical literature," BULLETIN, AM. MATH. SOC. Historical appendix. Bibliography. 18 figures. viii + 558pp. 5⅜ x 8. S349 Paperbound **$2.55**

THEORY OF DIFFERENTIAL EQUATIONS, A. R. Forsyth. Out of print for over a decade, the complete 6 volumes (now bound as 3) of this monumental work represent the most comprehensive treatment of differential equations ever written. Historical presentation includes in 2500 pages every substantial development. Vol. 1, 2: EXACT EQUATIONS, PFAFF'S PROBLEM; ORDINARY EQUATIONS, NOT LINEAR: methods of Grassmann, Clebsch, Lie, Darboux; Cauchy's theorem; branch points; etc. Vol. 3, 4: ORDINARY EQUATIONS, NOT LINEAR; ORDINARY LINEAR EQUATIONS: Zeta Fuchsian functions, general theorems on algebraic integrals, Brun's theorem, equations with uniform periodic coffiecients, etc. Vol. 4, 5: PARTIAL DIFFERENTIAL EQUATIONS: 2 existence-theorems, equations of theoretical dynamics, Laplace transformations, general transformation of equations of the 2nd order, much more. Indexes. Total of 2766pp. 5⅜ x 8. S576-7-8 Clothbound: the set **$15.00**

DIFFERENTIAL AND INTEGRAL EQUATIONS OF MECHANICS AND PHYSICS (DIE DIFFERENTIAL- UND INTEGRALGLEICHUNGEN DER MECHANIK UND PHYSIK), edited by P. Frank and R. von Mises. Most comprehensive and authoritative work on the mathematics of mathematical physics available today in the United States: the standard, definitive reference for teachers, physicists, engineers, and mathematicians—now published (in the original German) at a relatively inexpensive price for the first time! Every chapter in this 2,000-page set is by an expert in his field: Caratheodory, Courant, Frank, Mises, and a dozen others. Vol. I, on mathematics, gives concise but complete coverages of advanced calculus, differential equations, integral equations, and potential, and partial differential equations. Index. xxiii + 916pp. Vol. II (physics): classical mechanics, optics, continuous mechanics, heat conduction and diffusion, the stationary and quasi-stationary electromagnetic field, electromagnetic oscillations, and wave mechanics. Index. xxiv + 1106pp. Two volume set. Each volume available separately. 5⅝ x 8⅜. S787 Vol I Clothbound **$7.50** S788 Vol II Clothbound **$7.50** The set **$15.00**

MATHEMATICAL ANALYSIS OF ELECTRICAL AND OPTICAL WAVE-MOTION, Harry Bateman. Written by one of this century's most distinguished mathematical physicists, this is a practical introduction to those developments of Maxwell's electromagnetic theory which are directly connected with the solution of the partial differential equation of wave motion. Methods of solving wave-equation, polar-cylindrical coordinates, diffraction, transformation of coordinates, homogeneous solutions, electromagnetic fields with moving singularities, etc. Index. 168pp. 5⅜ x 8. S14 Paperbound **$1.60**

See also: **THE ANALYTICAL THEORY OF HEAT, J. Fourier; INTRODUCTION TO BESSEL FUNC- TIONS, F. Bowman.**

Statistics

ELEMENTARY STATISTICS, WITH APPLICATIONS IN MEDICINE AND THE BIOLOGICAL SCIENCES, F. E. Croxton. A sound introduction to statistics for anyone in the physical sciences, assuming no prior acquaintance and requiring only a modest knowledge of math. All basic formulas carefully explained and illustrated; all necessary reference tables included. From basic terms and concepts, the study proceeds to frequency distribution, linear, non-linear, and multiple correlation, skewness, kurtosis, etc. A large section deals with reliability and significance of statistical methods. Containing concrete examples from medicine and biology, this book will prove unusually helpful to workers in those fields who increasingly must evaluate, check, and interpret statistics. Formerly titled "Elementary Statistics with Applications in Medicine." 101 charts. 57 tables. 14 appendices. Index. iv + 376pp. 5⅜ x 8. S506 Paperbound **$1.95**

METHODS OF STATISTICS, L. H. C. Tippett. A classic in its field, this unusually complete systematic introduction to statistical methods begins at beginner's level and progresses to advanced levels for experimenters and poll-takers in all fields of statistical research. Supplies fundamental knowledge of virtually all elementary methods in use today by sociologists, psychologists, biologists, engineers, mathematicians, etc. Explains logical and mathematical basis of each method described, with examples for each section. Covers frequency distributions and measures, inference from random samples, errors in large samples, simple analysis of variance, multiple and partial regression and correlation, etc. 4th revised (1952) edition. 16 charts. 5 significance tables. 152-item bibliography. 96 tables. 22 figures. 395pp. 6 x 9. S228 Clothbound **$7.50**

STATISTICS MANUAL, E. L. Crow, F. A. Davis, M. W. Maxfield. Comprehensive collection of classical, modern statistics methods, prepared under auspices of U. S. Naval Ordnance Test Station, China Lake, Calif. Many examples from ordnance will be valuable to workers in all fields. Emphasis is on use, with information on fiducial limits, sign tests, Chi-square runs, sensitivity, quality control, much more. "Well written . . . excellent reference work," Operations Research. Corrected edition of NAVORD Report 3360 NOTS 948. Introduction. Appendix of 32 tables, charts. Index. Bibliography. 95 illustrations. 306pp. 5⅜ x 8. S599 Paperbound **$1.55**

ANALYSIS & DESIGN OF EXPERIMENTS, H. B. Mann. Offers a method for grasping the analysis of variance and variance design within a short time. Partial contents: Chi-square distribution and analysis of variance distribution, matrices, quadratic forms, likelihood ration tests and tests of linear hypotheses, power of analysis, Galois fields, non-orthogonal data, interblock estimates, etc. 15pp. of useful tables. x + 195pp. 5 x 7⅜. S180 Paperbound **$1.45**

Numerical analysis, tables

PRACTICAL ANALYSIS, GRAPHICAL AND NUMERICAL METHODS, F. A. Willers. Translated by R. T. Beyer. Immensely practical handbook for engineers, showing how to interpolate, use various methods of numerical differentiation and integration, determine the roots of a single algebraic equation, system of linear equations, use empirical formulas, integrate differential equations, etc. Hundreds of shortcuts for arriving at numerical solutions. Special section on American calculating machines, by T. W. Simpson. 132 illustrations. 422pp. 5⅜ x 8.
S273 Paperbound **$2.00**

NUMERICAL SOLUTIONS OF DIFFERENTIAL EQUATIONS, H. Levy & E. A. Baggott. Comprehensive collection of methods for solving ordinary differential equations of first and higher order. All must pass 2 requirements: easy to grasp and practical, more rapid than school methods. Partial contents: graphical integration of differential equations, graphical methods for detailed solution. Numerical solution. Simultaneous equations and equations of 2nd and higher orders. "Should be in the hands of all in research in applied mathematics, teaching," NATURE. 21 figures. viii + 238pp. 5⅜ x 8. S168 Paperbound **$1.75**

NUMERICAL INTEGRATION OF DIFFERENTIAL EQUATIONS, Bennett, Milne & Bateman. Unabridged republication of original monograph prepared for National Research Council. New methods of integration of differential equations developed by 3 leading mathematicians: THE INTERPOLATIONAL POLYNOMIAL and SUCCESSIVE APPROXIMATIONS by A. A. Bennett; STEP-BY-STEP METHODS OF INTEGRATION by W. W. Milne; METHODS FOR PARTIAL DIFFERENTIAL EQUATIONS by H. Bateman. Methods for partial differential equations, transition from difference equations to differential equations, solution of differential equations to non-integral values of a parameter will interest mathematicians and physicists. 288 footnotes, mostly bibliographic; 235-item classified bibliography. 108pp. 5⅜ x 8. S305 Paperbound **$1.35**

INTRODUCTION TO RELAXATION METHODS, F. S. Shaw. Fluid mechanics, design of electrical networks, forces in structural frameworks, stress distribution, buckling, etc. Solve linear simultaneous equations, linear ordinary differential equations, partial differential equations, Eigen-value problems by relaxation methods. Detailed examples throughout. Special tables for dealing with awkwardly-shaped boundaries. Indexes. 253 diagrams. 72 tables. 400pp. 5⅜ x 8. S244 Paperbound **$2.45**

TABLES OF INDEFINITE INTEGRALS, G. Petit Bois. Comprehensive and accurate, this orderly grouping of over 2500 of the most useful indefinite integrals will save you hours of laborious mathematical groundwork. After a list of 49 common transformations of integral expressions, with a wide variety of examples, the book takes up algebraic functions, irrational monomials, products and quotients of binomials, transcendental functions, natural logs, etc. You will rarely or never encounter an integral of an algebraic or transcendental function not included here; any more comprehensive set of tables costs at least $12 or $15. Index. 2544 integrals. xii + 154pp. 6⅛ x 9¼. S225 Paperbound **$1.65**

A TABLE OF THE INCOMPLETE ELLIPTIC INTEGRAL OF THE THIRD KIND, R. G. Selfridge, J. E. Maxfield. The first complete 6 place tables of values of the incomplete integral of the third kind, prepared under the auspices of the Research Department of the U.S. Naval Ordnance Test Station. Calculated on an IBM type 704 calculator and thoroughly verified by echo-checking and a check integral at the completion of each value of **a**. Of inestimable value in problems where the surface area of geometrical bodies can only be expressed in terms of the incomplete integral of the third and lower kinds; problems in aero-, fluid-, and thermodynamics involving processes where nonsymmetrical repetitive volumes must be determined; various types of seismological problems; problems of magnetic potentials due to circular current; etc. Foreword. Acknowledgment. Introduction. Use of table. xiv + 805pp. 5⅝ x 8⅜. S501 Clothbound **$7.50**

MATHEMATICAL TABLES, H. B. Dwight. Unique for its coverage in one volume of almost every function of importance in applied mathematics, engineering, and the physical sciences. Three extremely fine tables of the three trig functions and their inverse functions to thousandths of radians; natural and common logarithms; squares, cubes; hyperbolic functions and the inverse hyperbolic functions; $(a^2 + b^2)$ exp. ½a; complete elliptic integrals of the 1st and 2nd kind; sine and cosine integrals; exponential integrals $Ei(x)$ and $Ei(-x)$; binomial coefficients; factorials to 250; surface zonal harmonics and first derivatives; Bernoulli and Euler numbers and their logs to base of 10; Gamma function; normal probability integral; over 60 pages of Bessel functions; the Riemann Zeta function. Each table with formulae generally used, sources of more extensive tables, interpolation data, etc. Over half have columns of differences, to facilitate interpolation. Introduction. Index. viii + 231pp. 5⅜ x 8. S445 Paperbound **$1.75**

TABLES OF FUNCTIONS WITH FORMULAE AND CURVES, E. Jahnke & F. Emde. The world's most comprehensive 1-volume English-text collection of tables, formulae, curves of transcendent functions. 4th corrected edition, new 76-page section giving tables, formulae for elementary functions—not in other English editions. Partial contents: sine, cosine, logarithmic integral; factorial function; error integral; theta functions; elliptic integrals, functions; Legendre, Bessel, Riemann, Mathieu, hypergeometric functions, etc. Supplementary books. Bibliography. Indexed. "Out of the way functions for which we know no other source," SCIENTIFIC COMPUTING SERVICE, Ltd. 212 figures. 400pp. 5⅜ x 8.   S133 Paperbound **$2.00**

JACOBIAN ELLIPTIC FUNCTION TABLES, L. M. Milne-Thomson. An easy to follow, practical book which gives not only useful numerical tables, but also a complete elementary sketch of the application of elliptic functions. It covers Jacobian elliptic functions and a description of their principal properties; complete elliptic integrals; Fourier series and power series expansions; periods, zeros, poles, residues, formulas for special values of the argument; transformations, approximations, elliptic integrals, conformal mapping, factorization of cubic and quartic polynomials; application to the pendulum problem; etc. Tables and graphs form the body of the book: Graph, 5 figure table of the elliptic function sn (u m); cn (u m); dn (u m). 8 figure table of complete elliptic integrals K, K′, E, E′, and the nome q. 7 figure table of the Jacobian zeta-function Z(u). 3 figures. xi + 123pp. 5⅜ x 8.
S194 Paperbound **$1.35**

PHYSICS

General physics

FOUNDATIONS OF PHYSICS, R. B. Lindsay & H. Margenau. Excellent bridge between semi-popular works & technical treatises. A discussion of methods of physical description, construction of theory; valuable for physicist with elementary calculus who is interested in ideas that give meaning to data, tools of modern physics. Contents include symbolism, mathematical equations; space & time foundations of mechanics; probability; physics & continua; electron theory; special & general relativity; quantum mechanics; causality. "Thorough and yet not overdetailed. Unreservedly recommended," NATURE (London). Unabridged, corrected edition. List of recommended readings. 35 illustrations. xi + 537pp. 5⅜ x 8.
S377 Paperbound **$2.45**

FUNDAMENTAL FORMULAS OF PHYSICS, ed. by D. H. Menzel. Highly useful, fully inexpensive reference and study text, ranging from simple to highly sophisticated operations. Mathematics integrated into text—each chapter stands as short textbook of field represented. Vol. 1: Statistics, Physical Constants, Special Theory of Relativity, Hydrodynamics, Aerodynamics, Boundary Value Problems in Math. Physics; Viscosity, Electromagnetic Theory, etc. Vol. 2: Sound, Acoustics, Geometrical Optics, Electron Optics, High-Energy Phenomena, Magnetism, Biophysics, much more. Index. Total of 800pp. 5⅜ x 8.   Vol. 1 S595 Paperbound **$2.00**
Vol. 2 S596 Paperbound **$2.00**

MATHEMATICAL PHYSICS, D. H. Menzel. Thorough one-volume treatment of the mathematical techniques vital for classic mechanics, electromagnetic theory, quantum theory, and relativity. Written by the Harvard Professor of Astrophysics for junior, senior, and graduate courses, it gives clear explanations of all those aspects of function theory, vectors, matrices, dyadics, tensors, partial differential equations, etc., necessary for the understanding of the various physical theories. Electron theory, relativity, and other topics seldom presented appear here in considerable detail. Scores of definitions, conversion factors, dimensional constants, etc. "More detailed than normal for an advanced text . . . excellent set of sections on Dyadics, Matrices, and Tensors," JOURNAL OF THE FRANKLIN INSTITUTE. Index. 193 problems, with answers. x + 412pp. 5⅜ x 8.   S56 Paperbound **$2.00**

THE SCIENTIFIC PAPERS OF J. WILLARD GIBBS. All the published papers of America's outstanding theoretical scientist (except for "Statistical Mechanics" and "Vector Analysis"). Vol I (thermodynamics) contains one of the most brilliant of all 19th-century scientific papers—the 300-page "On the Equilibrium of Heterogeneous Substances," which founded the science of physical chemistry, and clearly stated a number of highly important natural laws for the first time; 8 other papers complete the first volume. Vol II includes 2 papers on dynamics, 8 on vector analysis and multiple algebra, 5 on the electromagnetic theory of light, and 6 miscellaneous papers. Biographical sketch by H. A. Bumstead. Total of xxxvi + 718pp. 5⅝ x 8⅜.
S721 Vol I Paperbound **$2.00**
S722 Vol II Paperbound **$2.00**
The set **$4.00**

Relativity, quantum theory, nuclear physics

THE PRINCIPLE OF RELATIVITY, A. Einstein, H. Lorentz, M. Minkowski, H. Weyl. These are the 11 basic papers that founded the general and special theories of relativity, all translated into English. Two papers by Lorentz on the Michelson experiment, electromagnetic phenomena. Minkowski's SPACE & TIME, and Weyl's GRAVITATION & ELECTRICITY. 7 epoch-making papers by Einstein: ELECTROMAGNETICS OF MOVING BODIES, INFLUENCE OF GRAVITATION IN PROPAGATION OF LIGHT, COSMOLOGICAL CONSIDERATIONS, GENERAL THEORY, and 3 others. 7 diagrams. Special notes by A. Sommerfeld. 224pp. 5⅜ x 8.
S81 Paperbound **$1.75**

SPACE TIME MATTER, Hermann Weyl. "The standard treatise on the general theory of relativity," (Nature), written by a world-renowned scientist, provides a deep clear discussion of the logical coherence of the general theory, with introduction to all the mathematical tools needed: Maxwell, analytical geometry, non-Euclidean geometry, tensor calculus, etc. Basis is classical space-time, before absorption of relativity. Partial contents: Euclidean space, mathematical form, metrical continuum, relativity of time and space, general theory. 15 diagrams. Bibliography. New preface for this edition. xviii + 330pp. 5⅜ x 8.
S267 Paperbound **$1.85**

PRINCIPLES OF QUANTUM MECHANICS, W. V. Houston. Enables student with working knowledge of elementary mathematical physics to develop facility in use of quantum mechanics, understand published work in field. Formulates quantum mechanics in terms of Schroedinger's wave mechanics. Studies evidence for quantum theory, for inadequacy of classical mechanics, 2 postulates of quantum mechanics; numerous important, fruitful applications of quantum mechanics in spectroscopy, collision problems, electrons in solids; other topics. "One of the most rewarding features . . . is the interlacing of problems with text," Amer. J. of Physics. Corrected edition. 21 illus. Index. 296pp. 5⅜ x 8.   S524 Paperbound **$1.85**

PHYSICAL PRINCIPLES OF THE QUANTUM THEORY, Werner Heisenberg. A Nobel laureate discusses quantum theory; Heisenberg's own work, Compton, Schroedinger, Wilson, Einstein, many others. Written for physicists, chemists who are not specialists in quantum theory, only elementary formulae are considered in the text; there is a mathematical appendix for specialists. Profound without sacrifice of clarity. Translated by C. Eckart, F. Hoyt. 18 figures. 192pp. 5⅜ x 8.   S113 Paperbound **$1.25**

SELECTED PAPERS ON QUANTUM ELECTRODYNAMICS, edited by J. Schwinger. Facsimiles of papers which established quantum electrodynamics, from initial successes through today's position as part of the larger theory of elementary particles. First book publication in any language of these collected papers of Bethe, Bloch, Dirac, Dyson, Fermi, Feynman, Heisenberg, Kusch, Lamb, Oppenheimer, Pauli, Schwinger, Tomonoga, Weisskopf, Wigner, etc. 34 papers in all, 29 in English, 1 in French, 3 in German, 1 in Italian. Preface and historical commentary by the editor. xvii + 423pp. 6⅛ x 9¼.   S444 Paperbound **$2.45**

THE FUNDAMENTAL PRINCIPLES OF QUANTUM MECHANICS, WITH ELEMENTARY APPLICATIONS, E. C. Kemble. An inductive presentation, for the graduate student or specialist in some other branch of physics. Assumes some acquaintance with advanced math; apparatus' necessary beyond differential equations and advanced calculus is developed as needed. Although a general exposition of principles, hundreds of individual problems are fully treated, with applications of theory being interwoven with development of the mathematical structure. The author is the Professor of Physics at Harvard Univ. "This excellent book would be of great value to every student . . . a rigorous and detailed mathematical discussion of all of the principal quantum-mechanical methods . . . has succeeded in keeping his presentations clear and understandable," Dr. Linus Pauling, J. of the American Chemical Society. Appendices: calculus of variations, math. notes, etc. Indexes. 611pp. 5⅜ x 8.
S472 Paperbound **$2.95**

ATOMIC SPECTRA AND ATOMIC STRUCTURE, G. Herzberg. Excellent general survey for chemists, physicists specializing in other fields. Partial contents: simplest line spectra and elements of atomic theory, building-up principle and periodic system of elements, hyperfine structure of spectral lines, some experiments and applications. Bibliography. 80 figures. Index. xii + 257pp. 5⅜ x 8.   S115 Paperbound **$1.95**

THE THEORY AND THE PROPERTIES OF METALS AND ALLOYS, N. F. Mott, H. Jones. Quantum methods used to develop mathematical models which show interrelationship of basic chemical phenomena with crystal structure, magnetic susceptibility, electrical, optical properties. Examines thermal properties of crystal lattice, electron motion in applied field, cohesion, electrical resistance, noble metals, para-, dia-, and ferromagnetism, etc. "Exposition . . . clear . . . mathematical treatment . . . simple," Nature. 138 figures. Bibliography. Index. xiii + 320pp. 5⅜ x 8.   S456 Paperbound **$1.85**

FOUNDATIONS OF NUCLEAR PHYSICS, edited by R. T. Beyer. 13 of the most important papers on nuclear physics reproduced in facsimile in the original languages of their authors: the papers most often cited in footnotes, bibliographies. Anderson, Curie, Joliot, Chadwick, Fermi, Lawrence, Cockcroft, Hahn, Yukawa. UNPARALLELED BIBLIOGRAPHY. 122 double-columned pages, over 4,000 articles, books classified. 57 figures. 288pp. 6⅛ x 9¼.
S19 Paperbound **$1.75**

MESON PHYSICS, R. E. Marshak. Traces the basic theory, and explicity presents results of experiments with particular emphasis on theoretical significance. Phenomena involving mesons as virtual transitions are avoided, eliminating some of the least satisfactory predictions of meson theory. Includes production and study of π mesons at nonrelativistic nucleon energies, contrasts between π and μ mesons, phenomena associated with nuclear interaction of π mesons, etc. Presents early evidence for new classes of particles and indicates theoretical difficulties created by discovery of heavy mesons and hyperons. Name and subject indices. Unabridged reprint. viii + 378pp. 5⅜ x 8.   S500 Paperbound **$1.95**

See also: **STRANGE STORY OF THE QUANTUM, B. Hoffmann; FROM EUCLID TO EDDINGTON, E. Whittaker; MATTER AND LIGHT, THE NEW PHYSICS, L. de Broglie; THE EVOLUTION OF SCIENTIFIC THOUGHT FROM NEWTON TO EINSTEIN, A. d'Abro; THE RISE OF THE NEW PHYSICS, A. d'Abro; THE THEORY OF GROUPS AND QUANTUM MECHANICS, H. Weyl; SUBSTANCE AND FUNCTION, & EINSTEIN'S THEORY OF RELATIVITY, E. Cassirer; FUNDAMENTAL FORMULAS OF PHYSICS, D. H. Menzel.**

Hydrodynamics

HYDRODYNAMICS, H. Dryden, F. Murnaghan, Harry Bateman. Published by the National Research Council in 1932 this enormous volume offers a complete coverage of classical hydrodynamics. Encyclopedic in quality. Partial contents: physics of fluids, motion, turbulent flow, compressible fluids, motion in 1, 2, 3 dimensions; viscous fluids rotating, laminar motion, resistance of motion through viscous fluid, eddy viscosity, hydraulic flow in channels of various shapes, discharge of gases, flow past obstacles, etc. Bibliography of over 2,900 items. Indexes. 23 figures. 634pp. 5⅜ x 8.   S303 Paperbound **$2.75**

A TREATISE ON HYDRODYNAMICS, A. B. Basset. Favorite text on hydrodynamics for 2 generations of physicists, hydrodynamical engineers, oceanographers, ship designers, etc. Clear enough for the beginning student, and thorough source for graduate students and engineers on the work of d'Alembert, Euler, Laplace, Lagrange, Poisson, Green, Clebsch, Stokes, Cauchy, Helmholtz, J. J. Thomson, Love, Hicks, Greenhill, Besant, Lamb, etc. Great amount of documentation on entire theory of classical hydrodynamics. Vol I: theory of motion of frictionless liquids, vortex, and cyclic irrotational motion, etc. 132 exercises. Bibliography. 3 Appendixes. xii + 264pp. Vol II: motion in viscous liquids, harmonic analysis, theory of tides, etc. 112 exercises. Bibliography. 4 Appendixes. xv + 328pp. Two volume set. 5⅜ x 8.
S724 Vol I Paperbound **$1.75**
S725 Vol II Paperbound **$1.75**
The set **$3.50**

HYDRODYNAMICS, Horace Lamb. Internationally famous complete coverage of standard reference work on dynamics of liquids & gases. Fundamental theorems, equations, methods, solutions, background, for classical hydrodynamics. Chapters include Equations of Motion, Integration of Equations in Special Gases, Irrotational Motion, Motion of Liquid in 2 Dimensions, Motion of Solids through Liquid-Dynamical Theory, Vortex Motion, Tidal Waves, Surface Waves, Waves of Expansion, Viscosity, Rotating Masses of liquids. Excellently planned, arranged; clear, lucid presentation. 6th enlarged, revised edition. Index. Over 900 footnotes, mostly bibliographical. 119 figures. xv + 738pp. 6⅛ x 9¼.   S256 Paperbound **$2.95**

See also: **FUNDAMENTAL FORMULAS OF PHYSICS, D. H. Menzel; THEORY OF FLIGHT, R. von Mises; FUNDAMENTALS OF HYDRO- AND AEROMECHANICS, L. Prandtl and O. G. Tietjens; APPLIED HYDRO- AND AEROMECHANICS, L. Prandtl and O. G. Tietjens; HYDRAULICS AND ITS APPLICATIONS, A. H. Gibson; FLUID MECHANICS FOR HYDRAULIC ENGINEERS, H. Rouse.**

Acoustics, optics, electromagnetics

ON THE SENSATIONS OF TONE, Hermann Helmholtz. This is an unmatched coordination of such fields as acoustical physics, physiology, experiment, history of music. It covers the entire gamut of musical tone. Partial contents: relation of musical science to acoustics, physical vs. physiological acoustics, composition of vibration, resonance, analysis of tones by sympathetic resonance, beats, chords, tonality, consonant chords, discords, progression of parts, etc. 33 appendixes discuss various aspects of sound, physics, acoustics, music, etc. Translated by A. J. Ellis. New introduction by Prof. Henry Margenau of Yale. 68 figures. 43 musical passages analyzed. Over 100 tables. Index. xix + 576pp. 6⅛ x 9¼.
S114 Paperbound **$2.95**

THE THEORY OF SOUND, Lord Rayleigh. Most vibrating systems likely to be encountered in practice can be tackled successfully by the methods set forth by the great Nobel laureate, Lord Rayleigh. Complete coverage of experimental, mathematical aspects of sound theory. Partial contents: Harmonic motions, vibrating systems in general, lateral vibrations of bars, curved plates or shells, applications of Laplace's functions to acoustical problems, fluid friction, plane vortex-sheet, vibrations of solid bodies, etc. This is the first inexpensive edition of this great reference and study work. Bibliography. Historical introduction by R. B. Lindsay. Total of 1040pp. 97 figures. 5⅜ x 8.

S292, S293, Two volume set, paperbound, **$4.00**

THE DYNAMICAL THEORY OF SOUND, H. Lamb. Comprehensive mathematical treatment of the physical aspects of sound, covering the theory of vibrations, the general theory of sound, and the equations of motion of strings, bars, membranes, pipes, and resonators. Includes chapters on plane, spherical, and simple harmonic waves, and the Helmholtz Theory of Audition. Complete and self-contained development for student and specialist; all fundamental differential equations solved completely. Specific mathematical details for such important phenomena as harmonics, normal modes, forced vibrations of strings, theory of reed pipes, etc. Index. Bibliography. 86 diagrams. viii + 307pp. 5⅜ x 8.
S655 Paperbound **$1.50**

WAVE PROPAGATION IN PERIODIC STRUCTURES, L. Brillouin. A general method and application to different problems: pure physics, such as scattering of X-rays of crystals, thermal vibration in crystal lattices, electronic motion in metals; and also problems of electrical engineering. Partial contents: elastic waves in 1-dimensional lattices of point masses. Propagation of waves along 1-dimensional lattices. Energy flow. 2 dimensional, 3 dimensional lattices. Mathieu's equation. Matrices and propagation of waves along an electric line. Continuous electric lines. 131 illustrations. Bibliography. Index. xii + 253pp. 5⅜ x 8.
S34 Paperbound **$1.85**

THEORY OF VIBRATIONS, N. W. McLachlan. Based on an exceptionally successful graduate course given at Brown University, this discusses linear systems having 1 degree of freedom, forced vibrations of simple linear systems, vibration of flexible strings, transverse vibrations of bars and tubes, transverse vibration of circular plate, sound waves of finite amplitude, etc. Index. 99 diagrams. 160pp. 5⅜ x 8.
S190 Paperbound **$1.35**

LOUD SPEAKERS: THEORY, PERFORMANCE, TESTING AND DESIGN, N. W. McLachlan. Most comprehensive coverage of theory, practice of loud speaker design, testing; classic reference, study manual in field. First 12 chapters deal with theory, for readers mainly concerned with math. aspects; last 7 chapters will interest reader concerned with testing, design. Partial contents: principles of sound propagation, fluid pressure on vibrators, theory of moving-coil principle, transients, driving mechanisms, response curves, design of horn type moving coil speakers, electrostatic speakers, much more. Appendix. Bibliography. Index. 165 illustrations, charts. 411pp. 5⅜ x 8.
S588 Paperbound **$2.25**

MICROWAVE TRANSMISSION, J. S. Slater. First text dealing exclusively with microwaves, brings together points of view of field, circuit theory, for graduate student in physics, electrical engineering, microwave technician. Offers valuable point of view not in most later studies. Uses Maxwell's equations to study electromagnetic field, important in this area. Partial contents: infinite line with distributed parameters, impedance of terminated line, plane waves, reflections, wave guides, coaxial line, composite transmission lines, impedance matching, etc. Introduction. Index. 76 illus. 319pp. 5⅜ x 8.
S564 Paperbound **$1.50**

THE ANALYSIS OF SENSATIONS, Ernst Mach. Great study of physiology, psychology of perception, shows Mach's ability to see material freshly, his "incorruptible skepticism and independence." (Einstein). Relation of problems of psychological perception to classical physics, supposed dualism of physical and mental, principle of continuity, evolution of senses, will as organic manifestation, scores of experiments, observations in optics, acoustics, music, graphics, etc. New introduction by T. S. Szasz, M. D. 58 illus. 300-item bibliography. Index. 404pp. 5⅜ x 8.
S525 Paperbound **$1.75**

APPLIED OPTICS AND OPTICAL DESIGN, A. E. Conrady. With publication of vol. 2, standard work for designers in optics is now complete for first time. Only work of its kind in English; only detailed work for practical designer and self-taught. Requires, for bulk of work, no math above trig. Step-by-step exposition, from fundamental concepts of geometrical, physical optics, to systematic study, design, of almost all types of optical systems. Vol. 1: all ordinary ray-tracing methods; primary aberrations; necessary higher aberration for design of telescopes, low-power microscopes, photographic equipment. Vol. 2: (Completed from author's notes by R. Kingslake, Dir. Optical Design, Eastman Kodak.) Special attention to high-power microscope, anastigmatic photographic objectives. "An indispensable work," J., Optical Soc. of Amer. "As a practical guide this book has no rival," Transactions, Optical Soc. Index. Bibliography. 193 diagrams. 852pp. 6⅛ x 9¼.
Vol. 1 T611 Paperbound **$2.95**
Vol. 2 T612 Paperbound **$2.95**

THE THEORY OF OPTICS, Paul Drude. One of finest fundamental texts in physical optics, classic offers thorough coverage, complete mathematical treatment of basic ideas. Includes fullest treatment of application of thermodynamics to optics; sine law in formation of images, transparent crystals, magnetically active substances, velocity of light, apertures, effects depending upon them, polarization, optical instruments, etc. Introduction by A. A. Michelson. Index. 110 illus. 567pp. 5⅜ x 8.
S532 Paperbound **$2.45**

OPTICKS, Sir Isaac Newton. In its discussions of light, reflection, color, refraction, theories of wave and corpuscular theories of light, this work is packed with scores of insights and discoveries. In its precise and practical discussion of construction of optical apparatus, contemporary understandings of phenomena it is truly fascinating to modern physicists, astronomers, mathematicians. Foreword by Albert Einstein. Preface by I. B. Cohen of Harvard University. 7 pages of portraits, facsimile pages, letters, etc. cxvi + 414pp. 5⅜ x 8.
S205 Paperbound **$2.00**

OPTICS AND OPTICAL INSTRUMENTS: AN INTRODUCTION WITH SPECIAL REFERENCE TO PRACTICAL APPLICATIONS, B. K. Johnson. An invaluable guide to basic practical applications of optical principles, which shows how to set up inexpensive working models of each of the four main types of optical instruments—telescopes, microscopes, photographic lenses, optical projecting systems. Explains in detail the most important experiments for determining their accuracy, resolving power, angular field of view, amounts of aberration, all other necessary facts about the instruments. Formerly "Practical Optics." Index. 234 diagrams. Appendix. 224pp. 5⅜ x 8.
S642 Paperbound **$1.65**

PRINCIPLES OF PHYSICAL OPTICS, Ernst Mach. This classical examination of the propagation of light, color, polarization, etc. offers an historical and philosophical treatment that has never been surpassed for breadth and easy readability. Contents: Rectilinear propagation of light. Reflection, refraction. Early knowledge of vision. Dioptrics. Composition of light. Theory of color and dispersion. Periodicity. Theory of interference. Polarization. Mathematical representation of properties of light. Propagation of waves, etc. 279 illustrations, 10 portraits. Appendix. Indexes. 324pp. 5⅜ x 8.
S178 Paperbound **$1.75**

FUNDAMENTALS OF ELECTRICITY AND MAGNETISM, L. B. Loeb. For students of physics, chemistry, or engineering who want an introduction to electricity and magnetism on a higher level and in more detail than general elementary physics texts provide. Only elementary differential and integral calculus is assumed. Physical laws developed logically, from magnetism to electric currents, Ohm's law, electrolysis, and on to static electricity, induction, etc. Covers an unusual amount of material; one third of book on modern material: solution of wave equation, photoelectric and thermionic effects, etc. Complete statement of the various electrical systems of units and interrelations. 2 Indexes. 75 pages of problems with answers stated. Over 300 figures and diagrams. xix +669pp. 5⅜ x 8.
S745 Paperbound **$2.75**

THE ELECTROMAGNETIC FIELD, Max Mason & Warren Weaver. Used constantly by graduate engineers. Vector methods exclusively: detailed treatment of electrostatics, expansion methods, with tables converting any quantity into absolute electromagnetic, absolute electrostatic, practical units. Discrete charges, ponderable bodies, Maxwell field equations, etc. Introduction. Indexes. 416pp. 5⅜ x 8.
S185 Paperbound **$2.00**

ELECTRICAL THEORY ON THE GIORGI SYSTEM, P. Cornelius. A new clarification of the fundamental concepts of electricity and magnetism, advocating the convenient m.k.s. system of units that is steadily gaining followers in the sciences. Illustrating the use and effectiveness of his terminology with numerous applications to concrete technical problems, the author here expounds the famous Giorgi system of electrical physics. His lucid presentation and well-reasoned, cogent argument for the universal adoption of this system form one of the finest pieces of scientific exposition in recent years. Index. Conversion tables for translating earlier data into modern units. Translated from 3rd Dutch edition by L. J. Jolley. x + 187pp. 5½ x 8¾.
S909 Clothbound **$6.00**

THEORY OF ELECTRONS AND ITS APPLICATION TO THE PHENOMENA OF LIGHT AND RADIANT HEAT, H. Lorentz. Lectures delivered at Columbia University by Nobel laureate Lorentz. Unabridged, they form a historical coverage of the theory of free electrons, motion, absorption of heat, Zeeman effect, propagation of light in molecular bodies, inverse Zeeman effect, optical phenomena in moving bodies, etc. 109 pages of notes explain the more advanced sections. Index. 9 figures. 352pp. 5⅜ x 8.
S173 Paperbound **$1.85**

TREATISE ON ELECTRICITY AND MAGNETISM, James Clerk Maxwell. For more than 80 years a seemingly inexhaustible source of leads for physicists, mathematicians, engineers. Total of 1082pp. on such topics as Measurement of Quantities, Electrostatics, Elementary Mathematical Theory of Electricity, Electrical Work and Energy in a System of Conductors, General Theorems, Theory of Electrical Images, Electrolysis, Conduction, Polarization, Dielectrics, Resistance, etc. "The greatest mathematical physicist since Newton," Sir James Jeans. 3rd edition. 107 figures, 21 plates. 1082pp. 5⅜ x 8.   S636-7, 2 volume set, paperbound **$4.00**

See also: **FUNDAMENTAL FORMULAS OF PHYSICS, D. H. Menzel; MATHEMATICAL ANALYSIS OF ELECTRICAL & OPTICAL WAVE MOTION, H. Bateman.**

Mechanics, dynamics, thermodynamics, elasticity

MECHANICS VIA THE CALCULUS, P. W. Norris, W. S. Legge. Covers almost everything, from linear motion to vector analysis: equations determining motion, linear methods, compounding of simple harmonic motions, Newton's laws of motion, Hooke's law, the simple pendulum, motion of a particle in 1 plane, centers of gravity, virtual work, friction, kinetic energy of rotating bodies, equilibrium of strings, hydrostatics, sheering stresses, elasticity, etc. 550 problems. 3rd revised edition. xii + 367pp. 6 x 9.
S207 Clothbound **$3.95**

MECHANICS, J. P. Den Hartog. Already a classic among introductory texts, the M.I.T. professor's lively and discursive presentation is equally valuable as a beginner's text, an engineering student's refresher, or a practicing engineer's reference. Emphasis in this highly readable text is on illuminating fundamental principles and showing how they are embodied in a great number of real engineering and design problems: trusses, loaded cables, beams, jacks, hoists, etc. Provides advanced material on relative motion and gyroscopes not usual in introductory texts. "Very thoroughly recommended to all those anxious to improve their real understanding of the principles of mechanics." MECHANICAL WORLD. Index. List of equations. 334 problems, all with answers. Over 550 diagrams and drawings. ix + 462pp. 5⅜ x 8.
S754 Paperbound **$2.00**

THEORETICAL MECHANICS: AN INTRODUCTION TO MATHEMATICAL PHYSICS, J. S. Ames, F. D. Murnaghan. A mathematically rigorous development of theoretical mechanics for the advanced student, with constant practical applications. Used in hundreds of advanced courses. An unusually thorough coverage of gyroscopic and baryscopic material, detailed analyses of the Corilis acceleration, applications of Lagrange's equations, motion of the double pendulum, Hamilton-Jacobi partial differential equations, group velocity and dispersion, etc. Special relativity is also included. 159 problems. 44 figures. ix + 462pp. 5⅜ x 8.
S461 Paperbound **$2.00**

THEORETICAL MECHANICS: STATICS AND THE DYNAMICS OF A PARTICLE, W. D. MacMillan. Used for over 3 decades as a self-contained and extremely comprehensive advanced undergraduate text in mathematical physics, physics, astronomy, and deeper foundations of engineering. Early sections require only a knowledge of geometry; later, a working knowledge of calculus. Hundreds of basic problems, including projectiles to the moon, escape velocity, harmonic motion, ballistics, falling bodies, transmission of power, stress and strain, elasticity, astronomical problems. 340 practice problems plus many fully worked out examples make it possible to test and extend principles developed in the text. 200 figures. xvii + 430pp. 5⅜ x 8.
S467 Paperbound **$2.00**

THEORETICAL MECHANICS: THE THEORY OF THE POTENTIAL, W. D. MacMillan. A comprehensive, well balanced presentation of potential theory, serving both as an introduction and a reference work with regard to specific problems, for physicists and mathematicians. No prior knowledge of integral relations is assumed, and all mathematical material is developed as it becomes necessary. Includes: Attraction of Finite Bodies; Newtonian Potential Function; Vector Fields, Green and Gauss Theorems; Attractions of Surfaces and Lines; Surface Distribution of Matter; Two-Layer Surfaces; Spherical Harmonics; Ellipsoidal Harmonics; etc. "The great number of particular cases . . . should make the book valuable to geophysicists and others actively engaged in practical applications of the potential theory," Review of Scientific Instruments. Index. Bibliography. xiii + 469pp. 5⅜ x 8.
S486 Paperbound **$2.25**

THEORETICAL MECHANICS: DYNAMICS OF RIGID BODIES, W. D. MacMillan. Theory of dynamics of a rigid body is developed, using both the geometrical and analytical methods of instruction. Begins with exposition of algebra of vectors, it goes through momentum principles, motion in space, use of differential equations and infinite series to solve more sophisticated dynamics problems. Partial contents: moments of inertia, systems of free particles, motion parallel to a fixed plane, rolling motion, method of periodic solutions, much more. 82 figs. 199 problems. Bibliography. Indexes. xii + 476pp. 5⅜ x 8.
S641 Paperbound **$2.00**

MATHEMATICAL FOUNDATIONS OF STATISTICAL MECHANICS, A. I. Khinchin. Offering a precise and rigorous formulation of problems, this book supplies a thorough and up-to-date exposition. It provides analytical tools needed to replace cumbersome concepts, and furnishes for the first time a logical step-by-step introduction to the subject. Partial contents: geometry & kinematics of the phase space, ergodic problem, reduction to theory of probability, application of central limit problem, ideal monatomic gas, foundation of thermo-dynamics, dispersion and distribution of sum functions. Key to notations. Index. viii + 179pp. 5⅜ x 8.
S147 Paperbound **$1.35**

ELEMENTARY PRINCIPLES IN STATISTICAL MECHANICS, J. W. Gibbs. Last work of the great Yale mathematical physicist, still one of the most fundamental treatments available for advanced students and workers in the field. Covers the basic principle of conservation of probability of phase, theory of errors in the calculated phases of a system, the contributions of Clausius, Maxwell, Boltzmann, and Gibbs himself, and much more. Includes valuable comparison of statistical mechanics with thermodynamics: Carnot's cycle, mechanical definitions of entropy, etc. xvi + 208pp. 5⅜ x 8.
S707 Paperbound **$1.45**

THE DYNAMICS OF PARTICLES AND OF RIGID, ELASTIC, AND FLUID BODIES; BEING LECTURES ON MATHEMATICAL PHYSICS, A. G. Webster. The reissuing of this classic fills the need for a comprehensive work on dynamics. A wide range of topics is covered in unusually great depth, applying ordinary and partial differential equations. Part I considers laws of motion and methods applicable to systems of all sorts; oscillation, resonance, cyclic systems, etc. Part 2 is a detailed study of the dynamics of rigid bodies. Part 3 introduces the theory of potential; stress and strain, Newtonian potential functions, gyrostatics, wave and vortex motion, etc. Further contents: Kinematics of a point; Lagrange's equations; Hamilton's principle; Systems of vectors; Statics and dynamics of deformable bodies; much more, not easily found together in one volume. Unabridged reprinting of 2nd edition. 20 pages of notes on differential equations and the higher analysis. 203 illustrations. Selected bibliography. Index. xi + 588pp. 5⅜ x 8.
S522 Paperbound **$2.35**

THE THEORY OF HEAT RADIATION, Max Planck. A pioneering work in thermodynamics, providing basis for most later work. Nobel Laureate Planck writes on Deductions from Electrodynamics and Thermodynamics, Entropy and Probability, Irreversible Radiation Processes, etc. Starts with simple experimental laws of optics, advances to problems of spectral distribution of energy and irreversibility. Bibliography. 7 illustrations, xiv + 224pp. 5⅜ x 8.
S546 Paperbound **$1.50**

A HISTORY OF THE THEORY OF ELASTICITY AND THE STRENGTH OF MATERIALS, I. Todhunter and K. Pearson. For over 60 years a basic reference, unsurpassed in scope or authority. Both a history of the mathematical theory of elasticity from Galileo, Hooke, and Mariotte to Saint Venant, Kirchhoff, Clebsch, and Lord Kelvin and a detailed presentation of every important mathematical contribution during this period. Presents proofs of thousands of theorems and laws, summarizes every relevant treatise, many unavailable elsewhere. Practically a book apiece is devoted to modern founders: Saint Venant, Lame, Boussinesq, Rankine, Lord Kelvin, F. Neumann, Kirchhoff, Clebsch. Hundreds of pages of technical and physical treatises on specific applications of elasticity to particular materials. Indispensable for the mathematician, physicist, or engineer working with elasticity. Unabridged, corrected reprint of original 3-volume 1886-1893 edition. Three volume set. Two indexes. Appendix to Vol. I. Total of 2344pp. 5⅜ x 8⅜.
S914-916 The set, Clothbound **$12.50**

THE MATHEMATICAL THEORY OF ELASTICITY, A. E. H. Love. A wealth of practical illustration combined with thorough discussion of fundamentals—theory, application, special problems and solutions. Partial Contents: Analysis of Strain & Stress, Elasticity of Solid Bodies, Elasticity of Crystals, Vibration of Spheres, Cylinders, Propagation of Waves in Elastic Solid Media, Torsion, Theory of Continuous Beams, Plates. Rigorous treatment of Volterra's theory of dislocations, 2-dimensional elastic systems, other topics of modern interest. "For years the standard treatise on elasticity," AMERICAN MATHEMATICAL MONTHLY. 4th revised edition. Index. 76 figures. xviii + 643pp. 6⅛ x 9¼.
S174 Paperbound **$2.95**

RAYLEIGH'S PRINCIPLE AND ITS APPLICATIONS TO ENGINEERING, G. Temple & W. Bickley. Rayleigh's principle developed to provide upper and lower estimates of true value of fundamental period of a vibrating system, or condition of stability of elastic systems. Illustrative examples; rigorous proofs in special chapters. Partial contents: Energy method of discussing vibrations, stability. Perturbation theory, whirling of uniform shafts. Criteria of elastic stability. Application of energy method. Vibrating systems. Proof, accuracy, successive approximations, application of Rayleigh's principle. Synthetic theorems. Numerical, graphical methods. Equilibrium configurations, Ritz's method. Bibliography. Index. 22 figures. ix + 156pp. 5⅜ x 8.
S307 Paperbound **$1.50**

INVESTIGATIONS ON THE THEORY OF THE BROWNIAN MOVEMENT, Albert Einstein. Reprints from rare European journals. 5 basic papers, including the Elementary Theory of the Brownian Movement, written at the request of Lorentz to provide a simple explanation. Translated by A. D. Cowper. Annotated, edited by R. Fürth. 33pp. of notes elucidate, give history of previous investigations. Author, subject indexes. 62 footnotes. 124pp. 5⅜ x 8.
S304 Paperbound **$1.25**

See also: **FUNDAMENTAL FORMULAS OF PHYSICS, D. H. Menzel.**

ENGINEERING

THEORY OF FLIGHT, Richard von Mises. Remains almost unsurpassed as balanced, well-written account of fundamental fluid dynamics, and situations in which air compressibility effects are unimportant. Stressing equally theory and practice, avoiding formidable mathematical structure, it conveys a full understanding of physical phenomena and mathematical concepts. Contains perhaps the best introduction to general theory of stability. "Outstanding," Scientific, Medical, and Technical Books. New introduction by K. H. Hohenemser. Bibliographical, historical notes. Index. 408 illustrations. xvi + 620pp. 5⅜ x 8⅜.
S541 Paperbound **$2.85**

THEORY OF WING SECTIONS, I. H. Abbott, A. E. von Doenhoff. Concise compilation of subsonic aerodynamic characteristics of modern NASA wing sections, with description of their geometry, associated theory. Primarily reference work for engineers, students, it gives methods, data for using wing-section data to predict characteristics. Particularly valuable: chapters on thin wings, airfoils; complete summary of NACA's experimental observations, system of construction families of airfoils. 350pp. of tables on Basic Thickness Forms, Mean Lines, Airfoil Ordinates, Aerodynamic Characteristics of Wing Sections. Index. Bibliography. 191 illustrations. Appendix. 705pp. 5⅜ x 8.
S558 Paperbound **$2.95**

SUPERSONIC AERODYNAMICS, E. R. C. Miles. Valuable theoretical introduction to the supersonic domain, with emphasis on mathematical tools and principles, for practicing aerodynamicists and advanced students in aeronautical engineering. Covers fundamental theory, divergence theorem and principles of circulation, compressible flow and Helmholtz laws, the Prandtl-Busemann graphic method for 2-dimensional flow, oblique shock waves, the Taylor-Maccoll method for cones in supersonic flow, the Chaplygin method for 2-dimensional flow, etc. Problems range from practical engineering problems to development of theoretical results. "Rendered outstanding by the unprecedented scope of its contents . . . has undoubtedly filled a vital gap," AERONAUTICAL ENGINEERING REVIEW. Index. 173 problems, answers. 106 diagrams. 7 tables. xii + 255pp. 5⅜ x 8.
S214 Paperbound **$1.45**

WEIGHT-STRENGTH ANALYSIS OF AIRCRAFT STRUCTURES, F. R. Shanley. Scientifically sound methods of analyzing and predicting the structural weight of aircraft and missiles. Deals directly with forces and the distances over which they must be transmitted, making it possible to develop methods by which the minimum structural weight can be determined for any material and conditions of loading. Weight equations for wing and fuselage structures. Includes author's original papers on inelastic buckling and creep buckling. "Particularly successful in presenting his analytical methods for investigating various optimum design principles," AERONAUTICAL ENGINEERING REVIEW. Enlarged bibliography. Index. 199 figures. xiv + 404pp. 5⅝ x 8⅜.   S660 Paperbound **$2.45**

INTRODUCTION TO THE STATISTICAL DYNAMICS OF AUTOMATIC CONTROL SYSTEMS, V. V. Solodovnikov. First English publication of text-reference covering important branch of automatic control systems—random signals; in its original edition, this was the first comprehensive treatment. Examines frequency characteristics, transfer functions, stationary random processes, determination of minimum mean-squared error, of transfer function for a finite period of observation, much more. Translation edited by J. B. Thomas, L. A. Zadeh. Index. Bibliography. Appendix. xxii + 308pp. 5⅜ x 8.   S420 Paperbound **$2.25**

TENSORS FOR CIRCUITS, Gabriel Kron. A boldly original method of analysing engineering problems, at center of sharp discussion since first introduced, now definitely proved useful in such areas as electrical and structural networks on automatic computers. Encompasses a great variety of specific problems by means of a relatively few symbolic equations. "Power and flexibility . . . becoming more widely recognized," Nature. Formerly "A Short Course in Tensor Analysis." New introduction by B. Hoffmann. Index. Over 800 diagrams. xix + 250pp. 5⅜ x 8.   S534 Paperbound **$1.85**

DESIGN AND USE OF INSTRUMENTS AND ACCURATE MECHANISM, T. N. Whitehead. For the instrument designer, engineer; how to combine necessary mathematical abstractions with independent observation of actual facts. Partial contents: instruments & their parts, theory of errors, systematic errors, probability, short period errors, erratic errors, design precision, kinematic, semikinematic design, stiffness, planning of an instrument, human factor, etc. Index. 85 photos, diagrams. xii + 288pp. 5⅜ x 8.   S270 Paperbound **$1.95**

APPLIED ELASTICITY, J. Prescott. Provides the engineer with the theory of elasticity usually lacking in books on strength of materials, yet concentrates on those portions useful for immediate application. Develops every important type of elasticity problem from theoretical principles. Covers analysis of stress, relations between stress and strain, the empirical basis of elasticity, thin rods under tension or thrust, Saint Venant's theory, transverse oscillations of thin rods, stability of thin plates, cylinders with thin walls, vibrations of rotating disks, elastic bodies in contact, etc. "Excellent and important contribution to the subject, not merely in the old matter which he has presented in new and refreshing form, but also in the many original investigations here published for the first time," NATURE. Index. 3 Appendixes. vi + 672pp. 5⅜ x 8.   S726 Paperbound **$2.95**

STRENGTH OF MATERIALS, J. P. Den Hartog. Distinguished text prepared for M.I.T. course, ideal as introduction, refresher, reference, or self-study text. Full clear treatment of elementary material (tension, torsion, bending, compound stresses, deflection of beams, etc.), plus much advanced material on engineering methods of great practical value: full treatment of the Mohr circle, lucid elementary discussions of the theory of the center of shear and the "Myosotis" method of calculating beam deflections, reinforced concrete, plastic deformations, photoelasticity, etc. In all sections, both general principles and concrete applications are given. Index. 186 figures (160 others in problem section). 350 problems, all with answers. List of formulas. viii + 323pp. 5⅜ x 8.   S755 Paperbound **$1.95**

PHOTOELASTICITY: PRINCIPLES AND METHODS, H. T. Jessop, F. C. Harris. For the engineer, for specific problems of stress analysis. Latest time-saving methods of checking calculations in 2-dimensional design problems, new techniques for stresses in 3 dimensions, and lucid description of optical systems used in practical photoelasticity. Useful suggestions and hints based on on-the-job experience included. Partial contents: strained and stress-strain relations, circular disc under thrust along diameter, rectangular block with square hole under vertical thrust, simply supported rectangular beam under central concentrated load, etc. Theory held to minimum, no advanced mathematical training needed. Index. 164 illustrations. viii + 184pp. 6⅛ x 9¼.   S137 Clothbound **$3.75**

MECHANICS OF THE GYROSCOPE, THE DYNAMICS OF ROTATION, R. F. Deimel, Professor of Mechanical Engineering at Stevens Institute of Technology. Elementary general treatment of dynamics of rotation, with special application of gyroscopic phenomena. No knowledge of vectors needed. Velocity of a moving curve, acceleration to a point, general equations of motion, gyroscopic horizon, free gyro, motion of discs, the damped gyro, 103 similar topics. Exercises. 75 figures. 208pp. 5⅜ x 8.   S66 Paperbound **$1.65**
                                                                              S144 Paperbound **$1.98**

A TREATISE ON GYROSTATICS AND ROTATIONAL MOTION: THEORY AND APPLICATIONS, Andrew Gray. Most detailed, thorough book in English, generally considered definitive study. Many problems of all sorts in full detail, or step-by-step summary. Classical problems of Bour, Lottner, etc.; later ones of great physical interest. Vibrating systems of gyrostats, earth as a top, calculation of path of axis of a top by elliptic integrals, motion of unsymmetrical top, much more. Index. 160 illus. 550pp. 5⅜ x 8.   S589 Paperbound **$2.75**

FUNDAMENTALS OF HYDRO- AND AEROMECHANICS, L. Prandtl and O. G. Tietjens. The well-known standard work based upon Prandtl's lectures at Goettingen. Wherever possible hydrodynamics theory is referred to practical considerations in hydraulics, with the view of unifying theory and experience. Presentation is extremely clear and though primarily physical, mathematical proofs are rigorous and use vector analysis to a considerable extent. An Enginering Society Monograph, 1934. 186 figures. Index. xvi + 270pp. 5⅜ x 8.
S374 Paperbound **$1.85**

APPLIED HYDRO- AND AEROMECHANICS, L. Prandtl and O. G. Tietjens. Presents, for the most part, methods which will be valuable to engineers. Covers flow in pipes, boundary layers, airfoil theory, entry conditions, turbulent flow in pipes, and the boundary layer, determining drag from measurements of pressure and velocity, etc. "Will be welcomed by all students of aerodynamics," NATURE. Unabridged, unaltered. An Engineering Society Monograph, 1934. Index. 226 figures, 28 photographic plates illustrating flow patterns. xvi + 311pp. 5⅜ x 8.
S375 Paperbound **$1.85**

HYDRAULICS AND ITS APPLICATIONS, A. H. Gibson. Excellent comprehensive textbook for the student and thorough practical manual for the professional worker, a work of great stature in its area. Half the book is devoted to theory and half to applications and practical problems met in the field. Covers modes of motion of a fluid, critical velocity, viscous flow, eddy formation, Bernoulli's theorem, flow in converging passages, vortex motion, form of effluent streams, notches and weirs, skin friction, losses at valves and elbows, siphons, erosion of channels, jet propulsion, waves of oscillation, and over 100 similar topics. Final chapters (nearly 400 pages) cover more than 100 kinds of hydraulic machinery: Pelton wheel, speed regulators, the hydraulic ram, surge tanks, the scoop wheel, the Venturi meter, etc. A special chapter treats methods of testing theoretical hypotheses: scale models of rivers, tidal estuaries, siphon spillways, etc. 5th revised and enlarged (1952) edition. Index. Appendix. 427 photographs and diagrams. 95 examples, answers. xv + 813pp. 6 x 9.
S791 Clothbound **$8.00**

FLUID MECHANICS FOR HYDRAULIC ENGINEERS, H. Rouse. Standard work that gives a coherent picture of fluid mechanics from the point of view of the hydraulic engineer. Based on courses given to civil and mechanical engineering students at Columbia and the California Institute of Technology, this work covers every basic principle, method, equation, or theory of interest to the hydraulic engineer. Much of the material, diagrams, charts, etc., in this self-contained text are not duplicated elsewhere. Covers irrotational motion, conformal mapping, problems in laminar motion, fluid turbulence, flow around immersed bodies, transportation of sediment, general charcteristics of wave phenomena, gravity waves in open channels, etc. Index. Appendix of physical properties of common fluids. Frontispiece + 245 figures and photographs. xvi + 422pp. 5⅜ x 8.
S729 Paperbound **$2.25**

THE MEASUREMENT OF POWER SPECTRA FROM THE POINT OF VIEW OF COMMUNICATIONS ENGINEERING, R. B. Blackman, J. W. Tukey. This pathfinding work, reprinted from the "Bell System Technical Journal," explains various ways of getting practically useful answers in the measurement of power spectra, using results from both transmission theory and the theory of statistical estimation. Treats: Autocovariance Functions and Power Spectra; Direct Analog Computation; Distortion, Noise, Heterodyne Filtering and Pre-whitening; Aliasing; Rejection Filtering and Separation; Smoothing and Decimation Procedures; Very Low Frequencies; Transversal Filtering; much more. An appendix reviews fundamental Fourier techniques. Index of notation. Glossary of terms. 24 figures. XII tables. Bibliography. General index. 192pp. 5⅜ x 8.
S507 Paperbound **$1.85**

MICROWAVE TRANSMISSION DESIGN DATA, T. Moreno. Originally classified, now rewritten and enlarged (14 new chapters) for public release under auspices of Sperry Corp. Material of immediate value or reference use to radio engineers, systems designers, applied physicists, etc. Ordinary transmission line theory; attenuation; capacity; parameters of coaxial lines; higher modes; flexible cables; obstacles, discontinuities, and injunctions; tuneable wave guide impedance transformers; effects of temperature and humidity; much more. "Enough theoretical discussion is included to allow use of data without previous background," Electronics. 324 circuit diagrams, figures, etc. Tables of dielectrics, flexible cable, etc., data. Index. Ix + 248pp. 5⅜ x 8.
S459 Paperbound **$1.50**

A CATALOGUE OF SELECTED DOVER BOOKS
IN ALL FIELDS OF INTEREST

A CATALOGUE OF SELECTED DOVER BOOKS
IN ALL FIELDS OF INTEREST

AMERICA'S OLD MASTERS, James T. Flexner. Four men emerged unexpectedly from provincial 18th century America to leadership in European art: Benjamin West, J. S. Copley, C. R. Peale, Gilbert Stuart. Brilliant coverage of lives and contributions. Revised, 1967 edition. 69 plates. 365pp. of text.

21806-6 Paperbound $3.00

FIRST FLOWERS OF OUR WILDERNESS: AMERICAN PAINTING, THE COLONIAL PERIOD, James T. Flexner. Painters, and regional painting traditions from earliest Colonial times up to the emergence of Copley, West and Peale Sr., Foster, Gustavus Hesselius, Feke, John Smibert and many anonymous painters in the primitive manner. Engaging presentation, with 162 illustrations. xxii + 368pp.

22180-6 Paperbound $3.50

THE LIGHT OF DISTANT SKIES: AMERICAN PAINTING, 1760-1835, James T. Flexner. The great generation of early American painters goes to Europe to learn and to teach: West, Copley, Gilbert Stuart and others. Allston, Trumbull, Morse; also contemporary American painters—primitives, derivatives, academics—who remained in America. 102 illustrations. xiii + 306pp. 22179-2 Paperbound $3.50

A HISTORY OF THE RISE AND PROGRESS OF THE ARTS OF DESIGN IN THE UNITED STATES, William Dunlap. Much the richest mine of information on early American painters, sculptors, architects, engravers, miniaturists, etc. The only source of information for scores of artists, the major primary source for many others. Unabridged reprint of rare original 1834 edition, with new introduction by James T. Flexner, and 394 new illustrations. Edited by Rita Weiss. 6⅝ x 9⅝.

21695-0, 21696-9, 21697-7 Three volumes, Paperbound $15.00

EPOCHS OF CHINESE AND JAPANESE ART, Ernest F. Fenollosa. From primitive Chinese art to the 20th century, thorough history, explanation of every important art period and form, including Japanese woodcuts; main stress on China and Japan, but Tibet, Korea also included. Still unexcelled for its detailed, rich coverage of cultural background, aesthetic elements, diffusion studies, particularly of the historical period. 2nd, 1913 edition. 242 illustrations. lii + 439pp. of text.

20364-6, 20365-4 Two volumes, Paperbound $6.00

THE GENTLE ART OF MAKING ENEMIES, James A. M. Whistler. Greatest wit of his day deflates Oscar Wilde, Ruskin, Swinburne; strikes back at inane critics, exhibitions, art journalism; aesthetics of impressionist revolution in most striking form. Highly readable classic by great painter. Reproduction of edition designed by Whistler. Introduction by Alfred Werner. xxxvi + 334pp.

21875-9 Paperbound $3.00

VISUAL ILLUSIONS: THEIR CAUSES, CHARACTERISTICS, AND APPLICATIONS, Matthew Luckiesh. Thorough description and discussion of optical illusion, geometric and perspective, particularly; size and shape distortions, illusions of color, of motion; natural illusions; use of illusion in art and magic, industry, etc. Most useful today with op art, also for classical art. Scores of effects illustrated. Introduction by William H. Ittleson. 100 illustrations. xxi + 252pp.

21530-X Paperbound $2.00

A HANDBOOK OF ANATOMY FOR ART STUDENTS, Arthur Thomson. Thorough, virtually exhaustive coverage of skeletal structure, musculature, etc. Full text, supplemented by anatomical diagrams and drawings and by photographs of undraped figures. Unique in its comparison of male and female forms, pointing out differences of contour, texture, form. 211 figures, 40 drawings, 86 photographs. xx + 459pp. 5⅜ x 8⅜.

21163-0 Paperbound $3.50

150 MASTERPIECES OF DRAWING, Selected by Anthony Toney. Full page reproductions of drawings from the early 16th to the end of the 18th century, all beautifully reproduced: Rembrandt, Michelangelo, Dürer, Fragonard, Urs, Graf, Wouwerman, many others. First-rate browsing book, model book for artists. xviii + 150pp. 8⅜ x 11¼.

21032-4 Paperbound' $3.50

THE LATER WORK OF AUBREY BEARDSLEY, Aubrey Beardsley. Exotic, erotic, ironic masterpieces in full maturity: Comedy Ballet, Venus and Tannhauser, Pierrot, Lysistrata, Rape of the Lock, Savoy material, Ali Baba, Volpone, etc. This material revolutionized the art world, and is still powerful, fresh, brilliant. With *The Early Work*, all Beardsley's finest work. 174 plates, 2 in color. xiv + 176pp. 8⅛ x 11.

21817-1 Paperbound $3.75

DRAWINGS OF REMBRANDT, Rembrandt van Rijn. Complete reproduction of fabulously rare edition by Lippmann and Hofstede de Groot, completely reedited, updated, improved by Prof. Seymour Slive, Fogg Museum. Portraits, Biblical sketches, landscapes, Oriental types, nudes, episodes from classical mythology—All Rembrandt's fertile genius. Also selection of drawings by his pupils and followers. "Stunning volumes," *Saturday Review*. 550 illustrations. lxxviii + 552pp. 9⅛ x 12¼.

21485-0, 21486-9 Two volumes, Paperbound $10.00

THE DISASTERS OF WAR, Francisco Goya. One of the masterpieces of Western civilization—83 etchings that record Goya's shattering, bitter reaction to the Napoleonic war that swept through Spain after the insurrection of 1808 and to war in general. Reprint of the first edition, with three additional plates from Boston's Museum of Fine Arts. All plates facsimile size. Introduction by Philip Hofer, Fogg Museum. v + 97pp. 9⅜ x 8¼.

21872-4 Paperbound $2.50

GRAPHIC WORKS OF ODILON REDON. Largest collection of Redon's graphic works ever assembled: 172 lithographs, 28 etchings and engravings, 9 drawings. These include some of his most famous works. All the plates from *Odilon Redon: oeuvre graphique complet,* plus additional plates. New introduction and caption translations by Alfred Werner. 209 illustrations. xxvii + 209pp. 9⅛ x 12¼.

21966-8 Paperbound **$5.00**

DESIGN BY ACCIDENT; A BOOK OF "ACCIDENTAL EFFECTS" FOR ARTISTS AND DESIGNERS, James F. O'Brien. Create your own unique, striking, imaginative effects by "controlled accident" interaction of materials: paints and lacquers, oil and water based paints, splatter, crackling materials, shatter, similar items. Everything you do will be different; first book on this limitless art, so useful to both fine artist and commercial artist. Full instructions. 192 plates showing "accidents," 8 in color. viii + 215pp. 8⅜ x 11¼. 21942-9 Paperbound $3.75

THE BOOK OF SIGNS, Rudolf Koch. Famed German type designer draws 493 beautiful symbols: religious, mystical, alchemical, imperial, property marks, runes, etc. Remarkable fusion of traditional and modern. Good for suggestions of timelessness, smartness, modernity. Text. vi + 104pp. 6⅛ x 9¼. 20162-7 Paperbound $1.25

HISTORY OF INDIAN AND INDONESIAN ART, Ananda K. Coomaraswamy. An unabridged republication of one of the finest books by a great scholar in Eastern art. Rich in descriptive material, history, social backgrounds; Sunga reliefs, Rajput paintings, Gupta temples, Burmese frescoes, textiles, jewelry, sculpture, etc. 400 photos. viii + 423pp. 6⅜ x 9¾. 21436-2 Paperbound $5.00

PRIMITIVE ART, Franz Boas. America's foremost anthropologist surveys textiles, ceramics, woodcarving, basketry, metalwork, etc.; patterns, technology, creation of symbols, style origins. All areas of world, but very full on Northwest Coast Indians. More than 350 illustrations of baskets, boxes, totem poles, weapons, etc. 378 pp. 20025-6 Paperbound $3.00

THE GENTLEMAN AND CABINET MAKER'S DIRECTOR, Thomas Chippendale. Full reprint (third edition, 1762) of most influential furniture book of all time, by master cabinetmaker. 200 plates, illustrating chairs, sofas, mirrors, tables, cabinets, plus 24 photographs of surviving pieces. Biographical introduction by N. Bienenstock. vi + 249pp. 9⅞ x 12¾. 21601-2 Paperbound $4.00

AMERICAN ANTIQUE FURNITURE, Edgar G. Miller, Jr. The basic coverage of all American furniture before 1840. Individual chapters cover type of furniture—clocks, tables, sideboards, etc.—chronologically, with inexhaustible wealth of data. More than 2100 photographs, all identified, commented on. Essential to all early American collectors. Introduction by H. E. Keyes. vi + 1106pp. 7⅞ x 10¾. 21599-7, 21600-4 Two volumes, Paperbound $11.00

PENNSYLVANIA DUTCH AMERICAN FOLK ART, Henry J. Kauffman. 279 photos, 28 drawings of tulipware, Fraktur script, painted tinware, toys, flowered furniture, quilts, samplers, hex signs, house interiors, etc. Full descriptive text. Excellent for tourist, rewarding for designer, collector. Map. 146pp. 7⅞ x 10¾. 21205-X Paperbound $2.50

EARLY NEW ENGLAND GRAVESTONE RUBBINGS, Edmund V. Gillon, Jr. 43 photographs, 226 carefully reproduced rubbings show heavily symbolic, sometimes macabre early gravestones, up to early 19th century. Remarkable early American primitive art, occasionally strikingly beautiful; always powerful. Text. xxvi + 207pp. 8⅜ x 11¼. 21380-3 Paperbound $3.50

ALPHABETS AND ORNAMENTS, Ernst Lehner. Well-known pictorial source for decorative alphabets, script examples, cartouches, frames, decorative title pages, calligraphic initials, borders, similar material. 14th to 19th century, mostly European. Useful in almost any graphic arts designing, varied styles. 750 illustrations. 256pp. 7 x 10.   21905-4 Paperbound $4.00

PAINTING: A CREATIVE APPROACH, Norman Colquhoun. For the beginner simple guide provides an instructive approach to painting: major stumbling blocks for beginner; overcoming them, technical points; paints and pigments; oil painting; watercolor and other media and color. New section on "plastic" paints. Glossary. Formerly *Paint Your Own Pictures*. 221pp.   22000-1 Paperbound $1.75

THE ENJOYMENT AND USE OF COLOR, Walter Sargent. Explanation of the relations between colors themselves and between colors in nature and art, including hundreds of little-known facts about color values, intensities, effects of high and low illumination, complementary colors. Many practical hints for painters, references to great masters. 7 color plates, 29 illustrations. x + 274pp.
20944-X Paperbound $2.75

THE NOTEBOOKS OF LEONARDO DA VINCI, compiled and edited by Jean Paul Richter. 1566 extracts from original manuscripts reveal the full range of Leonardo's versatile genius: all his writings on painting, sculpture, architecture, anatomy, astronomy, geography, topography, physiology, mining, music, etc., in both Italian and English, with 186 plates of manuscript pages and more than 500 additional drawings. Includes studies for the Last Supper, the lost Sforza monument, and other works. Total of xlvii + 866pp. 7⅞ x 10¾.
22572-0, 22573-9 Two volumes, Paperbound $11.00

MONTGOMERY WARD CATALOGUE OF 1895. Tea gowns, yards of flannel and pillow-case lace, stereoscopes, books of gospel hymns, the New Improved Singer Sewing Machine, side saddles, milk skimmers, straight-edged razors, high-button shoes, spittoons, and on and on . . . listing some 25,000 items, practically all illustrated. Essential to the shoppers of the 1890's, it is our truest record of the spirit of the period. Unaltered reprint of Issue No. 57, Spring and Summer 1895. Introduction by Boris Emmet. Innumerable illustrations. xiii + 624pp. 8½ x 11⅝.
22377-9 Paperbound $6.95

THE CRYSTAL PALACE EXHIBITION ILLUSTRATED CATALOGUE (LONDON, 1851). One of the wonders of the modern world—the Crystal Palace Exhibition in which all the nations of the civilized world exhibited their achievements in the arts and sciences—presented in an equally important illustrated catalogue. More than 1700 items pictured with accompanying text—ceramics, textiles, cast-iron work, carpets, pianos, sleds, razors, wall-papers, billiard tables, beehives, silverware and hundreds of other artifacts—represent the focal point of Victorian culture in the Western World. Probably the largest collection of Victorian decorative art ever assembled—indispensable for antiquarians and designers. Unabridged republication of the Art-Journal Catalogue of the Great Exhibition of 1851, with all terminal essays. New introduction by John Gloag, F.S.A. xxxiv + 426pp. 9 x 12.
22503-8 Paperbound $5.00

A History of Costume, Carl Köhler. Definitive history, based on surviving pieces of clothing primarily, and paintings, statues, etc. secondarily. Highly readable text, supplemented by 594 illustrations of costumes of the ancient Mediterranean peoples, Greece and Rome, the Teutonic prehistoric period; costumes of the Middle Ages, Renaissance, Baroque, 18th and 19th centuries. Clear, measured patterns are provided for many clothing articles. Approach is practical throughout. Enlarged by Emma von Sichart. 464pp. 21030-8 Paperbound $3.50

Oriental Rugs, Antique and Modern, Walter A. Hawley. A complete and authoritative treatise on the Oriental rug—where they are made, by whom and how, designs and symbols, characteristics in detail of the six major groups, how to distinguish them and how to buy them. Detailed technical data is provided on periods, weaves, warps, wefts, textures, sides, ends and knots, although no technical background is required for an understanding. 11 color plates, 80 halftones, 4 maps. vi + 320pp. $6\frac{1}{8}$ x $9\frac{1}{8}$. 22366-3 Paperbound $5.00

Ten Books on Architecture, Vitruvius. By any standards the most important book on architecture ever written. Early Roman discussion of aesthetics of building, construction methods, orders, sites, and every other aspect of architecture has inspired, instructed architecture for about 2,000 years. Stands behind Palladio, Michelangelo, Bramante, Wren, countless others. Definitive Morris H. Morgan translation. 68 illustrations. xii + 331pp. 20645-9 Paperbound $3.00

The Four Books of Architecture, Andrea Palladio. Translated into every major Western European language in the two centuries following its publication in 1570, this has been one of the most influential books in the history of architecture. Complete reprint of the 1738 Isaac Ware edition. New introduction by Adolf Placzek, Columbia Univ. 216 plates. xxii + 110pp. of text. $9\frac{1}{2}$ x $12\frac{3}{4}$. 21308-0 Clothbound $12.50

Sticks and Stones: A Study of American Architecture and Civilization, Lewis Mumford.One of the great classics of American cultural history. American architecture from the medieval-inspired earliest forms to the early 20th century; evolution of structure and style, and reciprocal influences on environment. 21 photographic illustrations. 238pp. 20202-X Paperbound $2.00

The American Builder's Companion, Asher Benjamin. The most widely used early 19th century architectural style and source book, for colonial up into Greek Revival periods. Extensive development of geometry of carpentering, construction of sashes, frames, doors, stairs; plans and elevations of domestic and other buildings. Hundreds of thousands of houses were built according to this book, now invaluable to historians, architects, restorers, etc. 1827 edition. 59 plates. 114pp. $7\frac{7}{8}$ x $10\frac{3}{4}$. 22236-5 Paperbound $3.50

Dutch Houses in the Hudson Valley Before 1776, Helen Wilkinson Reynolds. The standard survey of the Dutch colonial house and outbuildings, with constructional features, decoration, and local history associated with individual homesteads. Introduction by Franklin D. Roosevelt. Map. 150 illustrations. 469pp. $6\frac{5}{8}$ x $9\frac{1}{4}$. 21469-9 Paperbound $5.00

THE ARCHITECTURE OF COUNTRY HOUSES, Andrew J. Downing. Together with Vaux's *Villas and Cottages* this is the basic book for Hudson River Gothic architecture of the middle Victorian period. Full, sound discussions of general aspects of housing, architecture, style, decoration, furnishing, together with scores of detailed house plans, illustrations of specific buildings, accompanied by full text. Perhaps the most influential single American architectural book. 1850 edition. Introduction by J. Stewart Johnson. 321 figures, 34 architectural designs. xvi + 560pp.

22003-6 Paperbound $4.00

LOST EXAMPLES OF COLONIAL ARCHITECTURE, John Mead Howells. Full-page photographs of buildings that have disappeared or been so altered as to be denatured, including many designed by major early American architects. 245 plates. xvii + 248pp. 7⅞ x 10¾.

21143-6 Paperbound $3.50

DOMESTIC ARCHITECTURE OF THE AMERICAN COLONIES AND OF THE EARLY REPUBLIC, Fiske Kimball. Foremost architect and restorer of Williamsburg and Monticello covers nearly 200 homes between 1620-1825. Architectural details, construction, style features, special fixtures, floor plans, etc. Generally considered finest work in its area. 219 illustrations of houses, doorways, windows, capital mantels. xx + 314pp. 7⅞ x 10¾.

21743-4 Paperbound $4.00

EARLY AMERICAN ROOMS: 1650-1858, edited by Russell Hawes Kettell. Tour of 12 rooms, each representative of a different era in American history and each furnished, decorated, designed and occupied in the style of the era. 72 plans and elevations, 8-page color section, etc., show fabrics, wall papers, arrangements, etc. Full descriptive text. xvii + 200pp. of text. 8⅜ x 11¼.

21633-0 Paperbound $5.00

THE FITZWILLIAM VIRGINAL BOOK, edited by J. Fuller Maitland and W. B. Squire. Full modern printing of famous early 17th-century ms. volume of 300 works by Morley, Byrd, Bull, Gibbons, etc. For piano or other modern keyboard instrument; easy to read format. xxxvi + 938pp. 8⅜ x 11.

21068-5, 21069-3 Two volumes, Paperbound $10.00

KEYBOARD MUSIC, Johann Sebastian Bach. Bach Gesellschaft edition. A rich selection of Bach's masterpieces for the harpsichord: the six English Suites, six French Suites, the six Partitas (Clavierübung part I), the Goldberg Variations (Clavierübung part IV), the fifteen Two-Part Inventions and the fifteen Three-Part Sinfonias. Clearly reproduced on large sheets with ample margins; eminently playable. vi + 312pp. 8⅛ x 11.

22360-4 Paperbound $5.00

THE MUSIC OF BACH: AN INTRODUCTION, Charles Sanford Terry. A fine, nontechnical introduction to Bach's music, both instrumental and vocal. Covers organ music, chamber music, passion music, other types. Analyzes themes, developments, innovations. x + 114pp.

21075-8 Paperbound $1.50

BEETHOVEN AND HIS NINE SYMPHONIES, Sir George Grove. Noted British musicologist provides best history, analysis, commentary on symphonies. Very thorough, rigorously accurate; necessary to both advanced student and amateur music lover. 436 musical passages. vii + 407 pp.

20334-4 Paperbound $2.75

JOHANN SEBASTIAN BACH, Philipp Spitta. One of the great classics of musicology, this definitive analysis of Bach's music (and life) has never been surpassed. Lucid, nontechnical analyses of hundreds of pieces (30 pages devoted to St. Matthew Passion, 26 to B Minor Mass). Also includes major analysis of 18th-century music. 450 musical examples. 40-page musical supplement. Total of xx + 1799pp.

(EUK) 22278-0, 22279-9 Two volumes, Clothbound $17.50

MOZART AND HIS PIANO CONCERTOS, Cuthbert Girdlestone. The only full-length study of an important area of Mozart's creativity. Provides detailed analyses of all 23 concertos, traces inspirational sources. 417 musical examples. Second edition. 509pp.
21271-8 Paperbound $3.50

THE PERFECT WAGNERITE: A COMMENTARY ON THE NIBLUNG'S RING, George Bernard Shaw. Brilliant and still relevant criticism in remarkable essays on Wagner's Ring cycle, Shaw's ideas on political and social ideology behind the plots, role of Leitmotifs, vocal requisites, etc. Prefaces. xxi + 136pp.
(USO) 21707-8 Paperbound $1.75

DON GIOVANNI, W. A. Mozart. Complete libretto, modern English translation; biographies of composer and librettist; accounts of early performances and critical reaction. Lavishly illustrated. All the material you need to understand and appreciate this great work. Dover Opera Guide and Libretto Series; translated and introduced by Ellen Bleiler. 92 illustrations. 209pp.
21134-7 Paperbound $2.00

BASIC ELECTRICITY, U. S. Bureau of Naval Personel. Originally a training course, best non-technical coverage of basic theory of electricity and its applications. Fundamental concepts, batteries, circuits, conductors and wiring techniques, AC and DC, inductance and capacitance, generators, motors, transformers, magnetic amplifiers, synchros, servomechanisms, etc. Also covers blue-prints, electrical diagrams, etc. Many questions, with answers. 349 illustrations. x + 448pp. 6½ x 9¼.
20973-3 Paperbound $3.50

REPRODUCTION OF SOUND, Edgar Villchur. Thorough coverage for laymen of high fidelity systems, reproducing systems in general, needles, amplifiers, preamps, loudspeakers, feedback, explaining physical background. "A rare talent for making technicalities vividly comprehensible," R. Darrell, *High Fidelity*. 69 figures. iv + 92pp.
21515-6 Paperbound $1.35

HEAR ME TALKIN' TO YA: THE STORY OF JAZZ AS TOLD BY THE MEN WHO MADE IT, Nat Shapiro and Nat Hentoff. Louis Armstrong, Fats Waller, Jo Jones, Clarence Williams, Billy Holiday, Duke Ellington, Jelly Roll Morton and dozens of other jazz greats tell how it was in Chicago's South Side, New Orleans, depression Harlem and the modern West Coast as jazz was born and grew. xvi + 429pp.
21726-4 Paperbound $3.00

FABLES OF AESOP, translated by Sir Roger L'Estrange. A reproduction of the very rare 1931 Paris edition; a selection of the most interesting fables, together with 50 imaginative drawings by Alexander Calder. v + 128pp. 6½x9¼.
21780-9 Paperbound $1.50

AGAINST THE GRAIN (A REBOURS), Joris K. Huysmans. Filled with weird images, evidences of a bizarre imagination, exotic experiments with hallucinatory drugs, rich tastes and smells and the diversions of its sybarite hero Duc Jean des Esseintes, this classic novel pushed 19th-century literary decadence to its limits. Full unabridged edition. Do not confuse this with abridged editions generally sold. Introduction by Havelock Ellis. xlix + 206pp.   22190-3 Paperbound $2.50

VARIORUM SHAKESPEARE: HAMLET. Edited by Horace H. Furness; a landmark of American scholarship. Exhaustive footnotes and appendices treat all doubtful words and phrases, as well as suggested critical emendations throughout the play's history. First volume contains editor's own text, collated with all Quartos and Folios. Second volume contains full first Quarto, translations of Shakespeare's sources (Belleforest, and Saxo Grammaticus), Der Bestrafte Brudermord, and many essays on critical and historical points of interest by major authorities of past and present. Includes details of staging and costuming over the years. By far the best edition available for serious students of Shakespeare. Total of xx + 905pp.
21004-9, 21005-7, 2 volumes, Paperbound $7.00

A LIFE OF WILLIAM SHAKESPEARE, Sir Sidney Lee. This is the standard life of Shakespeare, summarizing everything known about Shakespeare and his plays. Incredibly rich in material, broad in coverage, clear and judicious, it has served thousands as the best introduction to Shakespeare. 1931 edition. 9 plates. xxix + 792pp.   21967-4 Paperbound $4.50

MASTERS OF THE DRAMA, John Gassner. Most comprehensive history of the drama in print, covering every tradition from Greeks to modern Europe and America, including India, Far East, etc. Covers more than 800 dramatists, 2000 plays, with biographical material, plot summaries, theatre history, criticism, etc. "Best of its kind in English," New Republic. 77 illustrations. xxii + 890pp.
20100-7 Clothbound $10.00

THE EVOLUTION OF THE ENGLISH LANGUAGE, George McKnight. The growth of English, from the 14th century to the present. Unusual, non-technical account presents basic information in very interesting form: sound shifts, change in grammar and syntax, vocabulary growth, similar topics. Abundantly illustrated with quotations. Formerly Modern English in the Making. xii + 590pp.
21932-1 Paperbound $4.00

AN ETYMOLOGICAL DICTIONARY OF MODERN ENGLISH, Ernest Weekley. Fullest, richest work of its sort, by foremost British lexicographer. Detailed word histories, including many colloquial and archaic words; extensive quotations. Do not confuse this with the Concise Etymological Dictionary, which is much abridged. Total of xxvii + 830pp. 6½ x 9¼.
21873-2, 21874-0 Two volumes, Paperbound $7.90

FLATLAND: A ROMANCE OF MANY DIMENSIONS, E. A. Abbott. Classic of science-fiction explores ramifications of life in a two-dimensional world, and what happens when a three-dimensional being intrudes. Amusing reading, but also useful as introduction to thought about hyperspace. Introduction by Banesh Hoffmann. 16 illustrations. xx + 103pp.   20001-9 Paperbound $1.25

POEMS OF ANNE BRADSTREET, edited with an introduction by Robert Hutchinson. A new selection of poems by America's first poet and perhaps the first significant woman poet in the English language. 48 poems display her development in works of considerable variety—love poems, domestic poems, religious meditations, formal elegies, "quaternions," etc. Notes, bibliography. viii + 222pp.

22160-1 Paperbound $2.50

THREE GOTHIC NOVELS: THE CASTLE OF OTRANTO BY HORACE WALPOLE; VATHEK BY WILLIAM BECKFORD; THE VAMPYRE BY JOHN POLIDORI, WITH FRAGMENT OF A NOVEL BY LORD BYRON, edited by E. F. Bleiler. The first Gothic novel, by Walpole; the finest Oriental tale in English, by Beckford; powerful Romantic supernatural story in versions by Polidori and Byron. All extremely important in history of literature; all still exciting, packed with supernatural thrills, ghosts, haunted castles, magic, etc. xl + 291pp.

21232-7 Paperbound $2.50

THE BEST TALES OF HOFFMANN, E. T. A. Hoffmann. 10 of Hoffmann's most important stories, in modern re-editings of standard translations: Nutcracker and the King of Mice, Signor Formica, Automata, The Sandman, Rath Krespel, The Golden Flowerpot, Master Martin the Cooper, The Mines of Falun, The King's Betrothed, A New Year's Eve Adventure. 7 illustrations by Hoffmann. Edited by E. F. Bleiler. xxxix + 419pp. 21793-0 Paperbound $3.00

GHOST AND HORROR STORIES OF AMBROSE BIERCE, Ambrose Bierce. 23 strikingly modern stories of the horrors latent in the human mind: The Eyes of the Panther, The Damned Thing, An Occurrence at Owl Creek Bridge, An Inhabitant of Carcosa, etc., plus the dream-essay, Visions of the Night. Edited by E. F. Bleiler. xxii + 199pp. 20767-6 Paperbound $1.50

BEST GHOST STORIES OF J. S. LEFANU, J. Sheridan LeFanu. Finest stories by Victorian master often considered greatest supernatural writer of all. Carmilla, Green Tea, The Haunted Baronet, The Familiar, and 12 others. Most never before available in the U. S. A. Edited by E. F. Bleiler. 8 illustrations from Victorian publications. xvii + 467pp. 20415-4 Paperbound $3.00

MATHEMATICAL FOUNDATIONS OF INFORMATION THEORY, A. I. Khinchin. Comprehensive introduction to work of Shannon, McMillan, Feinstein and Khinchin, placing these investigations on a rigorous mathematical basis. Covers entropy concept in probability theory, uniqueness theorem, Shannon's inequality, ergodic sources, the E property, martingale concept, noise, Feinstein's fundamental lemma, Shanon's first and second theorems. Translated by R. A. Silverman and M. D. Friedman. iii + 120pp. 60434-9 Paperbound $2.00

SEVEN SCIENCE FICTION NOVELS, H. G. Wells. The standard collection of the great novels. Complete, unabridged. *First Men in the Moon, Island of Dr. Moreau, War of the Worlds, Food of the Gods, Invisible Man, Time Machine, In the Days of the Comet.* Not only science fiction fans, but every educated person owes it to himself to read these novels. 1015pp. (USO) 20264-X Clothbound $6.00

LAST AND FIRST MEN AND STAR MAKER, TWO SCIENCE FICTION NOVELS, Olaf Stapledon. Greatest future histories in science fiction. In the first, human intelligence is the "hero," through strange paths of evolution, interplanetary invasions, incredible technologies, near extinctions and reemergences. Star Maker describes the quest of a band of star rovers for intelligence itself, through time and space: weird inhuman civilizations, crustacean minds, symbiotic worlds, etc. Complete, unabridged. v + 438pp. (USO) 21962-3 Paperbound $2.50

THREE PROPHETIC NOVELS, H. G. WELLS. Stages of a consistently planned future for mankind. *When the Sleeper Wakes*, and *A Story of the Days to Come*, anticipate *Brave New World* and *1984*, in the 21st Century; *The Time Machine*, only complete version in print, shows farther future and the end of mankind. All show Wells's greatest gifts as storyteller and novelist. Edited by E. F. Bleiler. x + 335pp. (USO) 20605-X Paperbound $2.50

THE DEVIL'S DICTIONARY, Ambrose Bierce. America's own Oscar Wilde—Ambrose Bierce—offers his barbed iconoclastic wisdom in over 1,000 definitions hailed by H. L. Mencken as "some of the most gorgeous witticisms in the English language." 145pp. 20487-1 Paperbound $1.25

MAX AND MORITZ, Wilhelm Busch. Great children's classic, father of comic strip, of two bad boys, Max and Moritz. Also Ker and Plunk (Plisch und Plumm), Cat and Mouse, Deceitful Henry, Ice-Peter, The Boy and the Pipe, and five other pieces. Original German, with English translation. Edited by H. Arthur Klein; translations by various hands and H. Arthur Klein. vi + 216pp. 20181-3 Paperbound $2.00

PIGS IS PIGS AND OTHER FAVORITES, Ellis Parker Butler. The title story is one of the best humor short stories, as Mike Flannery obfuscates biology and English. Also included, That Pup of Murchison's, The Great American Pie Company, and Perkins of Portland. 14 illustrations. v + 109pp. 21532-6 Paperbound $1.25

THE PETERKIN PAPERS, Lucretia P. Hale. It takes genius to be as stupidly mad as the Peterkins, as they decide to become wise, celebrate the "Fourth," keep a cow, and otherwise strain the resources of the Lady from Philadelphia. Basic book of American humor. 153 illustrations. 219pp. 20794-3 Paperbound $2.00

PERRAULT'S FAIRY TALES, translated by A. E. Johnson and S. R. Littlewood, with 34 full-page illustrations by Gustave Doré. All the original Perrault stories—Cinderella, Sleeping Beauty, Bluebeard, Little Red Riding Hood, Puss in Boots, Tom Thumb, etc.—with their witty verse morals and the magnificent illustrations of Doré. One of the five or six great books of European fairy tales. viii + 117pp. 8⅛ x 11. 22311-6 Paperbound $2.00

OLD HUNGARIAN FAIRY TALES, Baroness Orczy. Favorites translated and adapted by author of the *Scarlet Pimpernel*. Eight fairy tales include "The Suitors of Princess Fire-Fly," "The Twin Hunchbacks," "Mr. Cuttlefish's Love Story," and "The Enchanted Cat." This little volume of magic and adventure will captivate children as it has for generations. 90 drawings by Montagu Barstow. 96pp. (USO) 22293-4 Paperbound $1.95

THE RED FAIRY BOOK, Andrew Lang. Lang's color fairy books have long been children's favorites. This volume includes Rapunzel, Jack and the Bean-stalk and 35 other stories, familiar and unfamiliar. 4 plates, 93 illustrations x + 367pp.
21673-X Paperbound $2.50

THE BLUE FAIRY BOOK, Andrew Lang. Lang's tales come from all countries and all times. Here are 37 tales from Grimm, the Arabian Nights, Greek Mythology, and other fascinating sources. 8 plates, 130 illustrations. xi + 390pp.
21437-0 Paperbound $2.75

HOUSEHOLD STORIES BY THE BROTHERS GRIMM. Classic English-language edition of the well-known tales — Rumpelstiltskin, Snow White, Hansel and Gretel, The Twelve Brothers, Faithful John, Rapunzel, Tom Thumb (52 stories in all). Translated into simple, straightforward English by Lucy Crane. Ornamented with head-pieces, vignettes, elaborate decorative initials and a dozen full-page illustrations by Walter Crane. x + 269pp.
21080-4 Paperbound **$2.00**

THE MERRY ADVENTURES OF ROBIN HOOD, Howard Pyle. The finest modern versions of the traditional ballads and tales about the great English outlaw. Howard Pyle's complete prose version, with every word, every illustration of the first edition. Do not confuse this facsimile of the original (1883) with modern editions that change text or illustrations. 23 plates plus many page decorations. xxii + 296pp.
22043-5 Paperbound $2.75

THE STORY OF KING ARTHUR AND HIS KNIGHTS, Howard Pyle. The finest children's version of the life of King Arthur; brilliantly retold by Pyle, with 48 of his most imaginative illustrations. xviii + 313pp. 6⅛ x 9¼.
21445-1 Paperbound $2.50

THE WONDERFUL WIZARD OF OZ, L. Frank Baum. America's finest children's book in facsimile of first edition with all Denslow illustrations in full color. The edition a child should have. Introduction by Martin Gardner. 23 color plates, scores of drawings. iv + 267pp.
20691-2 Paperbound $2.50

THE MARVELOUS LAND OF OZ, L. Frank Baum. The second Oz book, every bit as imaginative as the Wizard. The hero is a boy named Tip, but the Scarecrow and the Tin Woodman are back, as is the Oz magic. 16 color plates, 120 drawings by John R. Neill. 287pp.
20692-0 Paperbound $2.50

THE MAGICAL MONARCH OF MO, L. Frank Baum. Remarkable adventures in a land even stranger than Oz. The best of Baum's books not in the Oz series. 15 color plates and dozens of drawings by Frank Verbeck. xviii + 237pp.
21892-9 Paperbound $2.25

THE BAD CHILD'S BOOK OF BEASTS, MORE BEASTS FOR WORSE CHILDREN, A MORAL ALPHABET, Hilaire Belloc. Three complete humor classics in one volume. Be kind to the frog, and do not call him names . . . and 28 other whimsical animals. Familiar favorites and some not so well known. Illustrated by Basil Blackwell. 156pp.
(USO) 20749-8 Paperbound $1.50

EAST O' THE SUN AND WEST O' THE MOON, George W. Dasent. Considered the best of all translations of these Norwegian folk tales, this collection has been enjoyed by generations of children (and folklorists too). Includes True and Untrue, Why the Sea is Salt, East O' the Sun and West O' the Moon, Why the Bear is Stumpy-Tailed, Boots and the Troll, The Cock and the Hen, Rich Peter the Pedlar, and 52 more. The only edition with all 59 tales. 77 illustrations by Erik Werenskiold and Theodor Kittelsen. xv + 418pp. 22521-6 Paperbound $3.50

GOOPS AND HOW TO BE THEM, Gelett Burgess. Classic of tongue-in-cheek humor, masquerading as etiquette book. 87 verses, twice as many cartoons, show mischievous Goops as they demonstrate to children virtues of table manners, neatness, courtesy, etc. Favorite for generations. viii + 88pp. 6½ x 9¼.
22233-0 Paperbound $1.50

ALICE'S ADVENTURES UNDER GROUND, Lewis Carroll. The first version, quite different from the final *Alice in Wonderland,* printed out by Carroll himself with his own illustrations. Complete facsimile of the "million dollar" manuscript Carroll gave to Alice Liddell in 1864. Introduction by Martin Gardner. viii + 96pp. Title and dedication pages in color. 21482-6 Paperbound $1.25

THE BROWNIES, THEIR BOOK, Palmer Cox. Small as mice, cunning as foxes, exuberant and full of mischief, the Brownies go to the zoo, toy shop, seashore, circus, etc., in 24 verse adventures and 266 illustrations. Long a favorite, since their first appearance in St. Nicholas Magazine. xi + 144pp. 6⅝ x 9¼.
21265-3 Paperbound $1.75

SONGS OF CHILDHOOD, Walter De La Mare. Published (under the pseudonym Walter Ramal) when De La Mare was only 29, this charming collection has long been a favorite children's book. A facsimile of the first edition in paper, the 47 poems capture the simplicity of the nursery rhyme and the ballad, including such lyrics as I Met Eve, Tartary, The Silver Penny. vii + 106pp. (USO) 21972-0 Paperbound $2.00

THE COMPLETE NONSENSE OF EDWARD LEAR, Edward Lear. The finest 19th-century humorist-cartoonist in full: all nonsense limericks, zany alphabets, Owl and Pussycat, songs, nonsense botany, and more than 500 illustrations by Lear himself. Edited by Holbrook Jackson. xxix + 287pp. (USO) 20167-8 Paperbound $2.00

BILLY WHISKERS: THE AUTOBIOGRAPHY OF A GOAT, Frances Trego Montgomery. A favorite of children since the early 20th century, here are the escapades of that rambunctious, irresistible and mischievous goat—Billy Whiskers. Much in the spirit of *Peck's Bad Boy,* this is a book that children never tire of reading or hearing. All the original familiar illustrations by W. H. Fry are included: 6 color plates, 18 black and white drawings. 159pp. 22345-0 Paperbound $2.00

MOTHER GOOSE MELODIES. Faithful republication of the fabulously rare Munroe and Francis "copyright 1833" Boston edition—the most important Mother Goose collection, usually referred to as the "original." Familiar rhymes plus many rare ones, with wonderful old woodcut illustrations. Edited by E. F. Bleiler. 128pp. 4½ x 6⅜. 22577-1 Paperbound $1.00

TWO LITTLE SAVAGES; BEING THE ADVENTURES OF TWO BOYS WHO LIVED AS INDIANS AND WHAT THEY LEARNED, Ernest Thompson Seton. Great classic of nature and boyhood provides a vast range of woodlore in most palatable form, a genuinely entertaining story. Two farm boys build a teepee in woods and live in it for a month, working out Indian solutions to living problems, star lore, birds and animals, plants, etc. 293 illustrations. vii + 286pp.

20985-7 Paperbound $2.50

PETER PIPER'S PRACTICAL PRINCIPLES OF PLAIN & PERFECT PRONUNCIATION. Alliterative jingles and tongue-twisters of surprising charm, that made their first appearance in America about 1830. Republished in full with the spirited woodcut illustrations from this earliest American edition. 32pp. $4\frac{1}{2}$ x $6\frac{3}{8}$.

22560-7 Paperbound $1.00

SCIENCE EXPERIMENTS AND AMUSEMENTS FOR CHILDREN, Charles Vivian. 73 easy experiments, requiring only materials found at home or easily available, such as candles, coins, steel wool, etc.; illustrate basic phenomena like vacuum, simple chemical reaction, etc. All safe. Modern, well-planned. Formerly *Science Games for Children*. 102 photos, numerous drawings. 96pp. $6\frac{1}{8}$ x $9\frac{1}{4}$.

21856-2 Paperbound $1.25

AN INTRODUCTION TO CHESS MOVES AND TACTICS SIMPLY EXPLAINED, Leonard Barden. Informal intermediate introduction, quite strong in explaining reasons for moves. Covers basic material, tactics, important openings, traps, positional play in middle game, end game. Attempts to isolate patterns and recurrent configurations. Formerly *Chess*. 58 figures. 102pp. (USO) 21210-6 Paperbound $1.25

LASKER'S MANUAL OF CHESS, Dr. Emanuel Lasker. Lasker was not only one of the five great World Champions, he was also one of the ablest expositors, theorists, and analysts. In many ways, his Manual, permeated with his philosophy of battle, filled with keen insights, is one of the greatest works ever written on chess. Filled with analyzed games by the great players. A single-volume library that will profit almost any chess player, beginner or master. 308 diagrams. xli x 349pp.

20640-8 Paperbound $2.75

THE MASTER BOOK OF MATHEMATICAL RECREATIONS, Fred Schuh. In opinion of many the finest work ever prepared on mathematical puzzles, stunts, recreations; exhaustively thorough explanations of mathematics involved, analysis of effects, citation of puzzles and games. Mathematics involved is elementary. Translated by F. Göbel. 194 figures. xxiv + 430pp. 22134-2 Paperbound $3.50

MATHEMATICS, MAGIC AND MYSTERY, Martin Gardner. Puzzle editor for Scientific American explains mathematics behind various mystifying tricks: card tricks, stage "mind reading," coin and match tricks, counting out games, geometric dissections, etc. Probability sets, theory of numbers clearly explained. Also provides more than 400 tricks, guaranteed to work, that you can do. 135 illustrations. xii + 176pp.

20335-2 Paperbound $1.75

MATHEMATICAL PUZZLES FOR BEGINNERS AND ENTHUSIASTS, Geoffrey Mott-Smith. 189 puzzles from easy to difficult—involving arithmetic, logic, algebra, properties of digits, probability, etc.—for enjoyment and mental stimulus. Explanation of mathematical principles behind the puzzles. 135 illustrations. viii + 248pp.

20198-8 Paperbound $1.75

PAPER FOLDING FOR BEGINNERS, William D. Murray and Francis J. Rigney. Easiest book on the market, clearest instructions on making interesting, beautiful origami. Sail boats, cups, roosters, frogs that move legs, bonbon boxes, standing birds, etc. 40 projects; more than 275 diagrams and photographs. 94pp.

20713-7 Paperbound $1.00

TRICKS AND GAMES ON THE POOL TABLE, Fred Herrmann. 79 tricks and games— some solitaires, some for two or more players, some competitive games—to entertain you between formal games. Mystifying shots and throws, unusual caroms, tricks involving such props as cork, coins, a hat, etc. Formerly *Fun on the Pool Table*. 77 figures. 95pp.

21814-7 Paperbound $1.25

HAND SHADOWS TO BE THROWN UPON THE WALL: A SERIES OF NOVEL AND AMUSING FIGURES FORMED BY THE HAND, Henry Bursill. Delightful picturebook from great-grandfather's day shows how to make 18 different hand shadows: a bird that flies, duck that quacks, dog that wags his tail, camel, goose, deer, boy, turtle, etc. Only book of its sort. vi + 33pp. 6½ x 9¼.   21779-5 Paperbound $1.00

WHITTLING AND WOODCARVING, E. J. Tangerman. 18th printing of best book on market. "If you can cut a potato you can carve" toys and puzzles, chains, chessmen, caricatures, masks, frames, woodcut blocks, surface patterns, much more. Information on tools, woods, techniques. Also goes into serious wood sculpture from Middle Ages to present, East and West. 464 photos, figures. x + 293pp.

20965-2 Paperbound $2.00

HISTORY OF PHILOSOPHY, Julián Marías. Possibly the clearest, most easily followed, best planned, most useful one-volume history of philosophy on the market; neither skimpy nor overfull. Full details on system of every major philosopher and dozens of less important thinkers from pre-Socratics up to Existentialism and later. Strong on many European figures usually omitted. Has gone through dozens of editions in Europe. 1966 edition, translated by Stanley Appelbaum and Clarence Strowbridge. xviii + 505pp.

21739-6 Paperbound $3.50

YOGA: A SCIENTIFIC EVALUATION, Kovoor T. Behanan. Scientific but non-technical study of physiological results of yoga exercises; done under auspices of Yale U. Relations to Indian thought, to psychoanalysis, etc. 16 photos. xxiii + 270pp.

20505-3 Paperbound $2.50

Prices subject to change without notice.
Available at your book dealer or write for free catalogue to Dept. GI, Dover Publications, Inc., 180 Varick St., N. Y., N. Y. 10014. Dover publishes more than 150 books each year on science, elementary and advanced mathematics, biology, music, art, literary history, social sciences and other areas.